Classical Electromagnetic Theory
Second Edition

AF531584

Fundamental Theories of Physics

An International Book Series on The Fundamental Theories of Physics: Their Clarification, Development and Application

Editor:
ALWYN VAN DER MERWE, *University of Denver, U.S.A.*

Editorial Advisory Board:
GIANCARLO GHIRARDI, *University of Trieste, Italy*
LAWRENCE P. HORWITZ, *Tel-Aviv University, Israel*
BRIAN D. JOSEPHSON, *University of Cambridge, U.K.*
CLIVE KILMISTER, *University of London, U.K.*
PEKKA J. LAHTI, *University of Turku, Finland*
ASHER PERES, *Israel Institute of Technology, Israel*
EDUARD PRUGOVECKI, *University of Toronto, Canada*
FRANCO SELLERI, *Università di Bara, Italy*
TONY SUDBERY, *University of York, U.K.*
HANS-JÜRGEN TREDER, *Zentralinstitut für Astrophysik der Akademie der Wissenschaften, Germany*

Classical Electromagnetic Theory

Second Edition

by

Jack Vanderlinde

University of New Brunswick,
Fredericton, NB, Canada

A C.I.P. Catalogue record for this book is available from the Library of Congress.

ISBN 10 1-4020-2699-4 (HB)
ISBN 13 978-1-4020-2699-7 (HB)
ISBN 10 1-4020-2700-1 (e-book)
ISBN 13 978-1-4020-2700-0 (e-book)

springer.com

All rights reserved. No part of this work may be reproduced, stored in a retrieval system, or transmitted in any form or by any means, electronic, mechanical, photocopying, microfilming, recording or otherwise, without written permission from the publisher, with the exception of any material supplied specifically for the purpose of being entered and executed on a computer system, for exclusive use by the purchaser of the work.

Vanderlinde: Classical Electromagnetic Theory, Second Edition

© 2004 Springer

All rights reserved. No part of this publication may be reproduced, stored in any electronic or mechanical form, including photocopy, recording or otherwise, without the prior written permission of the publisher.

First Indian Reprint 2007

ISBN 978-81-8128-746-5

This edition is manufactured in India for sale only in India, Pakistan, Bangladesh, Nepal and Sri Lanka and any other country as authorized by the publisher.

This edition is published by Springer (India) Private Limited,
A part of Springer Science+Business Media, Registered Office: 906-907,
Akash Deep Building, Barakhamba Road, New Delhi – 110 001, India.

Printed in India by Raj Press, New Delhi.

Preface

In questions of science, the authority of a thousand is not worth the humble reasoning of a single individual.
Galileo Galilei, physicist and astronomer (1564-1642)

This book is a second edition of "Classical Electromagnetic Theory" which derived from a set of lecture notes compiled over a number of years of teaching electromagnetic theory to fourth year physics and electrical engineering students. These students had a previous exposure to electricity and magnetism, and the material from the first four and a half chapters was presented as a review. I believe that the book makes a reasonable transition between the many excellent elementary books such as Griffith's *Introduction to Electrodynamics* and the obviously graduate level books such as Jackson's *Classical Electrodynamics* or Landau and Lifshitz' *Electrodynamics of Continuous Media.* If the students have had a previous exposure to Electromagnetic theory, all the material can be reasonably covered in two semesters. Neophytes should probable spend a semester on the first four or five chapters as well as, depending on their mathematical background, the Appendices B to F. For a shorter or more elementary course, the material on spherical waves, waveguides, and waves in anisotropic media may be omitted without loss of continuity.

In this edition I have added a segment on Schwarz-Christoffel transformations to more fully explore conformal mappings. There is also a short heuristic segment on Cherenkov radiation and Bremstrahlung. In Appendix D there is a brief discussion of orthogonal function expansions. For greater completeness, Appendices E and F have been expanded to include the solution of the Bessel equation and Legendre's equation as well as obtaining the generating function of each of the solutions. This material is not intended to supplant a course in mathematical methods but to provide a ready reference provide a backstop for those topics missed elsewhere. Frequently used vector identities and other useful formulas are found on the inside of the back cover and referred to inside the text by simple number (1) to (42).

Addressing the complaint "*I don't know where to start, although I understand all the theory*", from students faced with a non-transparent problem, I have included a large number of examples of varying difficulty, worked out in detail. This edition has been enriched with a number of new examples. These examples illustrate both the theory and the techniques used in solving problems. Working through these examples should equip the student with both the confidence and the knowledge to solve realistic problems. In response to suggestions by my colleagues I have numbered all equations for ease of referencing and more clearly delineated examples from the main text.

Because students appear generally much less at ease with magnetic effects than with electrical phenomena, the theories of electricity and magnetism are developed in parallel. From the demonstration of the underlying interconvertability of the fields in Chapter One to the evenhanded treatment of electrostatic and magnetostatic problems to the covariant formulation, the treatment emphasizes the relation between the electric and magnetic fields. No attempt has been made to follow the historical development of the theory.

An extensive chapter on the solution of Laplace's equation explores most of the techniques used in electro- and magnetostatics, including conformal mappings and separation of variable in Cartesian, cylindrical polar, spherical polar and oblate ellipsoidal coordinates. The magnetic scalar potential is exploited in many examples to demonstrate the equivalence of methods used for the electric and magnetic potentials. The next chapter explores the use of image charges in solving Poisson's equation and then introduces Green's functions, first heuristically, then more formally. As always, concepts introduced are put to use in examples and exercises. A fairly extensive treatment of radiation is given in the later portions of this book. The implications of radiation reaction on causality and other limitations of the theory are discussed in the final chapter.

I have chosen to sidestep much of the tedious vector algebra and vector calculus by using the much more efficient tensor methods, although, on the advice of colleagues, delaying their first use to chapter 4 in this edition. Although it almost universally assumed that students have some appreciation of the concept of a tensor, in my experience this is rarely the case. Appendix B addresses this frequent gap with an exposition of the rudiments of tensor analysis. Although this appendix cannot replace a course in differential geometry, I strongly recommend it for self-study or formal teaching if students are not at ease with tensors. The latter segments of this appendix are particularly recommended as an introduction to the tensor formulation of Special Relativity.

The exercises at the end of each chapter are of varying difficulty but all should be within the ability of strong senior students. In some problems, concepts not elaborated in the text are explored. A number of new problems have been added to the text both as exercises and as examples. As every teacher knows, it is essential that students consolidate their learning by solving problems on a regular basis. A typical regimen would consist of three to five problems weekly.

I have attempted to present clearly and concisely the reasoning leading to inferences and conclusions without excessive rigor that would make this a book in Mathematics rather than Physics. Pathological cases are generally dismissed. In an attempt to have the material transfer more easily to notes or board, I have labelled vectors by overhead arrows rather than the more usual bold face. As the material draws fairly heavily on mathematics I have strived to make the book fairly self sufficient by including much of the relevant material in appendices.

Rationalized SI units are employed throughout this book, having the advantage of yielding the familiar electrical units used in everyday life. This connection to reality tends to lessen the abstractness many students impute to electromagnetic theory. It is an added advantage of SI units that it becomes easier to maintain a clear distinction between B and H, a distinction frequently lost to users of gaussian units.

I am indebted to my students and colleagues who provided motivation for this book, and to Dr. Matti Stenroos and Dr. E.G. Jones who class tested a number of chapters and provided valuable feedback. Lastly, recognizing the unfortunate number of errata that escaped me and my proofreaders in the first edition, I have made a significantly greater effort to assure the accuracy of this edition.

Jack Vanderlinde
email: jvdl@unb.ca

Table of Contents

Chapter 1

Static Electric and Magnetic Fields in Vacuum

1.1 Static Charges

Static electricity, produced by rubbing different materials against one another, was known to the early Greeks who gave it its name (derived from ἠλεκτρον, pronounced ēlectron, meaning amber). Experiments by Du Fay in the early 18th century established that there are two kinds of electricity, one produced by rubbing substances such as hard rubber and amber, called resinous, and another produced by rubbing glassy substances such as quartz, dubbed vitreous. Objects with like charge were found to repel one another, while objects with unlike charge were found to attract. Benjamin Franklin attempted to explain electricity in terms of an excess or deficiency of the vitreous electric fluid, leading to the designations *positive* and *negative.*

A report by Benjamin Franklin that a cork ball inside an electrically charged metal cup is not attracted to the inside surface of the cup led Joseph Priestly to infer that, like gravity, electrical forces obey an inverse square law. This hypothesis was almost immediately confirmed (to limited accuracy) by John Robison, but the results were not published for almost 50 years. Cavendish, in an elegant experiment, showed that if a power law holds,[1] the exponent of r in the force law could not differ from minus two by more than 1 part in 50, but he failed to publish his results. Charles Augustin de Coulomb, who, in the late 18th century, measured both the attractive and repulsive force between charges with a delicate torsion balance, is credited with the discovery of the force law bearing his name – he found that the force is proportional to the product of the charges, acts along the line joining the charges, and decreases inversely as the square of the distance between them. Charges of opposite sign attract one another, whereas charges of the same sign repel. It has been verified experimentally that the exponent of r varies from minus two by no more than 1 part in 10^{16} over distances of order one meter.

[1] A modern interpretation suggests that a test of the exponent is not appropriate because a power law is not the anticipated form. In line with considerations by Proca and Yukawa, the potential should take the form $e^{-\beta r}/r$ ($\beta = m_\gamma c/\hbar$) where m_γ is the rest mass (if any) of the photon. Astronomical measurements of Jupiter's magnetic field place an upper limit of 4×10^{-51} kg on the mass of the photon

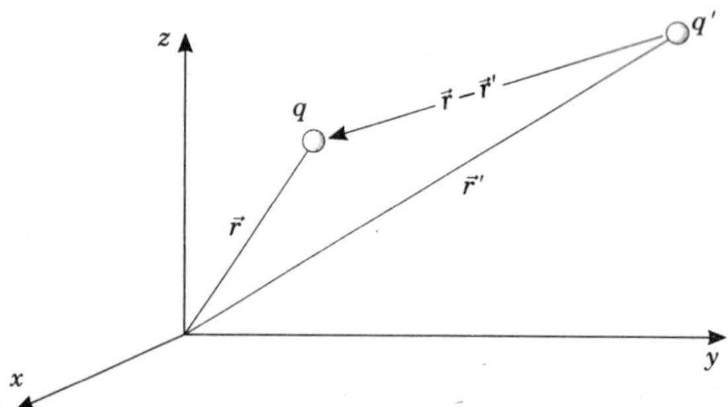

Figure 1.1: When q and q' are situated at r and r' respectively, the vector pointing from q' to q is $(\vec{r}-\vec{r}\,')$.

1.1.1 The Electrostatic Force

The inverse square electric force on a particle with charge q located at $\vec{r}$ due to a second charged particle with charge q' located at $\vec{r}\,'$ is (Figure 1.1) economically expressed by Coulomb's law:

$$\vec{F}_q = k_e \frac{qq'(\vec{r}-\vec{r}\,')}{|\vec{r}-\vec{r}\,'|^3} \tag{1–1}$$

Various system of units assign different values to k_e. *Gaussian Units* (Appendix A), used in many advanced texts, set $k_e \equiv 1$ thereby defining the unit of charge, the *esu*. In Gaussian units, the force is measured in *dynes* and r in cm. In this book we will uniformly use SI units, which have the advantage of dealing with ordinary electrical units such as volt and amperes at the cost of requiring k_e to take on a value of roughly 9×10^9 N-m^2/C^2. Anticipating later developments, we write $k = 1/4\pi\varepsilon_0$ to obtain:

$$\vec{F}_q = \frac{1}{4\pi\varepsilon_0}\frac{qq'(\vec{r}-\vec{r}\,')}{|\vec{r}-\vec{r}\,'|^3} \tag{1–2}$$

where ε_0, the *permittivity of free space*, is experimentally determined to be 8.84519×10^{-12} C^2/N-m^2. More properly, as we will see, ε_0 can be derived from the (defined) speed of light in vacuum, $c \equiv 299{,}792{,}458$ m/s and the (defined) *permeability of free space*, $\mu_0 \equiv 4\pi \times 10^{-7}$ kg-m/C^2, to give $\varepsilon_0 \equiv 1/\mu_0 c^2 = 8.85418781\cdots \times 10^{-12}$ C^2/N-m^2.

The force of several charges q_i' on q is simply the vector sum of the force q_1' exerts on q plus the force of q_2' on q and so on until the last charge q_n'. This statement may be summarized as

$$\vec{F}_q = \frac{q}{4\pi\varepsilon_0}\sum_{i=1}^{n}\frac{q_i'(\vec{r}-\vec{r}_i)}{|\vec{r}-\vec{r}_i|^3} \tag{1–3}$$

or, to translate it to the language of calculus, with the small element of source charge denoted by dq'

$$\vec{F}_q = \frac{q}{4\pi\varepsilon_0}\int\frac{(\vec{r}-\vec{r}\,')dq'}{|\vec{r}-\vec{r}\,'|^3} \tag{1–4}$$

Although we know that electric charges occur only in discrete quanta $\pm e = \pm 1.6021917 \times 10^{-19}$ coulomb (or $\pm\frac{1}{3}e$ and$\pm\frac{2}{3}e$ if quarks are considered), the elementary charge is so small that we normally deal with many thousands at a time and we replace the individual charges by a smeared-out *charge density.* Thus the charge distribution is described by a charge density $\rho(\vec{r}\,') = n(\vec{r}\,')e$, with n the *net* number (positive minus negative) of positive charges per unit volume centered on r'. For a distributed charge, we may generally write

$$\vec{F}_q = \frac{q}{4\pi\varepsilon_0}\int_\tau \frac{\rho(\vec{r}\,')(\vec{r}-\vec{r}\,')}{|\vec{r}-\vec{r}\,'|^3}\,d^3r' \tag{1-5}$$

where dq' has been replaced by$\rho(\vec{r}\,')d^3r'$. The differential d^3r' represents the three-dimensional differential volume in arbitrary coordinates. For example, in Cartesians $d^3r' \equiv dx'dy'dz'$, whereas in spherical polars, $d^3r' \equiv r'^2\sin\theta' dr' d\theta' d\varphi'$. For charges distributed over a surface, it suffices to replace dq' by $\sigma(\vec{r}\,')dA'$ and for line charges we write $dq' = \lambda(\vec{r}\,')d\ell'$.

If required, the lumpiness of a point charge q' can be accommodated in (1–5) by letting the charge density have the form of a three-dimensional Dirac δ function.[2] Line charges and surface charges can similarly be accommodated.

The original form (1–2) is easily recovered by setting $\rho(\vec{r}\,') = q'\delta(\vec{r}\,' - \vec{r}_q)$ with $\delta(\vec{r}\,' - \vec{r}_q) = \delta(x' - x_q)\delta(y' - y_q)\delta(z' - z_q)$ [3] and carrying out the integration called for in (1–5).

EXAMPLE 1.1: Find the force on a charge q lying on the z axis above the center of a circular hole of radius a in an infinite uniformly charged flat plate occupying the x-y plane, carrying surface charge density σ (Figure 1.2).

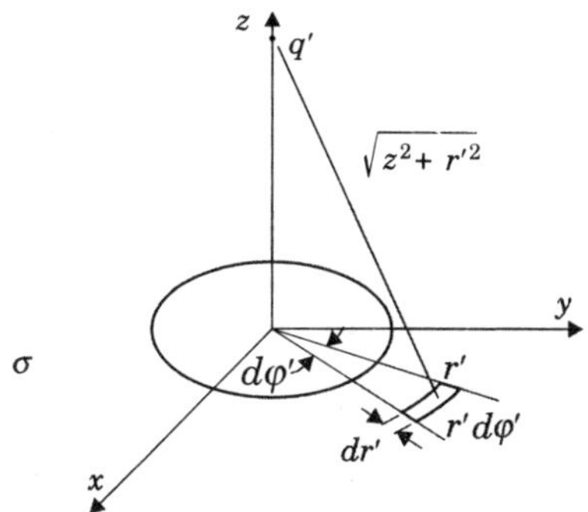

Figure 1.2: Example 1.1 – A uniformly distributed charge lies on the x-y plane surrounding the central hole in the plate.

[2]The δ function $\delta(x - a)$ is a sharply spiked function that vanishes everywhere except at $x = a$, where it is infinite. It is defined by $\int f(x)\delta(x-a)dx = f(a)$ when a is included in the region of integration; it vanishes otherwise. For further discussion, see Appendix C.

[3]In non-Cartesian coordinate systems the δ function may not be so obvious. In spherical polar coordinates, for example, $\delta(\vec{r} - \vec{r}\,') = r'^{-2}\delta(r - r')\delta(\cos\theta - \cos\theta')\,\delta(\varphi - \varphi')$.

Solution: The boundary of the integration over the charge distribution is most easily accommodated by working in cylindrical polar coordinates. The field point is located at $z\hat{k}$ and the source points have coordinates r' and φ' giving $\vec{r}-\vec{r}\,' = z\hat{k} - r'\cos\varphi'\hat{\imath} - r'\sin\varphi'\hat{\jmath}$. An element of charge dq' takes the form $dq' = \sigma(r', \varphi')dA' = \sigma r'\,dr'd\varphi'$. The distance $|\vec{r}-\vec{r}\,'|$ of the source charge element from the field point is $\sqrt{z^2+r'^2}$ leading us to write:

$$\vec{F}_q = \frac{q}{4\pi\varepsilon_0}\int_a^\infty \int_0^{2\pi} \frac{\sigma(z\hat{k} - r'\cos\varphi\,\hat{\imath} - r'\sin\varphi\,\hat{\jmath})}{(z^2+r'^2)^{3/2}} r'dr'd\varphi' \tag{Ex 1.1.1}$$

The integrations over $\sin\varphi$ and $\cos\varphi$ yield 0, reducing the integral to:

$$\vec{F}_q = \frac{q}{4\pi\varepsilon_0}\,2\pi\int_a^\infty \frac{\sigma z\hat{k}}{(z^2+r'^2)^{3/2}} r'dr'$$

$$= \frac{q}{2\varepsilon_0}\,\frac{-\sigma z\hat{k}}{\sqrt{z^2+r'^2}}\bigg|_a^\infty = \frac{q\sigma z\hat{k}}{2\varepsilon_0\sqrt{z^2+a^2}} \tag{Ex 1.1.2}$$

The force points upward above the plane and downward below. At large distances it tends to a constant, $\frac{1}{2\varepsilon_0}q\sigma\hat{k}$, exactly what it would be in the absence of the hole. We further verify that as $z \to 0$, in the center of the hole, the force vanishes. It is probably worth mentioning that the charge would not distribute itself uniformly on a conducting plate so that we have not solved the problem of a charged conducting plate with a hole.

Example 1.2: Find the force exerted on a point charge Q located at $\vec{r}$ in the x-y plane by a long (assume infinite) line charge λ, uniformly distributed along a thin wire lying along the z axis (Figure 1.3).

Solution: An element of charge along the wire is given by $dq' = \lambda dz'$ so that using (1–4) we can write the force on the charge

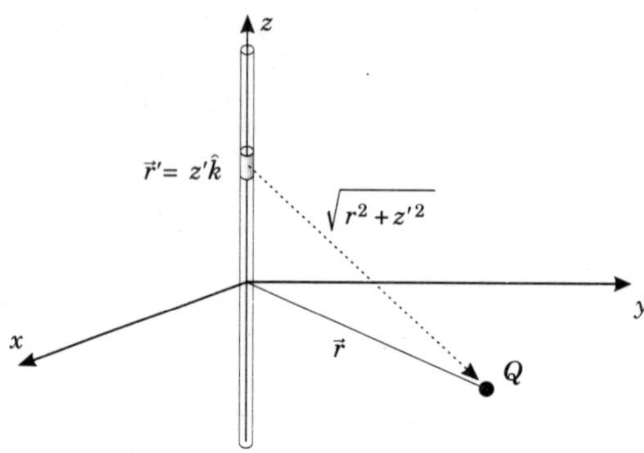

Figure 1.3: A line charge λ is distributed uniformly along the z axis.

$$\vec{F}_Q = \frac{Q}{4\pi\epsilon_0}\int_{-\infty}^{\infty}\frac{(r\hat{r} - z'\hat{k})\,\lambda dz'}{(r^2 + z'^2)^{3/2}} \tag{Ex 1.2.1}$$

where $\hat{r}$ is a unit vector in the x-y plane pointing from the origin to the charge Q. The integral is best evaluated in two parts. The second part,

$$\int_{-\infty}^{\infty}\frac{-\lambda z'\hat{k}dz'}{(r^2 + z'^2)^{3/2}} = 0 \tag{Ex 1.2.2}$$

because the integrand is odd. The remaining integral is then,

$$\vec{F}_Q = \frac{Q}{4\pi\epsilon_0}\int_{-\infty}^{\infty}\frac{\lambda r\hat{r}dz'}{(r^2 + z'^2)^{3/2}} \tag{Ex 1.2.3}$$

r and $\hat{r}$ are constants with respect to dz' so that we may use (28) to evaluate the integral:

$$\vec{F}_Q = \left.\frac{Q\lambda r\hat{r}z'}{4\pi\epsilon_0 r^2\sqrt{r^2 + z'^2}}\right|_{-\infty}^{\infty} = \frac{Q\lambda\hat{r}}{2\pi\epsilon_0 r} \tag{Ex 1.2.4}$$

This question will be revisited in example 1.5 where we will allow the charge carrying wire to have finite size.

1.1.2 The Electric Field

Although Coulomb's law does an adequate job of predicting the force one particle causes another to feel, there is something almost eerie about one particle pushing or pulling another without any physical contact. Somehow, it would be more satisfying if the charged particle felt a force due to some local influence, a *field*, created by all other charged particles. This field would presumably exist independent of the sensing particle q. (In quantum electrodynamics, even the notion of a field without a carrier [the photon] is held to be aphysical.)

The force on the sensing particle must be proportional to its charge; all the other properties of the force will be assigned to the *electric field*, $\vec{E}(\vec{r})$. Thus we define the electric field by

$$\vec{F} = q\vec{E}(\vec{r}) \tag{1–6}$$

where $\vec{F}$ is the force on the charge q situated at $\vec{r}$ and $\vec{E}(\vec{r})$ is the electric field at position $\vec{r}$ due to all other charges. (The source charge's coordinates will normally be distinguished from the *field point* by a prime [$'$].) One might well wonder why the sensing particle, q's field would not be a component of the field at its position. A simple answer in terms of the electric field's definition, (1–6) above, is that since a particle can exert no net force on itself, its own field cannot be part of the field it senses (in the same way that you cannot lift yourself by your bootstraps). Unfortunately, this appears to suggest that two point charges at the same position might well experience a different field. That argument, however, is somewhat academic, as two point charges at the same location would give rise to infinite interaction forces. One might also argue that, as a point particle's field must be spherically symmetric, it would, in fact make no difference whether we included the sensing particle's own field in computing the force on the particle. This particular point of view runs into trouble when we consider the no longer spherically symmetric fields of accelerated charges in Chapter 12. Whatever the best answer, this self field will continue to trouble us whenever we deal seriously with point particles.

Factoring the charge q from Coulomb's law (1–5), we find that the electric field produced by a charge distribution $\rho(\vec{r}')$ must be

$$\vec{E}(\vec{r}) = \frac{1}{4\pi\varepsilon_0}\int \frac{\rho(\vec{r}')\,(\vec{r}-\vec{r}')}{|\vec{r}-\vec{r}'|^3}\,d^3r' \tag{1–7}$$

where the integration is carried out over all space (ρ must of course vanish at sufficiently large r, making the volume of integration less than infinite). We reiterate that coordinates of the source of the field will be primed, while the *field* points are denoted by unprimed coordinates.

EXAMPLE 1.3: Find the electric field above the center of a flat, circular plate of radius R, bearing a charge Q uniformly distributed over the top surface (Figure 1.3).

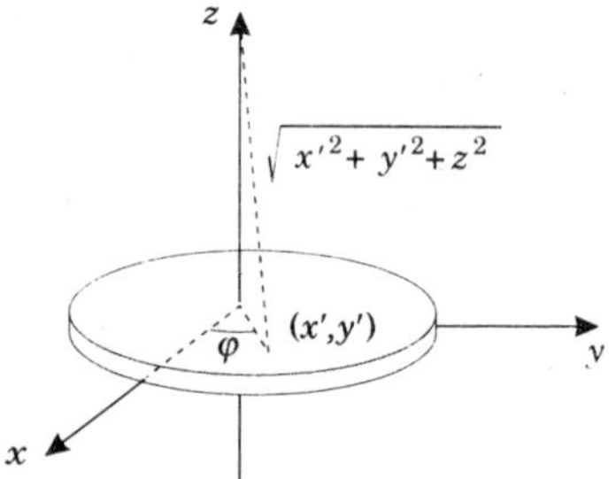

Figure 1.4: The field at height z above a uniformly charged disk.

Solution: The charge density on the plate takes the form $\rho(\vec{r}) = \frac{Q}{\pi R^2}\delta(z')$ for $x'^2+y'^2 \le R^2$ and 0 elsewhere. Using

$$\vec{E}(\vec{r}) = \frac{1}{4\pi\varepsilon_0}\int \frac{\rho(\vec{r}')\,(\vec{r}-\vec{r}')}{|\vec{r}-\vec{r}'|^3}\,d^3r' \tag{Ex 1.3.1}$$

we obtain the explicit expression

$$\vec{E}(0,0,z) = \frac{1}{4\pi\varepsilon_0}\frac{Q}{\pi R^2}\int_{-R}^{R}\int_{-\sqrt{R^2-y'^2}}^{\sqrt{R^2-y'^2}} \frac{z\hat{k}-x'\hat{\imath}-y'\hat{\jmath}}{(x'^2+y'^2+z^2)^{3/2}}\,dx'dy' \tag{Ex 1.3.2}$$

In cylindrical polar coordinates, $x' = r'\cos\varphi'$, $y' = r'\sin\varphi'$, and $dx'dy' = r'dr'd\varphi'$, giving

$$\vec{E}(0,0,z) = \frac{1}{4\pi\varepsilon_0}\frac{Q}{\pi R^2}\int_0^R\int_0^{2\pi} \frac{(z\hat{k}-r'\cos\varphi'\hat{\imath}-r'\sin\varphi'\hat{\jmath})\,d\varphi'\,r'\,dr'}{(r'^2+z^2)^{3/2}} \tag{Ex 1.3.3}$$

The integration over φ' eliminates the $\sin\varphi'$ and the $\cos\varphi'$ terms, leaving only

$$\vec{E}(0,0,z) = \frac{Q}{2\pi\varepsilon_0 R^2}\int_0^R \frac{z\hat{k}r'dr'}{(r'^2+z^2)^{3/2}} = -\frac{Q}{2\pi\varepsilon_0 R^2}\left.\frac{z\hat{k}}{(r'^2+z^2)^{1/2}}\right|_0^R$$

$$= \frac{Q}{2\pi\varepsilon_0 R^2}\left(\frac{z\hat{k}}{\sqrt{z^2}} - \frac{z\hat{k}}{\sqrt{R^2+z^2}}\right) = \frac{Q\hat{k}}{2\pi\varepsilon_0 R^2}\left(1 - \frac{z}{\sqrt{R^2+z^2}}\right) \quad \text{(Ex 1.3.4)}$$

When z is small compared to R the field reduces to $(\sigma/2\varepsilon_0)\hat{k}$, the value it would have above an infinite sheet, whereas at large distances it tends to $Q/(4\pi\varepsilon_0 z^2)$. It is worth noting that adding to Ex 1.3.4 the field of the plate with the hole deduced from Ex 1.1.2 gives precisely the field of the infinite plate with the hole filled in.

The invention of the electric field appears at this point no more than a response to a vague uneasiness about the action at a distance implicit in Coulomb's law. As we progress we will endow the field, $\vec{E}$, with properties such as energy and momentum, and the field will gain considerable reality. Whether $\vec{E}$ is merely a mathematical construct or has some independent objective reality cannot be settled until we discuss radiation in Chapter 10.

1.1.3 Gauss' Law

It is evident that the evaluation of $\vec{E}$, even for relatively simple source charge distributions, is fairly cumbersome. When problems present some symmetry, they can often be solved much more easily using the integral form of Gauss' law, which states

$$\oint_S \vec{E}(\vec{r})\cdot d\vec{S}(\vec{r}) = \frac{q'}{\varepsilon_0} \tag{1–8}$$

where S is any closed surface, q' is the charge enclosed within that surface, $d\vec{S}$ is surface element of S pointing in the direction of an outward-pointing normal, and $\vec{r}$ is the location of the element $d\vec{S}$ on the surface. Note that S need not be a physical surface.

To prove this result, we expand $\vec{E}$ in (1–8) using (1–7) to obtain

$$\oint_S \vec{E}(\vec{r})\cdot d\vec{S}(\vec{r}) = \frac{1}{4\pi\varepsilon_0}\oint\left[\int\frac{\rho(\vec{r}\,')(\vec{r}-\vec{r}\,')}{|\vec{r}-\vec{r}\,'|^3}\,d^3r'\right]\cdot d\vec{S}(\vec{r})$$

$$= \frac{1}{4\pi\varepsilon_0}\int\left[\oint\frac{(\vec{r}-\vec{r}\,')}{|\vec{r}-\vec{r}\,'|^3}\cdot d\vec{S}(\vec{r})\right]\rho(\vec{r}\,')\,d^3r' \tag{1–9}$$

We must now evaluate the surface integral $\oint \frac{(\vec{r}-\vec{r}\,')}{|\vec{r}-\vec{r}\,'|^3}\cdot d\vec{S}$.

We divide the source points into those lying inside the surface S and those lying outside. The divergence theorem (20) generally allows us to write

$$\oint\frac{(\vec{r}-\vec{r}\,')}{|\vec{r}-\vec{r}\,'|^3}\cdot d\vec{S} = \int\vec{\nabla}\cdot\frac{(\vec{r}-\vec{r}\,')}{|\vec{r}-\vec{r}\,'|^3}d^3r \tag{1–10}$$

For any fixed source point with coordinate $\vec{r}\,'$ outside the closed surface, S, the field point $\vec{r}$ (located inside S for the volume integration resulting from the application of the divergence theorem) never coincides with $\vec{r}\,'$, and it is easily verified by direct differentiation that the integrand on the right hand side of (1–10), the divergence of $(\vec{r}-\vec{r}\,')/|\vec{r}-\vec{r}\,'|^3$, vanishes identically. We conclude, therefore, that charges lying outside the surface make no contribution to the surface integral of the electric field.

When $\vec{r}\,'$ is inside the bounding surface S, the singularity at $\vec{r}=\vec{r}\,'$ prevents a similar conclusion because the divergence of (1–10) becomes singular. To deal with this circumstance, we exclude a small spherical region of radius R centered on $\vec{r}\,'$ from the $\oint$ and integrate over this spherical surface separately. In the remaining volume, excluding the small sphere, $(\vec{r}-\vec{r}\,')/|\vec{r}-\vec{r}\,'|^3$ again has a vanishing divergence and presents no singularities, allowing us to conclude that it too, makes no contribution to the surface integral of $\vec{E}$. Setting $\vec{R}=(\vec{r}-\vec{r}\,')$ and $\hat{R}\cdot d\vec{S}=R^2 d\Omega$, with $d\Omega=\sin\theta\, d\theta\, d\varphi$, an element of solid angle, we may write the integral

$$\oint_{sphere}\frac{(\vec{r}-\vec{r}\,')}{|\vec{r}-\vec{r}\,'|^3}\cdot d\vec{S}(\vec{r})=\oint_{sphere}\frac{(\vec{R}\cdot\hat{R})\,R^2\,d\Omega}{R^3}=4\pi \tag{1–11}$$

We substitute this result, (1–11) into (1–9) to get (1–8), the desired result

$$\oint_{sphere}\vec{E}(\vec{r})\cdot d\vec{S}(\vec{r})=\frac{1}{4\pi\varepsilon_0}\cdot 4\pi\int\rho(\vec{r}\,')d^3r'=\frac{q'}{\varepsilon_0}$$

A more geometric insight into the evaluation of (1–10) may be obtained by recognizing that the left hand side of (1–10) represents the solid angle covered by the surface as seen from $\vec{r}\,'$. When $\vec{r}\,'$ is inside the surface, S encloses the entire 4π solid angle, whereas when $\vec{r}\,'$ lies outside S, the contribution from near side of the surface makes the same solid angle as the back side, but the two bear opposite signs (because the normals point in opposite directions) and cancel one another.

To reiterate, only the charge inside the surface S enters into the integration. In words, Gauss' law states that the perpendicular component of the electric field integrated over a closed surface equals $1/\varepsilon_0$ times the charge enclosed within that surface, irrespective of the shape of the enclosing surface.

EXAMPLE 1.4: Find the electric field at a distance r from the center of a uniformly charged sphere of radius R and total charge Q.

Solution: The charge $Q(r)$ enclosed within a sphere of radius r centered on the charge center is

$$Q(r)=\begin{cases}\left(\dfrac{r}{R}\right)^3 Q & \text{for } r\le R\\ Q & \text{for } r>R\end{cases} \tag{Ex 1.4.1}$$

On a spherical surface of radius r, symmetry requires $\vec{E}=E_r(r)\hat{r}$ with no φ dependence. Gauss' law then gives us

$$\oint\vec{E}\cdot d\vec{S}=\oint\vec{E}\cdot\hat{r}\,r^2 d\Omega=4\pi r^2\,E_r \tag{Ex 1.4.2}$$

Equating this to $Q(r)/\varepsilon_0$ yields

$$E_r = \begin{cases} \dfrac{rQ}{4\pi\varepsilon_0 R^3} & r \le R \\ \dfrac{Q}{4\pi\varepsilon_0 r^2} & r > R \end{cases} \tag{Ex 1.4.3}$$

The electric field inside the sphere grows linearly with radius and falls off quadratically outside the sphere. As Gauss' law applies equally to gravity and electrostatics (it depends only on the r^{-2} nature of the force), the same field dependence pertains to gravitational fields within gravitating bodies.

EXAMPLE 1.5: Find the electric field near a long, uniformly charged cylindrical rod of radius a.

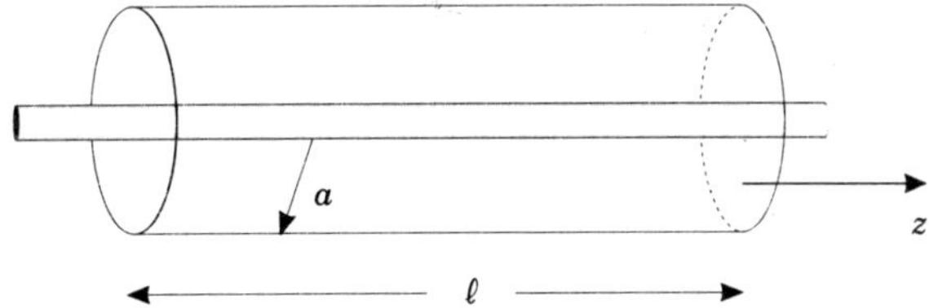

Figure 1.5: The cylinder about the rod forms a Gaussian surface perpendicular to the electric field. The end faces contribute nothing to the surface integral.

Solution: $E_z = 0$, as reversing the z axis, or translating the origin along the z axis does not change the problem. $E_\theta = 0$ as reversing θ or rotating the system about the z axis leaves the system invariant. Evidently $\vec{E}(r,\ \theta,\ z) = E_r(r)\,\hat{r}$. Drawing a cylinder about the rod as indicated in Figure 1.4 (the cylinder may be interior to the rod), we obtain from Gauss' law, (1–8),

$$\oint \vec{E}\cdot d\vec{S} = \int \frac{\rho}{\varepsilon_0}\, d^3r'$$

For $r > a$, this becomes

$$2\pi r\ell E_r = \frac{\rho\pi\, a^2\ell}{\varepsilon_0}, \qquad \text{or } E_r = \frac{\rho a^2}{2\varepsilon_0 r}, \tag{Ex 1.5.1}$$

while for $r < a$ we obtain

$$2\pi r\ell E_r = \frac{\rho\pi\, r^2\ell}{\varepsilon_0}, \qquad \text{or } E_r = \frac{\rho r}{2\varepsilon_0}. \tag{Ex 1.5.2}$$

We will make frequent use of Gaussian cylinders (and "pill-boxes" in the next example) throughout the remainder of this book.

EXAMPLE 1.6: Find the electric field between two conducting infinite parallel plates bearing surface charge densities σ and $-\sigma$.

Solution: For a Gaussian enclosing surface we now choose a very flat cylinder (commonly referred to as a *pillbox*) that includes one of the two charged surfaces, say the top surface, of the bottom plate, as illustrated in Figure 1.6. The charge enclosed within the pillbox is σA where A is the flat area included in the box. Because of the symmetry we anticipate an electric field whose only non vanishing component is E_z. Gauss' law then becomes:

$$\oint \vec{E} \cdot d\vec{S} = A(E_{top} - E_{bott}) = \frac{\sigma A}{\varepsilon_0} \qquad \text{(Ex 1.6.1)}$$

If the plates are conductors, then the electric field on the bottom surface of the pillbox lying inside the conductor must vanish (otherwise charges inside the metal would be subject to a Coulomb force and move until the field does vanish). We conclude, then, that the electric field at the top surface of the pillbox is $E_z = \sigma/\varepsilon_0$. Increasing the height of the pillbox straddling the bottom plate so that its top surface lies progressively closer to the top plate produces no variation of the enclosed charge; we infer that the field is uniform, (i.e., it does not vary with z.)

A similar argument could of course have been employed at the top plate, giving exactly the same result. This time below the plate, the surface of the pillbox along which $\vec{E}$ does not vanish points downward so that $-E_z = -\sigma\varepsilon_0$. As stated above, we shall make frequent use of the pillbox whenever we deal with the behavior of the field at surfaces, both conducting and nonconducting.

Gauss' law may be restated in terms of the local charge density by means of the divergence theorem. Writing the charge enclosed within the boundary S as the volume integral of the charge density enclosed, we find

$$\oint \vec{E} \cdot d\vec{S} = \int \frac{\rho}{\varepsilon_0} d^3r' \qquad (1\text{–}12)$$

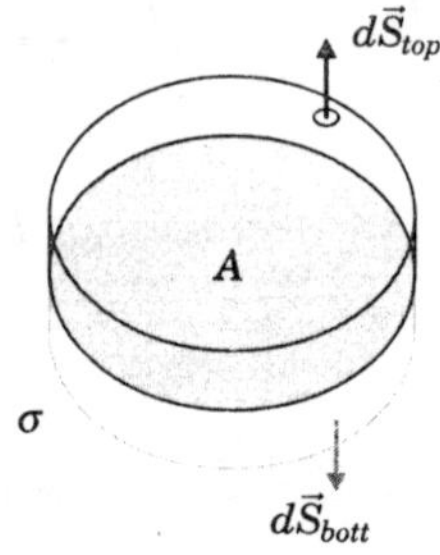

Figure 1.6: The "pillbox" encloses one of the charged surfaces. The electric field is parallel to the curved surface side so that the integral over the curved side makes no contribution to the surface integral.

With the aid of the divergence theorem, the surface integral of the electric field may be rewritten:

$$\int_{vol} \vec{\nabla} \cdot \vec{E}(\vec{r})d^3r = \int_{vol} \frac{\rho(\vec{r}\,')}{\varepsilon_0}\, d^3r' \tag{1–13}$$

Since the boundary and hence the volume of integration was arbitrary but the same on both sides, we conclude that the integrands must be equal

$$\vec{\nabla} \cdot \vec{E}(\vec{r}) = \frac{\rho(\vec{r})}{\varepsilon_0} \tag{1–14}$$

1.1.4 The Electric Potential

The expression for the electric field arising from a charge distribution may be usefully expressed as a gradient of a scalar integral as follows.

$$\begin{aligned}\vec{E}(\vec{r}) &= \frac{1}{4\pi\varepsilon_0}\int \rho(\vec{r}\,')\,\frac{(\vec{r}-\vec{r}\,')}{|\vec{r}-\vec{r}\,'|^3}\, d^3r' \\ &= -\frac{1}{4\pi\varepsilon_0}\int \rho(\vec{r}\,')\,\vec{\nabla}\left(\frac{1}{|\vec{r}-\vec{r}\,'|}\right) d^3r' \end{aligned} \tag{1–15}$$

As $\vec{\nabla}$ acts only on the unprimed coordinates, we may take it outside the integral (1–15) to obtain

$$\vec{E}(\vec{r}) = -\vec{\nabla}\left(\frac{1}{4\pi\varepsilon_0}\int \frac{\rho(\vec{r}\,')}{|\vec{r}-\vec{r}\,'|}\, d^3r'\right) \equiv -\vec{\nabla}V \tag{1–16}$$

We identify the *electric potential*, V, with the integral

$$V(\vec{r}) = \frac{1}{4\pi\varepsilon_0}\int \frac{\rho(\vec{r}\,')}{|\vec{r}-\vec{r}\,'|}\, d^3r' \tag{1–17}$$

Since the curl of any gradient vanishes, we have immediately

$$\vec{\nabla} \times \vec{E} = -\vec{\nabla} \times (\vec{\nabla}V) = 0 \tag{1–18}$$

There are obvious advantages to working with the scalar V instead of the vector field $\vec{E}$. First, the integral for V requires computing only one component rather than the three required for $\vec{E}$. Second, the electric field obtained from several localized sources would require taking the vector sum of the fields resulting from each source. Since taking the gradient is a linear operation, the electric field could as well be found from

$$\begin{aligned}\vec{E} &= -\vec{\nabla}V_1 - \vec{\nabla}V_2 - \vec{\nabla}V_3 - \ldots \\ &= -\vec{\nabla}(V_1 + V_2 + V_3 + \ldots)\end{aligned}$$

where now we need only find a scalar sum of potentials. These simplifications make it well worthwhile to use the electric scalar potential whenever possible.

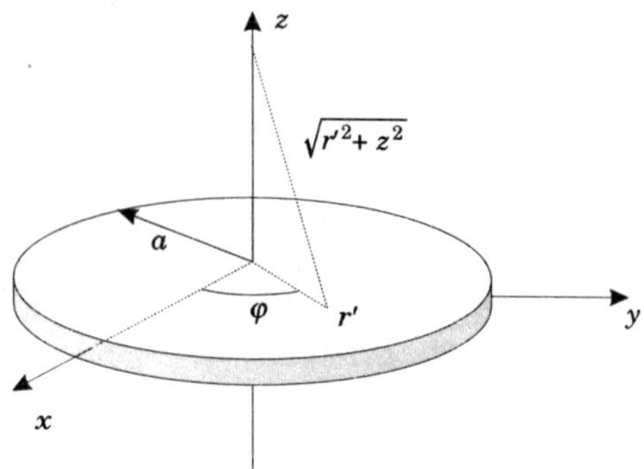

Figure 1.7: The circular plate is assumed to lie in the x-y plane.

EXAMPLE 1.7: Find the potential V at height z above the center of a disk of radius a carrying charge Q uniformly distributed over its top surface (Figure 1.7).

Solution: The charge density on the disk is

$$\rho = \begin{cases} \dfrac{Q}{\pi a^2}\,\delta(z') & \text{for } x'^2 + y'^2 \leq a^2 \\ 0 & \text{elsewhere} \end{cases} \tag{Ex 1.7.1}$$

The potential above the center of the plate is then

$$V(0,0,z) = \frac{1}{4\pi\varepsilon_0}\int \frac{\rho(\vec{r}\,')\,d^3r\,'}{|\vec{r}-\vec{r}\,'|} = \frac{1}{4\pi\varepsilon_0}\int_0^a \int_0^{2\pi} \frac{Q}{\pi a^2}\frac{r'dr'd\varphi'}{\sqrt{r'^2+z^2}}$$

$$= \frac{Q}{2\pi\varepsilon_0 a^2}\sqrt{r'^2+z^2}\,\Bigg|_0^a = \frac{Q}{2\pi\varepsilon_0 a^2}\left(\sqrt{a^2+z^2}-z\right) \tag{Ex 1.7.2}$$

To find the field $\vec{E}$ as we have already done in the example 1.3, we merely find the gradient of V giving

$$E_z(0,0,z) = -\frac{\partial V}{\partial z} = \frac{Q}{2\pi\varepsilon_0 a^2}\left(1 - \frac{z}{\sqrt{a^2+z^2}}\right) \tag{Ex 1.7.3}$$

At large distance the field reduces to that of a point charge. (Notice that because we have not calculated V as a function of x and y [or r and φ], no information about E_x or E_y can be obtained from V, however, symmetry dictates that they must vanish on the z axis.)

Like the electric field, the potential can also be expressed in terms of the local charge density by combining the differential form of Gauss' law with the definition of V, $-\vec{\nabla}V = \vec{E}$:

$$\vec{\nabla}\cdot\vec{E} = \vec{\nabla}\cdot(-\vec{\nabla}V) = -\nabla^2 V = \frac{\rho}{\varepsilon_0} \tag{1–19}$$

The differential equation solved by V, $\nabla^2 V = -\rho/\varepsilon_0$, is known as Poisson's equation. Its homogeneous counterpart ($\rho = 0$) is the Laplace equation. The Laplace equation will be discussed in considerable detail in Chapter 5 and the solution of Poisson's equation is the subject of Chapter 6.

1.2 Moving Charges

In the remainder of this chapter we consider the forces and fields due to slowly moving charges. Although the charges are allowed to move, we do insist that a steady state exist so that the forces are static. This restriction will be lifted in later chapters.

1.2.1 The Continuity Equation

Among the most fundamental conservation laws of physics is conservation of charge. There is no known interaction that creates or destroys charge (unlike mass, which can be created or annihilated). This conservation law is expressed by the equation of continuity (1–24).

We define the current flowing into some volume as the rate that charge accumulates in that volume:

$$I = \frac{dQ}{dt} \tag{1–20}$$

More usefully, we express I as the net amount of charge crossing the boundary into the volume τ with boundary S per unit time,

$$I = -\oint_S \rho\vec{v} \cdot d\vec{S} = -\oint_S \vec{J} \cdot d\vec{S} \tag{1–21}$$

where $\vec{J} \equiv \rho\vec{v}$ is the *current density*. (Recall that $d\vec{S}$ points outward from the volume so that $\vec{J} \cdot d\vec{S}$ is an outflow of current, hence the negative sign for current flowing into the volume.) Combining (1–20) and (1–21) and replacing Q with the volume integral of the charge density, we obtain

$$\oint_S \vec{J} \cdot d\vec{S} = -\frac{dQ}{dt} = -\frac{d}{dt}\int_\tau \rho\, d^3r \tag{1–22}$$

With the aid of the divergence theorem, the right and left hand side of this equation become

$$\int_\tau \vec{\nabla} \cdot \vec{J}\, d^3r = -\int_\tau \frac{\partial \rho}{\partial t}\, d^3r \tag{1–23}$$

Since the volume of integration τ was arbitrary, the integrands must be equal, giving the continuity equation

$$\vec{\nabla} \cdot \vec{J} + \frac{\partial \rho}{\partial t} = 0 \tag{1–24}$$

(It is perhaps useful to maintain this intuitive view of divergence as the outflow of a vector field from a point.) The equation of continuity states simply that an increase in charge density can only be achieved by having more current arrive at the point than leaves it. The continuity equation expresses the conservation of charge, one of the cornerstones of physics and is fundamental to the study of electromagnetic theory. When dealing with electrostatics we have $\vec{\nabla} \cdot \vec{J} = 0$ (note that this does not preclude current flows, only that $\partial\rho/\partial t = 0$).

1.2.2 Magnetic Forces

Magnetic forces were familiar to Arab navigators before 1000 A.D. who used lodestones as primitive compasses. Magnetic poles of the earth were postulated in the thirteenth century but it was not until about 1820 that Biot, Savart and Ampère discovered the interaction between currents and magnets.

As Biot, Savart and Ampère discovered, when charges are in motion, a force additional to the electrical force appears. We could merely postulate a force law (Equation (1–38), but it would be more satisfying to demonstrate the intimate connection between electricity and magnetism by obtaining magnetism as a consequence of electricity and relativistic covariance.

As a simple demonstration of why we expect currents to interact, let us consider two long line charges, each of length L with linear charge density λ_1 (consider them tending to infinity, requiring only that λL be a finite constant) lying along the x-axis and λ_2 parallel to the λ_1, at distance r from the x-axis. As seen by a stationary observer, the force on wire *2* is (Ex 1.2.4)

$$F = F_e = \frac{1}{2\pi\varepsilon_0}\frac{\lambda_1\lambda_2 L}{r^2}\vec{r} \tag{1–25}$$

A second observer moving with velocity v along the x axis sees the line charges in motion with velocity $-v$. According to special relativity, the transverse (to the motion) components of forces in a stationary and a moving (indicated by a prime) reference frame are related by $F = \gamma F'$ $[\gamma \equiv (1-v^2/c^2)^{-1/2}]$. We deduce, therefore, that in the moving observer's frame, the total force of one wire on the other should be $F' = \gamma^{-1}F$. [4]

Alternatively, we calculate a new length L' for the moving line charges using the length contraction formula $L' = \gamma^{-1}L$, and we deduce, assuming conservation of charge, an appropriately compressed charge density $\lambda_1' = \gamma\lambda_1$ and $\lambda_2' = \gamma\lambda_2$ in the moving frame. Thus, if the same laws of physics are to operate, the moving observer calculates an electric force

$$\vec{F}_e' = \frac{1}{2\pi\varepsilon_0}\frac{\lambda_1'\lambda_2'L'}{r^2}\vec{r} = \frac{1}{2\pi\varepsilon_0}\frac{(\gamma\lambda_1)(\gamma\lambda_2)\gamma^{-1}L}{r^2}\vec{r} = \frac{\gamma}{2\pi\varepsilon_0}\frac{\lambda_1\lambda_2 L}{r^2}\vec{r} = \gamma\vec{F}_e \tag{1–26}$$

clearly not the result anticipated above.

In fact, the moving observer, who of course believes the line charges to be in motion, must invent a second force, say F_m, in order to reconcile the results of the alternative calculations.

Thus

[4] More properly, the transverse force F, on a particle moving with velocity v in system Σ, is related to F' in Σ' where the particle has velocity v' by $\gamma F = \gamma' F'$, and transverse means perpendicular to the velocity of frame Σ' with respect to Σ.

$$\vec{F}' = \frac{\vec{F}_e}{\gamma} = \vec{F}_e' + \vec{F}_m' = \gamma \vec{F}_e + \vec{F}_m' \tag{1-27}$$

which we may solve for $\vec{F}_m'$:

$$\vec{F}_m' = \frac{\vec{F}_e}{\gamma} - \gamma \vec{F}_e = \gamma \vec{F}_e \left(\frac{1}{\gamma^2} - 1\right)$$

$$= \gamma \vec{F}_e \left(1 - \frac{v^2}{c^2} - 1\right) = -\frac{v^2}{c^2}\gamma \vec{F}_e$$

$$= -\vec{F}_e' \frac{v^2}{c^2} = \frac{(\lambda_1' v \lambda_2' v) L'}{2\pi\varepsilon_0 c^2 r^2}\vec{r}$$

$$= -\frac{I_1' I_2' L'}{2\pi\varepsilon_0 c^2 r^2}\vec{r} \equiv -\frac{\mu_0 I_1' I_2' L'}{2\pi r^2}\vec{r} \tag{1-28}$$

In the frame where the line charges move, we find a force opposite to the electrical force, proportional to the product of the currents. Parallel currents attract one another; antiparallel currents repel. The term $1/\varepsilon_0 c^2$, conventionally abbreviated as μ_0, is called the permeability of free space. The constant μ_0 has a defined value $4\pi \times 10^{-7}$ kg-m/C^2. This choice fixes the unit of charge, the coulomb, which we have conveniently left undefined to this point.

Again shying away from action at a distance we invent a field $\vec{B}$, produced by I_2 at the location of the current I_1 with which the current I_1 interacts. Since $d\vec{F}_m$, the magnetic force on a short segment $Id\vec{\ell}$ of current *1*, is perpendicular to I_1, it must be of the form

$$d\vec{F}_m = I_1 d\vec{\ell} \times \vec{B} \tag{1-29}$$

where $\vec{B}$ is a yet undetermined vector field known as the *magnetic induction field*, or alternatively the *magnetic flux density*.

The magnetic force on a moving charged point particle is easily deduced by identifying $\vec{v}dq$ with $Id\vec{\ell}$ in (1–29) to obtain $d\vec{F}_m = dq(\vec{v} \times \vec{B})$. The total force on a charged particle in a static electromagnetic field is known as the *Lorentz* force

$$\vec{F} = q(\vec{E} + \vec{v} \times \vec{B}) \tag{1-30}$$

Let us attempt to determine the magnetic induction field produced by the current I_2 assumed to run along the z axis. Equation (1–29) requires $\vec{B}$ to be perpendicular to $d\vec{F}_m$ (which is directed along $\vec{r}$, the cylindrical radial position vector of the current $I_1 d\ell$). If, in addition, we make the not unnatural assumption that $\vec{B}$ is also perpendicular to I_2, then $\vec{B}$ must be directed along $\vec{I}_2 \times \vec{r}$. Taking $\vec{B} = C(\vec{I}_2 \times \vec{r})$ and substituting this form into (1–29), we find

$$d\vec{F}_m = CI_1 d\vec{\ell} \times (\vec{I}_2 \times \vec{r}) = -CI_1 I_2 d\ell\, \vec{r}$$

which, when compared with the result from (1–28)

$$d\vec{F}_m = -\frac{\mu_0 I_1 d\ell\, I_2}{2\pi r^2}\vec{r} \tag{1-31}$$

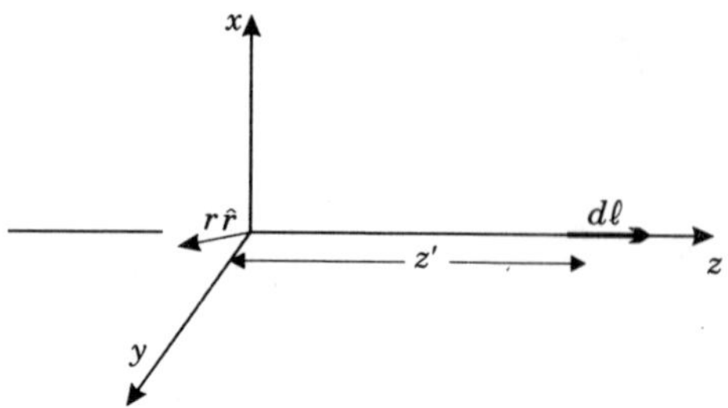

Figure 1.8: The current is assumed to run along the z axis, and we pick the observer in the x-y plane.

leads immediately to

$$\vec{B}(\vec{r}) = \frac{\mu_0}{2\pi}\frac{\vec{I}_2 \times \vec{r}}{r^2} \tag{1–32}$$

1.2.3 The Law of Biot and Savart

The magnetic induction field, B, of the long straight wire (1–32), must in fact be the sum of contributions from all parts of wire *2* stretching from $-\infty$ to $+\infty$. Since the magnetic force is related to the electric force by a Lorentz transformation that does not involve r, $\vec{E}$ and $\vec{B}$ must have the same r dependence, $(1/r^2)$. The field, $d\vec{B}$, generated by a short segment of wire, $d\vec{\ell}$ carrying current I_2 at the origin, must therefore be given by

$$d\vec{B} = \frac{\mu_0}{4\pi}\frac{I_2 d\vec{\ell}' \times \vec{r}}{r^3} \tag{1–33}$$

(The choice of numeric factor $[\mu_0/4\pi]$ will be confirmed below.) Equation (1–33) is easily generalized for current segments located at $\vec{r}'$, rather than at the origin, giving

$$d\vec{B}(\vec{r}) = \frac{\mu_0}{4\pi}\frac{I_2 d\vec{\ell}' \times (\vec{r} - \vec{r}')}{|\vec{r} - \vec{r}'|^3} \tag{1–34}$$

Integrating over the length of the current source, we obtain the Biot-Savart law:

$$\vec{B}(\vec{r}) = \frac{\mu_0}{4\pi}\int \frac{I_2 d\vec{\ell}' \times (\vec{r} - \vec{r}')}{|\vec{r} - \vec{r}'|^3} \tag{1–35}$$

We might verify that this expression does indeed give the field (1–32) of the infinite straight thin wire. Without loss of generality we may pick our coordinate system as in Figure 1.8, with the wire lying along the z-axis and the field point in the (x-y plane. Then $\vec{r} - \vec{r}' = r\hat{r} - z'\hat{k}$, $|\vec{r} - \vec{r}'| = \sqrt{r^2 + z'^2}$, and $d\vec{\ell}' = \hat{k}\, dz'$. The flux density, $\vec{B}$, may now be calculated:

$$\vec{B}(\vec{r}) = \frac{\mu_0 I_2}{4\pi}\int_{-\infty}^{+\infty} \frac{\hat{k} \times (r\hat{r} - z'\hat{k})}{(r^2 + z'^2)^{3/2}} dz' = \frac{\mu_0 I_2}{4\pi}\int_{-\infty}^{\infty} \frac{r\,(\hat{k} \times \hat{r})}{(r^2 + z'^2)^{3/2}} dz'$$

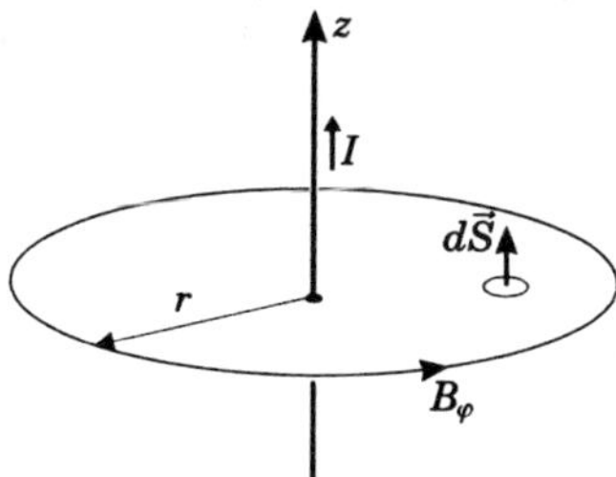

Figure 1.11: A circle of radius r, centered on the wire is used to calculate B_φ nearby.

Solution: From symmetry we expect the field to be independent of φ or z. The φ component of the field is then easily evaluated by integrating B_φ around a circle of radius r centered on the wire as in Figure 1.11. Taking B_φ to be constant on the circle, the contour integration is easily carried out:

$$\oint B_\varphi d\ell = \mu_0 \int_A \vec{J} \cdot d\vec{S} \tag{Ex 1.10.1}$$

$$2\pi r B_\varphi = \mu_0 I \tag{Ex 1.10.2}$$

leading to

$$B_\varphi = \frac{\mu_0 I}{2\pi r} \tag{Ex 1.10.3}$$

EXAMPLE 1.11: Find the magnetic field inside a long (assume infinite) closely wound solenoid with N with turn per unit length, each carrying the current I. Neglect the pitch of the windings.

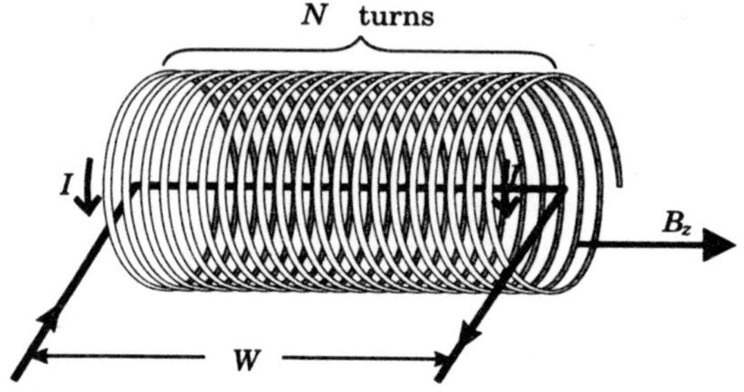

Figure 1.12: A current NI threads the rectangular loop of width W shown.

Solution: We construct a rectangular loop as shown in Figure 1.12. The inside segment lies inside the solenoid, whereas the segment completely outside the loop is placed sufficiently far from the solenoid that any field vanishes. Any supposed radial component of the field must be the same on both sides of the loop and, its contribution therefore cancels from the integral. Again applying Ampère's law,

$$\oint \vec{B} \cdot d\vec{\ell} = \mu_0 \int \vec{J} \cdot d\vec{S} \tag{Ex 1.11.1}$$

$$B_z W = \mu_0 NI \tag{Ex 1.11.2}$$

We conclude that for a unit length of solenoid ($W = 1$), $B_z = \mu_0 NI$.

If we move the segment of the rectangular loop inside the solenoid to the outside, we have zero current threading the loop, which leads to the conclusion that the field has no z-component outside the solenoid. The field inside the solenoid is evidently uniform, as the placement of the loop's inner segment does not change our result. We have not proved the radial component of the field zero; however, since the magnetic induction field is perpendicular to the current, we conclude that B_φ vanishes whence $\vec{\nabla} \cdot \vec{B} = 0$ requires B_r to vanish.

To gain some insight into the effect of nonzero pitch of the windings, we draw an Ampèrian loop around the coil in a plane perpendicular to the coil axis. When the loop is outside the coil, exactly I crosses the plane of the loop, leading us to conclude that the field outside no longer vanishes, instead, appealing to the rotational invariance, we deduce a field corresponding to a line current I along the axis. Inside the coil, no current crosses the loop, and the nonzero pitch has only minimal effect.

XAMPLE 1.12: Find the magnetic induction field inside a closely wound torus having a total of N turns. Neglect the pitch of the windings.

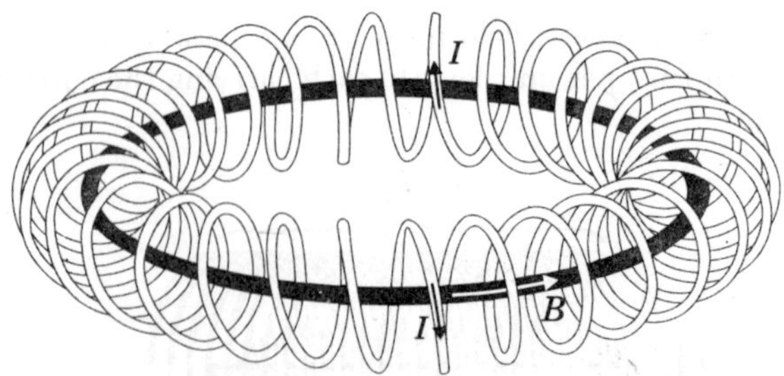

Figure 1.13: A circle drawn in the interior of the torus includes a current NI, whereas outside the torus the included (net) current is zero.

Solution: Assuming the torus has its midplane coinciding with the x-y plane, we construct a circle parallel to the x-y plane in the interior of the torus as shown in Figure 1.15. Integrating the azimuthal component of the field B_φ around the circle, we obtain $2\pi r B_\varphi = \mu_0 NI$, from which we conclude that

$$B_\varphi = \frac{\mu_0 NI}{2\pi r} \tag{Ex 1.12.1}$$

Interestingly, so long as the loop remains within the torus, up or down displacement makes no difference. The field is independent of z but decreases radially as $1/r$.

1.2.5 The Magnetic Vector Potential

It would be convenient if $\vec{B}$ could generally be derived from a scalar potential, much in the fashion of the electric field. Unfortunately, since the curl of $\vec{B}$ is not generally zero, $\vec{B}$ cannot generally be expressed as the gradient of a scalar function. (Notwithstanding this caution, wherever the current density vanishes, $\vec{B}$ can usefully be found as the gradient of a scalar function we will call the *scalar magnetic potential*, V_m.)

In fact, since, according to (1–43) and to the chagrin of many physicists whose theories predict the existence, there are no magnetic charges (monopoles) analogous to electric charges, $\vec{B}$ has zero divergence.

A field without divergence (no sources or sinks) can always be expressed as the curl of a vector. We define the *magnetic vector potential*, $\vec{A}(\vec{r})$, by

$$\vec{B}(\vec{r}) = \vec{\nabla} \times \vec{A}(\vec{r}) \tag{1–51}$$

In much the same manner as the derivation of the electrical scalar potential where we expressed $\vec{E}$ as the gradient of a function, we now seek to express $\vec{B}$ as the curl of a vector field. Glancing back at (1–42), we find $\vec{B}$ expressed in exactly this form. We conclude that

$$\vec{A}(\vec{r}) \equiv \frac{\mu_0}{4\pi} \int \frac{\vec{J}(\vec{r}\,')}{|\vec{r} - \vec{r}\,'|}\, d^3r' \tag{1–52}$$

serves as an expression for the magnetic vector potential

For many years it was thought that the vector potential $\vec{A}$ was merely a mathematical construct since all forces and apparently physical effects depend on $\vec{B}$. In fact, the phase of a charged particle's wave function depends on the line integral of the vector potential along the particle's path. The *Bohm-Aharanov* experiment[5] has directly demonstrated a shift in the fringe pattern of electron waves diffracted by a long magnetized needle that, though presenting no appreciable magnetic flux density along the path, does have a nonzero vector potential outside the needle. Since forces play no central role in quantum mechanics, the disappearance of $\vec{E}$ and $\vec{B}$ from the formulation should come as no surprise. As we will see when discussing the covariant formulation in Chapter 11, the potentials V and $\vec{A}$ appear to be somewhat more fundamental than $\vec{E}$ and $\vec{B}$ in the relativistic formulation as well.

It is worth noting that according to (1–52), $\vec{A}$ is always parallel to the mean weighted current flow. We note also that from its definition, $\vec{A}$ is not unique. We can add the gradient of any function, say $\Lambda(\vec{r})$, to $\vec{A}$ without changing curl $\vec{A}$.

---

[5]See for example R. P. Feynman, R. B. Leighton, and M. Sands (1964) *The Feynman Lectures on Physics, Vol. 2*, Addison-Wesley Publishing Company, Reading (Mass.).

Although the evaluation of $\vec{A}$ from (1–52) appears quite straight-forward, in practice it is frequently necessary to calculate $\vec{B}$ first and then derive $\vec{A}$ from $\vec{B}$

The vector potential of a slowly moving point charge is obtained by setting $\vec{J}(\vec{r}\,') = \rho\vec{v}(\vec{r}\,') = q\,\delta\,(\vec{r}\,' - \vec{r}_q(t))\,\vec{v}$, giving on integration of (1–52)

$$\vec{A}(\vec{r}) = \frac{\mu_0 q\vec{v}}{4\pi\,|\vec{r} - \vec{r}_q(t)|} \tag{1–53}$$

Example 1.13: Find the vector potential $\vec{A}(0,\ 0,\ z)$ above a circular current loop of radius a in the x-y plane.

Solution: $\vec{J} = I\,\delta(r'-a)\,\delta(z')\,\hat{\varphi}$ where r' is the radial coordinate of a current element in cylindrical coordinates. Then

$$\vec{A}(0,0,z) = \frac{\mu_0}{4\pi}\int\frac{\vec{J}\,d^3r'}{|\vec{r}-\vec{r}\,'|} = \frac{\mu_0}{4\pi}\int\frac{I\,\delta(r'-a)\,\delta(z')\,\hat{\varphi}}{\sqrt{a^2+z^2}}\,r'dr'd\varphi'\,dz' \tag{Ex 1.13.1}$$

$$= \frac{\mu_0\,Ia}{4\pi\sqrt{a^2+z^2}}\int_0^{2\pi}\hat{\varphi}\,d\varphi' = \frac{\mu_0 Ia}{4\pi\sqrt{a^2+z^2}}\int_0^{2\pi}\frac{d\hat{r}}{d\varphi'}\,d\varphi' \tag{Ex 1.13.2}$$

$$= \frac{\mu_0 Ia}{4\pi\sqrt{a^2+z^2}}\,[\hat{r}(2\pi) - \hat{r}(0)] = 0 \tag{Ex 1.13.3}$$

Since we expect $\vec{B}$ to have no x or y component at this point, $\vec{A}$ should not have any z dependence. Clearly this result is not very useful. We will tackle this problem again using the magnetic scalar potential. Notwithstanding the successfully integration of the equations above using a curvilinear basis, the student is strongly advised to avoid using anything but a Cartesian basis in integrals.

Example 1.14: Find the vector potential at distance r from a long straight wire of radius $a \ll r$ carrying current I (Figure 1.13).

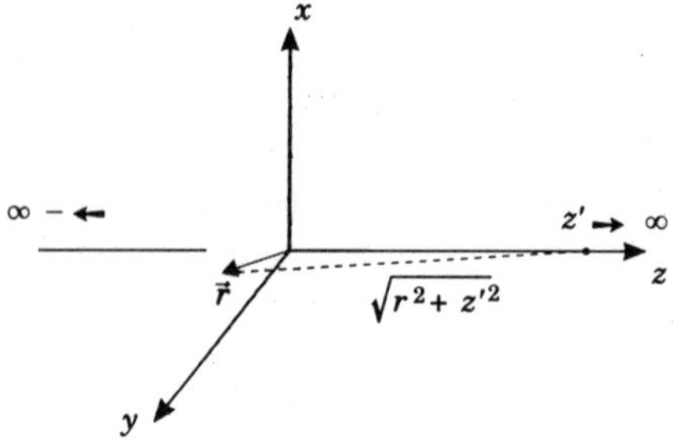

Figure 1.14: Since the wire is infinitely long, we may pick the origin so that $\vec{r}$ lies in the x-y plane.

Solution: Without loss of generality we can place the field point in the x-y plane. If we try

$$\vec{A}(r,\varphi,z) = \frac{\mu_0}{4\pi}\int \frac{\vec{J}(\vec{r}\,')\,d^3r'}{|\vec{r}-\vec{r}\,'|} \tag{Ex 1.14.1}$$

Neglecting a, we obtain for $\vec{A}$ outside the wire

$$\vec{A} = \frac{\mu_0}{4\pi}\int \frac{\vec{J}\,d^3r'}{|\vec{r}-\vec{r}\,'|} = \frac{\mu_0}{4\pi}\int_{-\infty}^{\infty} \frac{I\hat{k}dz'}{\sqrt{r^2+z'^2}} \tag{Ex 1.14.2}$$

giving

$$\vec{A} = \frac{\mu_0 I\hat{k}}{4\pi}\ln\left(z'+\sqrt{r^2+z'^2}\right)\Bigg|_{-\infty}^{\infty} \tag{Ex 1.14.3}$$

which is infinite. We can however evaluate $\vec{A}$ by equating its curl to the expression for $\vec{B}$ as found in (1–32) and integrating.

Setting $\vec{A} = A_z\hat{k}$ and noting that $\vec{B} = \frac{\mu_0 I}{2\pi r}\hat{\varphi}$, we replace $\vec{B}_\varphi$ by the φ component of the curl of $\vec{A}$ to write

$$\left(\vec{\nabla}\times\vec{A}\right)_\varphi = \left(\frac{\partial A_r}{\partial z} - \frac{\partial A_z}{\partial r}\right) = \frac{\mu_0 I}{2\pi r} \tag{Ex 1.14.4}$$

We conclude:

$$\frac{\partial A_z}{\partial r} = -\frac{\mu_0 I}{2\pi r} \quad\Rightarrow\quad A_z = -\frac{\mu_0 I}{2\pi}\ln r + \text{const.} \tag{Ex 1.14.5}$$

The vector potential $\vec{A}$ outside the wire then becomes

$$\vec{A} = -\frac{\mu_0 I\hat{k}}{2\pi}\ln r + \vec{\nabla}\Lambda \tag{Ex 1.14.6}$$

with Λ an arbitrary scalar function.

The expression (Ex 1.14.6) provides us with a useful form of the expression for the vector potential; for, if we consider that it governs the vector potential produced by a filament of current dI extending infinitely in the z-direction at $r = 0$,

$$dA_z(\vec{r}) = -\frac{\mu_0}{2\pi}\ln r\,dI \tag{1–54}$$

The vector potential of the entire current considered as a superposition of the filaments located ar $\vec{r}\,'$ becomes, with the replacement $dI = J_z(\vec{r}\,')dS'$,

$$A_z(\vec{r}) = -\frac{\mu_0}{2\pi}\int J_z(\vec{r}\,')\ln|r-r'|\,dS' \tag{1–55}$$

which may be integrated over arbitrary cross-sections.

The vector potential of a uniform magnetic induction field may be written as

$$\vec{A}(\vec{r}) = \frac{\vec{B} \times \vec{r}}{2} \tag{1–56}$$

which is easily verified as follows:

$$\vec{\nabla} \times \vec{A} = \vec{\nabla} \times \left(\frac{\vec{B} \times \vec{r}}{2}\right) = \frac{\vec{B}(\vec{\nabla} \cdot \vec{r}) - \vec{r}(\vec{\nabla} \cdot \vec{B}) + (\vec{r} \cdot \vec{\nabla})\vec{B} - (\vec{B} \cdot \vec{\nabla})\vec{r}}{2}$$

$$= \frac{3\vec{B} - \vec{B}}{2} = \vec{B} \tag{1–57}$$

1.2.6 The Magnetic Scalar Potential

Although not of the same theoretical importance as the magnetic vector potential, the magnetic scalar potential is extremely useful for solving problems involving magnetic fields. As we have already pointed out, in regions of space where the current density is zero (outside wires for instance), $\vec{\nabla} \times \vec{B} = 0$ (this holds only for static fields), which implies that $\vec{B}$ can be expressed as the gradient of a scalar potential:

$$\vec{B} = -\mu_0 \vec{\nabla} V_m \tag{1–58}$$

The reason for including the constant μ_0 in (1–58) will become clearer in Chapter 7 when we deal with fields in ponderable matter.

Merely knowing that V_m exists, does not give a prescription for finding this scalar potential. To obtain the magnetic scalar potential for a closed current loop we proceed as follows.

In general, the change in any scalar function, and in particular $V_m(\vec{r})$, in response to a small change $d\vec{r}$ of its argument, is given to first order by

$$dV_m = \vec{\nabla} V_m \cdot d\vec{r} = -\frac{\vec{B} \cdot d\vec{r}}{\mu_0} \tag{1–59}$$

Using the line current form of the Biot-Savart law, (1–35), to substitute an expression for $\vec{B}$ we obtain, with the abbreviation $\vec{R} = \vec{r} - \vec{r}\,'$,

$$dV_m = \left(-\frac{I}{4\pi}\oint \frac{d\vec{\ell} \times \vec{R}}{R^3}\right) \cdot d\vec{r} = -\frac{I}{4\pi}\oint \frac{(d\vec{\ell} \times \vec{R}) \cdot d\vec{r}}{R^3} \tag{1–60}$$

It is interesting to relate this expression to the solid angle, Ω, subtended by the current loop. With reference to Figure 1.15,

$$\Omega \equiv \int_A \frac{d\vec{A} \cdot (-\hat{R})}{R^2} = \int \frac{-d\vec{A} \cdot \vec{R}}{R^3} \tag{1–61}$$

where $-\vec{R}$ is a vector pointing from the observer at the *field* point to a point on the surface enclosed by the loop. It should be evident that the shape of the surface

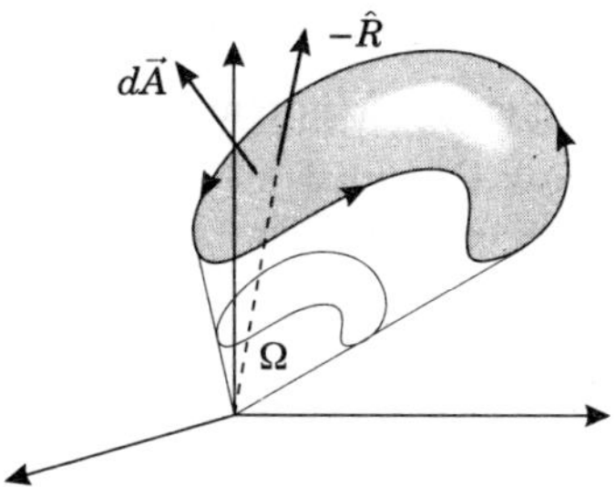

Figure 1.15: The area enclosed by the loop points in the direction shown, determined by the direction of the current.

enclosed is immaterial; only the boundary is significant. If the observer moves by an amount $d\vec{r}$ or, equivalently, the loop moves by $-d\vec{r}$, the solid angle subtended by the loop will change. In particular, the area gained by any segment $d\vec{\ell}$ of the loop is $-d\vec{r} \times d\vec{\ell}$, giving a change in solid angle subtended (see Figure 1.16):

$$d\Omega = \oint \frac{[(-d\vec{r}) \times d\vec{\ell}] \cdot (-\vec{R})}{R^3}$$

$$= \oint \frac{d\vec{r} \cdot (d\vec{\ell} \times \vec{R})}{R^3} \tag{1–62}$$

Comparing $d\Omega$, the integrand of (1–62) with dV_m, (1–60), we find that

$$dV_m = -\frac{I}{4\pi}\, d\Omega \tag{1–63}$$

Since we may in general write $dV_m = \vec{\nabla} V_m \cdot d\vec{r}$ and $d\Omega = \vec{\nabla}\Omega \cdot d\vec{r}$, we conclude that

$$V_m = -\frac{I}{4\pi}\, \Omega \tag{1–64}$$

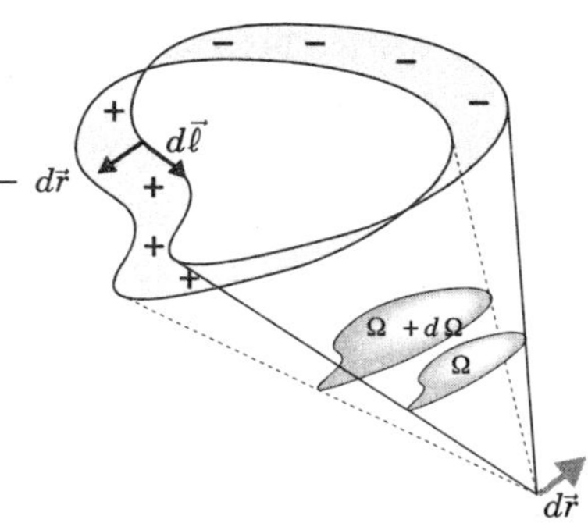

Figure 1.16: Although the area does not change under displacement, the solid angle does, because the mean R changes.

serves as a scalar potential for the magnetic induction field.

EXAMPLE 1.15: Find the magnetic scalar potential at a point below the center of a circular current loop of radius a (Figure 1.17). Use the scalar potential to find the magnetic induction field on the central axis.

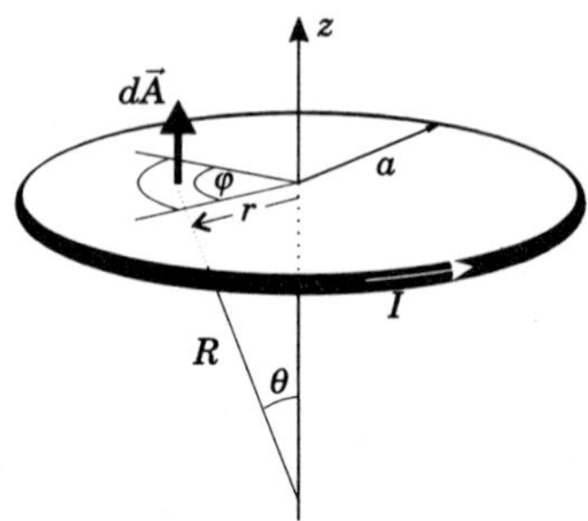

Figure 1.17: The observer is directly below the center of the loop.

Solution: We first find Ω. To this end, we note $\vec{R} = z\hat{k} - r'\hat{r}$, $R = \sqrt{z^2 + r'^2}$ and $\vec{R}\cdot d\vec{A} = zr'dr'd\varphi'$, so that

$$-\Omega = \int \frac{d\vec{A}\cdot\vec{R}}{R^3} = \int_0^a \int_0^{2\pi} \frac{zr'dr'd\varphi'}{(z^2 + r'^2)^{3/2}}$$

$$= \frac{-2\pi z}{\sqrt{z^2 + r'^2}}\Bigg|_0^a = 2\pi\left(1 - \frac{z}{\sqrt{z^2 + a^2}}\right) \qquad \text{(Ex 1.15.1)}$$

from which we conclude

$$V_m = -\frac{I}{2}\left(\frac{z}{\sqrt{z^2 + a^2}} - 1\right) \qquad \text{(Ex 1.15.2)}$$

(the constant term may of course be dropped without loss). As shown in the next example, (1.16), the scalar potential of the current loop becomes a building block for a number of other problems whose currents can be decomposed into current loops.

The magnetic induction field along the z axis previously obtained in example 1.8 is now easily found:

$$B_z(0,0,z) = -\mu_0 \frac{\partial}{\partial z} V_m$$

$$= \frac{I\mu_0}{2}\left[\frac{1}{(z^2 + a^2)^{1/2}} - \frac{z^2}{(z^2 + a^2)^{3/2}}\right]$$

$$= \frac{I\mu_0}{2}\frac{a^2}{(z^2 + a^2)^{3/2}} \qquad \text{(Ex 1.15.3)}$$

EXAMPLE 1.16: Find the scalar magnetic potential along the axis of a solenoid of length L with N closely spaced turns, carrying current I. Assume the pitch may be neglected.

Solution: We choose our coordinate system so that the z axis runs along the axis of the solenoid and the origin is at the center of the solenoid. Using the preceding example 1.15, the magnetic scalar potential of a coil at the origin is

$$V_m = -\frac{I}{2}\frac{z}{\sqrt{z^2+a^2}} \tag{Ex 1.16.1}$$

If the loop were located at position z' instead of the origin, the expression for V_m would become

$$V_m(z) = -\frac{I}{2}\frac{(z-z')}{\sqrt{(z-z')^2+a^2}} \tag{Ex 1.16.2}$$

The scalar potential from N turns at varying locations z' is then obtained by summing the potential from each of the loops:

$$\begin{aligned} V_m(z) &= -\frac{I}{2}\int\frac{(z-z')dN}{\sqrt{(z-z')^2+a^2}} = -\frac{NI}{2L}\int_{-L/2}^{L/2}\frac{(z-z')\,dz'}{\sqrt{(z-z')^2+a^2}} \\ &= \frac{NI}{2L}\sqrt{(z-z')^2+a^2}\,\Bigg|_{-L/2}^{L/2} \\ &= -\frac{NI}{2L}\left(\sqrt{\left(z+\tfrac{1}{2}L\right)^2+a^2}-\sqrt{\left(z-\tfrac{1}{2}L\right)^2+a^2}\right) \end{aligned} \tag{Ex 1.16.3}$$

The calculation of the magnetic induction field along the axis of the solenoid is now a simple matter. (Exercise 1-23)

Exercises and Problems

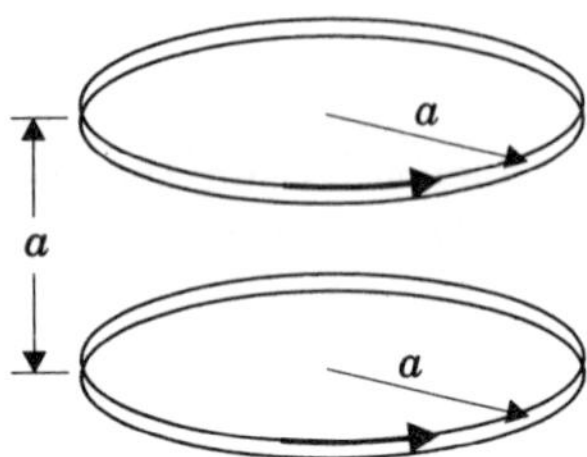

Figure 1.18: A Helmholtz coil consists of two identical parallel coils, spaced at their common radius.

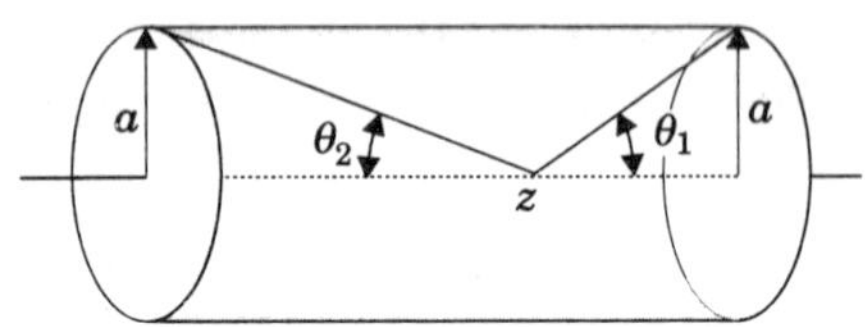

Figure 1.19: Geometry of the solenoid of question 21.

1-1 Verify that Ex 1.3.4 yields the limiting form $Q/4\pi\varepsilon_0 z^2$ when $z \to 0$.

1-2 Find the electric field along the axis of a charged ring of radius a lying in the x-y plane when the charge density on the ring varies sinusoidally around the ring as

$$\rho = \lambda_0(1 + \sin\varphi)\,\delta(r - a)\delta(z)$$

1-3 Find the electric field above the center of a flat circular plate of radius a when the charge distribution is $\rho = br^2\delta(z)$ when $r \le a$ and 0 elsewhere.

1-4 Find the electrostatic field established by two long concentric cylinders having radii a and b, bearing respective surface charge densities σ_a and σ_b, in the region inside the inner cylinder, between the cylinders, and outside the two cylinders.

1-5 Find the electrostatic field produced by a spherical charge distribution with charge density $\rho_0 e^{-kr}$.

1-6 Find the electric field produced by a spherically symmetric charge distribution with charge density

$$\rho = \begin{cases} \rho_0\left(1 - \dfrac{r}{a}\right)^2 & \text{for } r \le a \text{ and} \\ 0 & \text{for } r \ge a \end{cases}$$

1-7 Two large parallel flat plates bear uniform surface charge densities σ and $-\sigma$. Find the force on one of the plates due to the other. Neglect the fringing fields. Note that just using the electric field between the plates to calculate the force as $\sigma\vec{E}A$ gives twice the correct result. (Why?)

1-8 Find the electric field at any point (not on the charge) due to a line charge with charge density

$$\rho = \begin{cases} bz\delta(x)\delta(y) & z \in (-a,\ a) \\ 0 & \text{elsewhere} \end{cases}$$

lying along the z axis between a and $-a$. (Fairly messy integral!)

1-9 Within a conductor, charges move freely until the remaining electric field is zero. Show that this implies that the electric field near the surface of a conductor is perpendicular to the conductor.

1-10 Find the electric potential for the charge distribution of problem 1-8.

1-11 Find the electric potential for the charge distribution of problem 1-2.

1-12 Find the electric potential for the charge distribution of problem 1-3.

1-13 A fine needle emits electrons isotropically at a steady rate. Find the divergence of the current density and the resulting current flux at distance r from the point in the steady state.

1-14 Using the magnetic scalar potential, find the magnetic induction field along the spin axis of a uniformly charged spinning disk.

1-15 Using the results from 1-14, find the magnetic induction field along the spin axis of a uniformly charged spinning thin spherical shell of radius R at a point outside the shell.

1-16 Using the results from 1-15, find the magnetic induction field along the spin axis of a uniformly charged spinning sphere of radius R at a point outside the sphere.

1-17 A Helmholtz coil consists of two parallel plane coils each of radius a and spaced by a (Figure 1.18). Find the magnetic induction field at the center ($a/2$ from the plane of each coil).

1-18 Use the Biot-Savart law to find the axial magnetic induction field of a closely wound solenoid having length L and N closely spaced turns. (Ignore the pitch of the windings.)

1-19 Find the magnetic induction field of a long coaxial cable carrying a uniformly distributed current on its inner conductor (radius a) and an equal countercurrent along its outer conductor of radius b and thickness $<< b$.

1-20 Expresses the vector potential of a filamentary current loop as a line integral and show that the magnetic vector potential of the current loop vanishes.

1-21 Show that the magnetic vector potential satisfies $\nabla^2 \vec{A} = -\mu_0 \vec{J}$.

1-22 Find the vector potential of two long parallel wires each of radius a, spaced a distance h apart, carrying equal currents I in opposing directions.

1-23 A low-velocity beam of charged particles will spread out because of mutual repulsion. By what factor is the mutual repulsion reduced when the particles are accelerated to $.99c$?

1-24 Differentiate the axial scalar magnetic potential of a solenoid of length L with N closely spaced turns to obtain the axial magnetic induction field. Show that this field can be conveniently written

$$B_z = \frac{NI\mu_0}{2L} (\cos\theta_1 + \cos\theta_2)$$

where θ_1 and θ_2 are respectively the angles subtended by the solenoid's radius at the left and the right end of the coil (Figure 1.19).

1-25 A charged particle in a crossed electric and magnetic induction field experiences no net force when it moves through the fields with an appropriate velocity known as the *plasma drift velocity*. Find an expression for the plasma drift velocity.

Chapter 2

Charge and Current Distributions

2.1 Multipole Moments

When the charge and current distribution does not exhibit considerable symmetry or, alternatively, the potential is required at some point not on a symmetry axis, the evaluation of the integrals in Coulomb's law or in the Biot-Savart law becomes rather formidable. It becomes useful to approximate the potentials by a power series in the field point coordinates. The coefficients of such a power series, called the (multipole) moments of the charge distribution, are independent of the field point coordinates. The utility of such an expansion rests on the fact that once the moments of a distribution have been determined, the evaluation of the potential at any point becomes merely a matter of substituting the field point coordinates into the power series.

Before embarking on the general expansion, let us briefly consider the potential of a pair of closely spaced charges of opposite sign. We place the positive charge at $\frac{1}{2}\vec{d}$ from the origin and the negative charge at $-\frac{1}{2}\vec{d}$ from the origin.

The potential at $\vec{r}$ due to the two charges q and $-q$ is

$$V(\vec{r}) = \frac{q}{4\pi\varepsilon_0}\left(\frac{1}{\left|\vec{r}-\frac{1}{2}\vec{d}\right|} - \frac{1}{\left|\vec{r}+\frac{1}{2}\vec{d}\right|}\right) \tag{2–1}$$

$$= \frac{q}{4\pi\varepsilon_0}\left(\frac{1}{\sqrt{r^2+(d/2)^2-\vec{r}\cdot\vec{d}}} - \frac{1}{\sqrt{r^2+(d/2)^2+\vec{r}\cdot\vec{d}}}\right) \tag{2–2}$$

Assuming $r \gg d$, we retain only the first order terms in d/r, to approximate the potential of the two charges by

$$V(\vec{r}) \approx \frac{q}{4\pi\varepsilon_0 r}\left(\frac{1}{\sqrt{1-\dfrac{\vec{r}\cdot\vec{d}}{r^2}}} - \frac{1}{\sqrt{1+\dfrac{\vec{r}\cdot\vec{d}}{r^2}}}\right) \tag{2–3}$$

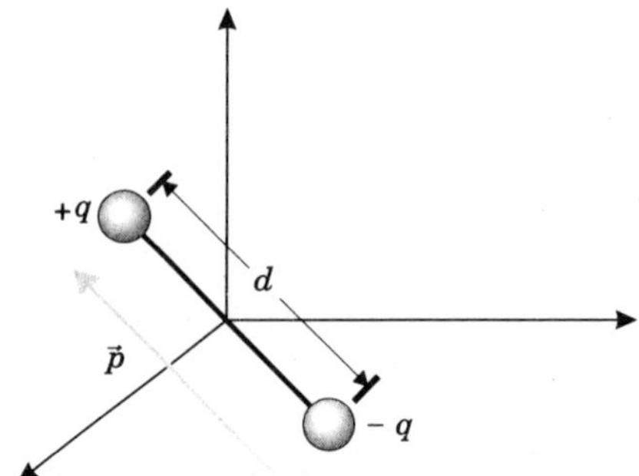

Figure 2.1: The dipole moment of a pair of equal but opposite charges points from $-q$ to q.

We expand (2–3) with the binomial expansion (33)[6] to obtain

$$V(\vec{r}) \approx \frac{q}{4\pi\varepsilon_0 r}\left(1 + \frac{\vec{r}\cdot\vec{d}}{2r^2} - 1 + \frac{\vec{r}\cdot\vec{d}}{2r^2} + \cdots\right)$$

$$= \frac{\vec{r}\cdot(q\vec{d})}{4\pi\varepsilon_0 r^3} = \frac{\vec{p}\cdot\vec{r}}{4\pi\varepsilon_0 r^3} \tag{2–4}$$

The bracketed term $(q\vec{d}) \equiv \vec{p}$ is known as the dipole moment of the charge pair (Figure 2.1). The neglected terms of order d^2 and higher involve the quadrupole moment, octopole moment, hexadecapole moment, and so forth. Once $\vec{p}$ is known, no further integrations are required to evaluate the potentials of an electric dipole at any (sufficiently distant) point.

EXAMPLE 2.1: Find the electric field resulting from an electric dipole $\vec{p}$ placed at the origin.

Solution: The potential of the dipole is (2–4)

$$V(\vec{r}) = \frac{\vec{p}\cdot\vec{r}}{4\pi\varepsilon_0 r^3} \tag{Ex 2.1.1}$$

leading to an electric field

$$\vec{E} = -\vec{\nabla}V = -\frac{\vec{\nabla}(\vec{p}\cdot\vec{r})}{4\,\pi\varepsilon_0 r^3} + \frac{3\,(\vec{p}\cdot\vec{r})\vec{r}}{4\,\pi\varepsilon_0 r^5}$$

$$= \frac{-1}{4\pi\varepsilon_0 r^3}\left(\vec{p} - \frac{3(\vec{p}\cdot\vec{r})\vec{r}}{r^2}\right) \tag{Ex 2.1.2}$$

[6]The binomial expansion,

$$(1+\varepsilon)^r = 1 + r\varepsilon + \frac{r(r-1)\varepsilon^2}{2!} + \frac{r(r-1)(r-2)\varepsilon^3}{3!} + \frac{r(r-1)(r-2)(r-3)\varepsilon^4}{4!} + \ldots$$

is finite when r is a positive integer. When r is negative or a noninteger, the infinite series converges only for $\varepsilon < 1$.

Had we placed the dipole at $\vec{r}'$ instead of at the origin, we need merely replace $\vec{r}$ by $(\vec{r}-\vec{r}')$ in Ex 2.1.2.

2.1.1 The Cartesian Multipole Expansion

In general the term $1/|\vec{r}-\vec{r}'|$ in the expression for the electric potential may be written as

$$\frac{1}{|\vec{r}-\vec{r}'|}=\frac{1}{\sqrt{r^2+r'^2-2\vec{r}\cdot\vec{r}'}}=\frac{1}{r_>\sqrt{1+\left(\dfrac{r_<}{r_>}\right)^2-\dfrac{2\vec{r}_>\cdot\vec{r}_<}{r_>^2}}} \tag{2–5}$$

where we have defined $r_>$ as the greater of r and r' and $r_<$ as the lesser of r and r'. This slightly cumbersome notation is necessary to ensure that the binomial expansion of the radical converges. We proceed to expand the radical by the binomial expansion (33) to obtain

$$\frac{1}{|\vec{r}-\vec{r}'|}=\frac{1}{r_>}\left[1-\frac{1}{2}\left(\frac{r_<^2}{r_>^2}-\frac{2\vec{r}_>\cdot\vec{r}_<}{r_>^2}\right)+\frac{3}{2}\frac{1}{2}\frac{1}{2!}\left(\frac{r_<^2}{r_>^2}-\frac{2\vec{r}_>\cdot\vec{r}_<}{r_>^2}\right)^2+\dots\right] \tag{2–6}$$

$$=\frac{1}{r_>}+\frac{\vec{r}_>\cdot\vec{r}_<}{r_>^3}-\frac{1}{2}\frac{r_<^2}{r_>^3}+\frac{3}{8r_>}\frac{4(\vec{r}_>\cdot\vec{r}_<)^2}{r_>^4}+\dots \tag{2–7}$$

where the neglected terms are of order $(r_</r_>)^3$ or higher. We evaluate the two terms in $r_<^2$ in such a manner as to separate the $r_<$ and $r_>$ terms.

$$\frac{3}{8r_>}\frac{4(\vec{r}_>\cdot\vec{r}_<)^2}{r_>^4}-\frac{1}{2}\frac{r_<^2}{r_>^3}=\frac{\left[3(\vec{r}_>\cdot\vec{r}_<)^2-r_<^2r_>^2\right]}{2r_>^5}=\frac{3(\vec{r}\cdot\vec{r}')^2-r^2r'^2}{2r_>^5} \tag{2–8}$$

$$=\frac{1}{2r_>^5}\sum_{\substack{i=1\\j=1}}^{3,3}3(x_ix_i')(x_jx_j')-x_ix_ix_j'x_j' \tag{2–9}$$

$$=\frac{1}{2r_>^5}\sum_{\substack{i=1\\j=1}}^{3,3}x_ix_j(3x_i'x_j'-\delta_{ij}r'^2) \tag{2–10}$$

The primed and unprimed coordinates could, of course, have been exchanged. To be explicit we assume that the source coordinates r' are consistently smaller than r (only then does the multipole expansion make sense), and we can write

$$\frac{1}{|\vec{r}-\vec{r}'|}=\frac{1}{r}+\frac{\vec{r}\cdot\vec{r}'}{r^3}+\frac{1}{2r^5}\sum_{\substack{i=1\\j=1}}^{3,3}x_ix_j\left(3x_i'x_j'-\delta_{ij}r'^2\right)+\dots \tag{2–11}$$

leading to

$$V(\vec{r})=\frac{1}{4\pi\varepsilon_0}\int\frac{\rho(\vec{r}')}{|\vec{r}-\vec{r}'|}d^3r' \tag{2–12}$$

$$= \frac{1}{4\pi\varepsilon_0}\left(\frac{\int \rho(\vec{r}\,')d^3r'}{r} + \frac{\vec{r}\cdot\int \rho(\vec{r}\,')\vec{r}\,'\,d^3r'}{r^3}\right.$$

$$\left. + \frac{\sum x_i x_j \int \rho(\vec{r}\,')\left[3x_i'x_j' - \delta_{ij}r'^2\right]d^3r'}{2r^5} + \dots\right) \qquad (2\text{–}13)$$

$$= \frac{1}{4\,\pi\varepsilon_0}\left(\frac{q}{r} + \frac{\vec{r}\cdot\vec{p}}{r^3} + \frac{\sum x_i x_j Q_{ij}}{2r^5} + \dots\right) \qquad (2\text{–}14)$$

Here q is the total charge of the source,

$$\vec{p} = \int \rho(\vec{r}\,')\vec{r}\,'d^3r' \qquad (2\text{–}15)$$

is the dipole moment of the distribution and

$$Q_{ij} \equiv \int \rho(\vec{r}\,')(3x_i'x_j' - \delta_{ij}r'^2)d^3r' \qquad (2\text{–}16)$$

is the ij component of the Cartesian (electric) quadrupole moment tensor. In principle this expansion could be continued to higher order, but it is rarely fruitful to do so. The *trace* (or sum of diagonal elements) of the quadrupole moment tensor is easily seen to vanish:

$$Q_{11} + Q_{22} + Q_{33} = \int \rho\left[(3x'^2 - r'^2) + (3y'^2 - r'^2) + (3z'^2 - r'^2)\right]d^3r'$$

$$= \int \rho(3x'^2 + 3y'^2 + 3z'^2 - 3r'^2)d^3r' = 0 \qquad (2\text{–}17)$$

To confuse matters, when a quadrupole has azimuthal symmetry, the zz component, $Q_{33} = -2Q_{11} = -2Q_{22}$, is often referred to as *the* quadrupole moment.

The multipole moment of a charge distribution generally depends on the choice of origin. Nonetheless, the lowest order nonzero moment is unique, independent of the choice of origin.

EXAMPLE 2.2: Find the dipole moment of a line charge of length a and charge density $\rho(\vec{r}\,') = \lambda z'\delta(x')\delta(y')$ for $z' \in (-a/2, a/2)$ and 0 elsewhere.

Solution: The x and y components of the dipole moment clearly vanish, and the z component is easily obtained from the definition (2–15).

$$p_x = p_y = 0 \qquad (\text{Ex } 2.2.1)$$

$$p_z = \int_{-a/2}^{a/2} \lambda z'^2\,dz' = \left.\frac{\lambda z'^3}{3}\right|_{-a/2}^{a/2} = \frac{\lambda a^3}{12} \qquad (\text{Ex } 2.2.2)$$

We note that as the net charge is zero, the dipole moment should be unique. Shifting the origin to $-\frac{1}{2}a$ along the z axis, for instance, gives us a new charge density, $\rho = \lambda(z' - \frac{1}{2}a)\delta(x')\delta(y')$ and the dipole moment with the shifted origin becomes

$$p_z = \int_0^a \lambda(z' - \tfrac{1}{2}a)z'dz' = \frac{\lambda a^3}{12} \tag{Ex 2.2.3}$$

EXAMPLE 2.3: Find the quadrupole moment of the charge distribution shown in Figure 2.2.

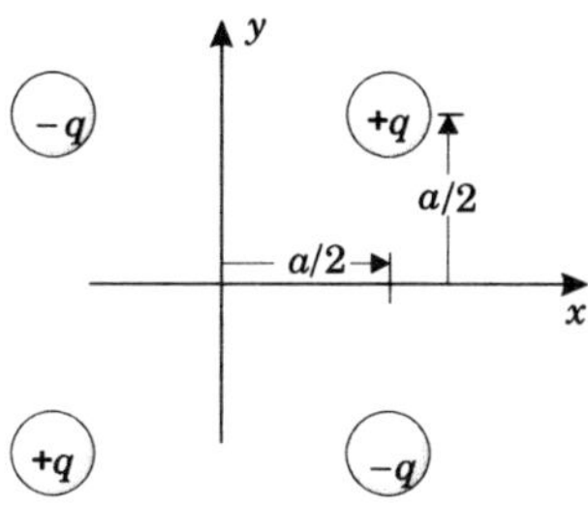

Figure 2.2: Two positive charges occupy diagonally opposite corners, and two equal, negative charges occupy the remaining corners.

Solution: The xx (1,1) component of the quadrupole tensor is again easily evaluated from its definition (2–16) replacing the integral over the source by a sum over the source particles:

$$Q_{xx} = (+q)\left[3\left(\frac{a}{2}\right)^2 - \left(\frac{a^2}{2}\right)\right] + (-q)\left[3\left(\frac{-a}{2}\right)^2 - \left(\frac{a^2}{2}\right)\right]$$

$$+(+q)\left[3\left(\frac{-a}{2}\right)^2 - \left(\frac{a^2}{2}\right)\right] + (-q)\left[3\left(\frac{a}{2}\right)^2 - \left(\frac{a^2}{2}\right)\right] = 0 \tag{Ex 2.3.1}$$

In the same manner, Q_{yy} and Q_{zz} are found to be 0. $Q_{xy} = Q_{yx}$ is

$$Q_{xy} = 3q\left(\frac{a}{2}\cdot\frac{a}{2}\right) - 3q\left(\frac{a}{2}\cdot\frac{-a}{2}\right) + 3q\left(\frac{-a}{2}\cdot\frac{-a}{2}\right) - 3q\left(\frac{-a}{2}\cdot\frac{a}{2}\right) \tag{Ex 2.3.2}$$

$$= 3qa^2 \tag{Ex 2.3.3}$$

and $Q_{xz} = Q_{zx}$ vanishes.

EXAMPLE 2.4: Find the quadrupole moment of a uniformly charged ellipsoid of revolution, with semi-major axis a along the z axis and semi-minor axis b bearing total charge Q.

Solution: We begin by finding the zz component of the quadrupole moment tensor. The equation of the ellipsoid can be expressed as

$$\frac{s^2}{b^2} + \frac{z^2}{a^2} = 1 \tag{Ex 2.4.1}$$

where $s = \sqrt{x^2 + y^2}$ is the cylindrical radius. Then

$$Q_{zz} = \int_{volume} (3z'^2 - r'^2)\rho d^3r' = \int_{z'=-a}^{a} \int_{s'=0}^{b\sqrt{1-(z'/a)^2}} \rho(2z'^2 - s'^2) 2\pi s' ds' dz' \quad \text{(Ex 2.4.2)}$$

$$= 2\pi\rho \int_{-a}^{a} \left[z'^2 b^2 \left(1 - \frac{z'^2}{a^2}\right) - \frac{1}{4} b^4 \left(1 - \frac{z'^2}{a^2}\right)^2 \right] dz' \quad \text{(Ex 2.4.3)}$$

$$= \tfrac{8}{15}\pi\rho a b^2 (a^2 - b^2) \quad \text{(Ex 2.4.4)}$$

The total charge in the volume is $\frac{4}{3}\pi\rho a b^2$, so that $Q_{zz} = \frac{2}{5}Q(a^2 - b^2)$. Rather than compute the xx and yy components, we make use of the fact that the trace of the quadrupole tensor vanishes, requiring $Q_{xx} = Q_{yy} = -\frac{1}{2}Q_{zz}$. The off-diagonal elements vanish.

The net charge on the ellipsoid is not zero, hence we conclude that the computed quadrupole moment is not unique. The identical calculation has considerable application in gravitational theory with mass replacing charge.

The vector potential $\vec{A}$ can be expanded in precisely the same fashion as V. Let us consider the expression for the vector potential of a plane current loop (1–52)

$$\vec{A}(\vec{r}) = \frac{\mu_0}{4\pi} \int \frac{\vec{J}(\vec{r}\,')d^3r'}{|\vec{r} - \vec{r}\,'|}$$

$$= \frac{\mu_0 I}{4\pi} \oint \frac{d\vec{\ell}\,'}{|\vec{r} - \vec{r}\,'|} \quad (2\text{–}18)$$

in which we expand the denominator using the binomial theorem (33) as

$$\vec{A}(\vec{r}) = \frac{\mu_0 I}{4\pi r} \oint d\vec{\ell}\,' + \frac{\mu_0 I}{4\pi r^3} \oint d\vec{\ell}\,'(\vec{r} \cdot \vec{r}\,') + \ldots \quad (2\text{–}19)$$

The first term vanishes identically, as the sum of vector displacements around a loop is zero. The second term is the contribution from the magnetic dipole moment of the loop. Using identity (17),

$$\oint (\vec{r} \cdot \vec{r}\,') d\vec{\ell}\,' = \int d\vec{S}' \times \vec{\nabla}'(\vec{r} \cdot \vec{r}\,') = -\int \vec{r} \times d\vec{S}' \quad (2\text{–}20)$$

we find for the vector potential of a small current loop

$$\vec{A}(\vec{r}) = \frac{\mu_0 I}{4\pi r^3} \int d\vec{S}' \times \vec{r} = \frac{\mu_0}{4\pi r^3} \left(I \int d\vec{S}' \right) \times \vec{r} \quad (2\text{–}21)$$

The bracketed term,

$$I \int d\vec{S}' \equiv \vec{m} \quad (2\text{–}22)$$

is the magnetic dipole moment of the loop. We have then, to first order, the magnetic vector potential

$$\vec{A}(\vec{r}) = \frac{\mu_0}{4\pi r^3}\vec{m} \times \vec{r} \tag{2–23}$$

Other useful forms for the magnetic moments are easily obtained by generalizing to not necessarily planar current loops:

$$\vec{m} = I\int d\vec{S}' = \frac{I}{2}\oint \vec{r}\,' \times d\vec{\ell}\,' = \frac{1}{2}\int \vec{r}\,' \times \vec{J}(\vec{r}\,')d^3r' \tag{2–24}$$

EXAMPLE 2.5: Find the magnetic dipole moment of a uniformly charged sphere of radius a rotating with angular velocity ω about the z axis.

Solution: The magnetic moment may be found from $\vec{m} = \frac{1}{2}\int(\vec{r}\,' \times \vec{J})\,d^3r'$. The integrand can be rearranged as follows.

$$\vec{r}\,' \times \vec{J} = \vec{r}\,' \times \rho\vec{v} = \rho\vec{r}\,' \times (\vec{\omega} \times \vec{r}\,') = \rho\left[r'^2\,\vec{\omega} - (\vec{r}\,' \cdot \vec{\omega})\vec{r}\,'\right] \tag{Ex 2.5.1}$$

$$= \rho\omega\left[r'^2\hat{k} - (r'\cos\theta)\vec{r}\,'\right] \tag{Ex 2.5.2}$$

$$= \rho\omega\left(r'^2\hat{k} - r'^2\cos^2\theta\hat{k} - r'^2\sin\theta\cos\theta\hat{s}\right) \tag{Ex 2.5.3}$$

where $\hat{s}$ is the unit cylindrical radial basis vector. The last term of the preceding expression (the $\hat{s}$ term) vanishes when integrated over φ, leaving

$$\vec{m} = \tfrac{1}{2}\rho\omega\hat{k}\Bigg(\int_0^a\int_0^{4\pi} r'^2 r'^2 dr'd\Omega' - \int_0^a\int_0^{\pi}\int_0^{2\pi} r'^2\cos^2\theta' r'^2 dr'\sin\theta'\,d\theta' d\varphi'\Bigg) \tag{Ex 2.5.4}$$

$$= \tfrac{1}{2}\rho\,\omega\hat{k}\left(\frac{4\pi a^5}{5} - \frac{2}{3}\frac{2\pi\,a^5}{5}\right) = \frac{4\pi\rho\omega a^5}{15}\hat{k} \tag{Ex 2.5.5}$$

To obtain the magnetic induction field of a magnetic dipole at the origin, we merely take the curl of the vector potential (2–23) yielding

$$\vec{B}(\vec{r}) = \vec{\nabla} \times \vec{A} = \frac{\mu_0}{4\pi}\left(-\frac{\vec{m}}{r^3} + \frac{3(\vec{m}\cdot\vec{r})\vec{r}}{r^5}\right) \tag{2–25}$$

The magnetic induction field of a small loop is illustrated in Figure 2.3. The dipole field is obtained as the area of the loop is shrunk to zero. The expression (2–25) can also be written as

$$\vec{B}(\vec{r}) = -\mu_0\vec{\nabla}\left(\frac{\vec{m}\cdot\vec{r}}{4\pi r^3}\right) \tag{2–26}$$

Recalling that outside current sources $\vec{B}$ can be expressed as $\vec{B} = -\mu_0\vec{\nabla}V_m$, we deduce that the scalar magnetic potential for a small current loop located at the origin is

$$V_m(\vec{r}) = \frac{\vec{m}\cdot\vec{r}}{4\pi r^3} \tag{2–27}$$

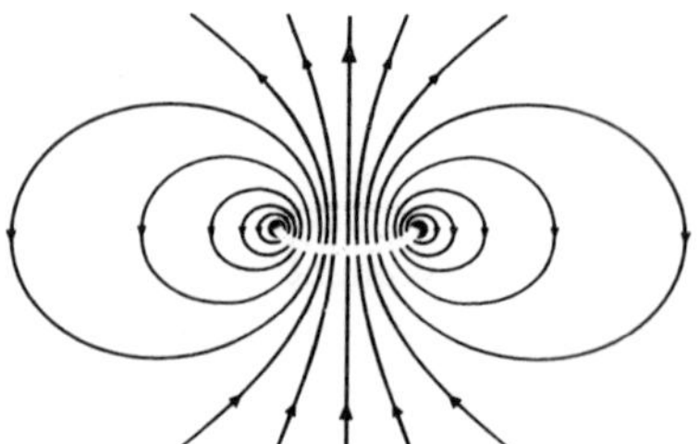

Figure 2.3: The magnetic induction field of a small current loop.

This form of the magnetic scalar potential allows us to easily generate our earlier result (1–64) for the arbitrary current loop. If the loop is considered as a sum of small bordering loops (each of which is adequately described as a dipole) situated at r' (Figure 2.4), we find

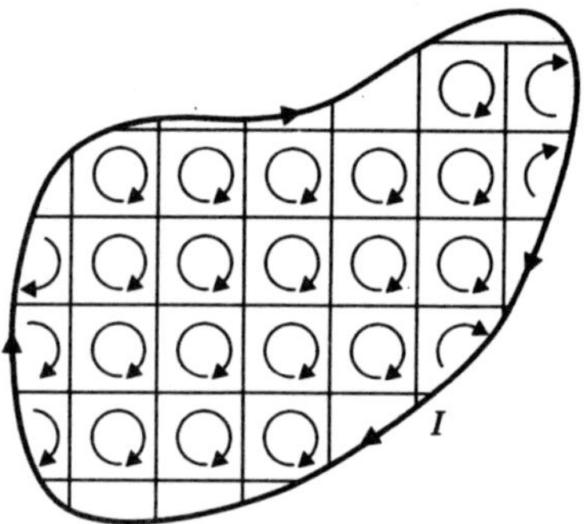

Figure 2.4: The scalar magnetic potential of the loop may be found as the sum of the potentials of each of the small loops. The countercurrents along the boundary of the inner loops are fictitious and cancel exactly.

$$V_m = \int dV_m = \int \frac{d\vec{m}(\vec{r}\,') \cdot (\vec{r} - \vec{r}\,')}{4\pi\, |\vec{r} - \vec{r}\,'|^3} = \int \frac{I d\vec{A} \cdot \vec{R}}{4\pi R^3} = -\frac{I\Omega}{4\pi} \tag{2–28}$$

Although only the current along the outside of the loop is real, we supply cancelling currents along the boundaries of contiguous loops. The fact that we need look at no further terms than those of the magnetic dipole to generate the correct form of the magnetic scalar potential suggests there is little to be gained from considering higher order terms.

2.1.2 The Spherical Polar Multipole Expansion

An alternative expansion for the potential of a charge distribution is offered by the generating function, (F–34), for *Legendre polynomials*:

$$\frac{1}{\sqrt{1 - 2t\cos\gamma + t^2}} = \sum_{n=0}^{\infty} t^n \mathrm{P}_n(\cos\gamma) \tag{2–29}$$

The term

$$\frac{1}{|\vec{r}-\vec{r}'|} = \frac{1}{r_> \sqrt{1 - \dfrac{2r_< \cos\gamma}{r_>} + \left(\dfrac{r_<}{r_>}\right)^2}} \tag{2-30}$$

where γ is the angle between $\vec{r}$ and $\vec{r}'$, has almost exactly this form. Thus

$$\frac{1}{|\vec{r}-\vec{r}'|} = \frac{1}{r_>} \sum_{n=0}^{\infty} \left(\frac{r_<}{r_>}\right)^n \mathrm{P}_n(\cos\gamma) \tag{2-31}$$

and

$$V(\vec{r}) = \frac{1}{4\pi\varepsilon_0} \int \frac{\rho(\vec{r}')d^3r'}{|\vec{r}-\vec{r}'|} = \frac{1}{4\pi\varepsilon_0} \int \sum_{n=0}^{\infty} \frac{r_<^n}{r_>^{n+1}} \mathrm{P}_n(\cos\gamma)\rho(\vec{r}')\, d^3r' \tag{2-32}$$

Two special cases arise. If the charge distribution lies along the z axis—that is, $\rho(\vec{r}') = \lambda(z')\delta(x')\delta(y')$ —then γ is the polar angle θ of $\vec{r}$, and we find for a line charge located on the z axis that

$$V(\vec{r}) = \frac{1}{4\pi\varepsilon_0} \sum_{n=0}^{\infty} \frac{\mathrm{P}_n(\cos\theta)}{r^{n+1}} \int (z')^n \lambda(z') dz' \tag{2-33}$$

when the source dimensions are smaller than r and

$$V(\vec{r}) = \frac{1}{4\pi\varepsilon_0} \sum r^n \mathrm{P}_n(\cos\theta) \int \frac{\lambda(z')}{z'^{n+1}} dz' \tag{2-34}$$

when r lies closer to the origin than any part of the source does. For any fixed r it is possible to divide the source into two parts, one part that lies inside a sphere defined by r and one that lies outside. The potential may then be found as the sum of the two contributions.

Alternatively, if we confine $\vec{r}$ to the z axis, then $\gamma = -\theta'$, the polar angle of $-\vec{r}'$, giving for a general $(r > r')$ charge distribution

$$V(0,0,z) = \frac{1}{4\pi\varepsilon_0} \sum_{n=0}^{\infty} \frac{1}{z^{n+1}} \int r'^n \mathrm{P}_n(\cos\theta')\rho(\vec{r}')r'^2 dr' \sin\theta' d\theta' d\varphi' \tag{2-35}$$

The unrestricted expansion in the "spherical basis" is obtained using the summation identity for spherical harmonics (F–47)

$$\mathrm{P}_\ell(\cos\gamma) = \frac{4\pi}{(2\ell+1)} \sum_{m=-\ell}^{\ell} \mathrm{Y}_\ell^m(\theta,\varphi)\mathrm{Y}_\ell^{*m}(\theta',\varphi') \tag{2-36}$$

Here γ is the angle between two vectors having polar angles (θ,φ) and (θ',φ') respectively, and Y_ℓ^{*m} is the complex conjugate of Y_ℓ^m, so that for $r > r'$

$$V(\vec{r}) = \frac{1}{\varepsilon_0} \sum_{\ell,m} \frac{\mathrm{Y}_\ell^m(\theta,\varphi)}{(2\ell+1)\, r^{\ell+1}} \int \rho(\vec{r}')r'^\ell \mathrm{Y}_\ell^{*m}(\theta',\varphi')d^3r' \tag{2-37}$$

The integral in the preceding expression, which we abbreviate $q_{\ell,m}$, is the m *component of the 2^ℓ-pole moment* of the charge distribution in the spherical basis. Given these numbers, characteristic of the distribution, it is easy to calculate the potential anywhere in space. The relationship between the components of the dipole and quadrupole moments in the spherical and Cartesian basis is

$$
\begin{aligned}
&q_{1,0} = \sqrt{\tfrac{3}{4\pi}}p_z \qquad q_{1,\pm1} = \mp\sqrt{\tfrac{3}{8\pi}}(p_x \mp ip_y) \\
&q_{2,0} = \sqrt{\tfrac{5}{16\pi}}Q_{zz} \quad q_{2,\pm1} = \sqrt{\tfrac{5}{24\pi}}(Q_{xz} \mp iQ_{yz}) \quad q_{2,\pm2} = \sqrt{\tfrac{5}{96\pi}}(Q_{xx} - Q_{yy} \mp 2iQ_{xy})
\end{aligned}
\tag{2–38}
$$

EXAMPLE 2.6: Find the dipole moment of a sphere of radius a centered on the origin bearing charge density $\rho = \rho_0 z'$ for $r' \le a$.

Solution: We exploit the convenient normalization of spherical harmonics (F–45) to simplify the integration. Writing $\rho(\vec{r}\,') = \rho_0 z'$ as a spherical harmonic,

$$\rho_0 z' = \rho_0 r' \cos\theta' = \sqrt{\frac{4\pi}{3}}\rho_0 r' \mathrm{Y}_1^0(\theta',\varphi') \tag{Ex 2.6.1}$$

using (2–38), we find that the integral for $q_{1,0}$ takes the form

$$q_{1,0} = \sqrt{\frac{4\pi}{3}}\int_0^a\int_0^{4\pi} \rho_0 r'^2 \mathrm{Y}_1^0(\theta',\varphi')\mathrm{Y}_1^0(\theta',\varphi') r'^2 d\Omega' dr' \tag{Ex 2.6.2}$$

the orthonormality relation (F–45) reduces the solid angle integral to unity so that the integral reduces to

$$q_{1,0} = \sqrt{\frac{4\pi}{3}}\int_0^a \rho_0 r'^4 dr' = \frac{\sqrt{4\pi}}{5\sqrt{3}}\rho_0 a^5 \tag{Ex 2.6.3}$$

finally then,

$$p_z = \sqrt{\tfrac{4\pi}{3}}q_{1,0} = \tfrac{4\pi}{15}\rho_0 a^5 \tag{Ex 2.6.4}$$

Symmetry dictates that the remaining components vanish. Of course we could have integrated $\rho_0 z'$ over the sphere directly to obtain the same results.

2.2 Interactions with the Field

For electric monopoles we have already seen that the force on a charge is given by $\vec{F} = q\vec{E}$. For the dipole of Section 2.1, it should be clear that in a uniform field the force on one of the charges is exactly balanced by the opposite force on the second charge, leaving us with zero net force. If the field is not parallel to the dipole axis, however, the two forces will not act along the axis, and a nonzero torque acts on the dipole. Thus we would conclude that in order to obtain a force on a dipole, a

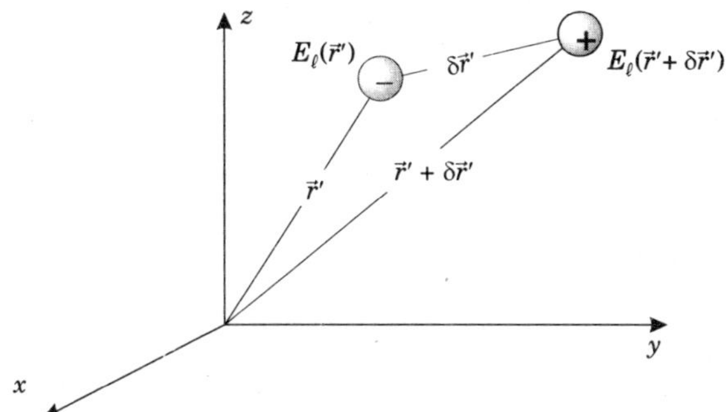

Figure 2.5: A small dipole in a nonhomogeneous electric field.

nonuniform field is required, whereas any field not parallel to the dipole suffices to produce a torque.

2.2.1 Electric Dipoles

Consider the small dipole composed of a negative charge situated at $\vec{r}'$ and an equal positive charge situated at $\vec{r}' + \delta\vec{r}'$ in a nonhomogeneous electric field (Figure 2.5). The ℓ component of the field at $\vec{r}' + \delta\vec{r}'$ can be expressed to first order in $\delta\vec{r}'$ as

$$\begin{aligned} E_\ell(\vec{r}' + \delta\vec{r}') &= E_\ell(\vec{r}') + \left.\frac{\partial E_\ell}{\partial x_j'}\right|_{\vec{r}'} \delta x_j' \\ &= E_\ell(\vec{r}') + (\delta\vec{r}' \cdot \vec{\nabla}')\, E_\ell(\vec{r}') \end{aligned} \tag{2–39}$$

The net force on the dipole is then

$$\begin{aligned} \vec{F}_{net} &= -q\vec{E}(\vec{r}') + q\vec{E}(\vec{r}' + \delta\vec{r}') \\ &= q(\delta\vec{r}' \cdot \vec{\nabla}')\vec{E}(\vec{r}') \\ &= (\vec{p} \cdot \vec{\nabla}')\vec{E}(\vec{r}') \end{aligned} \tag{2–40}$$

The torque on the dipole about the negative charge is easily found as $\vec{\tau} = q\,\delta\vec{r}' \times \vec{E}(\vec{r}' + \delta\vec{r}')$, which is to first order in $\delta\vec{r}'$

$$\vec{\tau} = \vec{p} \times \vec{E}(\vec{r}') \tag{2–41}$$

If the net force on the dipole vanishes, the torque calculated is independent of the choice of origin.

EXAMPLE 2.7: Find the force on an electric dipole $\vec{p} = (p_x,\, p_y,\, p_z)$ at distance r from a point charge q, located at the origin.

Solution: The electric field at the position of the dipole due to the point charge at the origin is $\frac{q\vec{r}}{4\pi\varepsilon_0 r^3}$. The force on the dipole is then

$$\vec{F} = (\vec{p} \cdot \vec{\nabla})\, \vec{E} = \left(p_x \frac{\partial}{\partial x} + p_y \frac{\partial}{\partial y} + p_z \frac{\partial}{\partial z}\right) \vec{E} \tag{Ex 2.7.1}$$

$$= \frac{q}{4\pi\varepsilon_0}\left(\frac{\vec{p}}{r^3} - \frac{3(\vec{p}\cdot\vec{r})\vec{r}}{r^5}\right) \tag{Ex 2.7.2}$$

We could, of course, just as well have found the force on q due to the electric field of the dipole, which gives the same result with only the sign reversed.

2.2.2 Magnetic Dipoles

The evaluation of the force on the magnetic dipole is slightly more complicated. We consider the small rectangular current loop of dimensions δx and δy in the x-y plane shown in Figure 2.6, threaded by a nonhomogeneous field $\vec{B}$. The net force in the x-direction is readily found to be

$$F_{net,x} = I\,\delta y\,B_z(x+\delta x) - I\,\delta y\,B_z(x) \tag{2–42}$$

$$= I\,\delta y\,\frac{\partial B_z}{\partial x}\,\delta x = m_z\,\frac{\partial B_z}{\partial x} \tag{2–43}$$

Obtaining the other two components of the net force in like fashion, we have for the force on a z-directed magnetic dipole

$$\vec{F}_{net} = m_z\vec{\nabla}\,B_z = \vec{\nabla}(m_z\,B_z) \tag{2–44}$$

Repeating the steps leading to (2–44) with similar loops having normals in the x and y direction easily generalizes (2–44) for a magnetic dipole with components (m_x, m_y, m_z) pointing in an arbitrary direction to give

$$\vec{F} = \vec{\nabla}(\vec{m}\cdot\vec{B}) \tag{2–45}$$

It is worth pointing out that the form of the force on the magnetic dipole is subtly different from that on the electric dipole. For example, if both dipoles are z-directed, then the force on the electric dipole is $\vec{F} = p_z\partial\vec{E}/\partial z$, whereas that on the magnetic dipole is $\vec{F} = m_z\vec{\nabla}B_z$.

The torque on a loop such as that of Figure 2.6 placed in a uniform magnetic induction field is fairly easily found. We compute the torque for each of three projections of an arbitrarily oriented loop. Since the monopole moment vanishes,

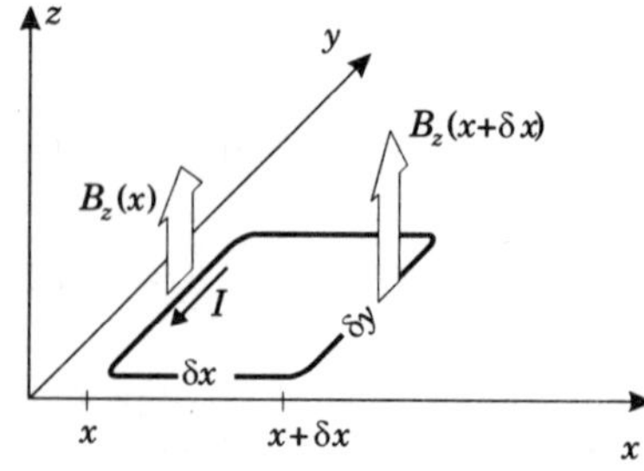

Figure 2.6: A small rectangular loop in the x-y plane has magnetic moment $\vec{m} = m_z\hat{k}$.

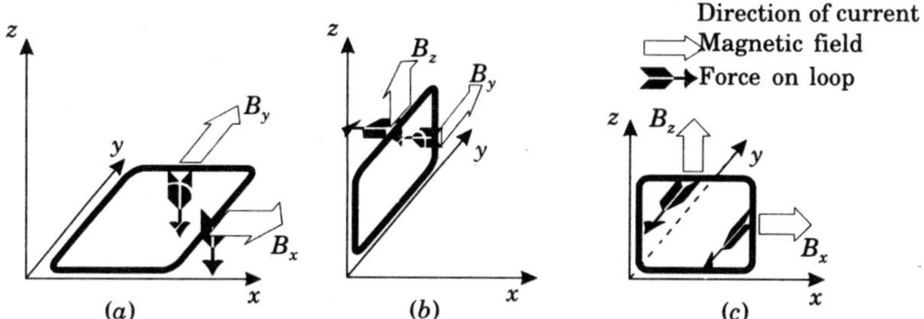

Figure 2.7: Forces on the sides of rectangular current loops in the (a) x-y plane, (b) y-z plane and (c) y-z plane.

we are free to choose the point about which we calculate the torque. For the loop of Figure 2.7, we calculate torques about the lower left corner of the loop in each of the three orientations. So long as the magnetic field intensity, $\vec{B}$, is uniform over the area of the loop, there is no net force, hence the torque is independent of origin.

x-y *plane, Figure* 2.7(a)	y-z *plane, Figure* 2.7(b)	x-z *plane, Figure* 2.7(c)
$\tau_y = +\delta_x I\delta_y B_x = +m_z B_x$	$\tau_z = +\delta_y I\delta_z B_y = +m_x B_y$	$\tau_z = +\delta_x I\delta_z B_x = -m_y B_x$
$\tau_x = -\delta_y I\delta_x B_y = -m_z B_y$	$\tau_y = -\delta_z I\delta_y B_z = -m_x B_z$	$\tau_x = -\delta_z I\delta_x B_z = +m_y B_z$

Considering each of these loops as the projection of an arbitrarily oriented loop, we combine these results to obtain generally

$$\begin{aligned}\tau_x &= m_y B_z - m_z B_y\\ \tau_y &= m_z B_x - m_x B_z\\ \tau_z &= m_x B_y - m_y B_x\end{aligned} \tag{2–46}$$

or, more briefly

$$\vec{\tau} = \vec{m} \times \vec{B} \tag{2–47}$$

2.3 Potential Energy

The potential energy W of a dipole in a field derives primarily from its orientation in the field. Not surprisingly, when the dipole is parallel to the field, its potential energy is lowest. Since, for a conservative force $\vec{F} = -\vec{\nabla}W$, we deduce immediately from (2–45) that the magnetic energy is given by

$$W = -\vec{m}\ \vec{B} \tag{2–48}$$

The case of the electric dipole in an electric field is not quite so obvious from our work. Suffice it to say that when $\vec{F}$ is conservative, curl $\vec{E}$ must vanish. If, in addition, $\vec{p}$ is independent of the coordinates, we have $(\vec{p}\cdot\vec{\nabla})\vec{E} = \vec{\nabla}(\vec{p}\cdot\vec{E})$, which leads to

$$W = -\vec{p}\cdot\vec{E} \tag{2–49}$$

A more instructive approach to the potential energy of a charge distribution in an external field is obtained by considering the potential energy, $W = \sum q_i V(\vec{r}^{(i)})$ of a collection of point charges q_i. Expanding the potential energy as a Taylor series about the origin, we obtain

$$\sum q_i V(\vec{r}^{(i)}) = \sum q_i V(0) + \sum q_i r_\ell^{(i)} \frac{\partial V}{\partial x_\ell} + \frac{1}{2!} \sum q_i r_\ell^{(i)} r_m^{(i)} \frac{\partial^2 V}{\partial x_\ell \, \partial x_m} + \cdots \quad (2\text{–}50)$$

where the superscript (i) on the coordinates indicates the charge to which the coordinate belongs. Since the external field cannot have charges at the locations $\vec{r}^{(i)}$ as these positions are already occupied by the charges subjected to the field, we have $\vec{\nabla} \cdot \vec{E} = \nabla^2 V = 0$, which we use to rewrite W as follows:

$$W = QV + \vec{p} \cdot \vec{\nabla} V + \tfrac{1}{6} \sum 3 q_i x_\ell^{(i)} x_m^{(i)} \frac{\partial^2 V}{\partial x_\ell \partial x_m} + \cdots \quad (2\text{–}51)$$

$$= QV - \vec{p} \cdot \vec{E} + \tfrac{1}{6} \sum q_i \left(3 x_\ell^{(i)} x_m^{(i)} \frac{\partial^2 V}{\partial x_\ell \partial x_m} - r^{(i)2} \, \nabla^2 V \right) \cdots \quad (2\text{–}52)$$

$$= QV - \vec{p} \cdot \vec{E} + \tfrac{1}{6} \sum q_i \left(3 x_\ell^{(i)} x_m^{(i)} - \delta_{\ell m} r^{(i)2} \right) \frac{\partial^2 V}{\partial x_\ell \partial x_m} \cdots \quad (2\text{–}53)$$

$$= QV - \vec{p} \cdot \vec{E} + \tfrac{1}{6} Q_{\ell m} \frac{\partial^2 V}{\partial x_\ell \, \partial x_m} \cdots \quad (2\text{–}54)$$

We assume summation over repeated indices in (2–54).

EXAMPLE 2.8: Find the energy of the quadrupole of Figure 2.2 in the potential $V(x, y) = V_0 xy$.

Solution: Using the form of the quadrupole energy given in (2-54), we have

$$W = \tfrac{1}{6} \left(Q_{xx} \frac{\partial^2 V}{\partial x^2} + Q_{xy} \frac{\partial^2 V}{\partial x \, \partial y} + Q_{xz} \frac{\partial^2 V}{\partial x \, \partial z} + Q_{yx} \frac{\partial^2 V}{\partial y \partial x} + Q_{yy} \frac{\partial^2 V}{\partial y^2} + \cdots \right) \quad (\text{Ex } 2.8.1)$$

$$= \tfrac{1}{6} \left(3qa^2 V_0 + 3qa^2 V_0 \right) = qa^2 V_0 \quad (\text{Ex } 2.8.2)$$

Exercises and Problems

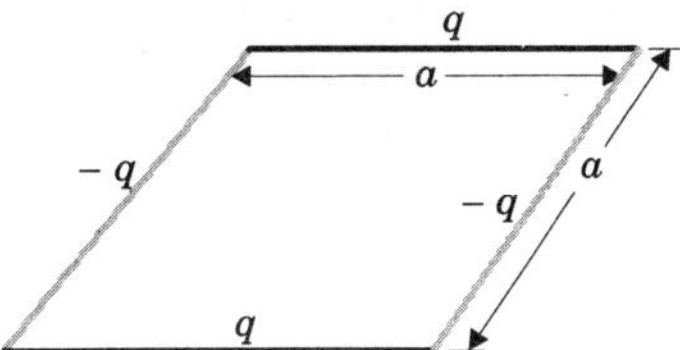

Figure 2.8: The square has alternating positive and negative charges along the edges.

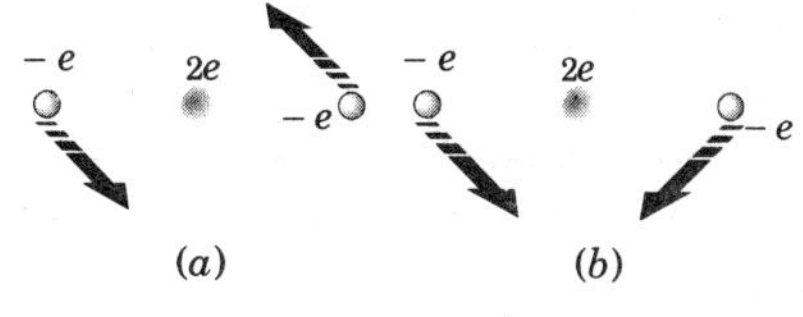

Figure 2.9: Classical model of helium.

2-1 Find the electric dipole moment of a thin ring lying in the x-y plane centered on the origin bearing line charge $\rho = \lambda\,\delta(r-a)\,\delta(z)\,\cos\varphi$.

2-2 Find the dipole moment of a thin, charged rod bearing charge density $\rho = \lambda\,z\,\delta(x)\,\delta(y)$ for $z \in (-a, a)$.

2-3 Compute the curl of (2–23) to obtain (2–25).

2-4 Show that the dipole moment of a charge distribution is unique when the monopole (charge) vanishes.

2-5 Find the quadrupole moment of two concentric coplanar ring charges q and $-q$, having radii a and b respectively.

2-6 Find the quadrupole moment of a square whose edges, taken in turn, have alternating charges $\pm\, q$ uniformly distributed over each as illustrated in Figure 2.8.

2-7 Find the *gyromagnetic ratio*, g ($\vec{m} = g\vec{L}$), for a charged, spinning object whose mass has the same distribution as its charge.

2-8 Find the gyromagnetic ratio of a charged spinning sphere whose mass is uniformly distributed through the volume and whose entire charge is uniformly distributed on the surface.

2-9 Find the quadrupole moment of a rod of length L bearing charge density $\rho = \eta\,(z^2 - L^2/12)$, with z measured from the midpoint of the rod.

2-10 Show that the potential generated by a cylindrically symmetric quadrupole at the origin is

$$V = \frac{Q_{zz}}{16\pi\varepsilon_0 r^3}(3\cos^2\theta - 1)$$

2-11 Find the charge Q contained in a sphere of radius a centered on the origin, whose charge density varies as $\rho_0 z^2$.

2-12 Find the quadrupole moment of the sphere of radius a of problem 2-11.

2-13 Show that the dipole term of the multipole expansion of the potential can be written

$$V_2 = -\frac{1}{4\pi\varepsilon_0}\sum q_i \vec{r}^{\,(i)} \cdot \vec{\nabla}\left(\frac{1}{r}\right)$$

2-14 Show that the quadrupole term of the multipole expansion of the potential can be written

$$V_3 = \frac{1}{8\pi\varepsilon_0}\sum q_i \vec{r}^{\,(i)} \cdot \vec{\nabla}\left[\vec{r}^{\,(i)} \cdot \vec{\nabla}\left(\frac{1}{r}\right)\right]$$

2-15 Use the "multipole expansion" (2–32) to find the potential due to the rod of (2-2) at points $|\vec{r}| > a$.

2-16 Use (2–32) to find the potential due to the charged rod of (2-2) at points not on the rod having $|\vec{r}| < a$.

2-17 A classical model of the helium atom has two electrons orbiting the nucleus. Assuming the electrons have coplanar circular orbits of radius a and rotate with angular frequency ω, find the electric dipole and quadrupole moments as a function of time in each of the following cases. (a) The electrons co-rotate diametrically opposed, (b) the electrons counter-rotate (see Figure 2.9).

2-18 Find an expression for the force between two dipoles $\vec{p}_1$, and $\vec{p}_2$, separated by $\vec{r}$.

2-19 Find an expression for the force between a quadrupole with components $Q_{\ell m}$ and a dipole $\vec{p}$ separated by r.

2-20 If magnetic monopoles existed, their scalar magnetic potential would be given by

$$V_m(\vec{r}) = \frac{1}{4\pi}\frac{q_m}{|\vec{r} - \vec{r}\,'|}$$

Obtain the magnetic scalar potential for two such hypothetical monopoles of opposite sign separated by a small distance $\vec{a}$. Show that, if we set $q_m = m/a$, the magnetic scalar potential becomes, in the limit of $a \to 0$, that of a magnetic dipole of strength $q_m\vec{a}$.

2-21 The proton has a *Landé g-factor* of 5.58 (the magnetic moment is $\vec{m} = 2.79\, e\hbar/2m$). When a proton is placed in a magnetic induction field, its spin precesses about the field axis. Find the frequency of precession.

2-22 Obtain an expression for the potential arising from a sheet of dipoles distributed over some surface. Assume a dipole layer density $\vec{D} = n\langle\vec{p}\rangle$, where n is the number of dipoles per unit area and $\langle\vec{p}\rangle$ is the mean dipole moment of these dipoles.

2-23 In the Stern-Gerlach experiment, atoms with differently oriented magnetic moments are separated in passage through a nonhomogeneous magnetic field produced by a wedge-shaped magnet. Assuming the field has

$$\frac{\partial\vec{B}}{\partial z} = \alpha\hat{k}$$

find (classically) the transverse force on atoms whose magnetic moment makes angle θ with the z axis.

2-24 Clearly, one could separate electric quadrupoles using an approach similar to that of the Stern-Gerlach experiment. Given that molecules have quadrupole moments of order 10^{-39} C-m^2, find the electric field gradient required to impart an impulse of 10^{-26} kgm/s to a molecule travelling at 100 m/s through a 1 mm region containing the gradient.

Chapter 3

Slowly Varying Fields in Vacuum

3.1 Magnetic Induction

In this chapter, we consider the effect of a slow variation in the electromagnetic fields. By slow, we mean that the sources do not change significantly during the time it takes for their fields to propagate to any point in the region of interest.

We have seen that charged particles experience a force when moving through a magnetic field, so it should come as no surprise that a moving source of magnetic field exerts a force on a stationary charged particle. At the position of the particle, a moving source of field is perceived as a temporally varying magnetic field. Any local field interpretation would therefore require that the force on the particle depend on $\partial\vec{B}/\partial t$. It will evolve that the force felt by such a stationary particle must be reinterpreted as resulting from an electric field.

We begin our consideration of time varying-fields with a short discussion of electromotive force.

3.1.1 Electromotive Force

When a (long) wire is connected between the terminals of a battery, it is not surprising that a current will flow through the circuit just completed. On further reflection, this obvious physical effect seems at odds with $\oint \vec{E}\cdot d\vec{\ell} = 0$ for static fields, a result that follows from $\vec{\nabla}\times\vec{E} = 0$ (equation 1–18). Clearly, there must be a yet-unaccounted-for force driving the charges around the circuit. The force responsible for the motion of charges must be nonelectrostatic in nature and may be mechanical or chemical. In this particular case, of course, the battery provides the "motive force," which is communicated throughout the wire by the electric field. The line integral of the electric field along the wire is precisely cancelled by the line integral of the field through the battery. The force communicated to the charges is well defined only as a line integral. The line integral around the loop of the total force per charge is called the *electromotive force*, or *EMF*, which we give the symbol $\mathcal{E}$.

$$\mathcal{E} = \oint \frac{\vec{F}}{q}\cdot d\vec{\ell} \tag{3–1}$$

The name is rather a misnomer because the EMF is certainly not a force, rather, it is the work that would be performed on a unit charge in travelling around the loop. In electrostatics it makes no difference whether forces arising from the static electric field $\vec{E}$ are included in $\oint \vec{F}\cdot d\vec{\ell}$, because this contribution would sum to zero in any case.

To clarify how chemistry might give rise to nonelectrostatic forces, we digress briefly to the specific example of a dilute solution of an electrolyte such as HCl, whose concentration varies spatially. The electrolyte will be almost entirely dissociated into H^+ and Cl^- ions. H^+, being much lighter, diffuses more rapidly than Cl^-; therefore, more H^+ ions than Cl^- diffuse into regions of low concentration. If the concentration gradient is maintained, a net positive current will flow into the low-concentration region until the accumulation of excess charge produces an electric field large enough to counter the differential diffusion. We might usefully think of the diffusion resulting from a force $\vec{F}$ causing the movement of the ions. In terms of this force, the equilibrium condition becomes $\vec{F} + e\vec{E} = 0$. Clearly, inside the medium, $e\vec{E} = -\vec{F}$. We could build a battery on this principle–separating two halves of a container with a permeable membrane, filling one side of the container with HCl and filling the other with clear water.

3.1.2 Magnetically Induced Motional EMF

When a charge is forced to move through a magnetic induction field, it is subjected to a force (1–30) due to motion through the field:

$$\vec{F} = q\vec{v} \times \vec{B} \tag{3–2}$$

Although the motion produced by this force gives the charged particle the capacity to do work, it is important to recognize that the magnetic field does not do any work on the charge; instead, whatever agent produces or maintains $\vec{v}$ does the work. Let us consider the EMF for a mobile loop placed in a static electric and magnetic field. In particular, we allow the loop to stretch and deform with velocity $\vec{v}(\vec{r},t)$. The force on a charge attached to the moving loop is then $q(\vec{E} + \vec{v} \times \vec{B})$,

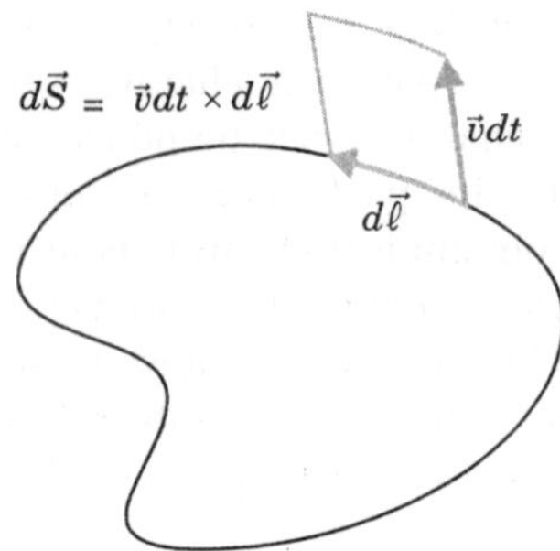

Figure 3.1: A small segment of the loop increases the area at a rate $\vec{v} \times d\vec{\ell}$.

leading to an EMF

$$\mathcal{E} = \oint_{\Gamma(t)} \vec{E} \cdot d\vec{\ell} + \oint_{\Gamma(t)} (\vec{v} \times \vec{B}) \cdot d\vec{\ell} \tag{3–3}$$

For static fields, the first integral vanishes, and the triple product in the second integral may be rearranged to give

$$\mathcal{E} = \oint_{\Gamma(t)} (d\vec{\ell} \times \vec{v}) \cdot \vec{B} \tag{3–4}$$

During a time dt, a segment of the loop of length $d\ell$ moves to increase the area within the loop by $d\vec{S} = \vec{v}dt \times d\vec{\ell}$ (Figure 3.1). Thus we write $d\vec{\ell} \times \vec{v} = -d\vec{S}/dt$, and, instead of summing over the length of the loop, we sum the area increments:

$$\mathcal{E} = -\int_{\Sigma(t)} \frac{d\vec{S}}{dt} \cdot \vec{B} \tag{3–5}$$

where $\Sigma(t)$ is the area included in the loop.

Defining the magnetic *flux*, Φ, as

$$\Phi \equiv \int_{\Sigma} \vec{B} \cdot d\vec{S} \tag{3–6}$$

we see that in the case of static fields,

$$\mathcal{E} = -\frac{d\Phi}{dt} \tag{3–7}$$

(It should, incidently, now be clear why B is sometimes called the *magnetic flux density.*)

EXAMPLE 3.1: A flat circular coil of N turns and radius a travels in a direction parallel to the plane of the coil through a uniform magnetic induction field perpendicular to the plane of the coil. Find the EMF generated.

Solution: The flux through the loop is constant; hence, no EMF will be generated. The EMF generated at the leading edge (semicircle) is precisely cancelled by the EMF at the trailing edge.

EXAMPLE 3.2: A nova sheds a ring of ionized gas expanding radially through a uniform magnetic induction field $B_z\hat{k}$ with velocity $\vec{v}$. Find the tangential acceleration of the charged particles in the ring.

Solution: The EMF generated around a loop of radius r is

$$\oint \frac{\vec{F}}{q} \cdot d\vec{\ell} = -\frac{d\Phi}{dt} = -2\pi r B_z \frac{dr}{dt} \tag{Ex 3.2.1}$$

The line integral of the tangential force is just $2\pi r F_\varphi$, giving $F_\varphi = qv_r B_z$ (a result we might have anticipated) and finally

$$a_\varphi = -\frac{q}{m} v B_z \tag{Ex 3.2.2}$$

negative charges are accelerated in the $\hat{\varphi}$ direction whereas positive charges are accelerated in the opposite direction. This conclusion could, of course, have been reached much more easily from a direct application of the Lorentz force $\vec{F} = q(d\vec{r}/dt) \times \vec{B}$.

3.1.3 Time-Dependent Magnetic Fields

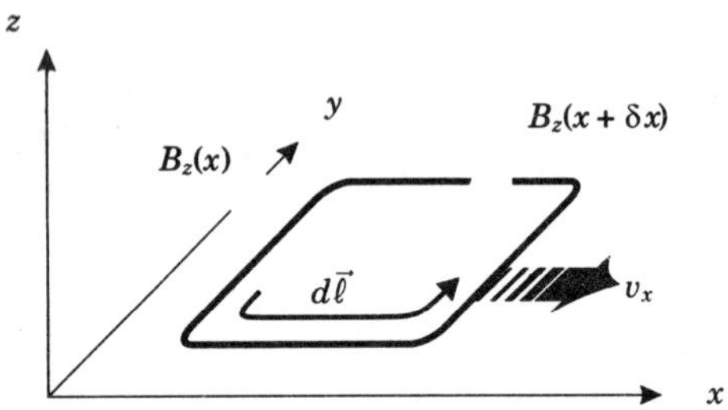

Figure 3.2: The rectangular loop moves in the x direction through a non-homogeneous magnetic induction field.

Let us now consider a loop whose area does not change but instead moves through a magnetic induction field whose strength varies with position. The sides of the moving loop will evidently experience a time-dependent field. To simplify matters, consider a small rectangular loop of dimensions δx and δy in the x-y plane, moving in the x direction through a magnetic induction field whose z component varies (to first order) linearly with x (Figure 3.2). The EMF generated around the moving loop is generally

$$\mathcal{E} = \oint \frac{\vec{F}}{q} \cdot d\vec{\ell} = \oint (\vec{v} \times \vec{B}) \cdot d\vec{\ell} \tag{3–8}$$

If the field is not homogeneous, having value $B_z(x, y)$ at x and $B_z(x + \delta x, y) = B_z(x, y) + (\partial B_z/\partial x)\,\delta x$ on the other side of the loop, we expand the integral as

$$\mathcal{E} = \int_{\delta y} \Big(\vec{v} \times \vec{B}(x, y)\Big) \cdot -\hat{\jmath}\, dy + \int_{\delta y} \Big(\vec{v} \times \vec{B}(x + \delta x, y)\Big) \cdot \hat{\jmath}\, dy \tag{3–9}$$

The integrals over the sides parallel to the velocity make no contribution to the EMF and have therefore been neglected. Gathering the two terms, we have

$$\mathcal{E} = \int_{\delta y} \vec{v} \times \frac{\partial \vec{B}(x, y)}{\partial x} \delta x \cdot \hat{\jmath} dy \tag{3–10}$$

which becomes, on putting in the explicit directions of $\vec{v}$ and $\vec{B}$,

$$\begin{aligned} \mathcal{E} &= \int_{\delta y} \hat{\imath} \frac{dx}{dt} \times \hat{k} \frac{\partial B_z(x, y)}{\partial x} \delta x \cdot \hat{\jmath}\, dy \\ &= -\int_{\delta y} \frac{\partial B_z}{\partial t} \delta x\, dy \end{aligned} \tag{3–11}$$

We conclude that

$$\mathcal{E} = -\int\int \frac{\partial B_z}{\partial t}\, dS_z \tag{3–12}$$

It is not difficult to generalize this result to fields and motions in arbitrary directions, to obtain for a loop of arbitrary, but constant, area

$$\mathcal{E} = -\int\int \frac{\partial \vec{B}}{\partial t} \cdot d\vec{S} = -\frac{d}{dt}\int\int \vec{B} \cdot d\vec{S} \tag{3–13}$$

We note that this result can again, as in (3–7), be written

$$\mathcal{E} = -\frac{d}{dt}\Phi$$

If instead of moving the loop we move the magnet responsible for the field above, special relativity would require the same EMF, but since now the velocity $v = 0$, implying there can be no contribution from $\vec{v} \times \vec{B}$. Charges within the wire of the loop have no way to tell whether the loop is moving or some other means is used to vary the field temporally. The conclusion must then be that the first integral of (3–3) cannot vanish when we have temporally varying fields. Instead, we must have

$$\mathcal{E} = \oint \vec{E} \cdot d\vec{\ell} = -\frac{d\Phi}{dt} \tag{3–14}$$

We note that as a consequence of (3–14), when $\vec{B}$ varies in time, we cannot maintain a vanishing curl of $\vec{E}$.

EXAMPLE 3.3: An electron with speed v executes cyclotron motion between the parallel faces of an electromagnet whose field is increased at a rate of dB_z/dt. Determine the tangential acceleration of the electron.

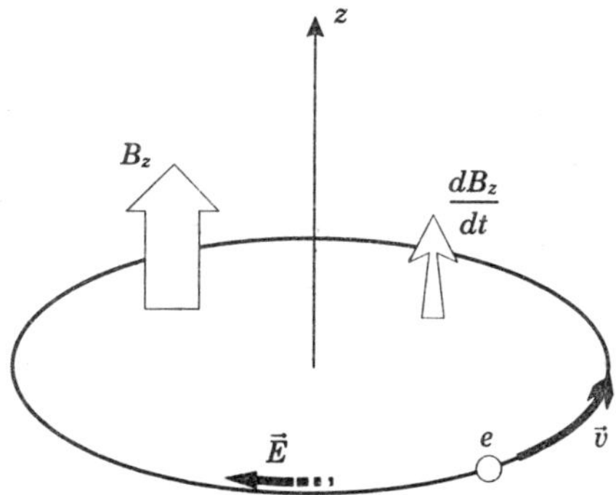

Figure 3.3: When the cyclotron field $\vec{B}$ and its derivative $d\vec{B}/dt$ are parallel, the charged particle increases its speed.

Solution: The radius of the electron's orbit (Figure 3.3) is readily obtained from

$$-evB_z = \frac{mv^2}{r} \quad \Rightarrow \quad r = \frac{mv}{-eB_z} \tag{Ex 3.3.1}$$

The flux included in the orbit is just $\Phi = \pi r^2 B_z$, so that the EMF is given by

$$\oint \vec{E} \cdot d\vec{\ell} = 2\pi r E_\varphi = -\pi r^2 \frac{dB_z}{dt} \qquad \text{(Ex 3.3.2)}$$

giving rise to a tangential acceleration

$$\vec{a} = \frac{e\vec{E}}{m} = -\frac{er}{2m}\frac{dB_z}{dt}\hat{\varphi} = \frac{v}{2B_z}\frac{dB_z}{dt}\hat{\varphi} \qquad \text{(Ex 3.3.3)}$$

The electron is accelerated in the direction of its motion.

3.1.4 Faraday's Law

Moving the magnet is, of course, just one means of changing $\vec{B}$ within the loop as a function of time. We might equally well decrease or increase the current to an electromagnet or use other means of changing the field. We postulate that, under all conditions,

$$\oint_\Gamma \vec{E} \cdot d\vec{\ell} = -\frac{d}{dt}\int_\Sigma \vec{B} \cdot d\vec{S} \qquad (3\text{–}15)$$

The relation (3–15) is Faraday's law in integral form. We can obtain the differential form by applying Stokes' theorem to the leftmost integral:

$$\int_\Sigma (\vec{\nabla} \times \vec{E}) \cdot d\vec{S} = -\frac{d}{dt}\int_\Sigma \vec{B} \cdot d\vec{S} = -\int_\Sigma \frac{\partial \vec{B}}{\partial t} \cdot d\vec{S} \qquad (3\text{–}16)$$

Since Σ was arbitrary, the integrands must be equal, giving

$$\vec{\nabla} \times \vec{E} = -\frac{\partial \vec{B}}{\partial t} \qquad (3\text{–}17)$$

for the required result. We note that because $\vec{\nabla} \cdot \vec{B}$ vanishes, $\vec{\nabla} \cdot (\vec{\nabla} \times \vec{E})$ vanishes, as of course it must.

3.2 Displacement Current

In this section, we will see that just as a time-varying magnetic induction field causes a curl of the electric field, so a time-varying electric field causes a curl of the magnetic field.

Consider the current flowing along a wire terminated by a capacitor plate that charges in response to the current, as shown in Figure 3.4. Drawing an Ampèrian loop Γ around the wire, as in the figure, we have from (1–39)

$$\oint_\Gamma \vec{B} \cdot d\vec{\ell} = \mu_0 \int\!\!\int_S \vec{J} \cdot d\vec{S} \qquad (3\text{–}18)$$

Now, although there is no ambiguity about what is meant by the path Γ, the included area could be either the flat surface S_1 or the bulbous surface S_2. If we use area S_1 to compute B from Ampère's law, then we get simply

$$\oint \vec{B} \cdot d\vec{\ell} = \mu_0 I \qquad (3\text{–}19)$$

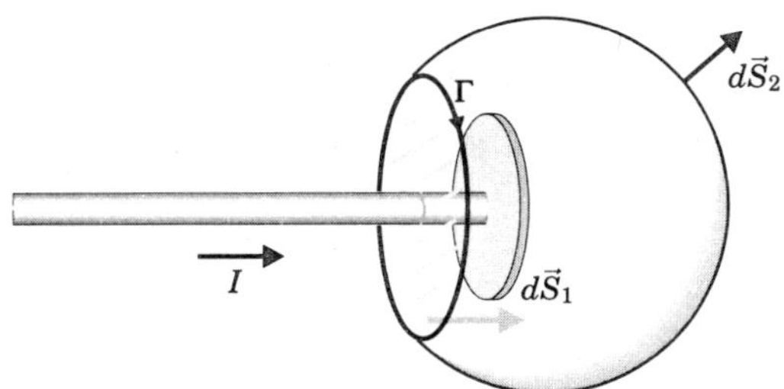

Figure 3.4: The surface included by the loop shown may be either the flat surface S_1 or the baglike surface S_2.

If, on the other hand, we use surface S_2, we find that $J = 0$ on the surface for a result quite inconsistent with (3–19). Since B cannot depend on which surface we use to compute it, (3–18) cannot be complete for nonstatic fields and we must postulate a second term to compensate for the discontinuity in the current, I.

With the Ampèrian loop indicated in Figure 3.4, the normal to the surface S_1 points into the volume enclosed by S_1 and S_2. The enclosing surface then becomes $S_2 - S_1$. (The minus sign in front of S_1 occurs because the sense in which the boundary Γ is followed indicates an inward orientation of the enclosed surface element S_1, while the enclosing surface of the volume must be taken with an outward-directed normal.) Because no current flows through surface S_2, we may without loss of generality replace

$$\iint_{S_1} \mu_0 \vec{J} \cdot d\vec{S} \quad \text{by minus} \quad \oint_{S_2 - S_1} \mu_0 \vec{J} \cdot d\vec{S} \tag{3–20}$$

and of course the right hand side of (3–20) can be converted to a volume integral using the divergence theorem (20). Thus

$$\iint_{S_1} \vec{J} \cdot d\vec{S} = -\iiint_{volume} (\vec{\nabla} \cdot \vec{J}) d^3r \tag{3–21}$$

We use the continuity equation to replace $\vec{\nabla} \cdot \vec{J}$ by $-\partial\rho/\partial t$ to get

$$\iint_{S_1} \vec{J} \cdot d\vec{S} = \iiint \frac{\partial \rho}{\partial t} d^3r \tag{3–22}$$

and with the aid of (1–14),

$$\iint_{S_1} \vec{J} \cdot d\vec{S} = \iiint \varepsilon_0 \frac{\partial}{\partial t} (\vec{\nabla} \cdot \vec{E}\,)\, d^3r = \oint_{S_2 - S_1} \varepsilon_0 \frac{\partial \vec{E}}{\partial t} \cdot d\vec{S} \tag{3–23}$$

Thus, $\oint \vec{B} \cdot d\vec{\ell}$ can be calculated equally well from integrating $\mu_0 \vec{J}$ over $\vec{S}_1$ or from integrating $\mu_0 \varepsilon_0 \partial \vec{E}/\partial t$ over the enclosing surface $S_2 - S_1$. This term, $\varepsilon_0 \partial \vec{E}/\partial t$, is called the displacement current, not a terribly descriptive name as it is not a current but only plays the role of one in Ampère's law. Since we can certainly imagine situations in which both $\vec{J}$ and $\partial\vec{E}/\partial t$ are nonzero, it seems reasonable to modify Ampère's law to read

$$\vec{\nabla}\times\vec{B} = \mu_0\vec{J} + \mu_0\varepsilon_0\frac{\partial\vec{E}}{\partial t} \tag{3–24}$$

This modification to Ampère's law is in fact necessary to preserve the general vector identity $\vec{\nabla}\cdot(\vec{\nabla}\times\vec{B}) = 0$, as shown below:

$$\vec{\nabla}\cdot(\vec{\nabla}\times\vec{B}) = \mu_0\left(\vec{\nabla}\cdot\vec{J} + \frac{\partial}{\partial t}(\varepsilon_0\vec{\nabla}\cdot\vec{E})\right) \tag{3–25}$$

$$= \mu_0\left(\vec{\nabla}\cdot\vec{J} + \frac{\partial\rho}{\partial t}\right) = 0 \tag{3–26}$$

It is this latter reasoning that rigorously yields (3–24).

EXAMPLE 3.4: A Van de Graaff generator with a spherical bowl of radius R is charged at a constant rate with a current I. Find the magnetic induction field at a distance a from the axis above the sphere. (Assume the current enters at the bottom of the bowl.)

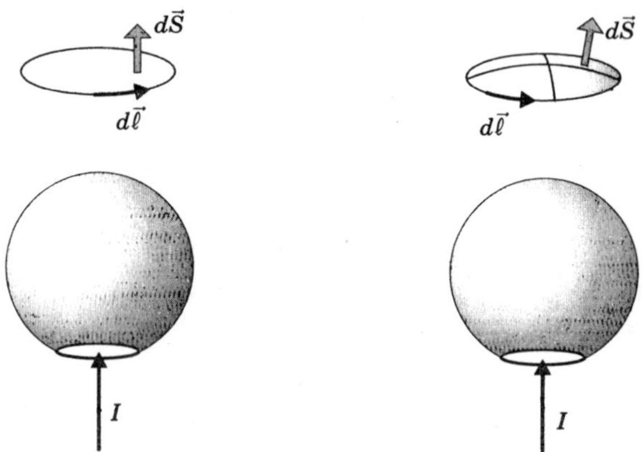

Figure 3.5: The surface enclosed by the loop can be taken to be either the flat surface on the left, or the spherical cap on the right.

Solution: As charge accumulates on the bowl, the electric field will increase at the following rate:

$$\frac{\partial\vec{E}}{\partial t} = \frac{1}{4\pi\varepsilon_0}\frac{\frac{dQ}{dt}\hat{r}}{r^2} = \frac{I}{4\pi\varepsilon_0}\frac{\hat{r}}{r^2} \tag{Ex 3.4.1}$$

We draw a loop of radius a about the z axis above the sphere where $\vec{J} = 0$ and integrate (3–24) over an arbitrary surface enclosed by the loop

$$\int_S(\vec{\nabla}\times\vec{B})\cdot d\vec{S} = \oint\vec{B}\cdot d\vec{\ell} = \mu_0\varepsilon_0\int\frac{\partial\vec{E}}{\partial t}\cdot d\vec{S} \tag{Ex 3.4.2}$$

$$2\pi a B_\varphi = \frac{\mu_0 I}{4\pi} \int \frac{\hat{r}}{r^2} \cdot d\vec{S} \tag{Ex 3.4.3}$$

The surface integral can be evaluated easily either over the flat surface included by the loop or alternatively over a spherical cap concentric with the sphere, as shown in Figure 3.5.

(a) For the flat surface, we need the z component of $\hat{r}$ in order to compute $\hat{r} \cdot d\vec{S}$. Calling the cylindrical radial coordinate ρ, we write $r = \sqrt{\rho^2 + z^2}$ and $(\hat{r})_z = \cos\theta = z/r$, to obtain

$$B_\varphi(z) = \frac{\mu_0 I}{8\pi^2 a} \int_0^a \frac{z}{r^3} 2\pi\rho \, d\rho \tag{Ex 3.4.4}$$

$$= \frac{\mu_0 I}{4\pi a} \left(\frac{z}{|z|} - \frac{z}{\sqrt{z^2 + a^2}} \right) \tag{Ex 3.4.5}$$

The reader is cautioned that lowering the surface below the top of the sphere will introduce a contribution from the current required to charge the sphere.

(b) The element of surface area for the spherical cap on the right of Figure 3.5 is $d\vec{S} = \hat{r} r^2 \, d\Omega = 2\pi \hat{r} r^2 \sin\theta \, d\theta$, leading for positive z to

$$B_\varphi = \frac{\mu_0 I}{8\pi^2 a} \int_0^{\cos^{-1}(z/r)} \frac{\hat{r} \cdot 2\pi \hat{r} r^2 \sin\theta \, d\theta}{r^2} \tag{Ex 3.4.6}$$

$$= \frac{\mu_0 I}{4\pi a} \int_0^{\cos^{-1}(z/r)} \sin\theta \, d\theta = \frac{\mu_0 I}{4\pi a} (-\cos\theta) \Big|_0^{\cos^{-1}(z/\sqrt{z^2+a^2})} \tag{Ex 3.4.7}$$

$$= \frac{\mu_0 I}{4\pi a} \left(1 - \frac{z}{\sqrt{z^2 + a^2}} \right) \tag{Ex 3.4.8}$$

Although the electric field grows spherically symmetrically about the bowl, the magnetic field induced has only axial symmetry. Presumably, the current entering at the bottom makes the z axis privileged. Interestingly, below the sphere, the magnetic induction field B_φ due to the temporally increasing electric field is replaced by the field from the real current running into the globe. It is in general impossible to produce a (non-trivial) zero divergence field that is spherically symmetric, a theorem that is most easily visualized by thinking of arrows representing the curling vector field as similar to the fuzz on a peach combed flat. It is not hard to convince oneself that the peach fuzz always presents at least two "crowns", meaning that the spherical symmetry is broken.

3.3 Maxwell's Equations

The results of the first chapter and this are very neatly summarized by four coupled differential equations first obtained by *James Clerk Maxwell.* (Maxwell did not have available to him the short hand notations for curl and div and wrote these equations out component by component, nor is the following the exact, final formulation, which will have to wait until chapter 8.) The equations (3–27) are appropriately known as **Maxwell's equations**.

$$\begin{aligned}
&\vec\nabla\cdot\vec E=\frac{\rho}{\varepsilon_0} && \text{Gauss' law}\\
&\vec\nabla\cdot\vec B=0 &&\\
&\vec\nabla\times\vec E=-\frac{\partial\vec B}{\partial t} && \text{Faraday's law}\\
&\vec\nabla\times\vec B=\mu_0\vec J+\mu_0\varepsilon_0\frac{\partial\vec E}{\partial t} && \text{Ampère's law}
\end{aligned}\tag{3–27}$$

Equations (3–27) are linear, meaning that superimposed solutions still solve the equations. It is worth noting that this linearity is not always true, in particular in materials the fields may not be simply additive. Even in vacuum, quantum mechanics would predict a nonlinearity, because virtual pairs of charged particles created by photons in the field can scatter other photons in the field.

Maxwell's equations and the Lorentz force equation together with Newton's second law constitute the basis of all classical electromagnetic interactions. Our development of electromagnetic theory rests almost entirely on Maxwell's equations, and we will use them frequently. The importance of these equations cannot be overstated, and it is earnestly recommended that they be committed to heart, as having to refer to this page each time we have need of Maxwell's equations will prove a considerable impediment to learning.

3.4 The Potentials

When the fields are time dependent, $\vec E$ can no longer be found simply as the gradient of a scalar potential since $\vec\nabla\times\vec E\neq 0$. In all cases $\vec\nabla\cdot\vec B\equiv 0$, so that we can still express $\vec B$ as $\vec B=\vec\nabla\times\vec A$. From Faraday's law we have

$$\begin{aligned}
\vec\nabla\times\vec E&=-\frac{\partial\vec B}{\partial t}=-\frac{\partial}{\partial t}(\vec\nabla\times\vec A)\\
&=-\vec\nabla\times\left(\frac{\partial\vec A}{\partial t}\right)
\end{aligned}\tag{3–28}$$

or

$$\vec\nabla\times\left(\vec E+\frac{\partial\vec A}{\partial t}\right)=0\tag{3–29}$$

We conclude that *not* $\vec E$, but instead $\vec E+\partial\vec A/\partial t$, is expressible as the gradient of a scalar potential. We have then $\vec E+\partial\vec A/\partial t=-\vec\nabla V$, or

$$\vec E=-\vec\nabla V-\frac{\partial\vec A}{\partial t}\tag{3–30}$$

Taking the divergence of (3–30), we obtain the equation

$$\vec\nabla\cdot\vec E=\vec\nabla\cdot\left(-\vec\nabla V-\frac{\partial\vec A}{\partial t}\right)$$

or, equating the right hand side to ρ/ε_0,

$$-\nabla^2 V - \frac{\partial}{\partial t}(\vec{\nabla}\cdot\vec{A}) = \frac{\rho}{\varepsilon_0} \tag{3–31}$$

Similarly, putting $\vec{B} = (\vec{\nabla}\times\vec{A})$ into Ampère's law we obtain the analogous equation for the vector potential $\vec{A}$.

$$\vec{\nabla}\times(\vec{\nabla}\times\vec{A}) = \vec{\nabla}(\vec{\nabla}\cdot\vec{A}) - \nabla^2\vec{A} = \mu_0\vec{J} - \mu_0\varepsilon_0\frac{\partial}{\partial t}\left(\vec{\nabla}V + \frac{\partial\vec{A}}{\partial t}\right)$$

or

$$\left(\nabla^2\vec{A} - \mu_0\varepsilon_0\frac{\partial^2\vec{A}}{\partial t^2}\right) - \vec{\nabla}\left(\vec{\nabla}\cdot\vec{A} + \mu_0\varepsilon_0\frac{\partial V}{\partial t}\right) = -\mu_0\vec{J} \tag{3–32}$$

3.4.1 The Lorentz Force and Canonical Momentum

In classical mechanics it is frequently useful to write conservative forces as gradients of potential energies. Although the forces associated with $\vec{B}$ are not conservative, we can nonetheless find a canonical momentum whose time derivative is the gradient of a quasipotential. We begin by writing the Lorentz force as the rate of change of momentum of a charge in an electromagnetic field,

$$\frac{d\vec{p}}{dt} = q(\vec{E} + \vec{v}\times\vec{B}) \tag{3–33}$$

$$= q\left[-\vec{\nabla}V - \frac{\partial\vec{A}}{\partial t} + \vec{v}\times(\vec{\nabla}\times\vec{A})\right] \tag{3–34}$$

The last term of (3–34) may be expanded using (9)

$$\vec{\nabla}(\vec{v}\cdot\vec{A}) = (\vec{v}\cdot\vec{\nabla})\vec{A} + (\vec{A}\cdot\vec{\nabla})\vec{v} + \vec{A}\times(\vec{\nabla}\times\vec{v}) + \vec{v}\times(\vec{\nabla}\times\vec{A})$$

If $\vec{v}$, although a function of time, is not explicitly a function of position, we may eliminate spatial derivatives of $\vec{v}$ from the expansion of $\vec{\nabla}(\vec{v}\cdot\vec{A})$ to obtain $\vec{v}\times(\vec{\nabla}\times\vec{A}) = \vec{\nabla}(\vec{v}\cdot\vec{A}) - (\vec{v}\cdot\vec{\nabla})\vec{A}$. Replacing the last term of (3–34) by this equality we find

$$\frac{d\vec{p}}{dt} = q\left[-\vec{\nabla}V - \frac{\partial\vec{A}}{\partial t} + \vec{\nabla}(\vec{v}\cdot\vec{A}) - (\vec{v}\cdot\vec{\nabla})\vec{A}\right] \tag{3–35}$$

The pair of terms

$$\frac{\partial\vec{A}}{\partial t} + (\vec{v}\cdot\vec{\nabla})\vec{A} \equiv \frac{d\vec{A}}{dt} \tag{3–36}$$

is called the *convective derivative*[7] of $\vec{A}$. Substituting $d\vec{A}/dt$ for the pair of terms

[7] As a charge moves through space, the change in $\vec{A}$ it experiences arises not only from the temporal change in $\vec{A}$ but also from the fact that it samples $\vec{A}$ in different locations.

$$\begin{aligned} d\vec{A} &= \vec{A}\,[\vec{r}(t+dt), t+dt] - \vec{A}\,[\vec{r}(t), t] \\ &= \left\{\vec{A}\,[\vec{r}(t)+\vec{v}dt, t+dt] - \vec{A}\,[\vec{r}(t)+\vec{v}dt, t]\right\} + \left\{\vec{A}\,[\vec{r}(t)+\vec{v}dt, t] - \vec{A}\,[\vec{r}(t), t]\right\} \\ &= \frac{\partial\vec{A}}{\partial t}dt + (\vec{v}\cdot\vec{\nabla})\vec{A}dt \end{aligned}$$

(3–36), equation (3–35) may be written

$$\frac{d\vec{p}}{dt} = q\left(-\vec{\nabla}(V - \vec{v}\cdot\vec{A}) - \frac{d\vec{A}}{dt}\right) \tag{3–37}$$

Grouping like terms gives the desired result:

$$\frac{d}{dt}\left(\vec{p} + q\vec{A}\right) = -\vec{\nabla}\left(qV - q\vec{v}\cdot\vec{A}\right) \tag{3–38}$$

The argument of $\vec{\nabla}$ on right hand side of (3–38) is the potential that enters Lagrange's equation as the potential energy of a charged particle in an electromagnetic field and the term $mv_i + qA_i$ on the left hand side is the momentum conjugate to the coordinate x_i (see Exercise 3.9).

3.4.2 Gauge Transformations

As we have mentioned previously, $\vec{A}$ is not unique since we can add any vector field whose curl vanishes without changing the physics. We see now from (3-30) that, concomitant with any change in A, we also require a compensating change in V in order to keep $\vec{E}$ (and hence the physics) unchanged.

Recall that a curl free field must be the gradient of a scalar field; hence we write as before, $\vec{A}' = \vec{A} + \vec{\nabla}\Lambda$. Let us denote the correspondingly changed potential as V'. The magnetic induction field is invariant under this change and we can use (3–30) to express the electric field in terms of the changed (′) potentials,

$$\vec{E} = -\vec{\nabla}V' - \frac{\partial\vec{A}'}{\partial t} \tag{3–39}$$

$$= -\vec{\nabla}V' - \frac{\partial}{\partial t}\left(\vec{A} + \vec{\nabla}\Lambda\right)$$

$$= -\vec{\nabla}\left(V' + \frac{\partial\Lambda}{\partial t}\right) - \frac{\partial\vec{A}}{\partial t} \tag{3–40}$$

or in terms of the unchanged potentials $\vec{E} = -\vec{\nabla}V - \partial\vec{A}/\partial t$. Comparison of the terms gives $V' = V - \partial\Lambda/\partial t$.

The pair of coupled transformations

$$\begin{aligned} \vec{A}' &= \vec{A} + \vec{\nabla}\Lambda \\ V' &= V - \frac{\partial\Lambda}{\partial t} \end{aligned} \tag{3–41}$$

given by (3–41) is called a *gauge transformation* and the invariance of the fields under such a transformation is called *gauge invariance*. The transformations are useful for recasting the somewhat awkward equations (3–31) and (3–32) into a more elegant form. Although many different choices of gauge can be made, the *Coulomb gauge* and the *Lorenz gauge* are of particular use.

In statics it is usually best to choose Λ so that $\vec{\nabla}\cdot\vec{A}$ vanishes, a choice known as the Coulomb gauge. With this choice (3–31) and (3–32) reduce to *Poisson's* equation,

$$\nabla^2 V = -\frac{\rho}{\varepsilon_0} \tag{3–42}$$

and

$$\nabla^2\vec{A} = -\mu_0\vec{J} \tag{3–43}$$

If the fields are not static it is frequently still useful to adopt the Coulomb gauge. The "wave" equation (3–32) for the vector potential now takes the form

$$\nabla^2\vec{A} - \mu_0\varepsilon_0\frac{\partial^2\vec{A}}{\partial t^2} = -\mu_0\vec{J} + \mu_0\varepsilon_0\vec{\nabla}\left(\frac{\partial V}{\partial t}\right) \tag{3–44}$$

Any vector can be written as a sum of curl free component (called *longitudinal*) and a divergence free component (called *solenoidal* or *transverse*). As we expect $\vec{\nabla}V$ to be curl free, such a decomposition may be useful. Labelling the components of such a resolution of $\vec{J}$ by subscripts l and s we write

$$\vec{J} = \vec{J}_l + \vec{J}_s \tag{3–45}$$

Substituting this into the general vector identity (13), $\vec{\nabla}\times(\vec{\nabla}\times\vec{J}) = \vec{\nabla}(\vec{\nabla}\cdot\vec{J}) - \nabla^2\vec{J}$, we obtain separate equations for J_l and J_s:

$$\nabla^2\vec{J}_s = -\vec{\nabla}\times(\vec{\nabla}\times\vec{J}) \tag{3–46}$$

and

$$\nabla^2\vec{J}_l = \vec{\nabla}(\vec{\nabla}\cdot\vec{J}) \tag{3–47}$$

Although (3–46) and (3–47) can be solved systematically, the reader is invited to consider (1–19) where V is shown to be the solution of $\nabla^2 V = -\rho/\varepsilon_0$. We know the solution to this equation to be (1–17). Inserting our form of the inhomogeneity instead of $-\rho/\varepsilon_0$ we obtain the solutions

$$\vec{J}_s(\vec{r}) = \frac{1}{4\pi}\int\frac{\vec{\nabla}'\times[\vec{\nabla}'\times\vec{J}(\vec{r}\,')]d^3r'}{|\vec{r}-\vec{r}\,'|} \tag{3–48}$$

$$\vec{J}_l(\vec{r}) = -\frac{1}{4\pi}\int\frac{\vec{\nabla}'[\vec{\nabla}'\cdot\vec{J}(\vec{r}\,')]d^3r'}{|\vec{r}-\vec{r}\,'|} \tag{3–49}$$

Focussing on (3–49), we integrate "by parts", noting from (5) that $A\vec{\nabla}B = \vec{\nabla}(AB) - B\vec{\nabla}A$ so that we write

$$\int\frac{\vec{\nabla}'[\vec{\nabla}'\cdot\vec{J}(\vec{r}\,')]d^3r'}{|\vec{r}-\vec{r}\,'|} = \int\vec{\nabla}'\left(\frac{\vec{\nabla}'\cdot\vec{J}(\vec{r}\,')}{|\vec{r}-\vec{r}\,'|}\right)d^3r' - \int(\vec{\nabla}'\cdot\vec{J})\vec{\nabla}'\left(\frac{1}{|\vec{r}-\vec{r}\,'|}\right)d^3r' \tag{3–50}$$

The first of the integrals in (3–50) may be integrated using (19) to get

$$\int\vec{\nabla}'\left(\frac{\vec{\nabla}'\cdot\vec{J}(\vec{r}\,')}{|\vec{r}-\vec{r}\,'|}\right)d^3r' = \oint\left(\frac{\vec{\nabla}'\cdot\vec{J}(\vec{r}\,')}{|\vec{r}-\vec{r}\,'|}\right)d\vec{S} \tag{3–51}$$

The volume integral (3–49) was to include all current meaning that zero current crosses the boundary allowing us to set the integral (3–51) to zero. Focussing now on the remaining integral on the right hand side of (3–50), we use $\vec{\nabla}'\frac{1}{|\vec{r}-\vec{r}'|} = -\vec{\nabla}\frac{1}{|\vec{r}-\vec{r}'|}$ to get

$$J_l = -\frac{1}{4\pi}\vec{\nabla}\int\frac{\vec{\nabla}'\cdot\vec{J}(\vec{r}')d^3r'}{|\vec{r}-\vec{r}'|} \tag{3–52}$$

Finally with the aid of the continuity equation (1–24) we replace $\vec{\nabla}\cdot\vec{J}$ by $-\partial\rho/\partial t$ to obtain

$$J_l = \frac{1}{4\pi}\vec{\nabla}\int\frac{(\partial\rho/\partial t)d^3r'}{|\vec{r}-\vec{r}'|} = \vec{\nabla}\frac{\partial}{\partial t}\int\frac{\rho(\vec{r}')d^3r'}{|\vec{r}-\vec{r}'|} = \varepsilon_0\vec{\nabla}\frac{\partial V}{\partial t} \tag{3–53}$$

Returning now to (3–44). we find that the last term, $\mu_0\varepsilon_0\vec{\nabla}(\partial V/\partial t)$ precisely cancels the longitudinal component of $-\mu_0\vec{J}$. Equation (3–4) may therefore be recast as

$$\nabla^2\vec{A} - \mu_0\varepsilon_0\frac{\partial^2\vec{A}}{\partial t^2} = -\mu_0\vec{J}_s \tag{3–54}$$

while the electric potential obeys

$$\nabla^2 V = -\frac{\rho}{\epsilon_0} \tag{3–55}$$

It is interesting to note that in the Coulomb gauge, V obeys the static equation giving instantaneous solutions (with no time lapse to account for the propagation time of changes in the charge density). The vector potential, on the other hand, obeys a wave equation which builds in the finite speed of propagation of disturbances in J_s. We will meet the solenoidal current again in Chapter 10 when we deal with multipole radiation.

In electrodynamics, it is also frequently useful to choose $\vec{\nabla}\cdot\vec{A} = -\mu_0\varepsilon_0\partial V/\partial t$, a choice known as the Lorenz gauge. With this choice, equations (3–31) and (3–32) take the form of a wave equation:

$$\nabla^2 V - \mu_0\varepsilon_0\frac{\partial^2 V}{\partial t^2} = -\frac{\rho}{\varepsilon_0} \tag{3–56}$$

$$\nabla^2\vec{A} - \mu_0\varepsilon_0\frac{\partial^2\vec{A}}{\partial t^2} = -\mu_0\vec{J} \tag{3–57}$$

It is clear that in the Lorenz gauge V and $\vec{A}$ obey manifestly similar equations that fit naturally into a relativistic framework.

The invariance of electromagnetism under gauge transformations has profound consequences in quantum electrodynamics. In particular, gauge symmetry permits the existence of a zero mass carrier of the electromagnetic field.

3.5 The Wave Equation in Vacuum

We have seen that a changing magnetic field engenders an electric field and, conversely, that a changing electric field generates a magnetic field. Taken together,

Faraday's law and Ampère's law give a wave equation whose solution we know as electromagnetic waves.

We can uncouple the curl equations using techniques analogous to those we use to decouple other pairs of differential equations. For example, the pair $dx/dt = -ky$ and $dy/dt = k'x$ is easily decoupled by differentiating each once more to give

$$\frac{d^2x}{dt^2} = -k\frac{dy}{dt} = -kk'x \quad \text{and} \quad \frac{d^2y}{dt^2} = k'\frac{dx}{dt} = -kk'y$$

In the same spirit, we take the curl of $(\vec{\nabla} \times \vec{E})$ in (3–27) to obtain, using (13)

$$\vec{\nabla} \times (\vec{\nabla} \times \vec{E}) = -\vec{\nabla} \times \frac{\partial \vec{B}}{\partial t}$$

$$\vec{\nabla}(\vec{\nabla} \cdot \vec{E}) - \nabla^2 \vec{E} = -\frac{\partial}{\partial t}(\vec{\nabla} \times \vec{B}) = -\frac{\partial}{\partial t}\left(\mu_0 \vec{J} + \mu_0\varepsilon_0 \frac{\partial \vec{E}}{\partial t}\right) \tag{3–58}$$

Eliminating $(\vec{\nabla} \cdot \vec{E})$ and $\mu_0\vec{J}$ as we wish to postpone consideration of sources of waves until Chapter 10, we obtain the homogeneous wave equation

$$\nabla^2 \vec{E} = \mu_0\varepsilon_0 \frac{\partial^2 \vec{E}}{\partial t^2} \tag{3–59}$$

and, in similar fashion

$$\nabla^2 \vec{B} = \mu_0\varepsilon_0 \frac{\partial^2 \vec{B}}{\partial t^2} \tag{3–60}$$

3.5.1 Plane Waves

We will now give the simplest solutions to the wave equations (3–59) and (3-60). In attempting to solve these equations, it is important to recognize that ∇^2, acting on a vector, is not merely ∇^2 acting on each of the vector's components.[8] Only when the basis is independent of the coordinates can we make the simplification $(\nabla^2\vec{A})_i = \nabla^2 A_i$. Fortunately this covers the important case of Cartesian coordinates. Thus (3–59) and (3–60) reduce to six identical, uncoupled *scalar* differential equations of the form

$$\nabla^2\psi(\vec{r},t) = \mu_0\varepsilon_0 \frac{\partial^2\psi(\vec{r},t)}{\partial t^2} \tag{3–61}$$

with Ψ representing any one of E_x, E_y, E_z, B_x, B_y, or B_z.

[8] As an example, the Laplacian of a vector $\vec{V}$ in terms of its polar coordinates is given below:

$$\begin{aligned}
(\nabla^2\vec{V})_r &= \nabla^2 V_r - \frac{2V_r}{r^2} - \frac{2V_r}{r^2}\frac{\partial V_\theta}{\partial\theta} - \frac{2\cos\theta\, V_\theta}{r^2\sin\theta} - \frac{2}{r^2\sin\theta}\frac{\partial V_\varphi}{\partial\varphi} \\
(\nabla^2\vec{V})_\theta &= \nabla^2 V_\theta - \frac{V_\theta}{r^2\sin^2\theta} + \frac{2}{r^2}\frac{\partial V_r}{\partial\theta} - \frac{2\cos\theta}{r^2\sin^2\theta}\frac{\partial V_\varphi}{\partial\varphi} \\
(\nabla^2\vec{V})_\varphi &= \nabla^2 V_\varphi - \frac{V_\varphi}{r^2\sin^2\theta} + \frac{2}{r^2\sin\theta}\frac{\partial V_r}{\partial\varphi} + \frac{2\cos\theta}{r^2\sin^2\theta}\frac{\partial V_\theta}{\partial\varphi}
\end{aligned}$$

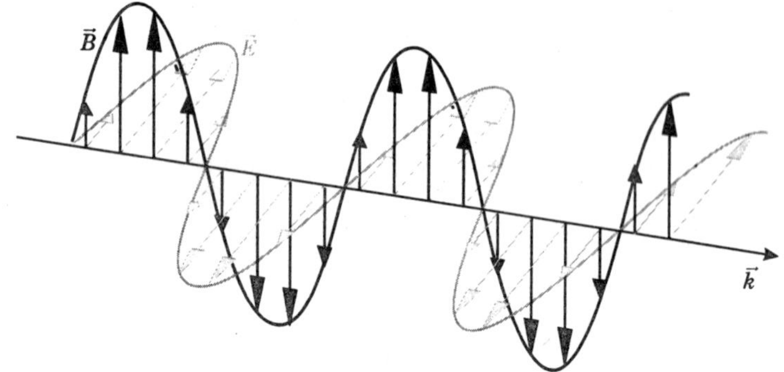

Figure 3.6: $\vec{E}$ and $\vec{B}$ oscillate in phase and are directed perpendicular to one-another as well as to $\vec{k}$.

Among the solutions of (3–61) are plane-wave solutions of the form

$$\psi(\vec{r}, t) = \psi_0 e^{i(\vec{k}\cdot\vec{r}\pm\omega t)} \tag{3–62}$$

We back-substitute ψ into the scalar wave equation (3–61) and find, with the help of $\nabla^2\psi = -k^2\psi$ and $\partial^2\psi/\partial t^2 = -\omega^2\psi$, that $k^2\psi = \omega^2\mu_0\varepsilon_0\psi$. We conclude that both the phase velocity ω/k and the group velocity $\partial\omega/\partial k$ are equal to $c = 1/\sqrt{\varepsilon_0\mu_0}$.

In deriving (3–59) and (3–60), we differentiated to remove the coupling between $\vec{E}$ and $\vec{B}$. The fields $\vec{E}$ and $\vec{B}$ are not independent of each other, and we must now restore the lost coupling by insisting that $\vec{E}$ and $\vec{B}$ satisfy not only (3-59) and (3–60) but also Maxwell's equations (3–27) with $\rho = 0$ and $\vec{J} = 0$. Thus, noting that for a vector plane wave $\vec{\nabla}\cdot = i\vec{k}\cdot$ and $\vec{\nabla}\times = i\vec{k}\times$, we see that the fields satisfy

$$\begin{aligned}
\vec{\nabla}\cdot\vec{E} &= 0 & &\Rightarrow & i\vec{k}\cdot\vec{E} &= 0\\
\vec{\nabla}\cdot\vec{B} &= 0 & &\Rightarrow & i\vec{k}\cdot\vec{B} &= 0\\
\vec{\nabla}\times\vec{E} &= -\frac{\partial\vec{B}}{\partial t} & &\Rightarrow & i\vec{k}\times\vec{E} &= i\omega\vec{B}\\
\vec{\nabla}\times\vec{B} &= \mu_0\varepsilon_0\frac{\partial\vec{E}}{\partial t} & &\Rightarrow & i\vec{k}\times\vec{B} &= -i\omega\mu_0\varepsilon_0\vec{E}
\end{aligned} \tag{3–63}$$

The first two of these expressions imply that $\vec{E}$ and $\vec{B}$ are each perpendicular to the propagation vector $\vec{k}$, while the second two imply that $\vec{E}$ and $\vec{B}$ are perpendicular to each other. The relative orientations of $\vec{E}$, $\vec{B}$, and $\vec{k}$ are illustrated in Figure 3.6.

We reiterate that these conclusions are valid only for infinite plane waves and do not apply to spherical waves or bounded waves.

The curl equations in (3–63) give one other fact of physical importance. Rewriting the $\vec{\nabla}\times\vec{E}$ equation in terms of the magnitudes of the relevant vectors we have $k|E| = \omega|B|$. In other words, the freely propagating electromagnetic wave has an electric field with a magnitude c times greater than that of its magnetic induction field. The force (1–30) that the wave exerts on a charged particle is the sum of qE, due to the electric field, and $(v/c)qE$ due to the magnetic induction field associated with the electromagnetic wave. Inside atoms and molecules, the relative speed of electrons, v/c, is of order $\alpha \approx 1/137$, leading us to conclude that the interaction of matter with electromagnetic radiation is almost entirely through the electric field component of the wave. The alignment of $\vec{E}$ (perpendicular to the propagation vector, $\vec{k}$) is known as the *polarization* of the wave.

3.5.2 Spherical Waves

To obtain the solution to the vector wave equations (3–59) or (3–60), we employ a trick that allows us to generate the vector solutions from the solutions of the scalar equation, (3–61). For that reason we consider initially solutions to the scalar equation

$$\nabla^2\psi(r,\theta,\varphi) = \frac{1}{c^2}\frac{\partial^2\psi}{\partial t^2} \tag{3–64}$$

Using separation of variables (explored at length in Chapter 5) in spherical polar coordinates, we obtain the solution

$$\psi(r,\theta,\varphi) = \sum_\ell \left[A_\ell \mathrm{j}_\ell(kr) + B_\ell \mathrm{n}_\ell(kr)\right] \mathrm{Y}_\ell^m(\theta,\,\varphi)\, e^{-i\omega t} \tag{3–65}$$

where Y_ℓ^m is a spherical harmonic of order ℓ, m; $\mathrm{j}_\ell(z)$ and $\mathrm{n}_\ell(z)$ are spherical Bessel functions and spherical Neumann functions of order ℓ and A_ℓ and B_ℓ are arbitrary constants.

The vector solution can generally be generated from the scalar solution by the following stratagem. When ψ solves the scalar wave equation, then $\vec{\nabla}\psi$ and $\vec{r}\times\vec{\nabla}\psi$ both satisfy the corresponding vector equation (see problem 3–12). Of these solutions, only the second, $\vec{r}\times\vec{\nabla}\psi = -\vec{\nabla}\times(\vec{r}\psi)$, has zero divergence—a requirement for $\vec{B}$ and also for $\vec{E}$ in a source-free region of space. If we let $\vec{E} = \vec{r}\times\vec{\nabla}\psi$, we obtain one kind of solution for the electromagnetic wave, called *transverse electric* (TE), or M type. Alternatively, if we take $\vec{B} = \vec{r}\times\vec{\nabla}\psi$, we obtain a *transverse magnetic* (TM), or E type wave. The function ψ from which the fields may be derived is known as the *Debye Potential.* A superposition of the two types of waves is the most general possible.

For the case of TE waves we find $\vec{B}$ from $\vec{B} = \vec{\nabla}\times\vec{E}/i\omega$, while for TM waves we find $\vec{E}$ from $\vec{E} = ic^2\vec{\nabla}\times\vec{B}/\omega$. To summarize (after rescaling ψ),

$$\begin{aligned} \vec{E}_{\mathrm{TE}} &= ik\vec{\nabla}\times\vec{r}\psi \qquad & \vec{B}_{\mathrm{TE}} &= \frac{1}{c}\vec{\nabla}\times(\vec{\nabla}\times\vec{r}\psi) \\ \vec{E}_{\mathrm{TM}} &= \vec{\nabla}\times(\vec{\nabla}\times\vec{r}\psi) \qquad & \vec{B}_{\mathrm{TM}} &= -\frac{ik}{c}\vec{\nabla}\times\vec{r}\psi \end{aligned} \tag{3–66}$$

where the "constants" ik and $1/c$ have been introduced to give both the corresponding TE fields and TM fields the same dimensions (and the same energy flux).

We may write the components explicitly. Abbreviating the linear combination of spherical Bessel functions, $A\mathrm{j}_\ell + B\mathrm{n}_\ell = f_\ell$ (the combination $\mathrm{h}_\ell^{(1)} = \mathrm{j}_\ell + i\mathrm{n}_\ell$ is required to produce radially expanding waves), we obtain for $\mathrm{TE}_{\ell,m}$ waves

$$
\begin{aligned}
E_r &= 0 & B_r &= \frac{1}{c}\frac{\ell(\ell+1)}{r} f_\ell(kr) \mathrm{Y}_\ell^m(\theta,\varphi) \\
E_\theta &= -\frac{km}{\sin\theta} f_\ell(kr) \mathrm{Y}_\ell^m(\theta,\varphi) & B_\theta &= \frac{1}{cr}\frac{d[rf_\ell(kr)]}{dr}\frac{\partial \mathrm{Y}_\ell^m(\theta,\varphi)}{\partial\theta} \\
E_\varphi &= -ikf_\ell(kr)\frac{\partial \mathrm{Y}_\ell^m(\theta,\varphi)}{\partial\theta} & B_\varphi &= \frac{im}{cr\sin\theta}\frac{d[rf_\ell(kr)]}{dr}\mathrm{Y}_\ell^m(\theta,\varphi)
\end{aligned}
\tag{3–67}
$$

For $\mathrm{TM}_{\ell,m}$ waves, the components are:

$$
\begin{aligned}
E_r &= \frac{\ell(\ell+1)}{r} f_\ell(kr) \mathrm{Y}_\ell^m(\theta,\varphi) & B_r &= 0 \\
E_\theta &= \frac{1}{r}\frac{d\,[rf_\ell(kr)]}{dr}\frac{\partial \mathrm{Y}_\ell^m(\theta,\varphi)}{\partial\theta} & B_\theta &= \frac{km}{c\sin\theta} f_\ell(kr) \mathrm{Y}_\ell^m(\theta,\varphi) \\
E_\varphi &= \frac{im}{r\sin\theta}\frac{d\,[rf_\ell(kr)]}{dr}\mathrm{Y}_\ell^m(\theta,\varphi) & B_\varphi &= \frac{ik}{c} f_\ell(kr)\frac{\partial \mathrm{Y}_\ell^m(\theta,\varphi)}{\partial\theta}
\end{aligned}
\tag{3–68}
$$

The fields described above are the 2^ℓ-pole radiation fields. As we will see in Chapter 10, the TE waves are emitted by oscillating magnetic multipoles, while the TM waves are emitted by oscillating electric multipoles. Note that in contrast to the conclusion for plane waves that $\vec{E}$ and $\vec{B}$ are both perpendicular to $\vec{k}$, for spherical waves only one of the two can be perpendicular to $\vec{k}$.

Exercises and Problems

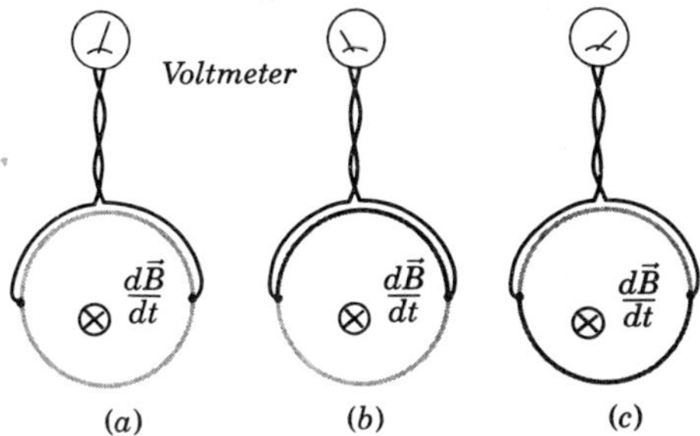

Figure 3.7: The voltmeter is attached to the loop ensuring that its leads do not add any area to the loop. (Exercise 3.18)

3-1 Show that the flux Φ threading a loop Γ may be written in terms of the vector potential as

$$\Phi = \oint_{\Gamma} \vec{A} \cdot d\vec{\ell}$$

3-2 The flux threading a single turn coil is gradually reduced, resulting in an induced current in the coil. Find the magnetic field produced at the center of the coil by the induced current of a coil with resistance R and radius a.

3-3 In a *Betatron* electrons are accelerated by an increasing magnetic flux density (as in example 3.3). How must the field at the orbit relate to the average field in order that the electron's orbit remain of constant radius?

3-4 Find the flux in a toroidal coil such as that illustrated in Figure 1.13 carrying current I. The mean radius of the coil is a, and the cross-sectional radius of the tube is b.

3-5 A current I charges a parallel plate capacitor made of two circular plates each of area A spaced at small distance d. Find the magnetic induction field between the plates.

3-6 Show that the magnetic induction field encircling the capacitor of problem 3-5 at a large enough distance for the electric field to vanish is given by

$$B_\varphi = \frac{\mu_0 I}{2\pi r}$$

Thus at a sufficient distance, it is impossible to tell from the magnetic field that the circuit is broken by the capacitor.

3-7 A pendulum consisting of a conducting loop of radius a and resistance R at the end of a massless string of length ℓ swings through an inhomogeneous magnetic induction field $\vec{B}$ perpendicular to the plane of the loop. Write an equation of motion for the pendulum.

3-8 A *flip coil* is a rectangular coil that is turned through 180° fairly rapidly in a static magnetic induction field. Show that the total charge transported through the coil as it is flipped is independent of the speed of flipping. (Assume a finite resistance.)

3-9 The Lagrangian of a charged particle moving in an electromagnetic field is $\mathcal{L} = \frac{1}{2}m\dot{r}^2 - qV(\vec{r}) + q\dot{\vec{r}} \cdot \vec{A}(\vec{r})$. Show that the i-component of $\vec{p}$, $p_i = \partial\mathcal{L}/\partial\dot{x}_i$ yields the canonical momentum on the left hand side of (3–38).

3-10 Show that, for plane waves, $\vec{\nabla} \times \vec{E} = i\vec{k} \times \vec{E}$ and $\vec{\nabla} \cdot \vec{E} = i\vec{k} \cdot \vec{E}$.

3-11 Obtain the wave equation for $\vec{B}$ (3–60) to verify that it is indeed identical to that for $\vec{E}$.

3-12 In order to ionize an atom, one needs to overcome the binding energy of the electron (of order 10 eV) in a distance of several atomic radii (say 10^{-9} m) in order to provide a reasonable tunnelling rate. Focused ruby pulse lasers can produce air sparks. Estimate the electric field strength and hence the irradiance of the laser light required to produce air sparks.

3-13 Show that if ψ solves the scalar wave equation, then $\vec{\nabla}\psi$ and $\vec{r} \times \vec{\nabla}\psi$ each solve the corresponding vector wave equation.

3-14 Fill in the missing steps in the last step of the footnote (page 59) to show that

$$\vec{A}\,[\vec{r}(t) + \vec{v}dt, t] - \vec{A}\,[\vec{r}(t), t] = (\vec{v} \cdot \vec{\nabla})\vec{A}dt$$

3-15 Show that a gauge transformation with Λ satisfying $\nabla^2\Lambda = 0$ preserves the Coulomb gauge.

3-16 Find the condition that a gauge function Λ needs to satisfy in order to preserve the Lorenz gauge.

3-17 A circular disk of conducting material spins about its axis in a magnetic field parallel to the axis. Find the EMF generated between the axis and a fixed point on the rim of the disk.

3-18 Resolve the following paradox. A voltmeter has its terminals attached to diametrically opposite points of a circular loop. From their points of attachment, the leads follow the curve of the loop and, after meeting, continue to the voltmeter as a twisted pair, as in Figure 3.7. A time-varying flux threads the loop. When the voltmeter leads are considered as part of the loop, the EMF measured appears to depend on whether the top half of the loop completes the circuit (Figure 3.7*b*) in which case the flux is zero, or the bottom half (Figure 3.7*c*) completes it, in which case the flux is just that through the original loop.

Chapter 4

Energy and Momentum

4.1 Energy of a Charge Distribution

To this point we have taken stationary or moving charge distributions as a given without any consideration of how such distributions were assembled. In this chapter we consider the work required to assemble the required distribution. The charges' gain in potential energy can be expressed in terms of their relative positions, but in analogy to a compressed spring's energy being stored in the strain of the spring rather than the position of the ends, we will also be able to identify the energy of the system with the strength of the fields created by the charge distribution.

4.1.1 Stationary Charges

Let us begin our calculation of the work required to assemble a given charge distribution by calculating the work required to move two charges, q_1 and q_2, to positions $\vec{r}_1$ and $\vec{r}_2$, respectively. For brevity we will write $r_{ij} = |\vec{r}_i - \vec{r}_j|$.

It requires no work at all to move the first charge to $\vec{r}_1$ since there is not yet any field to work against. To move the second charge into place from infinity to $\vec{r}_2$, the work

$$W_2 = \frac{q_1 q_2}{4\pi\varepsilon_0 r_{12}} \tag{4–1}$$

must be performed on the charge. To bring a third charge to $\vec{r}_3$, requires work

$$W_3 = \frac{1}{4\pi\varepsilon_0}\left(\frac{q_1 q_3}{r_{13}} + \frac{q_2 q_3}{r_{23}}\right) \tag{4–2}$$

and a fourth charge would require

$$W_4 = \frac{1}{4\pi\varepsilon_0}\left(\frac{q_1 q_4}{r_{14}} + \frac{q_2 q_4}{r_{24}} + \frac{q_3 q_4}{r_{34}}\right) \tag{4–3}$$

It will be convenient to symmetrize each of these terms by repeating each entry with indices reversed and halving the result. In other words, we rewrite (4–1) as

$$W_2 = \frac{1}{8\pi\varepsilon_0}\frac{q_1 q_2 + q_2 q_1}{r_{12}} \tag{4–4}$$

The succeeding terms are similarly rewritten

$$W_3 = \frac{1}{8\pi\varepsilon_0}\left(\frac{q_1q_3+q_3q_1}{r_{13}}+\frac{q_2q_3+q_3q_2}{r_{23}}\right) \tag{4–5}$$

$$W_4 = \frac{1}{8\pi\varepsilon_0}\left(\frac{q_1q_4+q_4q_1}{r_{14}}+\frac{q_2q_4+q_4q_3}{r_{24}}+\frac{q_3q_4+q_4q_3}{r_{34}}\right) \tag{4–6}$$

Finally, to put the nth charge into place,

$$W_n = \frac{1}{8\pi\varepsilon_0}\left(\frac{q_1q_n+q_nq_1}{r_{1n}}+\frac{q_2q_n+q_nq_2}{r_{2n}}+\frac{q_3q_n+q_nq_3}{r_{3n}}+\cdots+\frac{q_{n-1}q_n+q_nq_{n-1}}{r_{n-1,n}}\right) \tag{4–7}$$

We add the terms to obtain the work to assemble all the charges,

$$W = W_2+W_3+W_4+\cdots+W_n \tag{4–8}$$

$$\begin{aligned}= \frac{1}{8\pi\varepsilon_0}\Bigg[q_1&\left(\frac{q_2}{r_{12}}+\frac{q_3}{r_{13}}+\frac{q_4}{r_{14}}+\cdots+\frac{q_n}{r_{1,n}}\right)\\ &+q_2\left(\frac{q_1}{r_{21}}+\frac{q_3}{r_{23}}+\frac{q_4}{r_{24}}+\cdots+\frac{q_n}{r_{2,n}}\right)\\ &+q_3\left(\frac{q_1}{r_{31}}+\frac{q_2}{r_{32}}+\frac{q_4}{r_{34}}+\cdots+\frac{q_n}{r_{3,n}}\right)+\cdots\\ &\cdots+q_{n-1}\left(\frac{q_1}{r_{n-1,1}}+\frac{q_2}{r_{n-1,2}}+\cdots+\frac{q_n}{r_{n-1,n}}\right)\Bigg]\end{aligned} \tag{4–9}$$

$$= \frac{1}{8\pi\varepsilon_0}\sum_{i=1}^{n-1}q_i\sum_{\substack{j=1\\ j\neq i}}^{n}\frac{q_j}{r_{ij}} = \frac{1}{8\pi\varepsilon_0}\sum_{i=1}^{n}\sum_{\substack{j=1\\ j\neq i}}^{n}\frac{q_iq_j}{r_{ij}} \tag{4–10}$$

where in the last step we have increased the upper limit of the i summation by one which adds nothing as the $(n,\ n)$ term is eliminated from the sum by the $j \neq i$ requirement. It is useful to rewrite W as

$$W = \tfrac{1}{2}\sum_{i=1}^{n}q_i\sum_{\substack{j=1\\ j\neq i}}^{n}\frac{1}{4\pi\varepsilon_0}\frac{q_j}{r_{ij}} \tag{4–11}$$

$$= \tfrac{1}{2}\sum_{i=1}^{n}q_iV(\vec{r}_i) \tag{4–12}$$

where $V(\vec{r}_i)$ is the potential at $\vec{r}_i$ established by all the charges except q_i. Equation (4–12) leads to the obvious generalizations

$$W = \tfrac{1}{2}\int \rho(\vec{r})V(\vec{r})\,d^3r \tag{4–13}$$

and, replacing V by its integral form,

$$W = \frac{1}{8\pi\varepsilon_0}\int \rho(\vec{r})\left(\int \frac{\rho(\vec{r}\,')}{|\vec{r}-\vec{r}\,'|}\,d^3r'\right)d^3r \tag{4–14}$$

for continuous charge distributions. There is clearly a requirement that $\vec{r}$ not coincide with $\vec{r}\,'$ in equation (4–14).

Let us now express the energy of a charge distribution in terms of the field established. Starting with (4–13), we replace ρ by $\varepsilon_0\vec{\nabla}\cdot\vec{E}$. The energy of the assembled charges may then be written

$$W = \tfrac{1}{2}\int \varepsilon_0\left(\vec{\nabla}\cdot\vec{E}\right)V\,d^3r \tag{4–15}$$

which, with the use of the identity (7), $\vec{\nabla}\cdot(V\vec{E}) = V(\vec{\nabla}\cdot\vec{E}) + \vec{E}\cdot\vec{\nabla}V$, and the divergence theorem, gives

$$W = \frac{\varepsilon_0}{2}\int \vec{\nabla}\cdot(V\vec{E})d^3r - \frac{\varepsilon_0}{2}\int \vec{E}\cdot\vec{\nabla}V\,d^3r \tag{4–16}$$

$$= \frac{\varepsilon_0}{2}\oint V\vec{E}\cdot d\vec{S} + \frac{\varepsilon_0}{2}\int E^2\,d^3r \tag{4–17}$$

where the region of integration must extend over all regions of nonzero charge density. We extend the region of integration to large distances R. We note that if the volume of interest contains finite charge, the surface integral diminishes as 1 over R or faster, while the volume integral can only increase. Therefore, if we extend the integration to all space, we may write

$$W = \frac{\varepsilon_0}{2}\int_{\substack{\text{all}\\\text{space}}} E^2 d^3r \tag{4–18}$$

One might now reasonably reinterpret the work in (4–18) as the work required to produce the field E, or even as the energy of the field.

EXAMPLE 4.1: Find the energy of a conducting sphere of radius a carrying total charge Q uniformly distributed on its surface.

Solution: We will calculate the energy by each of the methods suggested by (4–13), (4–17), and (4–18).

(a) If we use (4–13),

$$W = \tfrac{1}{2}\int \rho V\,d^3r \tag{Ex 4.1.1}$$

$$= \tfrac{1}{2}\int_0^{4\pi}\frac{Q}{4\pi a^2}\frac{Q}{4\pi\varepsilon_0 a}a^2\,d\Omega = \frac{Q^2}{8\pi\varepsilon_0 a} \tag{Ex 4.1.2}$$

(b) Next we use (4–17). Since the charge resides entirely on the surface of the sphere we need only integrate over a very thin spherical shell including the surface. As the thickness of the volume of integration decreases to zero, the second integral of (4–17) vanishes, (the field interior to the sphere vanishes in any case, so even if the interior volume were included, the integral would vanish) leaving

$$W = \frac{\varepsilon_0}{2}\oint V\vec{E}\cdot d\vec{S}$$

$$= \frac{\varepsilon_0}{2}\int_0^{4\pi} \frac{Q}{4\pi\varepsilon_0 a}\frac{Q}{4\pi\varepsilon_0 a^2}\, a^2 d\Omega = \frac{Q^2}{8\pi\varepsilon_0 a} \tag{Ex 4.1.3}$$

(c) Finally we use (4-18):

$$W = \frac{\varepsilon_0}{2}\int E^2\, dV = \frac{\varepsilon_0}{2}\int_a^{\infty} \frac{Q^2}{(4\pi\varepsilon_0 r^2)^2} 4\pi r^2\, dr \tag{Ex 4.1.4}$$

$$= \frac{Q^2}{8\pi\varepsilon_0}\int_a^{\infty}\frac{dr}{r^2} = \frac{Q^2}{8\pi\varepsilon_0 a} \tag{Ex 4.1.5}$$

On further reflection (4–18) seems to lead to a rather uncomfortable paradox. An elementary calculation of the potential energy of two opposite sign charges shows that this energy is negative, yet (4–18), having a positive definite integrand, can never be less than zero. The problem arises in generalizing from (4–14) to (4–18). In (4–14), we carefully excluded the contribution from q_i to the potential $V(\vec{r}_i)$, but in (4–15) this exclusion is nowhere evident. This excluded energy is the energy it would take to assemble the charge q_i, which, for a point charge, would by our calculations be infinite! So long as the charge density ρ does not include point charges, the contribution $\rho(\vec{r})$ makes to $V(\vec{r})$ vanishes as $d^3r \to 0$, so that including this *self-energy* has no untoward effect. If, however, ρ includes point charges, (4-8) and its consequences cannot be correct. In fact, the integral will diverge so that the problem will be obvious. This self-energy will haunt us in later chapters, and even quantum electrodynamics does not satisfactorily resolve this problem.

4.1.2 Coefficients of Potential

In a system of discrete charges, or charges situated on disconnected conductors, the potential at any place bears a linear relationship to the charges in question. In particular, for a system of charge-bearing conductors, we may write the potential on conductor i in terms of the charges Q_j on conductor j:

$$V_i = \sum_j p_{ij} Q_j \tag{4–19}$$

The coefficients p_{ij} are known as the *coefficients of potential* and depend only on the geometry of the problem. Although the coefficients are not always calculable, the system is completely characterized by the p_{ij}, which may be determined experimentally.

The energy of the system of charged conductors is given by

$$W = \tfrac{1}{2} \sum_{i,j} p_{ij} Q_i Q_j \tag{4–20}$$

Alternatively, we can invert (4–19) to write

$$Q_j = \sum_i (P^{-1})_{ij} V_i \equiv \sum_i C_{ij} V_i \tag{4–21}$$

where the coefficients C_{ii} are known as the *coefficients of capacitance* and the C_{ij} $(i \neq j)$ are the *coefficients of induction.* In terms of the C_{ij} we can write the electrostatic energy as

$$W = \tfrac{1}{2} \sum_j Q_j V_j = \tfrac{1}{2} \sum_{i,j} C_{ij}\, V_i\, V_j \tag{4–22}$$

The coefficients p_{ij} and C_{ij} are symmetric in their indices (i.e., $p_{ij} = p_{ji}$) as we show below. Using the expression for the energy (4–20), we calculate the change in energy of the system as a small increment of charge ΔQ_k is added to the kth conductor while the charge on all others is maintained constant:

$$\begin{aligned} \Delta W &= \frac{\partial W}{\partial Q_k} \Delta Q_k = \tfrac{1}{2} \sum_{i,j} p_{ij} \left(Q_j \frac{\partial Q_i}{\partial Q_k} + Q_i \frac{\partial Q_j}{\partial Q_k} \right) \Delta Q_k \\ &= \tfrac{1}{2} \sum_{i,j} \left(p_{ij}\, \delta_{ik}\, Q_j \Delta Q_k + p_{ij} \delta_{jk},\, Q_i \Delta Q_k \right) \\ &= \tfrac{1}{2} \sum_j p_{kj} Q_j \Delta Q_k + \tfrac{1}{2} \sum_i p_{ik} Q_i \Delta Q_k \end{aligned} \tag{4–23}$$

Relabelling the dummy index i in the second sum as j, we obtain

$$\Delta W = \tfrac{1}{2} \sum_j \left(p_{kj} + p_{jk} \right) Q_j \Delta Q_k \tag{4–24}$$

The increase in energy of the system of charged conductors could equally have been written

$$\begin{aligned} \Delta W &= V_k \Delta Q_k \\ &= \sum_j p_{kj} Q_j \Delta Q_k \end{aligned} \tag{4–25}$$

Comparison of (4–24) and (4–25) leads immediately to $p_{jk} = p_{kj}$.

EXAMPLE 4.2: Two conductors of capacitance C_1 and C_2 are placed a large (compared to their dimensions) distance r apart. Find the coefficients C_{ij} to first order.

Solution: The approximate coefficients of potential for the system are easily found. If the charge on conductor *1* is q_1 and conductor *2* is not charged, we would have to first order

$$V_1 = \frac{q_1}{C_1} \quad \text{and} \quad V_2 = \frac{q_1}{4\pi\varepsilon_0 r} \tag{Ex 4.2.1}$$

where we have neglected the redistribution of charge over the second conductor. We conclude that the coefficients in the expression

$$V_i = \sum_j p_{ij} Q_j$$

take the values $p_{11} = 1/C_1$, $p_{12} = 1/4\pi\varepsilon_0 r$, and $p_{22} = 1/C_2$ or

$$p_{ij} = \begin{pmatrix} \dfrac{1}{C_1} & \dfrac{1}{4\pi\varepsilon_0 r} \\ \dfrac{1}{4\pi\varepsilon_0 r} & \dfrac{1}{C_2} \end{pmatrix} \tag{Ex 4.2.2}$$

Inverting the system to obtain the C_{ij} using

$$\begin{pmatrix} A & B \\ C & D \end{pmatrix}^{-1} = \begin{pmatrix} D & -B \\ -C & A \end{pmatrix} \Big/ \begin{vmatrix} A & B \\ C & D \end{vmatrix} \tag{Ex 4.2.3}$$

we find

$$C_{11} = \frac{1}{C_2}\left[\frac{1}{\dfrac{1}{C_1C_2} - \dfrac{1}{(4\pi\varepsilon_0 r)^2}}\right] \simeq C_1\left[1 + \frac{C_1C_2}{(4\pi\varepsilon_0 r)^2}\right] \quad C_{22} \simeq C_2\left[1 + \frac{C_1C_2}{(4\pi\varepsilon_0 r)^2}\right]$$

and

$$C_{12} = -\frac{1}{4\pi\varepsilon_0 r}\left[\frac{C_1C_2}{1 - \dfrac{C_1C_2}{(4\pi\varepsilon_0 r)^2}}\right] \simeq \frac{-C_1C_2}{4\pi\varepsilon_0 r} \tag{Ex 4.2.4}$$

4.1.3 Forces on Charge Distributions in Terms of Energy

Frequently the forces or torques on a component of a system of charges cannot easily be computed from Coulomb's law. If, however, the energy of the system can be evaluated in terms of the physical parameters, it is relatively easy to obtain the forces or torques.

Let us consider first an isolated system of charged components. If the net electrostatic force $\vec{F}^{(es)}$ were to produce a displacement $d\zeta$ in the $\vec{\zeta}$ direction, it would perform mechanical work $\vec{F}^{(es)} \cdot d\vec{\zeta}$. This work must be performed at the expense of the electrostatic energy $W^{(es)}$ (note that we have used W to denote energy rather than work) stored in the system. Conservation of energy in the isolated system then requires

$$F_\zeta^{(es)} d\zeta = -dW^{(es)} \tag{4–26}$$

Practically, the requirement of isolation of the system means that no charge enters or leaves the system. We therefore write the ζ component of the force on the mobile component as

$$F_\zeta^{(es)} = -\left.\frac{dW^{(es)}}{d\zeta}\right|_Q \tag{4–27}$$

a result entirely compatible with the usual form $\vec{F} = -\vec{\nabla}W$. A more interesting result obtains when the system is not isolated but charge is allowed to flow into or out of components of the system in order to maintain a constant potential on each of those components. Typically such an arrangement would have batteries maintaining the constant potentials. The electrostatic work done on the mobile component can now draw its energy from either a decrease of the stored electrostatic energy or from the batteries' energy $W^{(bat)}$, so that the expression of conservation of energy becomes

$$F_\zeta^{(\ )} d\zeta = -\,dW^{(es)}\Big|_V + dW^{(bat)}\Big|_V \tag{4–28}$$

The work done by the batteries in supplying charges dQ_i to each of the components labelled i at constant potential V_i is

$$dW^{(bat)}\Big|_V = \sum V_i dQ_i \tag{4–29}$$

The electrostatic energy of the system may be written in similar terms:

$$W^{(es)} = \tfrac{1}{2}\sum V_i Q_i \quad \Rightarrow \quad dW^{(es)}\big|_V = \tfrac{1}{2}\sum V_i dQ_i \tag{4–30}$$

We find, therefore,

$$F_\zeta^{(es)} \cdot d\zeta\Big|_V = +\,dW^{(es)}\Big|_V \tag{4–31}$$

leading to

$$F_\zeta^{(es)}\Big|_V = +\,\frac{dW^{(es)}}{d\zeta}\Bigg|_V \tag{4–32}$$

Example 4.3: A parallel plate capacitor of width h and spacing $d \ll h$ is held at potential V with respect to ground. A larger, thin, grounded conducting sheet is partially inserted midway between the capacitor plates, as shown in Figure 4.1. Find the force drawing the sheet into the capacitor. Ignore fringing fields.

Solution: Although the fringing fields are directly responsible for a lateral force on the sheet, because they don't change as sheet is inserted, they may be ignored in the calculation of the virtual change in energy of the system under lateral movement. The energy of the capacitor with the central plate inserted a distance ζ is given by

$$W^{(es)} = \frac{\varepsilon_0}{2}\int E^2\, d^3r \tag{Ex 4.3.1}$$

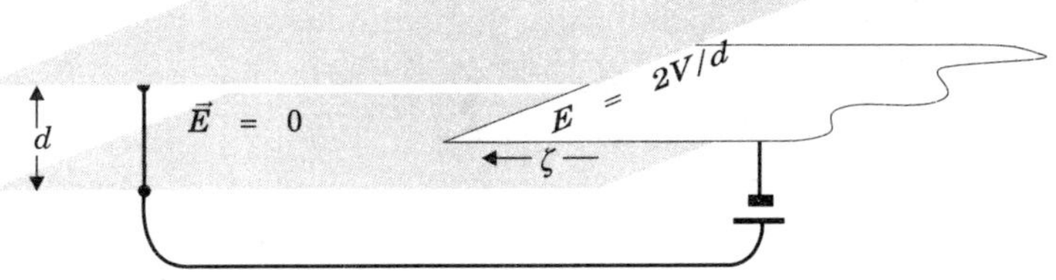

Figure 4.1: The field between the equipotential plates vanishes. Only where the grounded sheet is interposed is there a nonzero field.

$$= \frac{\varepsilon_0}{2}\left(\frac{2V}{d}\right)^2 dh\zeta \tag{Ex 4.3.2}$$

since the electric field is zero everywhere except in the volume ζhd.

The force on the inserted sheet is then given by

$$F_\zeta\big|_V = +\left.\frac{\partial W^{(es)}}{\partial\zeta}\right|_V = \frac{2\varepsilon_0 V^2 h}{d} \tag{Ex 4.3.3}$$

The displacements, of course, need not be linear; they could equally well be angular in which case the associated "force" becomes a torque eg. $\tau = \pm\,\partial W/\partial\theta$.

4.1.4 Potential Energy of Currents

While for charges we could merely separate the charges to infinity to reduce their potential energy to zero, it is far from clear that stretching current loops out to infinity would in fact diminish their energy. Instead we will use another strategy to find the energy of current loops. Unlike charge, we can make currents vanish by merely stopping the motion of charges. If we now calculate how much nondissipative work we have to do in gradually building the currents to their final value, we will have the potential energy of the currents.

When a current is established in a circuit, work must be done to overcome the induced EMF caused by the changing current. This work is fully recoverable, quite unlike the resistive losses, and does not depend on the rate that the current is increased. Only its final value and the geometry of the current loop(s) are important. Let us consider a system composed of a number n of current loops, each carrying a time dependent current I_k. Focusing our attention on the jth loop, we note that the magnetic flux Φ in the jth loop must have the form

$$\Phi_j = \sum_{k=1}^{n} M_{jk} I_k \tag{4–33}$$

since the magnetic field produced at a point inside the (jth loop by any current loop k (including the jth loop under consideration) is just proportional to the current in loop k. The coefficients M_{jk} are known as the mutual inductance of loop j and k, while $M_{jj} \equiv L_j$ is the self-inductance of loop j. The EMF generated around the loop is

$$\mathcal{E} = -\frac{d\Phi_j}{dt} = -\sum_{k=1}^{n} M_{jk}\frac{dI_k}{dt} \tag{4–34}$$

In the presence of this counter-EMF, the rate that work is done pushing I_j around the loop is just

$$\frac{dW_j}{dt} = -\,\mathcal{E} I_j = I_j(t)\sum_{k=1}^{n} M_{jk}\frac{dI_k}{dt} \tag{4–35}$$

The rate that the potential energy of the entire system is changed is the sum of (4–35) over all j:

$$\frac{dW}{dt} = \sum_{j=1}^{n}\sum_{k=1}^{n} I_j M_{jk}\frac{dI_k}{dt} \tag{4–36}$$

This is usefully rewritten as half the symmetric sum

$$\begin{aligned}\frac{dW}{dt} &= \tfrac{1}{2}\left(\sum_{j,k} M_{jk}I_j\frac{dI_k}{dt} + \sum_{j,k} M_{kj}I_k\frac{dI_j}{dt}\right)\\ &= \tfrac{1}{2}\sum_{j,k} M_{jk}\frac{d}{dt}(I_jI_k)\end{aligned} \tag{4–37}$$

where we have used the symmetry $M_{jk} = M_{kj}$.[9] The total work performed to establish all currents against the opposition of the counter EMF is now easily found by integrating (4–37)

$$\begin{aligned}W &= \tfrac{1}{2}\sum_j\sum_k M_{jk}I_kI_j\\ &= \tfrac{1}{2}\sum_j \Phi_j I_j\end{aligned} \tag{4–38}$$

As in the case of electric charges, we would like to express the energy in terms of the potential and the field. To this end, we replace Φ_j with

$$\Phi_j = \int \vec{B}\cdot d\vec{S}_j = \int(\vec{\nabla}\times\vec{A})\cdot d\vec{S}_j = \oint_{\Gamma_j}\vec{A}\cdot d\vec{\ell}_j \tag{4–39}$$

permitting us to write

$$W = \tfrac{1}{2}\sum_j \oint_{\Gamma_j}\vec{A}\cdot I_j d\vec{\ell}_j \tag{4–40}$$

This expression (4–40) is easily generalized for distributed currents to yield

$$W = \tfrac{1}{2}\int \vec{A}\cdot\vec{J}\,d^3r \tag{4–41}$$

EXAMPLE 4.4: Calculate the self-inductance per unit length of two parallel wires, each of radius a carrying equal currents in opposite directions (Figure 4.2).

[9] The symmetry of the mutual inductance under exchange of the indices is easily seen by writing the explicit expression for the contribution of the kth current to the flux in the jth loop as follows:

$$\begin{aligned}M_{jk}I_k &= \int \vec{B}_{jk}\cdot d\vec{S}_j = \int(\vec{\nabla}\times\vec{A}_{jk})\cdot d\vec{S}_j = \oint \vec{A}_{jk}\cdot d\vec{\ell}_j\\ &= \oint\left(\frac{\mu_0}{4\pi}\oint\frac{I_k d\vec{\ell}_k}{|\vec{r}-\vec{r}\,'|}\right)\cdot d\vec{\ell}_j = I_k\frac{\mu_0}{4\pi}\int\frac{d\vec{\ell}_k\cdot d\vec{\ell}_j}{|\vec{r}-\vec{r}\,'|}\end{aligned}$$

where B_{kj} and A_{jk} are, respectively, the magnetic induction field and vector potential due to loop k at the position of loop j. (Note that the subscripts are not coordinate indices; no implicit summation is intended.)

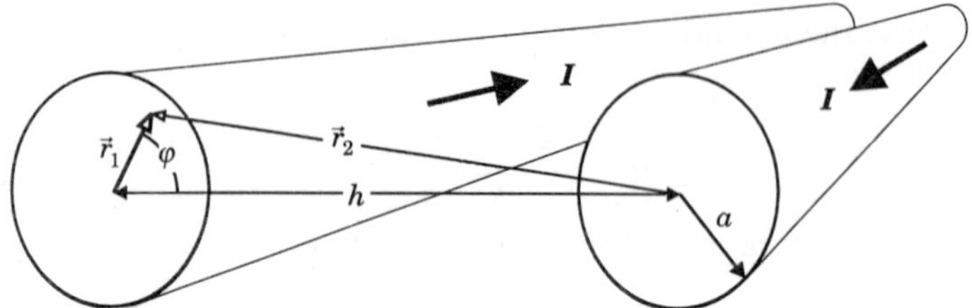

Figure 4.2: Two parallel wires carry equal currents in opposite directions.

Solution: We will calculate $\frac{1}{2}\int \vec{A}\cdot\vec{J}d^3r$ and equate this to $\frac{1}{2}LI^2$ to obtain the self-inductance L $(= M_{ii})$. For convenience we take the leftmost wire to be centered on the z-axis, while the center of the right wire lies at distance h from the origin.[10]

We begin by calculating the vector potential $\vec{A}$ both interior and exterior to a wire carrying current I in the z direction by solving $\nabla^2\vec{A} = -\mu_0\vec{J}$ (Equation 3–43). Since $\vec{A}$ is parallel to $\vec{J}$ in the Coulomb gauge, it suffices to solve for A_z.

Interior to the wire, the current density is

$$\vec{J} = \frac{I\hat{k}}{\pi a^2} \tag{Ex 4.4.1}$$

while it vanishes for $r \geq a$.

In cylindrical coordinates, $\nabla^2 A_z = -\mu_0 J_z$ becomes

$$\frac{1}{r}\frac{\partial}{\partial r}\left(r\frac{\partial A_z}{\partial r}\right) = \begin{cases} -\mu_0\, J_z & r \leq a \\ 0 & r > a \end{cases} \tag{Ex 4.4.2}$$

Integrating the expression in (Ex 4.2.2) for $r \leq a$ twice, we obtain

$$A_z(r \leq a) = -\frac{\mu_0 J_z\, r^2}{4} + C\ln r + D \tag{Ex 4.3.3}$$

Because A_z must be finite as $r \to 0$, we must set $C = 0$ leaving

$$A_z(r \leq a) = \frac{\mu_0 I}{4\pi}\left(D' - \frac{r^2}{a^2}\right) \tag{Ex 4.4.4}$$

For $r \geq a$, (Ex 4.4.2) implies that $A_z(r \geq a) = C\ln r + E$. We can determine C by the requirement that the exterior magnetic induction field, $\vec{\nabla}\times\vec{A}$, be given by (Ex 1.10.3)

$$\frac{\partial A_z}{\partial r} = \frac{C}{r} = -B_\varphi = -\frac{\mu_0 I}{2\pi r} \tag{Ex 4.4.5}$$

[10] We point out to the reader who was tempted to try to calculate the inductance from $L = d\Phi/dI$ that the boundary of the loop implied by Φ is ill defined. This problem recurs whenever finite-size wires are used.

We conclude that

$$C = \frac{-\mu_0 I}{2\pi} \tag{Ex 4.4.6}$$

Hence

$$A_z(r \geq a) = \frac{\mu_0 I}{4\pi}\left(E' - 2\ln\frac{r}{a}\right) \tag{Ex 4.4.7}$$

As B is finite at the boundary to the wire, A must be continuous. Matching the interior solution to the exterior solution at $r = a$, we obtain

$$\frac{\mu_0 I}{4\pi}\left(D' - \frac{a^2}{a^2}\right) = \frac{\mu_0 I}{4\pi}\left(E' - 2\ \ln\frac{a}{a}\right) \tag{Ex 4.4.8}$$

from which we conclude that $D' - 1 = E'$.

To summarize, for the vector potential of a long cylindrical wire, we have

$$A_z(r \geq a) = \frac{\mu_0 I}{4\pi}\left(D' - 1 - \ln\frac{r^2}{a^2}\right) \tag{Ex 4.4.9}$$

$$A_z(r \leq a) = \frac{\mu_0 I}{4\pi}\left(D' - \frac{r^2}{a^2}\right) \tag{Ex 4.4.10}$$

Returning to our problem, we note that interior to the wire at the origin, the vector potential is the superposition of its own interior A_z (Ex 4.4.10) and the exterior A_z, (Ex4.4.9), of the second wire carrying the same current in the opposite direction. Adding $A_z(r_2 \geq a, -I)$ to $A_z(r_1 \leq a, I)$ we get

$$A_z = \frac{\mu_0 I}{4\pi}\left(1 - \frac{r_1^2}{a^2} + \ln\frac{r_2^2}{a^2}\right) \tag{Ex 4.4.11}$$

From the geometry, $r_2^2 = r_1^2 + h^2 - 2r_1 h\cos\varphi$. With this substitution, we evaluate the volume integral of $\vec{A}\cdot\vec{J}$ over a length ℓ of the wire centered on the z axis:

$$\begin{aligned}
\int_{\text{wire 1}} \vec{J}\cdot\vec{A}\, d^3r &= \frac{\mu_0\, I^2\ell}{4\pi^2 a^2}\int_0^{2\pi}\int_0^a\left(1 - \frac{r_1^2}{a^2} + \ln\frac{r_1^2 + h^2 - 2r_1 h\cos\varphi}{a^2}\right) r_1 dr_1 d\varphi \\
&= \frac{\mu_0 I^2\ell}{2\,\pi^2 a^2}\int_0^a\left(\pi - \frac{\pi r_1^2}{a^2} + 2\pi\ln h - 2\pi\ln a\right) r_1 dr_1 \\
&= \frac{\mu_0\, I^2\ell}{2\pi a^2}\left(\frac{a^2}{2} - \frac{a^4}{4a^2} + a^2\ln\frac{h}{a}\right) = \frac{\mu_0 I^2\ell}{4\,\pi}\left(\frac{1}{2} + 2\ln\frac{h}{a}\right)
\end{aligned} \tag{Ex 4.4.12}$$

This result is just twice the energy integral (4–41) over one of the wires (or equal the integral over both). Equating the energy of the currents as expressed by the preceding integral to $\frac{1}{2}LI^2$, we find the inductance of the wires to be

$$L = \frac{\mu_0\ell}{4\pi}\left[1 + 4\ln(h/a)\right] \tag{Ex 4.4.13}$$

It is worth noting that for $h \gg a$, the logarithmic term of (Ex 4.4.13) dominates. The constant term, $\mu_0\ell/4\pi$, represents the self inductance of the current in each wire on itself.

As we previously did for electric charge distributions, we proceed now to express the potential energy of the current distribution in terms of the fields alone. Assuming there are no time-varying electric fields to contend with, we replace $\vec{J}$ in (4–41) by $(\vec{\nabla} \times \vec{B})/\mu_0$ to obtain

$$W = \frac{1}{2\mu_0}\int(\vec{\nabla} \times \vec{B}) \cdot \vec{A} d^3r \tag{4–42}$$

The argument of the integral in (4–42) may be rearranged with the aid of (8) as $(\vec{\nabla} \times \vec{B}) \cdot \vec{A} = \vec{B} \cdot (\vec{\nabla} \times \vec{A}) - \vec{\nabla} \cdot (\vec{A} \times \vec{B}) = B^2 - \vec{\nabla} \cdot (\vec{A} \times \vec{B})$, so that invoking the divergence theorem, the work expended to produce the field may be expressed as

$$W = \frac{1}{2\mu_0}\int B^2 d^3r - \frac{1}{2\mu_0}\oint(\vec{A} \times \vec{B}) \cdot d\vec{S} \tag{4–43}$$

B^2 is positive definite, implying that the volume integral of B^2 can only increase as the region of integration expands. By contrast, $(\vec{A} \times \vec{B})$ from a dipole field decreases as R^{-5} or faster, meaning that the surface integral goes to zero as $1/R^3$ or faster as $R \to \infty$. If, therefore, we include all space in the integral, we may write

$$W = \frac{1}{2\mu_0}\int_{\substack{all\\space}} B^2 d^3r \tag{4–44}$$

4.2 Poynting's Theorem

We will reconsider the energy required to produce a given electromagnetic field, this time less constrained by the construction of current loops or static charge distributions than the foregoing. This will give us not only the energy density of the fields but also the rate that energy is transported by the field.

We begin by considering the work dW done by the electromagnetic field on the charge dq contained in a small volume d^3r, moving through the field with velocity $\vec{v}$ when it is displaced though distance $d\vec{\ell}$. The work done on the charge is just the gain in mechanical energy (kinetic and potential) of the charge. If the field was the only agent acting on the charge, perhaps accelerating it, then the field must have supplied this energy and therefore decreased its own energy, or else there must have been a corresponding external input of energy.

The work performed by the Lorentz force is

$$\begin{aligned} dW &= dq\big(\vec{E} + \vec{v} \times \vec{B}\big) \cdot d\vec{\ell} \\ &= dq\,\big(\vec{E} + \vec{v} \times \vec{B}\big) \cdot \vec{v} dt \end{aligned} \tag{4–45}$$

Replacing dq with ρd^3r, we find the rate of increase of energy of all the charge contained in some volume τ as the volume integral

$$\frac{dW}{dt} = \int_\tau (\vec{E} + \vec{v} \times \vec{B}) \cdot \vec{v}\rho d^3r$$

$$= \int_\tau (\vec{E} \cdot \vec{J}) d^3r \tag{4–46}$$

We can express (4–46) using Ampère's law (3–24) in terms of the fields alone by replacing $\vec{J}$ with

$$\vec{J} = \frac{1}{\mu_0}(\vec{\nabla} \times \vec{B}) - \varepsilon_0 \frac{\partial \vec{E}}{\partial t} \tag{4–47}$$

With this substitution, the integrand of (4–46) can be rearranged as follows:

$$\begin{aligned}
\vec{E} \cdot \vec{J} &= \vec{E} \cdot \left(\frac{1}{\mu_0}(\vec{\nabla} \times \vec{B}) - \varepsilon_0 \frac{\partial \vec{E}}{\partial t} \right) \\
&= \frac{1}{\mu_0} \left[-\vec{\nabla} \cdot (\vec{E} \times \vec{B}) + \vec{B} \cdot (\vec{\nabla} \times \vec{E}) \right] - \varepsilon_0 \vec{E} \cdot \frac{\partial \vec{E}}{\partial t} \\
&= \frac{1}{\mu_0} \left(-\vec{\nabla} \cdot (\vec{E} \times \vec{B}) - \vec{B} \cdot \frac{\partial \vec{B}}{\partial t} \right) - \varepsilon_0 \vec{E} \cdot \frac{\partial \vec{E}}{\partial t} \\
&= -\frac{1}{2}\frac{\partial}{\partial t} \left(\varepsilon_0 E^2 + \frac{B^2}{\mu_0} \right) - \vec{\nabla} \cdot \frac{\vec{E} \times \vec{B}}{\mu_0}
\end{aligned} \tag{4–48}$$

With (4–48) and the divergence theorem (20), we cast (4-46) into the form

$$\frac{dW}{dt} = -\frac{d}{dt} \int \tfrac{1}{2} \left(\varepsilon_0 E^2 + \frac{B^2}{\mu_0} \right) d^3r - \oint \frac{(\vec{E} \times \vec{B})}{\mu_0} \cdot d\vec{S} \tag{4–49}$$

To interpret (4–49) we note that we have already met the terms in the volume integral in (4–18) and (4–44) as the energy density of the field of the electric and magnetic induction field respectively. If we suppose for the moment that the surface integral vanishes, (4–49) states that the rate at which the particles gain mechanical energy is just equal to the rate at which the fields lose energy. Now suppose that in spite of doing work on charges in the volume of interest, the fields remained constant. Clearly we would need an inflow of energy to allow this. The surface integral has exactly this form; including the minus sign, it measures the total $(\vec{E} \times \vec{B})/\mu_0$ crossing the surface *into* the volume. Clearly

$$\vec{S} \equiv (\vec{E} \times \vec{B})/\mu_0 \tag{4–50}$$

is the energy flux[11] crossing the surface out of the volume. $\vec{S}$ is known as Poynting's vector, and it is the rate that the electromagnetic field transports energy across a unit surface.

---

[11] A word of caution is advised. In reality, only the surface integral of $\vec{S}$ is associated with energy flow. It is easily verified that the field surrounding a charged magnetic dipole has a nonvanishing Poynting vector, but one would be most reluctant to associate an energy flow with this.

EXAMPLE 4.5: Find the energy flux (irradiance) of an electromagnetic plane wave with electric field amplitude $\vec{E}$.

Solution: Using $\vec{S} = (\vec{E} \times \vec{B})/\mu_0$, with $\vec{B} = (\hat{k} \times \vec{E})/c$, we find that

$$\vec{S} = \frac{\vec{E} \times (\hat{k} \times \vec{E})}{\mu_0 c} = \frac{E^2 \hat{k}}{\mu_0 c} \qquad \text{(Ex 4.5.1)}$$

This result is correct for the instantaneous flux. It should be noted, however, that one rarely measures an energy flux over time periods less than 10^{-14} seconds. Therefore it would be more correct to use the average value of the electric field in (Ex 4.5.1), namely $\langle E^2 \rangle = \frac{1}{2} E^2$.

4.3 Momentum of the Fields

As the electromagnetic field is capable of imparting momentum to charges, we would anticipate that the field must itself possess momentum. In the absence of external forces, we expect the total momentum (that of the charges and the electromagnetic field) to be conserved. If there are, on the other hand, external forces, they must present themselves at the boundary of the isolated region. Placing the volume of interest in an imaginary box, we may find the force on the volume by integrating the *stress* on the surface over the whole bounding surface (Figure 4.3).

4.3.1 The Cartesian Maxwell Stress Tensor for Electric Fields

As a preliminary to the development of the general Maxwell Stress Tensor in vacuum, we restrict ourselves to the stresses resulting from electric fields. Moreover, we sidestep questions of covariance by limiting this discussion to Cartesian coordinates. We seek in this section to express the force acting on an isolated system in terms of the fields at the bounding surface of the system. More quantitatively, the i component of the force acting on a surface element dS_j is given by $dF^i = -T^{ij} dS_j$, where the Cartesian stress tensor $\overleftrightarrow{T}$ has components T^{ij} and the summation over repeated indices is assumed. Familiar components of the mechanical stress tensor

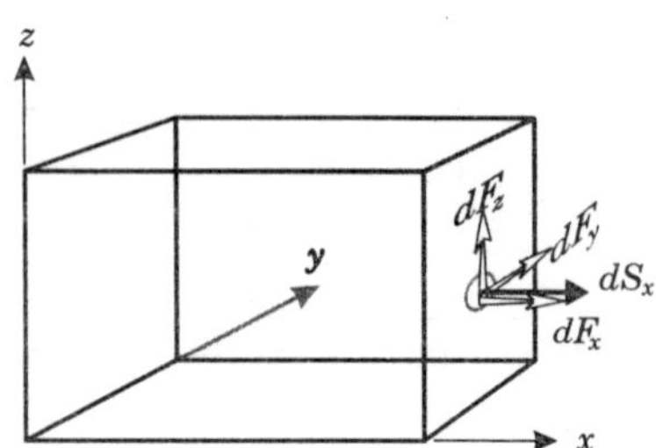

Figure 4.3: When a positive stress such as a pressure is applied to a surface, the resulting force is inward directed. The forces dF_y and dF_z shown on the right face result from shear stresses T_{xy} and T_{xz}.

are

$$\text{pressure} = \tfrac{1}{3}\sum_{i=1}^{3} T^{ii} \tag{4–51}$$

compressional stress, T^{jj}, and shear stress, T^{ij} $(i \neq j)$. The i component of the force on the entire volume is[12]

$$F^i = -\oint T^{ij}\, dS_j \tag{4–52}$$

where we have adopted the convention of summing over repeated indices (B.1.5) when they occur once as subscript and once as superscript. Expressing this force as the volume integral of the force per unit volume (or force density) $\vec{f}$,

$$F^i = \int_\tau f^i d^3r \tag{4–53}$$

we find, using a generalization of the divergence theorem for tensors,

$$\oint T^{ij} dS_j = \int \left(\partial_j T^{ij}\right) d^3r \tag{4–54}$$

meaning that the force density may be expressed as $f^i = -\partial_j T^{ij}$ ($\partial_j \equiv \partial/\partial x^j$ and j is not an exponent but only a label or *index*). Thus in analogy to the force being found as minus the gradient of the potential energy, we can find it as minus the divergence of the stress tensor (the comments in the footnote again apply).

In line with our earlier restriction, let us consider the stress tensor associated with a static electric field. We will express the force per unit volume on a given volume of charge as the divergence of a second rank tensor. The i component of the force density f is given by

$$\begin{aligned} f^i &= \rho E^i = (\varepsilon_0 \vec{\nabla} \cdot \vec{E}) E^i = \varepsilon_0 (\partial_j E^j) E^i \\ &= \varepsilon_0 \left[\partial_j (E^j E^i) - E^j \partial_j E^i\right] \end{aligned} \tag{4–55}$$

The first term of the (4–55) is already in the required form and we can transform the second term to the appropriate form using $\vec{\nabla} \times \vec{E} = 0$, or, in tensor form, $\partial_j E_i = \partial_i E_j$. Multiplying both sides by the "raising operator" g^{ik} we get $g^{ik}\partial_j E_i = g^{ik}\partial_i E_j$ which becomes $\partial_j E^k = \partial^k E_j$. Relabelling k as i and multiplying both sides by E^j, we have

$$\begin{aligned} E^j \partial_j E^i &= E^j \partial^i E_j \\ &= \tfrac{1}{2}\partial^i (E^j E_j) \\ &= \tfrac{1}{2}\delta^{ij} \partial_j E^2 \end{aligned} \tag{4–56}$$

[12]Some authors define the stress tensor with the opposite sign. As this would require the tensor elements representing an isotropic positive pressure to be negative, contrary to common use, we prefer the sign used. This choice of sign means that the Maxwell stress tensor's elements will be the negative of those defined by Griffiths, Wangsness, and others.

With this substitution, (4–55) may be written

$$f^i = -\partial_j \varepsilon_0 \left(-E^i E^j + \tfrac{1}{2}\delta^{ij}E^2\right) \tag{4–57}$$

The Maxwell stress tensor of a static electric field may now be written explicitly in matrix form:

$$\overleftrightarrow{T} = \varepsilon_0 \begin{pmatrix} \frac{1}{2}E^2 - E_x^2 & -E_xE_y & -E_xE_z \\ -E_yE_x & \frac{1}{2}E^2 - E_y^2 & -E_yE_z \\ -E_zE_x & -E_yE_z & \frac{1}{2}E^2 - E_z^2 \end{pmatrix} \tag{4–58}$$

This tensor is a symmetric tensor of second rank with eigenvalues $\lambda_1 = -\frac{1}{2}\varepsilon_0 E^2$ and $\lambda_2 = \lambda_3 = \frac{1}{2}\varepsilon_0 E^2$. The eigenvector of λ_1 is parallel to the field $\vec{E}$, whereas those of λ_2 and λ_3 are perpendicular to $\vec{E}$. In the principal axis system, with x chosen parallel to the electric field, $\overleftrightarrow{T}$ becomes

$$\overleftrightarrow{T} = \tfrac{1}{2}\varepsilon_0 \begin{pmatrix} -E^2 & 0 & 0 \\ 0 & E^2 & 0 \\ 0 & 0 & E^2 \end{pmatrix} \tag{4–59}$$

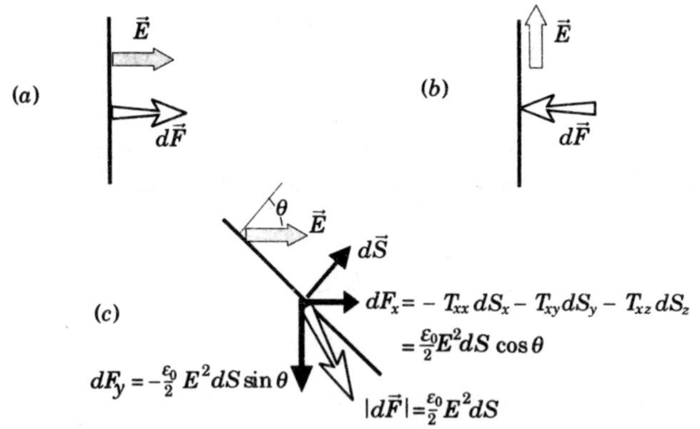

Figure 4.4: The force derived from the stress tensor when $\vec{E}$ is (a) perpendicular, (b) parallel, and (c) makes angle θ with the normal to the surface.

For various orientations of the field relative to a bounding surface this results in forces as illustrated in Figure 4.4. One can now use this formalism to calculate the force on a charge by drawing a box around the charge and integrating the stress tensor over the surface of the box.

EXAMPLE 4.6: Use the Maxwell stress tensor to find the forces two equal charges q of opposite sign exert on each other.

Solution: To make the problem somewhat more specific, we place the positive charge on the left, a distance $2d$ from the negative charge. We enclose one charge, say the

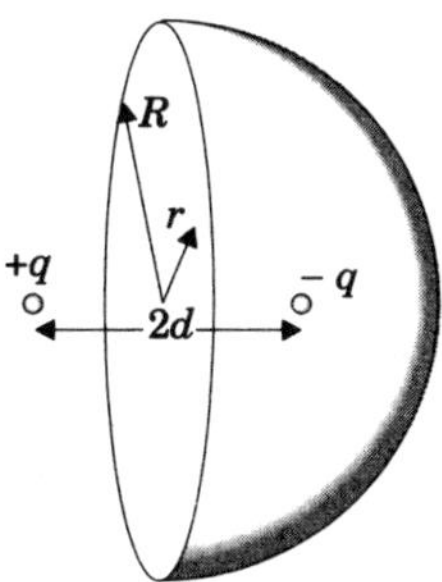

Figure 4.5: The stress tensor may be integrated of the hemisphere enclosing one of the two charges to find the force on the charge enclosed.

negative charge on the right of Figure 4.5 by a hemispherical "box" whose plane side lies along the midplane between the two charges. If we let the radius of the hemisphere grow large, the field at the hemispherical boundary becomes that of a dipole, $\propto R^{-3}$. The surface integral then varies as R^{-4}. Taking the hemisphere to ∞ makes this part of the surface integral vanish. At the midplane,

$$E_z = \frac{2}{4\pi\varepsilon_0}\frac{qd}{(d^2+r^2)^{3/2}} \tag{Ex 4.6.1}$$

$$\Rightarrow \qquad E^2 = \left(\frac{2qd}{4\pi\varepsilon_0}\right)^2 \frac{1}{(d^2+r^2)^3} \tag{Ex 4.6.2}$$

where the transverse components of the fields of the two charges just cancel. The zz component of the stress tensor T_{zz} is $\varepsilon_0(E_z^2 - \frac{1}{2}E^2)$, and the surface element is $dS_z = -2\pi r dr$. The force on the charge from the stress at the midplane is then

$$\begin{aligned} F_z &= -\int_0^\infty -\tfrac{1}{2}\varepsilon_0 E^2(-2\pi r dr) \\ &= -\tfrac{1}{2}\varepsilon_0 \frac{4q^2d^2}{(4\pi\varepsilon_0)^2}\int_0^\infty \frac{2\pi r\, dr}{(r^2+d^2)^3} \\ &= -\frac{q^2d^2}{4\pi\varepsilon_0}\int_d^\infty \frac{ds}{s^5} = -\frac{q^2}{4\pi\varepsilon_0(2d)^2} \end{aligned} \tag{Ex 4.6.3}$$

where we have defined $s^2 = r^2 + d^2$ to effect the integration. The result obtained is of course exactly the result given by Coulomb's law.

4.3.2 The Maxwell Stress Tensor and Momentum

We now consider the implications of conservation of momentum when the fields are not static. As before, we consider the force per unit volume ($= d\vec{\mathcal{P}}/dt$, where $\vec{\mathcal{P}}$ is the momentum density) exerted by the electromagnetic field on charges and currents contained in a volume τ and described by charge density ρ and current density $\vec{J}$. We will write this force as the divergence of a second rank tensor after eliminating ρ and $\vec{J}$ with the use of Maxwell's equations. The derivation requires considerable vector manipulation; therefore, to reduce the apparent arbitrariness of the manipulation and minimize the work, we use tensor notation and the *Levi-Cevita* symbol ϵ_{ijk} to express the cross products. To assure that these results will hold in any coordinate system, we employ full co– and contra-variant notation. If the reader is not familiar with these methods it is strongly advised that Appendix B.2.5 and B.3 be consulted.

The rate of change of the i component of the charges' momentum density, $\dot{\mathcal{P}}^i$, is given by the Lorentz force expression (1–30)

$$\begin{aligned}
\dot{\mathcal{P}}^i &= \rho E^i + \epsilon^{ijk} J_j B_k \\
&= \varepsilon_0(\partial_j E^j)E^i + \epsilon^{ijk}\left[\frac{1}{\mu_0}(\vec{\nabla}\times\vec{B})_j - \varepsilon_0 \dot{E}_j\right] B_k \\
&= \varepsilon_0(\partial_j E^j)E^i + \frac{1}{\mu_0}\epsilon^{ijk}\epsilon_{j\ell m}(\partial^\ell B^m)B_k - \varepsilon_0\epsilon^{ijk}\dot{E}_j B_k \qquad (4\text{–}60)\\
&= \varepsilon_0(\partial_j E^j)E^i + \frac{1}{\mu_0}(\delta^k_\ell\delta^i_m - \delta^k_m\delta^i_\ell)B_k\partial^\ell B^m - \varepsilon_0\epsilon^{ijk}\left[\frac{\partial}{\partial t}(E_j B_k) - E_j\dot{B}_k\right] \\
&= \varepsilon_0(\partial_j E^j)E^i + \frac{1}{\mu_0}(B_\ell\partial^\ell B^i - B_m\partial^i B^m) - \varepsilon_0\frac{\partial}{\partial t}(\vec{E}\times\vec{B})^i - \varepsilon_0\epsilon^{ijk}\left[E_j(\vec{\nabla}\times\vec{E})_k\right]
\end{aligned}$$

We group the time derivatives on the left and collect terms in E and B

$$\begin{aligned}
\dot{\mathcal{P}}^i + \varepsilon_0\frac{\partial}{\partial t}(\vec{E}\times\vec{B})^i &= \varepsilon_0\left[(\partial_j E^j)E^i - \epsilon^{ijk}E_j\epsilon_{k\ell m}\partial^\ell E^m\right] + \frac{1}{\mu_0}\left(B_\ell\partial^\ell B^i - B_m\partial^i B^m\right) \\
&= \varepsilon_0\left[(\partial_j E^j)E^i - (\delta^i_\ell\delta^j_m - \delta^i_m\delta^j_\ell)E_j\partial^\ell E^m\right] + \frac{1}{\mu_0}\left(B_\ell\partial^\ell B^i - B_m\partial^i B^m\right) \\
&= \varepsilon_0\left[(\partial_j E^j)E^i - E_m\partial^i E^m + E^j\partial_j E^i\right] + \frac{1}{\mu_0}\left(B_\ell\partial^\ell B^i - B_m\partial^i B^m\right) \qquad (4\text{–}61)
\end{aligned}$$

where we have rewritten $\delta^j_\ell E_j\partial^\ell = g^{jk}g_{k\ell}E_j\partial^\ell = E^k\partial_k$ and then relabelled k as j. We can add the inconsequential, identically vanishing term $B^i\partial_\ell B^\ell$ to the inside of the brackets containing the magnetic field terms of (4–61) to give it the same appearance as the electric field terms. With this addition we write

$$\dot{\mathcal{P}}^i + \varepsilon_0\frac{\partial}{\partial t}(\vec{E}\times\vec{B})^i = \partial_j\left[\varepsilon_0\left(E^jE^i - \tfrac{1}{2}\delta^{ij}E^2\right) + \frac{1}{\mu_0}\left(B^jB^i - \tfrac{1}{2}\delta^{ij}B^2\right)\right] \qquad (4\text{–}62)$$

The interpretation of (4–62) is simplified if, instead of considering the densities of (4–62), we integrate them over some finite volume. Anticipating our conclusions, we abbreviate the square bracketed term as $-T^{ij}$,

$$T^{ij} = -\left[\varepsilon_0\left(E^iE^j - \tfrac{1}{2}\delta^{ij}E^2\right) + \frac{1}{\mu_0}\left(B^iB^j - \tfrac{1}{2}\delta^{ij}B^2\right)\right] \qquad (4\text{–}63)$$

Then, integrating (4–62) over a volume τ, we obtain

$$\frac{d}{dt}\int_\tau\left[\mathcal{P}^i + \varepsilon_0(\vec{E}\times\vec{B})^i\right]d^3r = \int_\tau -\partial_j T^{ij}d^3r$$

$$= -\int_\Gamma T^{ij} dS_j \tag{4–64}$$

where Γ is the bounding surface to τ and we have used an obvious generalization of the divergence theorem (20).

Now, if the volume of integration is chosen large enough that the fields and hence the elements T^{ij} vanish on the boundary, we have

$$\frac{d}{dt}\int \left[\vec{\mathcal{P}} + \varepsilon_0(\vec{E} \times \vec{B})\right] d^3r = 0 \tag{4–65}$$

Since the system now has no external forces acting on it, the total momentum must be conserved. If the enclosed matter's momentum changes, the second term must change to compensate. Thus we are led to the conclusion that the term

$$\vec{\mathcal{P}}_{em} \equiv \varepsilon_0(\vec{E} \times \vec{B}) \tag{4–66}$$

is the momentum density of the electromagnetic field. This momentum density is, by inspection, equal to $\mu_0\varepsilon_0\vec{S}$, the energy flux divided by c^2. Since the energy flux is just c times the energy density, $U \times c$, we find the momentum density to be $1/c$ times the energy density, as might have been anticipated. This relation between U and $\mathcal{P}$ can of course hold only when there is in fact an energy flux and all the energy in the volume is being transported at velocity c in the same direction.

If the total momentum of the system does change, then an external force must be provided for by the term on the right of (4–64), which we have already seen to have the correct form.

The expression (4–64) lends itself to another interpretation, for we see the momentum in the volume decreasing at the rate that T^{ij} crosses the surface out of the volume. Clearly, then, T^{ij} also represents the momentum flux of the electromagnetic field exiting the volume.

A similar result can be obtained for the angular momentum of the fields. Writing the angular momentum density $\vec{\mathcal{L}}_{em} = \vec{r} \times \vec{\mathcal{P}}_{em}$, we can write

$$\frac{\partial}{\partial t}\left(\vec{\mathcal{L}}_{em} + \vec{\mathcal{L}}_{mech}\right) + \vec{\nabla} \cdot \overleftrightarrow{M} = 0 \tag{4–67}$$

where $\overleftrightarrow{M}$ is the *pseudo* tensor $\overleftrightarrow{M} = \vec{r} \times \overleftrightarrow{T}$ or $M_i^j = \epsilon_{ik\ell}x^k T^{j\ell}$.

At this point we might reflect on the question of the reality of the fields we posed early in Chapter 1. We have been able to associate with the fields energy, momentum, and angular momentum giving the fields considerably more substance than we had originally endowed them with. Nevertheless, as should be clear from the derivations, the momenta and energies could equally well be associated with the sources of the fields.

There are, however, cases of fields not obviously associated with their sources. The electromagnetic waves we discussed in the previous chapter carry an energy flux $\langle\vec{S}\rangle = \langle\vec{E} \times \vec{B}/\mu_0\rangle = E_0^2/2\mu_0 c$ independently of whatever sources produced them. If the fields are merely mathematical constructs, it is difficult to see how energy and momentum might be transported across space without the material sources of the fields. We are therefore forced to the view that the fields do indeed have some reality.

⋆ 4.4 Magnetic Monopoles

Although to date there have been no confirmed observations of magnetic monopoles, a number of arguments (advanced notably by *Dirac*) in favor of their existence has led to a continuing search for primitive magnetic monopoles.[13] Dirac found that an electron moving in the field of a hypothetical magnetic monopole would have a multivalued wave function unless the product of the electron charge and monopole charge were quantized. This quantization could provide an explanation for the apparently arbitrary quantization of charge. Most current *grand unified theories* (GUTs for short) require the existence of magnetic monopoles.

Taking a somewhat different approach from Dirac's, we may easily show that a magnetic monopole in the vicinity of an electric charge would produce an electro/-magneto–static field whose angular momentum is not zero. Remarkably, this angular momentum is independent of the spacing between the electric and magnetic monopoles. We produce below a demonstration of this assertion.

Let us consider the angular momentum, $\vec{L}$, of the fields of a magnetic monopole q_m located at the origin, and an electric charge e located at $a\hat{k}$:

$$\vec{L} = \int \vec{r} \times \varepsilon_0(\vec{E} \times \vec{B})d^3r = \varepsilon_0 \int \left[(\vec{r} \cdot \vec{B})\vec{E} - (\vec{r} \cdot \vec{E})\vec{B}\right] d^3r \tag{4–68}$$

where $\vec{E}$ is the electric field produced by the electric charge at $a\hat{k}$. For a magnetic monopole at the origin, the magnetic induction field $\vec{B}$ is given by

$$\vec{B} = \frac{\mu_0 \, q_m \vec{r}}{4\pi r^3} \tag{4–69}$$

which we substitute into the expression for $\vec{L}$ above to obtain

$$\vec{L} = \frac{\mu_0 \varepsilon_0 q_m}{4\pi} \int \left(\frac{\vec{E}}{r} - \frac{\vec{r}(\vec{E} \cdot \vec{r})}{r^3} \right) d^3r \tag{4–70}$$

We effect the integration of (4–70) one component at a time. To this end, we write

$$L^k = \frac{q_m}{4\pi c^2} \int \left[\frac{E^k}{r} - \frac{x^k(E^i x_i)}{r^3} \right] d^3r = \frac{q_m}{4\pi c^2} \int E^i \partial_i \left(\frac{x^k}{r} \right) d^3r \tag{4–71}$$

We integrate (4–71) by parts using

$$E^i \partial_i \left(\frac{x^k}{r} \right) = \partial_i \left(\frac{E^i x^k}{r} \right) - \frac{x^k}{r} \partial_i E^i \tag{4–72}$$

[13]The only positive report is a single uncorroborated candidate event reported by B. Cabrera, *Phys. Lett.* **48**, 1387 (1982). Among the more imaginative proposals for a search was the suggestion that because cosmic monopoles are more likely to be found in the oceans than on dry land, and oysters are prodigious filters of ocean water, we carefully distill many tons of oysters. The feasibility depends on the presumed massiveness of monopoles.

to obtain

$$L^k = \frac{q_m}{4\pi c^2}\int \partial_i\left(\frac{E^i x^k}{r}\right)d^3r - \frac{q_m}{4\pi c^2}\int \frac{x^k}{r}(\vec{\nabla}\cdot\vec{E})d^3r$$

$$= \frac{q_m}{4\pi c^2}\oint \frac{E^i x^k}{r}dS_i - \frac{q_m}{4\pi c^2}\int \frac{x^k}{r}\frac{\rho}{\varepsilon_0}d^3r \tag{4–73}$$

At sufficiently large distance from the charge, $|E|r^2 d\Omega = ed\Omega/4\pi\varepsilon_0$, a constant. Moreover, $\oint x^k d\Omega = 0$ for any spherical volume centered on the origin. We conclude that in the limit as the volume of integration tends to infinity, the surface integral vanishes. The charge density for the point charge e situated at $a\hat{k}$ is $\rho = e\delta(\vec{r} - a\hat{k})$. The remaining integral then becomes

$$L^k = -\frac{q_m e}{4\pi\varepsilon_0 c^2}\int \frac{x^k}{r}\delta(\vec{r} - a\hat{k})d^3r \tag{4–74}$$

The integration is now easily performed to give 0 for L_x and L_y and

$$L_z = -\frac{q_m e}{4\pi\varepsilon_0 c^2} \tag{4–75}$$

independent of a! Even if the magnetic and electric monopoles are separated by galactic distances, this intrinsic angular momentum of their fields remains.

We hypothesize that, consistent with quantum mechanics, the smallest permitted angular momentum is $\hbar/2$, leading to the smallest non-zero magnetic monopole charge

$$q_m = \frac{2\pi\varepsilon_0 c^2\hbar}{e} \tag{4–76}$$

4.5 Duality Transformations

We might see how Maxwell's equations would be altered to accommodate magnetic monopoles. Clearly we can no longer sustain $\vec{\nabla}\cdot\vec{B} = 0$, but have instead $\vec{\nabla}\cdot\vec{B} = \mu_0\rho_m$. Moving monopoles would constitute a magnetic current and presumably add a term to the $\vec{\nabla}\times\vec{E}$ equation. Postulating

$$\vec{\nabla}\times\vec{E} = -\frac{\partial\vec{B}}{\partial t} - \mu_0\vec{J}_m \tag{4–77}$$

we verify a continuity equation for magnetic monopoles:

$$0 = \vec{\nabla}\cdot(\vec{\nabla}\times\vec{E}) = -\frac{\partial}{\partial t}(\vec{\nabla}\cdot\vec{B}) - \mu_0\vec{\nabla}\cdot\vec{J}_m \tag{4–78}$$

or

$$\frac{\partial\rho_m}{\partial t} + \vec{\nabla}\cdot\vec{J}_m = 0 \tag{4–79}$$

The generalized Maxwell's equations then become

$$\vec{\nabla} \cdot \vec{E} = \frac{\rho_e}{\varepsilon_0} \qquad \vec{\nabla} \cdot \vec{B} = \mu_0 \rho_m$$
$$\vec{\nabla} \times \vec{E} = -\frac{\partial \vec{B}}{\partial t} - \mu_0 \vec{J}_m \qquad \vec{\nabla} \times \vec{B} = \frac{1}{c^2}\frac{\partial \vec{E}}{\partial t} + \mu_0 \vec{J}_e \tag{4–80}$$

The generalized equations present considerable symmetry. A *duality transformation* defined by

$$\vec{E}' = \vec{E}\cos\theta + c\vec{B}\sin\theta$$
$$c\vec{B}' = -\vec{E}\sin\theta + c\vec{B}\cos\theta \tag{4–81}$$

together with

$$c\rho_e' = c\rho_e\cos\theta + \rho_m\sin\theta \qquad \rho_m' = -c\rho_e\sin\theta + \rho_m\cos\theta \tag{4–82}$$

and the associated current transformations

$$cJ_e' = cJ_e\cos\theta + J_m\sin\theta \qquad J_m' = -cJ_e\sin\theta + J_m\cos\theta \tag{4–83}$$

leaves Maxwell's equations and therefore all ensuing physics invariant. Thus with the appropriate choice of *mixing* angle θ, magnetic monopoles may be made to appear or disappear at will. So long as all charges have the same ratio ρ_e/ρ_m (the same mixing angle), the existence or nonexistence of monopoles is merely a matter of convention. The Dirac monopole, on the other hand, would have a different mixing angle than customary charges.

Exercises and Problems

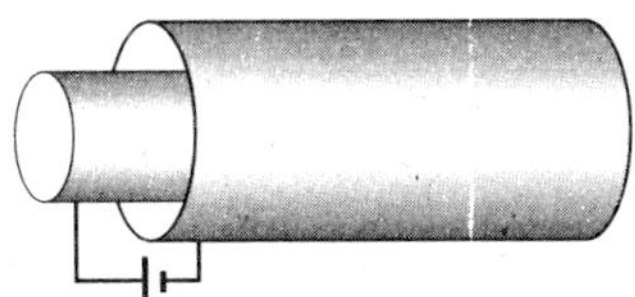

Figure 4.6: The concentric cylinders of problem 4-15.

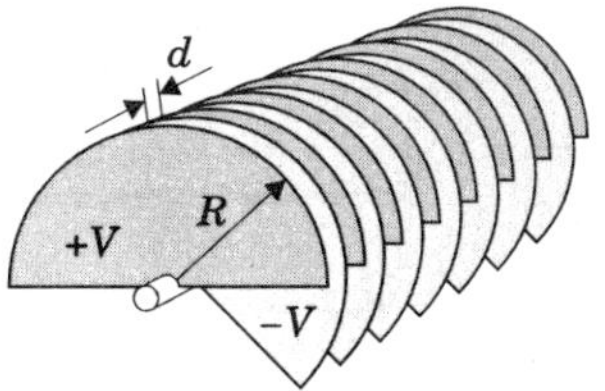

Figure 4.7: The variable capacitor of problem 4-16.

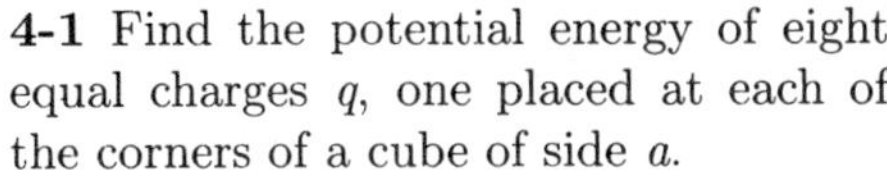

4-1 Find the potential energy of eight equal charges q, one placed at each of the corners of a cube of side a.

4-2 Find the inductance of a closely wound solenoid of radius R and length L having N turns when $R \ll L$.

4-3 Find the inductance of a closely wound toroidal coil of N turns with mean radius b and cross-sectional radius a using energy methods.

4-4 Find the inductance of a coaxial wire whose inner conductor has a radius a and whose outer conductor has inner radius b and the same cross-sectional area as the inner conductor. Assume the same current runs in opposite directions along the inner and outer conductor. (Hint: The magnetic induction field vanishes outside the outer conductor, meaning that the volume integral of B^2 is readily found.)

4-5 Find the magnetic flux Φ enclosed by a rectangular loop of dimensions $\ell \times (h-2a)$ placed between the two conductors of figure 4.2. Compare the inductance computed as $L = d\Phi/dI$ to the result of example 4.4.

4-6 Find the energy of a charge Q spread uniformly throughout the volume of a sphere of radius a.

4-7 Find the energy of a spherical charge whose density varies as

$$\rho = \rho_0\left(1 - \frac{r^2}{a^2}\right)$$

for $r \le a$ and vanishes when r exceeds a.

4-8 Find the electric field strength in a light beam emitted by a 5-watt laser if the beam has a 0.5 mm^2 (assume uniform) cross section.

4-9 Show that the surface integral of (4–43) vanishes for the parallel wire field at sufficiently large distance. Chose as volume of integration a large cylinder centered on one of the wires.

4-10 Estimate the mutual inductance of two parallel circular loops of radius a spaced by a small distance b ($b \ll a$).

4-11 A superconducting solenoid of length L and radius a carries current I in each of its N windings. Find the radial force on the windings and hence, the tensile strength required of the windings. (Note that the radial force resulting from the interior $\vec{B}$ is outward directed whereas the force on the end faces is inward.)

4-12 Find the force between two primitive magnetic monopoles, and compare this force to the force between two electric charges e.

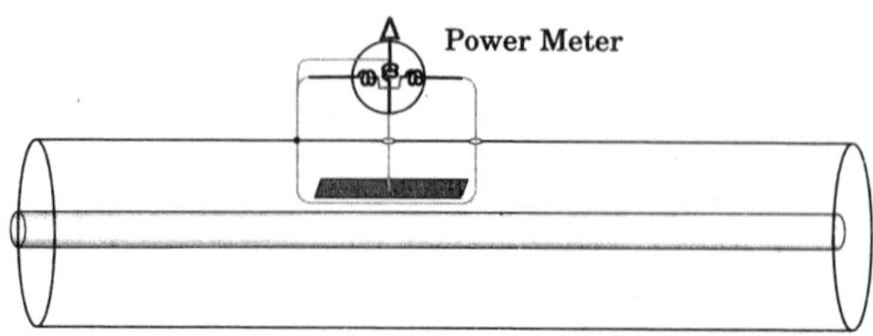

Figure 4.8: A directional Power meter may constructed by attaching the field coils of a power meter to an inductive loop and the movement coil to a capacitive element inside a coaxial conductor.

4-13 Show that the generalized Maxwell equations (4-44) are invariant under a duality transformation.

4-14 Find approximately the mutual capacitance of two 1-m radius spheres whose centers are separated by 10 m.

4-15 A pair of concentric cylinders of radii a and b, (Figure 4.6) are connected to the terminals of a battery supplying an EMF V. Find the force in the axial direction on the inner cylinder when it is partially extracted from the outer one.

4-16 A variable capacitor has 15 semicircular blades of radius R spaced at distance d (Figure 4.7). Alternate blades are charged to $\pm V$. The capacitance is varied by rotating one set of blades about an axis on the center of the diameter of the blades with respect to the other set. Find the electrostatic torque on the movable blades when partly engaged. Neglect any fringing fields. Could one reasonably design an electrostatic voltmeter using this principle?

4-17 A power meter may be constructed by using a signal proportional to the current to power the field (instead of the customary permanent magnet) of a galvanometer and a signal proportional to the voltage to power the moving coil of the galvanometer. A directional power meter may be constructed as illustrated in figure 4.8. Given that the torque on the needle is $\tau = kI_1I_2$ where I_1 is the current induced in the loop by the oscillating current in the coaxial wire and I_2 is the current required to capacitively charge the plate adjacent to the cental conductor, write an expression for the torque on the needle in terms of frequency, loop area and plate area, curved to maintain constant distance from the central conductor and the distance of each from the center.

4-18 A spherical soap bubble has a charge Q distributed over its surface. If the interior pressure and exterior pressure are the same, find the radius at which the compressive force from the surface tension balances the repulsive electrical force. Is the radius stable against perturbations? (Hint: the surface tension T on a spherical surface with radius of curvature r gives an inward pressure of $2T/r$.)

Chapter 5

Static Potentials in Vacuum – Laplace's Equation

5.1 Laplace's Equation

To this point we have assumed that all relevant charge and current distributions are known, allowing us to use Coulomb's law or the Biot-Savart law to obtain a closed form expression of the fields or potentials. In most cases, however, the charge distribution will not be known a priori. We might, for instance, know the total charge on the two plates of a finite size parallel plate capacitor, but the charge density, not being uniform, remains a mystery. Even when most of the charge density is known, induced charge densities on nearby conductors, or as we will see in the next chapter, polarizations or magnetizations of nearby materials, will generally contribute to the field. In such cases, the potentials or fields must be deduced from the *boundary conditions*, such as the requirement that the electric field be everywhere perpendicular to the surface of the conductor or, alternatively, that the surface of the conductor be a surface of constant potential. Finally, the unknown charge densities might be computed from the field at the surface of the conductor.

In this chapter we focus our attention on static fields resulting from charges or currents entirely on or outside the boundary of the region of interest. The lack of time dependence means that $\vec{E}$ has zero curl, and the absence of currents makes $\vec{B}$ curl free. Either $\vec{E}$ or $\vec{B}$ may therefore be expressed as the gradient of a potential V. The lack of sources inside the region makes the divergence of the fields vanish (of course $\vec{\nabla} \cdot \vec{B}$ vanishes identically in any case). Under these conditions, as has already been remarked in Chapter 1 (Section 1.1.3), the potential, V, satisfies the Laplace equation $\nabla^2 V = 0$. The particular solutions of Laplace's equation must satisfy the boundary conditions imposed by the physical situation.

The boundary conditions we will consider are of two general types. The first, the *Dirichlet* boundary condition, specifies the value of V everywhere on the boundary, while the second, the *Neumann* boundary condition, specifies the normal derivative $\partial V/\partial n$ (the electric field) at the surface. Of course it is entirely possible to have mixed boundary conditions with V specified on part of the boundary and the normal derivative on another part of the boundary. Before embarking on a detailed study of the solution of Laplace's equation, we prove a uniqueness theorem which asserts that the solution of Laplace's equation satisfying Dirichlet or Neumann (or mixed) boundary conditions over an enclosing surface is unique. This means that no matter how we arrive at a solution, if the solution satisfies $\nabla^2 V = 0$ and the boundary conditions we have *the* solution.

5.1.1 Uniqueness Theorem

Let us consider two proposed solutions V_1 and V_2, each satisfying the boundary conditions and $\nabla^2 V = 0$. The difference between these solutions, $\Phi = V_1 - V_2$, then also satisfies $\nabla^2\Phi = 0$ and in addition satisfies $\Phi = 0$ or $\partial\Phi/\partial n = 0$ according to the boundary conditions specified. In either case, the product of Φ and $(\vec{\nabla}\Phi)_n$ must vanish on the boundary. Thus, with the aid of the divergence theorem and making use of the vanishing of the Laplacian of Φ, we have

$$\begin{aligned} 0 &= \oint \Phi\vec{\nabla}\Phi \cdot d\vec{S} \\ &= \int \vec{\nabla}\cdot(\Phi\vec{\nabla}\Phi)d^3r \\ &= \int \left((\vec{\nabla}\Phi)^2 + \Phi\nabla^2\Phi\right) d^3r \\ &= \int (\vec{\nabla}\Phi)^2\, d^3r \end{aligned} \tag{5–1}$$

As $(\vec{\nabla}\Phi)^2$ is nowhere negative, the integral can vanish only if $(\vec{\nabla}\Phi)^2$ is identically zero. We conclude then that $\vec{\nabla}\Phi = 0$, implying Φ is constant. In the case that Φ was zero on the boundary (Dirichlet condition), it must be zero everywhere, implying $V_1 = V_2$. For the case that the normal derivative vanished (Neumann condition), V_1 and V_2 can differ only by a constant. This minor freedom in the latter case is not surprising; because we have essentially specified only the field at the boundary and we know that the field cannot determine the potential to better than an additive constant, we should have fully expected this inconsequential ambiguity in the solution.

The theorem we have just proved also proves incidently the useful observation that a charge-free region of space enclosed by a surface of constant potential has a constant potential and consequently a vanishing electric field within that boundary.

We proceed now to explore a number of techniques for solving Laplace's equation. Generally it will be best, if possible, to choose a coordinate system in which coordinate surfaces coincide with the boundary, as this makes the application of boundary conditions considerably easier. To clearly demonstrate the techniques we systematically explore one-, two- and three-dimensional solutions in a number of different coordinate systems.

5.1.2 ∇^2V = 0 in One Dimension

When the problem of interest is invariant to displacements in all but one of the coordinates, ∇^2 reduces to the one-dimensional ordinary differential operator. Despite calling this a one-dimensional problem, it is really a problem with one variable in three dimensional space. In the most frequently used coordinate systems the operators and the corresponding solutions are easily found as follows.

Cartesian Coordinates: Since V depends only on one of the coordinates, say z,

$$\frac{d^2V}{dz^2} = 0 \quad \Rightarrow \quad V = Az + B \tag{5–2}$$

Cylindrical Coordinates: Assuming that V depends only on r, we have

$$\frac{1}{r}\frac{d}{dr}\left(r\frac{dV}{dr}\right) = 0 \quad \Rightarrow \quad V = A\ln r + B \tag{5–3}$$

Spherical Polar Coordinates: Assuming that V depends only on r, we find

$$\frac{1}{r^2}\frac{d}{dr}\left(r^2\frac{dV}{dr}\right) = 0 \quad \Rightarrow \quad V = \frac{A}{r} + B \tag{5–4}$$

In each of these cases, boundary conditions must be applied to find the arbitrary constants A and B. It is worth noting that for $A \neq 0$, the one-dimensional solution in cylindrical coordinates diverges as $r \to \infty$ or $r \to 0$.

EXAMPLE 5.1: Find the potential between two concentric conducting spheres of radius a and b at potential V_a and V_b, respectively.

Solution: The general solution $V = \dfrac{A}{r} + B$ gives, at the boundaries,

$$\frac{A}{a} + B = V_a \qquad \frac{A}{b} + B = V_b \tag{Ex 5.1.1}$$

Solving the two equations simultaneously gives

$$A = \frac{ab(V_a - V_b)}{b - a} \quad \text{and} \quad B = \frac{aV_a - bV_b}{a - b} \tag{Ex 5.1.2}$$

5.2 $\nabla^2 V = 0$ in Two Dimensions

In this section we will explore a number of techniques for solving Laplace's equation in two dimensions. We include as two dimensional those cylindrical problems in three dimensions where there is no variation with z as well as those spherical problems where there is no dependence on the azimuthal angle. Since only the boundary conditions separate one problem from another, considerable attention will be given to the acquisition and application of boundary conditions.

One of the simplest techniques of reducing a partial differential equation in several variables to a set of ordinary differential equations is the separation of variables. In separating variables, we look for solutions V to Laplace's equation that can be expressed as a product of functions, each depending on only one of the coordinates, and obtain a differential equation for each of the "basis" functions. If the set of basis functions in each variable is *complete* in the space of that variable (in the sense that any reasonably behaved function in that space can be expanded in terms of the basis functions), then all solutions to Laplace's equation can be expressed as a linear combination of the products of the basis functions. The particular linear combination that is *the* solution for the potential V must be chosen to satisfy the boundary conditions. The separation and fitting to boundary conditions will be explicitly carried out in various coordinate systems.

We will also explore the use of conformal mappings in the solution of two-dimensional problems. Finally, a numerical approach, easily extended to three-dimensional problems, is briefly discussed.

5.2.1 Cartesian Coordinates in Two Dimensions

Let us consider the problem of finding the potential within a box of dimensions $a \times b$, extending infinitely in the z direction with the potential Φ_1, Φ_2, Φ_3 and Φ_4 on its four faces labelled 1, 2, 3, and 4. To solve $\nabla^2 V = 0$ we place the cart before the horse and begin by simplifying the boundary conditions before effecting the separation of variables. To simplify the boundary conditions we proceed as follows.

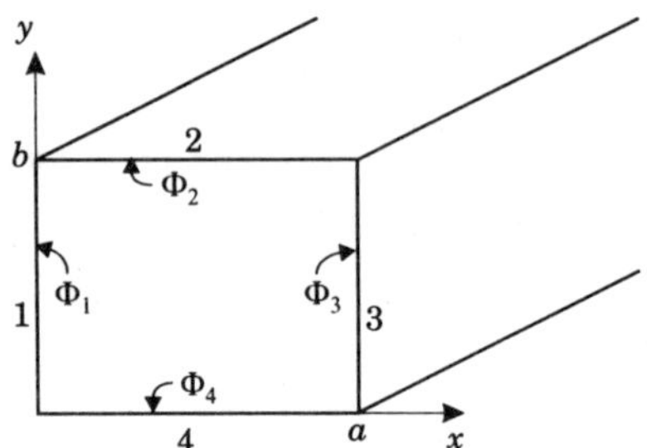

Figure 5.1: The rectangular box has the potential Φ as indicated along each of its sides.

(a) We temporarily set $\Phi_1 = \Phi_2 = \Phi_3 = 0$, keeping only $\Phi_4(x)$ as specified in the problem posed.

(b) We solve $\nabla^2 V_4(x, y) = 0$ subject to the boundary condition as simplified in (a).

(c) We repeat the procedure with $\Phi_1 = \Phi_2 = \Phi_4 = 0$, keeping only $\Phi_3(y)$ as specified, and solve for the corresponding V_3. We continue in similar fashion to obtain V_1 and V_2.

(d) The sum $V = V_1 + V_2 + V_3 + V_4$ satisfies $\nabla^2 V = 0$ and evidently also satisfies the boundary conditions.

To solve $\nabla^2 V_4 = 0$, we assume that V_4 can be factored into a product, $V_4(x, y) = \mathrm{X}(x)\mathrm{Y}(y)$ (a linear combination of such solutions will be required to satisfy the boundary conditions). Then

$$\frac{\nabla^2 V_4}{V_4} = \frac{\dfrac{\partial^2 V_4}{\partial x^2} + \dfrac{\partial^2 V_4}{\partial y^2}}{V_4} = \frac{\mathrm{Y}\dfrac{\partial^2 \mathrm{X}}{\partial x^2}}{\mathrm{XY}} + \frac{\mathrm{X}\dfrac{\partial^2 \mathrm{Y}}{\partial y^2}}{\mathrm{XY}} = 0 \tag{5-5}$$

or

$$\frac{1}{\mathrm{X}}\frac{\partial^2 \mathrm{X}}{\partial x^2} = -\frac{1}{\mathrm{Y}}\frac{\partial^2 \mathrm{Y}}{\partial y^2} \tag{5–6}$$

On the left of (5–6) we have a function depending on x only, which, according to the right side, is independent of x; therefore it must be constant. We can take this constant to be either positive or negative. Anticipating the requirements of the boundary condition, we take he separation contant to be negative and call it $-\lambda^2$ resulting in the equations,

$$\frac{1}{\mathrm{X}}\frac{\partial^2 \mathrm{X}}{\partial x^2} = -\lambda^2 \quad \text{and} \quad \frac{1}{\mathrm{Y}}\frac{\partial^2 \mathrm{Y}}{\partial y^2} = \lambda^2 \tag{5–7}$$

With this choice, we readily obtain the solution

$$\begin{aligned} \mathrm{X} &= A\cos\lambda x + B\sin\lambda x \\ \mathrm{Y} &= C\cosh\lambda y + D\sinh\lambda y \end{aligned} \tag{5–8}$$

Had we made the alternate choice for the sign of the separation constant $(+\lambda^2)$, we would have obtained solution

$$\begin{aligned} \mathrm{X} &= A\cosh\lambda x + B\sinh\lambda x \\ \mathrm{Y} &= C\cos\lambda y + D\sin\lambda y \end{aligned} \tag{5–9}$$

It is of course equally valid to write the cosh and the sinh solutions in terms of e^x and e^{-x}. The benefit of using cosh and sinh is that they have convenient values of 1 and 0 respectively at $x = 0$. Applying the boundary conditions, we note that V_4 is zero at boundary 1 and 3, so that the solution X must have two zeros. The cosine and sine readily yield any number of zeros as they are periodic. A nontrivial linear combination of cosh and sinh, on the other hand, has only one root. To illustrate, suppose we attempt to satisfy $\mathrm{X} = A\cosh\lambda x + B\sinh\lambda x = 0$ at $x = a$. Clearly $A = -B\tanh\lambda a$ is such a solution. Now applying this solution to evaluate X at 0, $A\cosh 0 + B\sinh 0 = A$. Requiring the solution to vanish at 0 now requires A to vanish meaning only the trivial solution $\mathrm{X} = 0 \Rightarrow V = 0$ can satisfy the boundary condition. Proceeding with the non-trivial solution, the boundary condition at $x = 0$ is

$$V_4(0, y) = A\big(C\cosh\lambda y + D\sinh\lambda y\big) = \Phi_1(y) = 0 \tag{5–10}$$

We conclude that either $C = D = 0$, or $A = 0$. To avoid a trivial solution, we pick $A = 0$. At the opposite boundary $(x = a)$,

$$V_4(a, y) = B\sin\lambda a\big(C\cosh\lambda y + D\sinh\lambda y\big) = \Phi_3(y) = 0 \tag{5–11}$$

This time, to avoid a trivial solution ($B = 0$ or $C = D = 0$), we must pick λ so that $\sin\lambda a$ vanishes; (in other words, $\lambda = n\pi/a$ with n an integer).

Along the side labelled 2 ($y = b$) in Figure 5.1, V_4 must also vanish, meaning that

$$V_4(x, b) = B\sin\frac{n\pi x}{a}\left(C\cosh\frac{n\pi b}{a} + D\sinh\frac{n\pi b}{a}\right) = \Phi_2(x) = 0 \tag{5–12}$$

which implies $C = -D\tanh\frac{n\pi b}{a}$.

Finally fitting the last boundary, $y = 0$, $V_4 = BC\sin(n\pi x/a)$. Clearly a single solution cannot fit an arbitrary $\Phi_4(x)$; we note, however, that n is undetermined other than being an integer. We therefore propose a linear combination of solutions $V_{4,n}$ that satisfies both the boundary conditions and Laplace's equation as the solution for V_4. Setting B to unity to avoid unnecessary arbitrary constants, we have

$$V_4(x,0) = \Phi_4(x) = \sum_{n=1}^{\infty} C_n \sin\frac{n\pi x}{a} \tag{5–13}$$

We immediately recognize the equation above as the Fourier expansion of $\Phi_4(x)$, allowing us to write

$$C_n = \frac{2}{a}\int_0^a \Phi_4(x)\sin\frac{n\pi x}{a}dx \tag{5–14}$$

Gathering terms, we have finally

$$V_4(x,y) = \sum_1^{\infty} C_n \sin\frac{n\pi x}{a}\left(\cosh\frac{n\pi y}{a} - \coth\frac{n\pi b}{a}\sinh\frac{n\pi y}{a}\right) \tag{5–15}$$

Repeating the procedure for each of the other sides taken nonzero one at a time yields V_1, V_2, and V_3 for the complete solution to the problem. It is noteworthy that although the solution of Laplace's equation is particularly simple in Cartesian coordinates, the fitting of boundary conditions is somewhat cumbersome compared to what we will experience when using other coordinate systems.

5.2.2 Plane Polar Coordinates

In plane polar coordinates, $\nabla^2 V$ assumes the form

$$\frac{1}{r}\frac{\partial}{\partial r}\left(r\frac{\partial V}{\partial r}\right) + \frac{1}{r^2}\frac{\partial^2 V}{\partial \varphi^2} = 0 \tag{5–16}$$

We assume a separable solution and try V of the form $V = \mathrm{R}(r)\Phi(\varphi)$ with R independent of φ and Φ independent of r. Substituting this form into (5–16), we get

$$\frac{\dfrac{1}{r}\dfrac{\partial}{\partial r}\left(r\dfrac{\partial \mathrm{R}}{\partial r}\right)}{\mathrm{R}} + \frac{\dfrac{1}{r^2}\dfrac{\partial^2\Phi}{\partial\varphi^2}}{\Phi} = 0 \tag{5–17}$$

or

$$\frac{r\dfrac{\partial}{\partial r}\left(r\dfrac{\partial \mathrm{R}}{\partial r}\right)}{\mathrm{R}} = -\frac{\dfrac{\partial^2\Phi}{\partial\varphi^2}}{\Phi} = \text{constant} = m^2 \tag{5–18}$$

where we have chosen a positive separation constant, m^2, in anticipation of a requirement by the boundary condition on Φ. The equation

$$\frac{\partial^2\Phi}{\partial\varphi^2} = -m^2\Phi \tag{5–19}$$

is easily solved to give

$$\Phi = A\cos m\varphi + B\sin m\varphi \tag{5–20}$$

The physical requirement that Φ be periodic with period 2π justifies our choice of a positive separation constant and further restricts m to integer values.

The remaining equation

$$r\frac{\partial}{\partial r}\left(r\frac{\partial \mathrm{R}}{\partial r}\right) = m^2\mathrm{R} \tag{5–21}$$

is easily solved. Clearly the equation suggests a monomial solution as differentiating r^ℓ lowers the exponent of r by one and multiplication by r promptly restores the original exponent. We try $\mathrm{R} = kr^\ell$, which, when substituted into (5–21), gives

$$\ell^2 kr^\ell = \ell^2\mathrm{R} = m^2\mathrm{R} \quad \Rightarrow \quad \ell = \pm m \tag{5–22}$$

We conclude that R takes the form $\mathrm{R} = Cr^m + Dr^{-m}$. The linear combination of the two linearly independent solutions is the general solution when $m \neq 0$. When $m = 0$, however, only one solution is obtained. To obtain the second, we integrate (5–21) directly with $m = 0$.

$$\frac{\partial}{\partial r}\left(r\frac{\partial \mathrm{R}}{\partial r}\right) = 0 \quad \Rightarrow \quad r\frac{\partial \mathrm{R}}{\partial r} = C \quad \Rightarrow \quad \frac{\partial \mathrm{R}}{\partial r} = \frac{C}{r} \tag{5–23}$$

Thus, when $m = 0$, $\mathrm{R} = C_0 \ln r + D_0$ is the solution.

Combining the two forms of the solution, we find the most general solution to $\nabla^2 V = 0$ in plane polar coordinates to be

$$V(r,\varphi) = C_0\ln r + D_0 + \sum_{m=1}^{\infty}\left(C_m r^m + \frac{D_m}{r^m}\right)(A_m\cos m\varphi + B_m \sin m\varphi) \tag{5–24}$$

It might be noted that there are more arbitrary constants in (5–24) than can be uniquely determined. The two trigonometric terms might, for instance, be added to give a single phase-shifted cosine at which point A_m and B_m become superfluous since the amplitude is already determined by C and D. In order to avoid a proliferation of constants, we will freely rename constants and product of constants as convenient.

Example 5.2: Find the potential when a neutral, long conducting circular cylinder is placed in an initially uniform electric field with its axis perpendicular to the field (Figure 5.2).

Solution: We choose the x axis along the initially uniform field so that at sufficiently large r, $\vec{E} = E_0\hat{\imath}$. To obtain this field at sufficiently large r, we must have $V \to -E_0 x = -E_0 r\cos\varphi$.

We now apply the boundary conditions to the general solution (5–24). Reflection symmetry about the x axis requires that $V(\varphi) = V(-\varphi)$, implying for $m \neq 0$, $B_m = 0$. The term $C_0 \neq 0$ implies a net charge on the cylinder. The terms $C_m r^m$

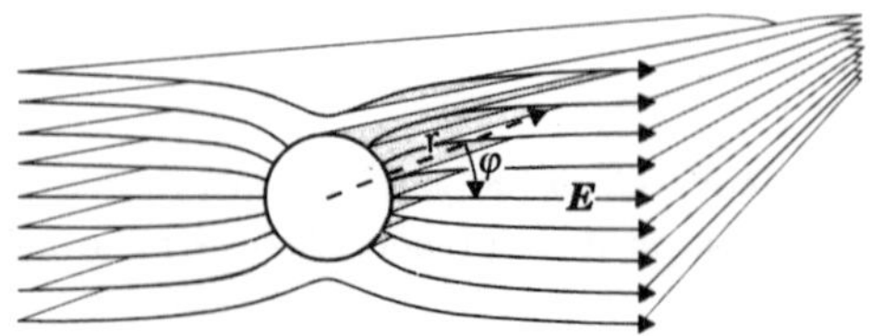

Figure 5.2: The coordinate system is chosen so that the axis of the cylinder lies along the z axis whereas the electric field at large distances is purely x directed.

all grow too rapidly as $r \to \infty$, requiring that $C_m = 0$ for $m \neq 1$. The general solution then reduces to

$$V(r, \varphi) = D_0 - E_0 r \cos\varphi + \sum_{m=1}^{\infty} \frac{D_m}{r^m} \cos m\varphi \qquad \text{(Ex 5.2.1)}$$

where the products of arbitrary constants, A_m and D_m, have been replaced by single constants, D_m.

On the surface of a conductor, the potential must be a constant, say V_0; therefore

$$V(a, \varphi) = V_0 = D_0 - E_0 a \cos\varphi + \sum_{m=1}^{\infty} \frac{D_m}{a^m} \cos m\varphi \qquad \text{(Ex 5.2.2)}$$

$$0 = (D_0 - V_0) + \left(\frac{D_1}{a} - E_0 a\right) \cos\varphi + \sum_{m=2}^{\infty} \frac{D_m}{a^m} \cos m\varphi \qquad \text{(Ex 5.2.3)}$$

As the $\{\cos m\varphi\}$ are linearly independent functions, each coefficient in the expansion must vanish (i.e., $D_0 = V_0$, $D_1 = E_0 a^2$, $D_m = 0$ for $m \geq 2$). The final result is then

$$V(r, \varphi) = V_0 - E_0 r \cos\varphi + \frac{E_0 a^2}{r} \cos\varphi \qquad \text{(Ex 5.2.4)}$$

It is straightforward to verify that (Ex 5.2.4) satisfies the boundary conditions $V = V_0$ at $r = a$ and $V \to -E_0 r \cos\varphi$ as $r \to \infty$.

More generally, the boundary conditions may be applied to two nested cylinders sharing a common axis. We sketch the procedure in the following example.

EXAMPLE 5.3: Two coaxial, nonconducting cylinders have surface charge densities $\sigma_a(\varphi)$ and $\sigma_b(\varphi)$ on the inner and outer cylinders, giving rise to potentials $V_a(\varphi)$ and $V_b(\varphi)$ on the two surfaces (Figure 5.3). Find the potential (a) for $r < a$, (b) for $r > b$, and (c) for $r \in (a, b)$.

Solution: We assume the general solution (5–24) and tailor it to the appropriate boundary conditions in each of the three regions.

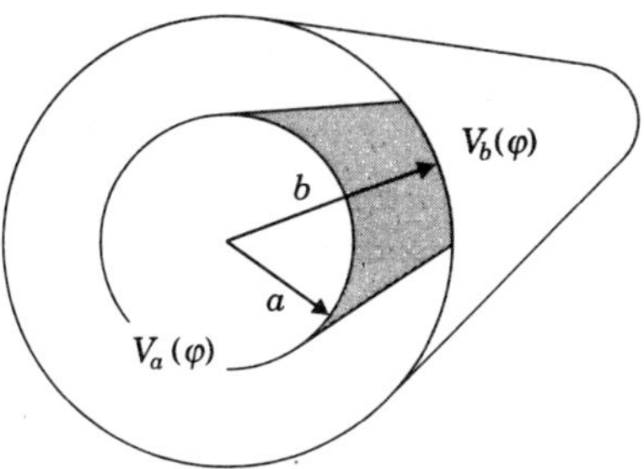

Figure 5.3: The two concentric cylinders of radius a and b are x at potential $V_a(\varphi)$ and $V_b(\varphi)$ respectively.

(a) For $r \le a$: terms in $\ln r$ or $1/r^m$ diverge at $r = 0$ and must be eliminated from the sum by setting their coefficients equal to 0. Hence

$$V = D_0 + \sum r^m (A_m \cos m\varphi + B_m \sin m\varphi) \tag{Ex 5.3.1}$$

At $r = a$, this specializes to

$$V_a(\varphi) = A_0 + \sum (a^m D_m \cos m\varphi + a^m B_m \sin m\varphi) \tag{Ex 5.3.2}$$

We recognize the righthand side as the Fourier expansion of $V_a(\varphi)$. The expansion coefficients are

$$A_0 = \frac{1}{2\pi}\int_0^{2\pi} V_a(\varphi)d\varphi \tag{Ex 5.3.3}$$

$$a^m A_m = \frac{1}{\pi}\int_0^{2\pi} V_a(\varphi)\cos m\varphi d\varphi \tag{Ex 5.3.4}$$

$$a^m B_m = \frac{1}{\pi}\int_0^{2\pi} V_a(\varphi)\sin m\varphi d\varphi \tag{Ex 5.3.5}$$

(b) For $r \ge b$: This time the terms in r^m diverge at ∞ ($\ln r$ also diverges, but such a term would be required by a nonzero net charge on the cylinders). Eliminating all the divergent terms except $\ln r$, we write the expansion of the potential as

$$V(r \ge b) = C_0 \ln r + D_0 + \sum_{m=1}^{\infty} \frac{1}{r^m}(A_m \cos m\varphi + B_m \sin m\varphi) \tag{Ex 5.3.6}$$

Matching the boundary condition at b, we have

$$V_b(\varphi) = (C_0 \ln b + D_0) + \sum_{m=1}^{\infty} \frac{1}{b^m}(A_m \cos m\varphi + B_m \sin m\varphi) \tag{Ex 5.3.7}$$

and we again identify the coefficients of the trigonometric terms with the coefficients of the Fourier series of $V_b(\varphi)$ to obtain

$$C_0 \ln b + D_0 = \frac{1}{2\pi} \int_0^{2\pi} V_b(\varphi) d\varphi \tag{Ex 5.3.8}$$

$$A_m = \frac{b^m}{\pi} \int_0^{2\pi} V_b(\varphi) \cos m\varphi d\varphi \tag{Ex 5.3.9}$$

$$B_m = \frac{b^m}{\pi} \int_0^{2\pi} V_b(\varphi) \sin m\varphi d\varphi \tag{Ex 5.3.10}$$

It is not possible to tell from V_b alone whether there is a net charge on the cylinders, hence we cannot distinguish between C_0 and B_0 without further information.

(c) For $a \le r \le b$: There is now no reason to eliminate any of the terms from (5–24). Fortunately we have twice as many boundary conditions; one at a and one at b. Equating the potential (5–24) to the values on the boundaries, we have

$$V_a = C_0 \ln a + D_0 + \sum_{m=1}^{\infty} \left(C_m a^m + \frac{D_m}{a^m} \right) (A_m \cos m\varphi + B_m \sin m\varphi) \tag{Ex 5.3.11}$$

and

$$V_b = C_0 \ln b + D_0 + \sum_{m=1}^{\infty} \left(C_m b^m + \frac{D_m}{b^m} \right) (A_m \cos m\varphi + B_m \sin m\varphi) \tag{Ex 5.3.12}$$

leading for $m = 0$ to

$$C_0 \ln a + D_0 = \frac{1}{2\pi} \int_0^{2\pi} V_a(\varphi) d\varphi \quad C_0 \ln b + D_0 = \frac{1}{2\pi} \int_0^{2\pi} V_b(\varphi) d\varphi \tag{Ex 5.3.13}$$

and for $m \neq 0$ we get

$$\left(C_m a^m + \frac{D_m}{a^m} \right) \begin{pmatrix} A_m \\ B_m \end{pmatrix} = \frac{1}{\pi} \int_0^{2\pi} V_a(\varphi) \begin{pmatrix} \cos m\varphi \\ \sin m\varphi \end{pmatrix} d\varphi \tag{Ex 5.3.14}$$

$$\left(C_m b^m + \frac{D_m}{b^m} \right) \begin{pmatrix} A_m \\ B_m \end{pmatrix} = \frac{1}{\pi} \int_0^{2\pi} V_b(\varphi) \begin{pmatrix} \cos m\varphi \\ \sin m\varphi \end{pmatrix} d\varphi \tag{Ex 5.3.15}$$

These equations may now be solved (two at a time) to obtain A_m, B_m, C_m, and D_m for a complete solution. For example, equations (Ex 5.3.13) yield

$$C_0 \ln \frac{b}{a} = \frac{1}{2\pi} \int_0^{2\pi} [V_b(\varphi) - V_a(\varphi)] \, d\varphi \tag{Ex 5.3.16}$$

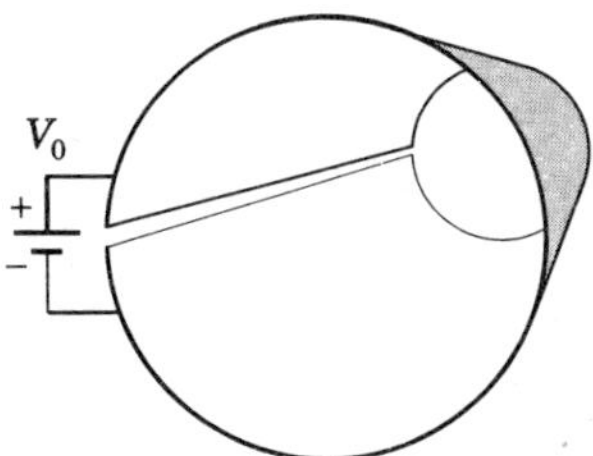

Figure 5.4: A current flowing around a resistive cut cylinder establishes a potential that varies linearly with φ around the cylinder.

and

$$D_0 \ln\frac{a}{b} = \frac{1}{2\pi}\int_0^{2\pi} [V_b(\varphi)\ln a - V_a(\varphi)\ln b].d\varphi \qquad \text{(Ex 5.3.17)}$$

EXAMPLE 5.4: A resistive cylinder of radius a with a narrow longitudinal gap at $\varphi = \pi$ carries a current in the azimuthal direction, giving rise to a linearly varying potential $V(a,\varphi) = V_0\varphi/2\pi$ for $-\pi < \varphi < \pi$ (Figure 5.4). Find the potential inside the cylinder.

Solution: Interior to the cylinder, the solution is of the form (we have redefined the constants A_ℓ and B_ℓ slightly in order to write r in normalized form):

$$V(r<a) = \sum_\ell \left(\frac{r}{a}\right)^\ell (A_\ell \cos \ell\varphi + \beta_\ell \cos\, \ell\varphi) \qquad \text{(Ex 5.4.1)}$$

giving, at $r = a$,

$$V(a,\varphi) = \sum A_\ell \cos \ell\varphi + B_\ell \sin \ell\varphi = \frac{V_0}{2\pi}\,\varphi \qquad \text{(Ex 5.4.2)}$$

We obtain the coefficients A_ℓ and B_ℓ in the usual fashion; multiplying both sides of the equation by $\cos m\varphi$ and integrating from $-\pi$ to $+\pi$, we get

$$\int_{-\pi}^{\pi} (A_\ell \cos \ell\varphi + B_\ell \sin \ell\varphi)\cos m\varphi\, d\varphi = \frac{V_0}{2\pi}\int_{-\pi}^{\pi} \varphi \cos m\varphi\, d\varphi \qquad \text{(Ex 5.4.3)}$$

or $A_m = 0$, which could have been predicted from the symmetry of V. Multiplying by $\sin\varphi$ and integrating, we have

$$\int_{-\pi}^{\pi} B_\ell \sin \ell\varphi \sin m\varphi\, d\varphi = \frac{V_0}{2\pi}\int_{-\pi}^{\pi} \varphi \sin m\varphi\, d\varphi \qquad \text{(Ex 5.4.4)}$$

or

$$\pi B_m = \frac{V_0}{2\pi}\int_{-\pi}^{\pi} \varphi \sin m\varphi\, d\varphi = -\frac{V_0 \cos m\pi}{m} = \frac{(-1)^{m+1}V_0}{m} \qquad \text{(Ex 5.4.5)}$$

The potential interior to the cylinder is then

$$V(r,\varphi) = V_0 \sum_{\ell=1}^{\infty} \frac{(-1)^{\ell+1}}{\ell\pi} \left(\frac{r}{a}\right)^{\ell} \sin \ell\varphi \tag{Ex 5.4.6}$$

Frequently, even when there are free charges or currents on some surface, it is feasible to solve for the potentials inside and outside the surface, and to convert the presence of charges or currents into a boundary condition that relates the difference between the inside and outside solutions to the source term.

EXAMPLE 5.5: A hollow cylindrical shell bearing no net charge of radius a has a surface charge $\sigma = \sigma_0 \cos\varphi$ distributed on it. Find the potential both inside and outside the shell.

Solution: The interior and exterior potentials are given by

$$V(r < a,\varphi) = \sum \left(\frac{r}{a}\right)^{\ell} (A_\ell \cos \ell\varphi + B_\ell \sin \ell\varphi) \tag{Ex 5.5.1}$$

and

$$V(r > a,\varphi) = \sum \left(\frac{a}{r}\right)^{\ell} (C_\ell \cos \ell\varphi + D_\ell \sin \ell\varphi) \tag{Ex 5.5.2}$$

The boundary conditions can be obtained from Maxwell's electric field equations, $\vec{\nabla} \times \vec{E} = 0$ and $\vec{\nabla} \cdot \vec{E} = \rho/\varepsilon_0$, as follows.

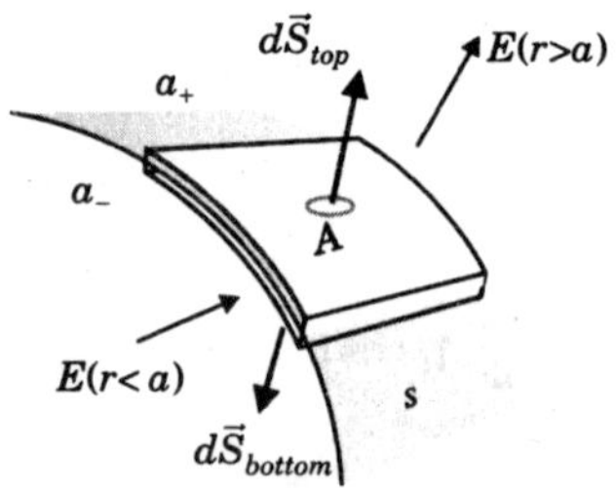

Figure 5.5: A thin box enclosing a small segment of the cylindrical surface.

(a) We integrate $\vec{\nabla} \cdot \vec{E}$ over the thin cylindrical shell segment of Figure 5.5 that encloses an area A of the cylinder:

$$\int \vec{\nabla} \cdot \vec{E} d^3r = \int \frac{\rho}{\varepsilon_0} d^3r = \frac{\bar{\sigma}A}{\varepsilon_0} \tag{Ex 5.5.3}$$

where $\bar{\sigma}$ is the average charge density on the surface A. With the aid of the divergence theorem, we recast this as

$$\oint \vec{E} \cdot d\vec{S} = \frac{\bar{\sigma}A}{\varepsilon_0} \tag{Ex 5.5.4}$$

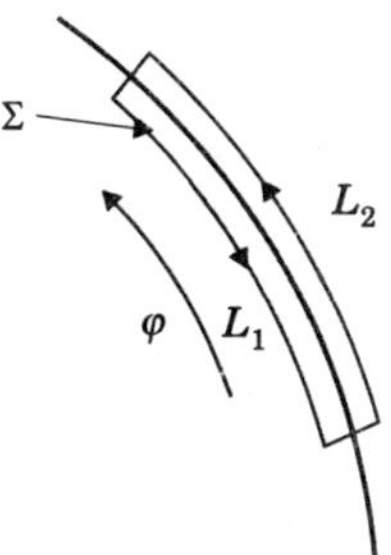

Figure 5.6: The loop encloses a section of the cylinder on a plane perpendicular to the cylinder.

We decompose the integral in terms of integrals along the top, the bottom, and the thin side

$$\oint \vec{E} \cdot d\vec{S} = \int_{\text{bottom}} -E_r dS + \int_{\text{top}} E_r dS + \int_{\text{sides}} \vec{E} \cdot d\vec{S} \qquad \text{(Ex 5.5.5)}$$

As we diminish the thickness of the box, the integral over the sides vanishes, leaving only $\bar{E}_r(a_+)A - \bar{E}_r(a_-)A = \bar{\sigma}A/\varepsilon_0$. Now, letting the area tend to zero, $\bar{E}_r \to E_r$ and $\bar{\sigma} \to \sigma$. Thus $E_r(a_+) = E_r(a_-) + \sigma/\varepsilon_0$.

(b) The second of Maxwell's $\vec{E}$ equations, $\vec{\nabla} \times \vec{E} = 0$, integrated over the area of the thin loop with long sides straddling the cylinder as shown in Figure 5.6, gives, using Stokes' theorem (18)

$$\begin{aligned}\int_\Sigma (\vec{\nabla} \times \vec{E}) \cdot d\vec{S} &= \oint \vec{E} \cdot d\vec{\ell} \\ &= \int_{L_1} -E_\varphi d\ell + \int_{L_2} E_\varphi d\ell + \int_{\substack{\text{short}\\ \text{sides}}} \vec{E} \cdot d\vec{\ell} \qquad \text{(Ex 5.5.6)}\end{aligned}$$

Shrinking the short sides to vanishingly small gives us

$$-\bar{E}_\varphi(a_-)L + \bar{E}_\varphi(a_+)L = 0 \qquad \text{(Ex 5.5.7)}$$

which, when $L \to 0$, becomes $E_\varphi(a_-) = E_\varphi(a_+)$.

To summarize, the boundary conditions are E_φ is continuous at a, or

$$-\frac{1}{a}\frac{\partial V(r,\varphi)}{\partial \varphi}\bigg|_{a_+} = -\frac{1}{a}\frac{\partial V(r,\varphi)}{\partial \varphi}\bigg|_{a_-} \qquad \text{(Ex 5.5.8)}$$

and E_r is discontinuous, increasing by σ/ε_0 as it crosses the surface out of the enclosed cylinder, or

$$-\frac{\partial V(r,\varphi)}{\partial r}\bigg|_{a_+} = -\frac{\partial V(r,\varphi)}{\partial r}\bigg|_{a_-} + \frac{\sigma}{\varepsilon_0} \qquad \text{(Ex 5.5.9)}$$

Applying the first of these conditions (Ex 5.5.8) to the series for V, we obtain

$$\sum (\ell A_\ell \sin \ell\varphi - \ell B_\ell \cos \ell\varphi) = \sum (\ell C_\ell \sin \ell\varphi - \ell D_\ell \cos \ell\varphi) \tag{Ex 5.5.10}$$

Grouping terms, we have

$$\sum \ell\,(A_\ell - C_\ell) \sin \ell\varphi + \sum \ell\,(B_\ell - D_\ell) \cos \ell\varphi = 0 \tag{Ex 5.5.11}$$

The linear independence of the trigonometric functions then requires that the coefficients of each term in the sum above equal zero; therefore $A_\ell = C_\ell$ and $B_\ell = D_\ell$.

The second boundary condition (Ex 5.5.9) applied to the series gives

$$\sum_\ell \frac{\ell}{a}\,(C_\ell \cos \ell\varphi + D_\ell \sin \ell\varphi) = -\sum \frac{\ell}{a}\,(A_\ell \cos \ell\varphi + B_\ell \sin \ell\varphi) + \frac{\sigma_0}{\varepsilon_0} \cos\varphi \tag{Ex 5.5.12}$$

We again equate the coefficient of each linearly independent function to zero, giving for the $\cos\varphi$ term

$$\left(C_1 + A_1 - \frac{\sigma_0 a}{\varepsilon_0}\right) \cos\varphi = 0 \tag{Ex 5.5.13}$$

while the remaining terms give $B_\ell + D_\ell = 0$ and $A_\ell + C_\ell = 0$. Together with the equations from (Ex 5.5.11) these latter imply that except for A_1 and C_1, the coefficients vanish. Substituting $A_1 = C_1$ in the equation above yields $A_1 = C_1 = \sigma_0 a/2\varepsilon_0$. Finally, then,

$$V(r < a) = \frac{\sigma_0 r}{2\varepsilon_0} \cos\varphi \tag{Ex 5.5.14}$$

and

$$V(r > a) = \frac{\sigma_0 a^2}{2\varepsilon_0 r} \cos\varphi \tag{Ex 5.5.15}$$

It might be noted that Exercise 5.5 is essentially a Neumann problem in that the derivatives of the potential are specified, albeit indirectly, on the boundary. It should be obvious that an arbitrary constant may be added to V without changing the fields.

5.2.3 Spherical Polar Coordinates with Axial Symmetry

When the potential is specified on a spherical boundary it is fairly clear that it would be advantageous to use spherical polar coordinates. It will develop, however, that even when there is only cylindrical symmetry in the problem it is still frequently advantageous to work in spherical polar coordinates. Laplace's equation in spherical polar coordinates when the solution is independent of φ reads as follows:

$$\frac{1}{r}\frac{\partial^2}{\partial r^2}(rV) + \frac{1}{r^2 \sin\theta}\frac{\partial}{\partial\theta}\left(\sin\theta \frac{\partial V}{\partial\theta}\right) = 0 \tag{5–25}$$

We assume a separable solution of the form $V = \mathrm{R}(r)\Theta(\theta)$, then $r^2\nabla^2 V/V = 0$ becomes

$$\frac{r^2}{r\mathrm{R}}\frac{d^2}{dr^2}(r\mathrm{R}) = -\frac{1}{\Theta\sin\theta}\frac{d}{d\theta}\left(\sin\theta\frac{d\Theta}{d\theta}\right) = \ell(\ell+1) \tag{5–26}$$

with no restrictions on ℓ yet. Our rather peculiar choice of writing the separation constant anticipates a well-behaved solution. The radial equation

$$r^2\frac{d^2}{dr^2}(r\mathrm{R}) = \ell(\ell+1)\,(r\mathrm{R}) \tag{5–27}$$

is readily solved to give $r\mathrm{R} = Ar^{\ell+1} + Br^{-\ell}$, whence we conclude

$$\mathrm{R}(r) = Ar^\ell + \frac{B}{r^{\ell+1}} \tag{5–28}$$

The remaining equation for Θ,

$$\frac{1}{\sin\theta}\frac{d}{d\theta}\left(\sin\theta\frac{d\Theta}{d\theta}\right) + \ell(\ell+1)\Theta = 0 \tag{5–29}$$

is may be converted to the Legendre equation (see Appendix F) by setting $x = \cos\theta \Rightarrow d/d\theta = -\sin\theta\cdot d/dx$, which results in

$$\frac{d}{dx}\left[(1-x^2)\frac{d\Theta}{dx}\right] + \ell(\ell+1)\Theta = 0 \tag{5–30}$$

The solutions to (5–30) are well known as the Legendre functions $\mathrm{P}_\ell(x)$ and $\mathrm{Q}_\ell(x)$. $\mathrm{Q}_\ell(\pm1) = \pm\infty$ and $\mathrm{P}_\ell(x)$ diverges as $x \to \pm1$ unless ℓ is an integer. The integral ℓ solutions $\mathrm{P}_\ell(x)$ are the Legendre Polynomials encountered earlier.

Recapitulating, we have found the well-behaved solutions to $\nabla^2 V = 0$ to be of the form

$$V(r,\theta) = \left(Ar^\ell + \frac{B}{r^{\ell+1}}\right)\mathrm{P}_\ell(\cos\theta) \tag{5–31}$$

and the general solution for problems including $\theta = 0$ and $\theta = \pi$ is

$$V(r,\theta) = \sum_{\ell=0}^{\infty}\left(A_\ell r^\ell + \frac{B_\ell}{r^{\ell+1}}\right)\mathrm{P}_\ell(\cos\theta) \tag{5–32}$$

Example 5.6: A grounded conducting sphere of radius a is placed in an initially uniform electric field $\vec{E} = E_0\hat{k}$. Find the resulting potential and electric field.

Solution: Besides the obvious boundary condition $V(a,\theta,\varphi) = 0$, we have also $V(r\to\infty) = -E_0 r\cos\theta$. In addition we expect the solution to be symmetric about the z axis . We therefore rewrite

$$V(r,\theta) = \sum_\ell\left(A_\ell r^\ell + \frac{B_\ell}{r^{\ell+1}}\right)\mathrm{P}_\ell(\cos\theta) \tag{Ex 5.6.1}$$

The boundary condition at ∞ implies that the potential can have no terms growing faster than r: $V(r) \to -E_0 r\cos\theta = A_1 r \mathrm{P}_1(\cos\theta)$ as $r \to \infty$ then gives $A_1 = -E_0$, and $A_\ell \neq 1 = 0$. With this restriction the solution at $r = a$ reduces to

$$0 = V(a,\theta) = -E_0 a\cos\theta + \sum_{\ell=0}^{\infty} \frac{B_\ell}{a^{\ell+1}} \mathrm{P}_\ell(\cos\theta) \tag{Ex 5.6.2}$$

$$= \frac{B_0}{a} + \left(\frac{B_1}{a^2} - E_0 a\right)\cos\theta + \sum_{\ell=2}^{\infty} \frac{B_\ell}{a^{\ell+1}} \mathrm{P}_\ell(\cos\theta) \tag{Ex 5.6.3}$$

As the Legendre polynomials are linearly independent functions, they can add to zero only if the coefficient of each vanishes. We conclude then that

$$\begin{aligned} \frac{B_0}{a} &= 0 &\Rightarrow\quad& B_0 = 0 \\ \frac{B_1}{a^2} - E_0 a &= 0 &\Rightarrow\quad& B_1 = E_0 a^3 \\ \frac{B_\ell}{a^{\ell+1}} &= 0 &\Rightarrow\quad& B_{\ell\neq 1} = 0 \end{aligned} \tag{Ex 5.6.4}$$

Recapitulating, we have found that the potential is

$$V(r,\theta) = \left(\frac{a^3}{r^2} - r\right) E_0 \cos\theta \tag{Ex 5.6.5}$$

from which we easily deduce the electric field $\vec{E} = -\vec{\nabla}V$:

$$\vec{E} = \left(1 + \frac{2a^3}{r^3}\right) E_0 \hat{r}\cos\theta + \left(\frac{a^3}{r^3} - 1\right) E_0 \hat{\theta}\sin\theta \tag{Ex 5.6.6}$$

It is worth noting that as $\hat{k} = \hat{r}\cos\theta - \hat{\theta}\sin\theta$,

$$\vec{E} = E_0\hat{k} + E_0\frac{a^3}{r^3}\left(3\hat{r}\cos\theta + \hat{\theta}\sin\theta - \hat{r}\cos\theta\right) \tag{Ex 5.6.7}$$

$$= E_0\hat{k} + \frac{E_0 a^3}{r^3}\left[3(\hat{r}\cdot\hat{k})\hat{r} - \hat{k}\right] \tag{Ex 5.6.8}$$

In other words, the field of the charge distribution on the sphere induced by the initially uniform external field $\vec{E}_0$ is just that of a dipole of strength $4\pi\varepsilon_0 a^3 E_0 \hat{k}$.

As a somewhat more powerful use of the spherical polar solution of Laplace's equation, we can use as boundary condition the known (by other means) potential along a symmetry axis. The solution thus obtained will give us the potential at any point in space.

Example 5.7: Obtain the magnetic scalar potential of a circular current loop of radius a at a point in the vicinity of the center ($r < a$).

Solution: The general solution for V_m that does not diverge as $r \to 0$ is

$$V_m(r,\theta) = \sum_\ell A_\ell r^\ell \mathrm{P}_\ell(\cos\theta) \tag{Ex 5.7.1}$$

For $\vec{r}$ along the z axis $\theta = 0$, allowing us to replace $\mathrm{P}_\ell(\cos\theta)$ by $\mathrm{P}_\ell(1) = 1$. Along the z axis, V_m becomes

$$V_m(z) = \sum_\ell A_\ell z^\ell \tag{Ex 5.7.2}$$

The magnetic scalar potential along the central axis of a circular current loop was found in (Ex 1.15.3) to be

$$V_m(z) = -\frac{I}{2}\frac{z}{\sqrt{z^2+a^2}} \tag{Ex 5.7.3}$$

which can be expanded by the binomial theorem as a power series in (z/a):

$$V_m = -\frac{Iz}{2a}\left(1+\frac{z^2}{a^2}\right)^{-1/2} \tag{Ex 5.7.4}$$

$$= -\frac{Iz}{2a}\left(1 - \frac{1}{2}\frac{z^2}{a^2} + \frac{3}{2}\frac{1}{2}\frac{1}{2!}\frac{z^4}{a^4} - \frac{5}{2}\frac{3}{2}\frac{1}{2}\frac{1}{3!}\frac{z^6}{a^6} + \dots\right) \tag{Ex 5.7.5}$$

$$= -\frac{I}{2a}z + \frac{I}{4a^3}z^3 - \frac{3I}{16a^5}z^5 + \frac{15}{96a^7}z^7 - \dots \tag{Ex 5.7.6}$$

A comparison of coefficients yields

$$A_1 = -\frac{I}{2a},\ A_3 = \frac{I}{4a^3},\ A_5 = -\frac{3I}{16a^5},\ A_7 = \frac{15I}{96a^7},\ \dots$$

The general result for the magnetic scalar potential of a circular current loop is then

$$V_m = -\frac{I}{2}\left[\frac{r}{a}\mathrm{P}_1(\cos\theta) - \frac{r^3}{2a^3}\mathrm{P}_3(\cos\theta) + \frac{3r^5}{8a^5}\mathrm{P}_5(\cos\theta) - \frac{15r^7}{48a^7}\mathrm{P}_7(\cos\theta) + \dots\right] \tag{Ex 5.7.7}$$

The magnetic induction field is now easily obtained as $\vec{B} = -\mu_0\vec{\nabla}V_m$.

The magnetic induction field of a plane coil is sufficiently important that we might try expressing the result above (Ex 5.8.5) in cylindrical coordinates in somewhat more elementary terms. Writing $\mathrm{P}_1(\cos\theta) = \cos\theta = z/r$, we expand

$$\mathrm{P}_3(\cos\theta) = \frac{1}{2}\left[5\left(\frac{z}{r}\right)^3 - 3\left(\frac{z}{r}\right)\right] \tag{5–33}$$

$$\mathrm{P}_5(\cos\theta) = \frac{1}{8}\left[63\left(\frac{z}{r}\right)^5 - 70\left(\frac{z}{r}\right)^3 + 15\left(\frac{z}{r}\right)\right] \tag{5–34}$$

Then, using the cylindrical coordinate ρ for distance from the axis, $r^2 = \rho^2 + z^2$ and we obtain

$$\frac{r}{a}P_1(\cos\theta) = \frac{r}{a}\cdot\frac{z}{r} = \frac{z}{a} \tag{5–35}$$

$$\left(\frac{r}{a}\right)^3 P_3(\cos\theta) = \frac{1}{2}\left(5\frac{z^3}{a^3} - 3\,\frac{r^2 z}{a^3}\right) = \frac{1}{2}\left(5\frac{z^3}{a^3} - \frac{3(z^2+\rho^2)z}{a^3}\right) = \frac{z^3}{a^3} - \frac{3}{2}\frac{\rho^2 z}{a^3} \tag{5–36}$$

$$\left(\frac{r}{a}\right)^5 P_5(\cos\theta) = \frac{1}{8}\left(63\frac{z^5}{a^5} - 70\,\frac{z^5}{a^5} - 70\frac{\rho^2 z^3}{a^5} + 15\frac{(\rho^2+z^2)^2 z}{a^5}\right)$$

$$= \frac{1}{8a^5}\left(8z^5 - 40z^3\rho^2 + 15z\rho^4\right) \tag{5–37}$$

Collecting terms, we obtain finally for $V_m(\rho, z)$

$$V_m(\rho, z) = -\frac{I}{2}\left[\frac{z}{a} - \frac{1}{2a^3}\left(z^3 - \frac{3}{2}\rho^2 z\right) + \frac{3}{64a^5}\left(8z^5 - 40\rho^2 z^3 + 15\rho^4 z\right)\cdots\right] \tag{5–38}$$

We see that centrally, the magnetic induction field takes the value $B_z = \mu_0 I/2a$, decreases quadratically along the z axis and increases quadratically as the wire is approached.

EXAMPLE 5.8: Obtain the magnetic scalar potential of a circular current loop of radius a at a point $r > a$.

Solution: The general expression for V_m that does not diverge as $r \to \infty$ is

$$V_m(r,\theta) = A_0 + \sum_\ell \frac{B_\ell}{r^{\ell+1}}P_\ell(\cos\theta) \tag{Ex 5.8.1}$$

giving along the z axis

$$V_m(z) = A_0 + \sum_\ell \frac{B_\ell}{z^{\ell+1}} \tag{Ex 5.8.2}$$

The previously found expression for the potential along the z axis can again be expanded in a power series, but this time in (a/z) to ensure convergence.

$$V_m(z) = -\frac{I}{2}\frac{z}{\sqrt{z^2+a^2}} = -\frac{I}{2}\left[1 + \left(\frac{a}{z}\right)^2\right]^{-\frac{1}{2}}$$

$$= -\frac{I}{2}\left[1 - \frac{1}{2}\left(\frac{a}{z}\right)^2 + \frac{1}{2}\frac{3}{2}\frac{1}{2!}\left(\frac{a}{z}\right)^4 - \frac{1}{2}\frac{3}{2}\frac{5}{2}\frac{1}{3!}\left(\frac{a}{z}\right)^6 + \cdots\right] \tag{Ex 5.8.3}$$

Comparing terms, we have $B_1 = \frac{1}{4}Ia^2$, $B_3 = -\frac{3}{16}Ia^4$, $B_6 = \frac{5}{32}Ia^6$, and so forth. The expansion for arbitrary $\vec{r}$, $r > a$, is then

$$V_m(r,\theta) = \frac{I}{2}\left[-1 + \frac{1}{2}\frac{a^2}{r^2}P_1(\cos\theta) - \frac{3}{8}\frac{a^4}{r^4}P_3(\cos\theta) + \frac{5}{16}\frac{a^6}{r^6}P_5(\cos\theta) + \cdots\right] \tag{Ex 5.8.4}$$

As a final example of the utility of this technique we compute the noncentral field of a pair of coils with parallel planes, separated by their radius. Such pairs of coils are known as Helmholtz coils and are useful for producing uniform magnetic induction fields over relatively large, open volumes. The spacing of the coils is exactly that required for the cubic terms of the scalar potential to cancel one-another.

EXAMPLE 5.9: Find the scalar magnetic potential and magnetic induction field at points near the axis of a pair of Helmholtz coils each of which has radius a and has one turn carrying current I.

Solution: The magnetic scalar potential of two coils centered at $z = \pm\frac{1}{2}a$ is easily written as

$$V_m(0,0,z) = -\frac{I}{2}\left(\frac{z-\frac{1}{2}a}{\sqrt{(z-\frac{1}{2}a)^2+a^2}} + \frac{z+\frac{1}{2}a}{\sqrt{(z+\frac{1}{2}a)^2+a^2}}\right) \tag{Ex 5.9.1}$$

We begin by expanding this axial potential as a power series in z/a. Some care is required in carrying all terms to order $(z/a)^6$. The result is

$$V_m(0,0,z) = \frac{-8I}{5^{3/2}}\left[\frac{z}{a} - \frac{144}{5^4}\left(\frac{z}{a}\right)^5 + \mathcal{O}\left(\frac{z}{a}\right)^7\right] \tag{Ex 5.9.2}$$

The general solution in spherical polar coordinates (assuming the obvious azimuthal symmetry) may be written

$$V_m(r,\theta) = \sum_\ell A_\ell\left(\frac{r}{a}\right)^\ell \mathrm{P}_\ell(\cos\theta) \tag{Ex 5.9.3}$$

for $r < a$. Along the z axis the general solution specializes to

$$V_m(z) = \sum_\ell A_\ell\left(\frac{z}{a}\right)^\ell \tag{Ex 5.9.4}$$

Comparison of series leads to $A_1 = -(8I/5)^{3/2}$, $A_2 = A_3 = A_4 = 0$, $A_5 = -(8I/5)^{3/2}(-144/5^4)$, $A_6 = 0, \cdots$; therefore

$$V_m(r,\theta) = \frac{-8I}{5^{3/2}}\left[\left(\frac{r}{a}\right)\mathrm{P}_1(\cos\theta) - \frac{144}{5^4}\left(\frac{r}{a}\right)^5\mathrm{P}_5(\cos\theta) + \cdots\right] \tag{Ex 5.9.5}$$

where $\mathrm{P}_5(x) = \frac{1}{8}(63x^5 - 70x^3 + 15x)$.

In cylindrical polar coordinates with $z = r\cos\theta$ and $r^2 = \rho^2 + z^2$, the scalar potential of the Helmholtz pair becomes

$$V_m(r,\theta) = -\frac{8I}{5^{3/2}}\left[\frac{z}{a} - \frac{144}{625}\left(\frac{z^5}{a^5} + \frac{15}{8}\frac{z\rho^4}{a^5} - 5\frac{z^3\rho^2}{a^5}\right) + \cdots\right] \tag{5–39}$$

The components of the magnetic induction field are now easily found, giving

$$B_z(\rho,z)=\frac{8\mu_0 I}{5^{3/2}a}\left\{1-\frac{144}{125}\left[\left(\frac{z}{a}\right)^4+\frac{3}{8}\left(\frac{\rho}{a}\right)^4-3\frac{z^2\rho^2}{a^4}\right]+\cdots\right\} \tag{5–40}$$

and

$$B_\rho(\rho,z)=\frac{8\mu_0 I}{5^{3/2}}\cdot\frac{144}{125}\left(\frac{4\rho z^3-3z\rho^3}{2a^5}\right)+\cdots \tag{5–41}$$

It will be noted that the central field is very homogeneous, a 10% of the radius displacement in any direction leads to only about a 10^{-4} deviation from the field's central value.

As mentioned before, when charges or currents reside entirely on the boundary, these may frequently be taken into account by the boundary conditions. In the next example we consider how we may account for a current on the bounding surface.

EXAMPLE 5.10: Obtain the magnetic induction field both inside and outside a uniformly charged rotating spherical shell.

Solution: The magnetic scalar potential will be of the form

$$V_m(\vec r)=\sum_{\ell=0}^{\infty}A_\ell r^\ell \mathrm{P}_\ell(\cos\theta)\quad \text{for } r<a \tag{Ex 5.10.1}$$

and

$$V_m(\vec r)=\sum_{\ell=0}^{\infty}\frac{B_\ell}{r^{\ell+1}}\mathrm{P}_\ell(\cos\theta)\quad \text{for } r>a \tag{Ex 5.10.2}$$

The boundary conditions are obtained by integrating Maxwell's equations at the surface of the sphere. Integrating $\vec\nabla\cdot\vec B=0$ over the thin volume of Figure 5.7, we have with the aid of the divergence theorem

$$0=\int_\tau \vec\nabla\cdot\vec B d^3r=\oint_S \vec B\cdot\vec S=\bar B_r^{ext}S_1-\bar B_r^{int}S_2+2\pi r\epsilon\langle\vec B_t\rangle \tag{Ex 5.10.3}$$

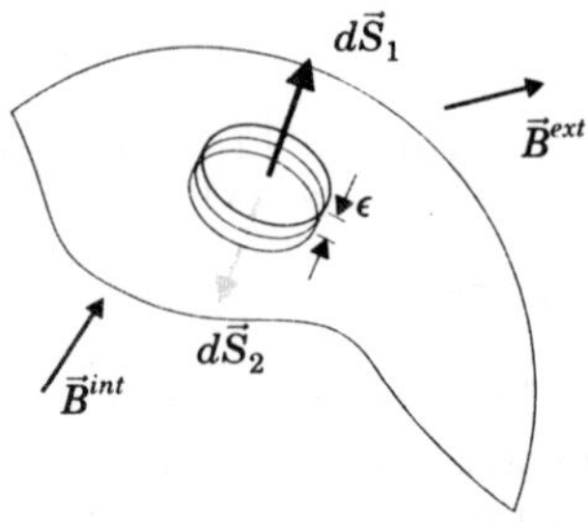

Figure 5.7: A thin pillbox shaped volume whose top and bottom surfaces are, respectively outside and inside the spherical boundary is used to obtain the boundary condition on B_r.

where $\langle \vec{B}_t \rangle$ is the mean value of the component of tangential field perpendicular to curved side of the pillbox. Letting the thickness ϵ approach zero, the last term vanishes, and S_1 approaches S_2. As the expression above is independent of the surface area S of the pillbox, we conclude $B_r^{ext} = B_r^{int}$. The perpendicular component of the induction field is continuous across the sphere.

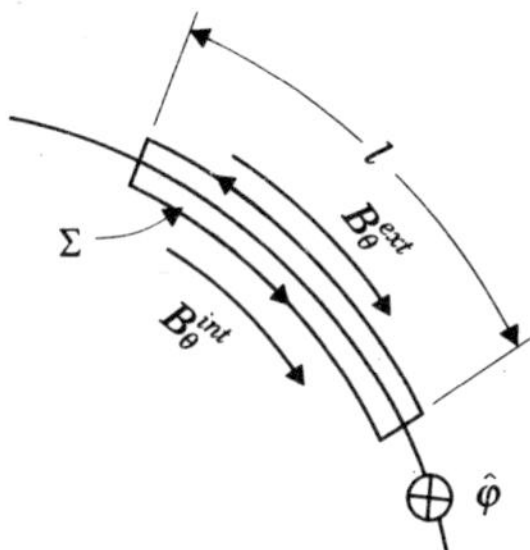

Figure 5.8: The loop lies in a plane perpendicular to the current and is traversed by current $l\sigma\omega\sin\theta$.

Integrating $\vec{\nabla} \times \vec{B} = \mu_0 \vec{J}$ over the area Σ illustrated in Figure 5.8 gives

$$\int_\Sigma (\vec{\nabla} \times \vec{B}) \cdot d\vec{S} = \int \mu_0 \vec{J} \cdot d\vec{S} \tag{Ex 5.10.4}$$

or, with the aid of Stokes' theorem ($d\vec{S} = -\hat{\varphi} dS$),

$$\oint \vec{B} \cdot d\vec{\ell} = -\mu_0 \int J_\varphi dS \tag{Ex 5.10.5}$$

Following the boundary in the direction indicated in Figure 5.16 we break the contour into four segments to get,

$$\oint \vec{B} \cdot d\vec{\ell} = l\bar{B}_\theta^{int} - l\bar{B}_\theta^{ext} + 2\epsilon \bar{B}_r \tag{Ex 5.10.6}$$

The bar indicates a mean value of the quantity under it. As we let ϵ tend to zero the last term vanishes. On the surface of the sphere, the current density is $\vec{J} = \sigma\delta(r-a)\vec{v}$ $= \sigma\delta(r-a)r\omega\sin\theta\hat{\varphi}$. The current passing through the thin rectangular area is then

$$\int \mu_0 \vec{J}_\varphi \cdot d\vec{S} = \mu_0 \int_{a-\epsilon/2}^{a+\epsilon/2} \int_{r\theta_0}^{r\theta_0 + l} \omega\sigma\delta(r-a) r \sin\theta \, dr d\ell$$

$$= \mu_0 \sigma a \omega l \sin\theta \tag{Ex 5.10.7}$$

Again, the relation must hold independent of l, meaning that

$$B_\theta^{ext} - B_\theta^{int} = \mu_0 \omega\sigma a \sin\theta \tag{Ex 5.10.8}$$

These boundary conditions are easily translated to boundary conditions on V_m. Since $\vec{B} = -\mu_0 \vec{\nabla} V_m$, these conditions reduce to

$$\left.\frac{\partial V_m}{\partial r}\right|_{r=a_+} = \left.\frac{\partial V_m}{\partial r}\right|_{r=a_-} \tag{Ex 5.10.9}$$

and

$$\left.\frac{1}{a}\frac{\partial V_m}{\partial \theta}\right|_{r=a_+} - \left.\frac{1}{a}\frac{\partial V_m}{\partial \theta}\right|_{r=a_-} = -\omega\sigma a \sin\theta \tag{Ex 5.10.10}$$

Using the general interior and exterior forms of the potential to evaluate the derivatives at a_- and a_+, respectively, we write for the latter equation (Ex 5.10.10)

$$\sum_{\ell=0}^{\infty}\left(\frac{B_\ell}{a^{\ell+2}} - A_\ell a^{\ell-1}\right)\mathrm{P}'_\ell(\cos\theta)\sin\theta = a\sigma\omega\sin\theta \tag{Ex 5.10.11}$$

which requires that

$$\frac{B_\ell}{a^{\ell+2}} = A_\ell a^{\ell-1} \tag{Ex 5.10.12}$$

except when $\mathrm{P}'_\ell(\cos\theta)$ is a constant in which case $\ell = 1$ and $P'_1 = 1$. Hence for $\ell = 1$, we obtain

$$\left(\frac{B_1}{a^3} - A_1\right) = a\sigma\omega \tag{Ex 5.10.13}$$

The boundary condition involving the radial derivatives (Ex 5.10.9) gives

$$-\frac{(\ell+1)B_\ell}{a^{\ell+2}}\mathrm{P}_\ell(\cos\theta) = A_\ell \ell a^{\ell-1}\mathrm{P}_\ell(\cos\theta) \tag{Ex 5.10.14}$$

which reduces to

$$-\frac{(\ell+1)B_\ell}{a^{\ell+2}} = A_\ell \ell a^{\ell-1} \tag{Ex 5.10.15}$$

Comparison of this term with the one above for which, for $\ell \neq 1$, shows that the two can be compatible only if ℓ equals $-(\ell+1)$. No integer solutions exist leading us to conclude that $A_\ell = B_\ell = 0$ for $\ell \neq 1$.

When $\ell = 1$, the relation (Ex 5.10.15) leads to

$$-\frac{2B_1}{a^3} = A_1 \tag{Ex 5.10.16}$$

Substituting for A_1 in the equation (Ex 5.10.13) above, we conclude

$$\left(\frac{B_1}{a^3} + \frac{2B_1}{a^3}\right) = a\sigma\omega \quad \Rightarrow \quad B_1 = \frac{a^4\sigma\omega}{3} \text{ and } A_1 = -\frac{2a\sigma\omega}{3} \tag{Ex 5.10.17}$$

The potentials are then

$$V_m(r > a) = \frac{a^4\sigma\omega}{3r^2}\cos\theta \tag{Ex 5.10.18}$$

and

$$V_m(r < a) = -\frac{2\sigma a\omega}{3} r\cos\theta = -\frac{2\sigma a\omega z}{3} \tag{Ex 5.10.19}$$

The magnetic induction field $\vec{B}$ is now readily evaluated as $\vec{B} = -\mu_0\vec{\nabla}V_m$,

$$\vec{B}(r < a) = \tfrac{2}{3}\mu_0 a\sigma\omega\hat{k} \tag{Ex 5.10.20}$$

$$B(r > a) = \tfrac{2}{3}\frac{\mu_0 a^4\sigma\omega}{r^3}\cos\theta\hat{r} + \tfrac{1}{3}\frac{\mu_0 a^4\sigma\omega}{r^3}\sin\theta\hat{\theta} \tag{Ex 5.10.21}$$

The magnetic induction field inside the sphere above is uniform throughout the entire sphere. While it is not practical to construct an apparatus inside a charged, rotating sphere, the same current density and hence the same magnetic field are achieved by a coil wound around the sphere with constant (angular) spacing between the windings.

5.2.4 Conformal Mappings

The idea behind conformal mappings is to take the solution to a very simple boundary condition problem and then to fold, stretch, or otherwise deform the boundary by a *conformal* mapping to match the boundary for a more complicated problem of interest. The same mapping that changes the boundary will also deform the field lines and constant potential lines of the simple problem to those of the more complicated problem. As the use of complex functions is central to the technique, we begin with a consideration of complex functions of a complex variable $z = x+iy$.

Generally, a complex function $f(z)$ may be written as the sum of a real part and an imaginary part, $f(z) = u(x, y) + iv(x, y)$. This means that for a point $x + iy$ in the (complex) z plane we can find a corresponding point $u + iv$ in the *image plane* (or, more briefly, the f plane) defined by f. The function f may be said to *map* the point $(x+iy)$ to $(u+iv)$. A series of points in the z plane will be mapped by f to a series of points in the f plane. If the function f is well behaved, adjacent points in the z plane are mapped to adjacent points in the f plane (i.e., a line in the z plane is mapped to a line in the f plane). It is such well-behaved complex functions or mappings that we will consider.

A function $f(z)$ is said to be *analytic* (the terms *regular* or *holomorphic* are also used) at a point z_0 if its derivative, defined by

$$\frac{df}{dz} = \lim_{\Delta z\to 0}\frac{f(z+\Delta z) - f(z)}{\Delta z} \tag{5–42}$$

exists (and has a unique value) in some neighborhood of z_0. A moment's reflection will show this to be a considerably stronger condition than the equivalent for functions of real variables because Δz can point in any direction in the z plane. The direction of taking this limit is immaterial for an analytic function. To investigate the consequence of this property, let us take the derivative first in the real direction and then in the imaginary direction.

Taking Δz along the real axis results in

$$\frac{df}{dz} = \lim_{\Delta x \to 0} \frac{u(x+\Delta x, y) - u(x,y) + i\big(v(x+\Delta x, y) - v(x,y)\big)}{\Delta x}$$
$$= \frac{\partial u}{\partial x} + i\frac{\partial v}{\partial x} \qquad (5\text{–}43)$$

whereas taking the derivative in the imaginary direction results in

$$\frac{df}{dz} = \lim_{\Delta y \to 0} \frac{u(x, y+\Delta y) - u(x,y) + i\big(v(x, y+\Delta y) - v(x,y)\big)}{i\Delta y}$$
$$= -i\frac{\partial u}{\partial y} + \frac{\partial v}{\partial y} \qquad (5\text{–}44)$$

Comparing the two expressions (5–41) and (5–42), we find that the real and imaginary parts of an analytic function f are not arbitrary but must be must be related by

$$\frac{\partial u}{\partial x} = \frac{\partial v}{\partial y} \quad \text{and} \quad \frac{\partial u}{\partial y} = -\frac{\partial v}{\partial x} \qquad (5\text{–}45)$$

These two equalities (5–45) are known as the *Cauchy-Riemann equations*. The validity of the Cauchy-Riemann equations is both a necessary and sufficient condition for f to be analytic.

To illustrate these ideas, we consider the function $f(z) = z^2 = (x+iy)^2 = (x^2 - y^2) + 2ixy$. Then $u = x^2 - y^2$ and $v = 2xy$. The derivatives are easily obtained

$$\frac{\partial u}{\partial x} = 2x, \quad \frac{\partial v}{\partial y} = 2x; \qquad \frac{\partial u}{\partial y} = -2y, \quad \frac{\partial v}{\partial x} = 2y$$

in accordance with the Cauchy-Riemann equations.

The Cauchy-Riemann equations can be differentiated to get

$$\frac{\partial^2 u}{\partial x^2} = \frac{\partial^2 v}{\partial x \partial y} \quad \text{and} \quad \frac{\partial^2 u}{\partial y^2} = -\frac{\partial^2 v}{\partial y \partial x} \qquad (5\text{–}46)$$

Since the order of differentiation should be immaterial, these can be added to give

$$\frac{\partial^2 u}{\partial x^2} + \frac{\partial^2 u}{\partial y^2} = \nabla^2 u = 0 \qquad (5\text{–}47)$$

In similar fashion

$$\frac{\partial^2 v}{\partial y^2} = \frac{\partial^2 u}{\partial y \partial x} \quad \text{and} \quad \frac{\partial^2 v}{\partial x^2} = -\frac{\partial^2 u}{\partial x \partial y} \quad \Rightarrow \quad \nabla^2 v = 0 \qquad (5\text{–}48)$$

The functions u and v each solve Laplace's equation. Any analytic function therefore supplies two solutions to Laplace's equation, suggesting that we might look for the solutions of static potential problems among analytic functions. The functions u and v are known as *conjugate harmonic functions*. It is easily verified that the curves $u =$ constant and $v =$ constant are perpendicular to one another, suggesting

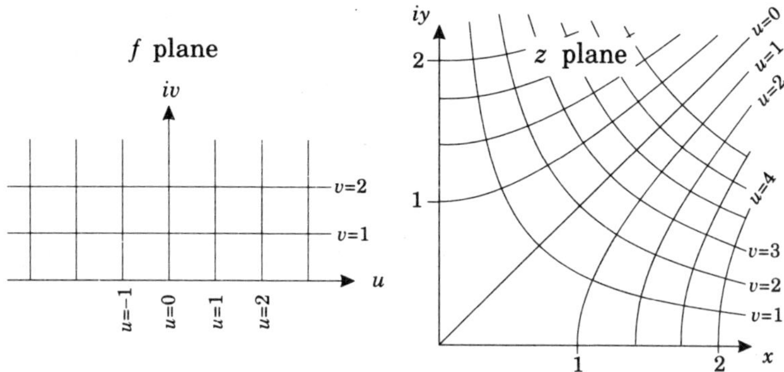

Figure 5.9: The image of the rectangular grid in the f plane under the mapping $z = \sqrt{f}$ is shown on the right in the z plane.

that if either u or v were to represent potential, then the other would represent electric field.

To return now to the notion of conformal mappings,[14] let us consider the mapping produced by the analytic function $f(z) = z^2$, (Figure 5.7). The inverse mapping, $z = \sqrt{f}$, is analytic everywhere except at $z = 0$. The image of the line $v = 0$ in the z plane is easily found: for $u > 0$, $z = \sqrt{u}$ produces an image line along the positive x axis, while for $u < 0$, $z = \sqrt{u} = i\sqrt{-u}$ produces an image line along the positive y axis of the z plane. The image of $v = 1$ is easily found from $(u + iv) = x^2 - y^2 + 2ixy$, implying that $2xy = 1$. Similarly, the line $u = c$ has image $x^2 - y^2 = c$.

Let us now consider the potential above an infinite flat conducting plate with potential $V = V_0$, lying along the u axis. Above the plate, the potential will be of the form $V = V_0 - av$. (A second plate at a different potential, parallel to the first would be required to determine the constant a. In a more general problem, V would be a function of both u and v. V may generally be considered to be the imaginary (or alternatively, real) part of an analytic function

$$\Phi(u, v) = U(u, v) + iV(u, v) = \Phi(f) \tag{5–49}$$

For this example, taking V to be the imaginary part of an analytic function Φ, $\Phi(u, v) = iV_0 - af = -au + i(V_0 - av)$. The corresponding electric field has components $E_v = a$ and $E_u = 0$.

[14]The mappings are called conformal because they preserve the angles between intersecting lines except when singular or having zero derivative. To see this, we consider two adjacent points z_0 and z on a line segment that makes angle φ with the x axis at z_0. Then $z - z_0$ may be written in polar form as $re^{i\varphi}$. The image of the segment $\Delta f = f(z) - f(z_0)$ can be expressed to first order as $f'(z_0)(z - z_0) = ae^{i\alpha}(z - z_0)$. If we write each term in polar form, Δf takes the form $Re^{i\theta} = ae^{i\alpha}re^{i\varphi}$. Thus the image line segment running from $f(z_0)$ to $f(z)$ makes angle $\theta = \alpha + \varphi$ with the u axis. This means that any line passing through z_0 is rotated through angle α in the mapping to the f plane providing that the derivative $f'(z_0)$ exists and that a unique polar angle α can be assigned to it. When f' is zero as it is for $f(z) = z^2$ at $z = 0$, conformality fails.

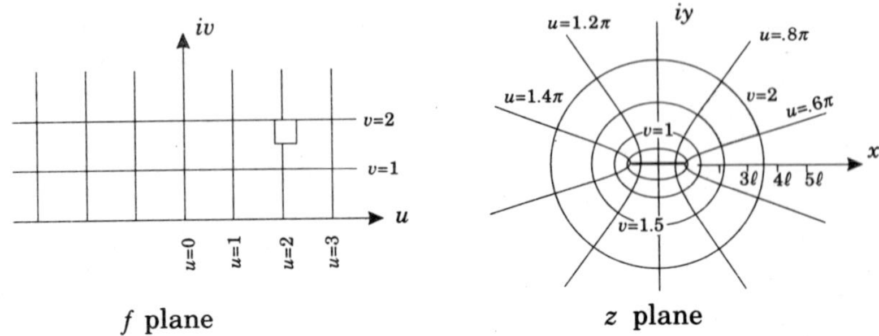

Figure 5.10: The figure on the left is the mapping of the rectangular grid on the right produced by the function $z = \ell \sin f$.

Now the mapping that takes the grid lines from the f plane to the z plane also takes Φ to the z plane. Expressed in terms of x and y

$$\Phi\left(f(z)\right) = iV_0 - az^2 = a(y^2 - x^2) + i(V_0 - 2axy) \tag{5–50}$$

Thus, in the z plane, $U(x,y) = a(y^2 - x^2)$ and $V(x,y) = V_0 - 2axy$. The latter is the potential produced by two conducting plates at potential V_0 intersecting at right angles. The field lines produced by v = constant in the u-v plane are now obtained from $u = y^2 - x^2$ = constant, while the constant potential lines are produced by $v = 2xy$ = constant. Figure 5.9 illustrates this mapping as it takes line segments and areas from the f plane to the z plane

A second example is offered by the mapping $z = \ell \sin f$, which maps the entire u axis onto a finite line segment of length 2ℓ in the z plane. We express z in terms of u and v by expanding $\sin f$ as

$$\sin f = \frac{e^{if} - e^{-if}}{2i} \tag{5–51}$$

to obtain

$$\begin{aligned}\sin f &= \frac{e^{i(u+iv)} - e^{-i(u+iv)}}{2i} = \frac{e^{-v}e^{iu} - e^{v}e^{-iu}}{2i}\\ &= \frac{e^{-v}\left(\cos u + i\sin u\right) - e^{v}\left(\cos u - i\sin u\right)}{2i}\\ &= i\sinh v \cdot \cos u + \cosh v \cdot \sin u\end{aligned} \tag{5–52}$$

Thus $x = \ell \cosh v \cdot \sin u$ and $y = \ell \sinh v \cdot \cos u$.

We again take for the potential above the $v = 0$ "plane" the imaginary part of $\Phi(u,v) = V_0 - af$. It is somewhat awkward, however, to find Φ in terms of x and y directly by substituting for f in the expression for Φ. Fortunately this is not really necessary. For a given point (u,v) in the f plane (where V is assumed to be known), the corresponding (x,y) value is readily found. In particular, the equipotentials at v = constant are easily obtained, for

$$\frac{x^2}{\ell^2 \cosh^2 v} + \frac{y^2}{\ell^2 \sinh^2 v} = 1 \tag{5–53}$$

In other words, the equipotentials around the flat strip are ellipses of major axis $\ell\cosh v$ and minor axis $\ell\sinh v$. At large distances $\cosh v$ and $\sinh v$ both tend to $\frac{1}{2}e^v$, meaning the equipotentials tend to circles. The potential and field lines are shown in Figure 5.10. (This problem can also be solved using separation of variables in elliptical coordinates.)

For a last example of somewhat more interest, let us consider the transformation

$$z = \frac{a}{2\pi}\left(1 + f + e^f\right) \tag{5–54}$$

We begin by considering some of the properties of this mapping. If f is real, then z is also real, and, as u varies from $-\infty$ to $+\infty$, x will take on values from $-\infty$ to $+\infty$. The real axis maps (albeit non-linearly) onto itself. Next consider the horizontal line $f = u + i\pi$. Substituting this into the expression for z gives

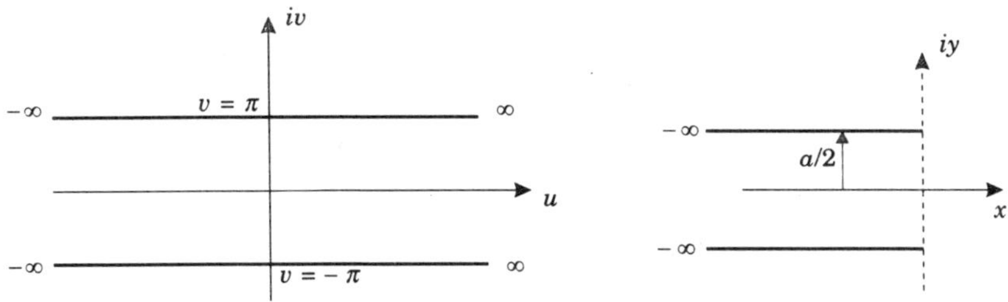

Figure 5.11: The infinite "planes" at $\pm i\pi$ in the f plane map to the semi-infinite "planes" in the z plane at $y = \pm a/2$ on the right.

$$z = \frac{a}{2\pi}\left(1 + u + i\pi + e^{u+i\pi}\right) = \frac{a}{2\pi}\left(1 + u + i\pi - e^u\right) \tag{5–55}$$

We conclude that

$$x = \frac{a}{2\pi}\left(1 + u - e^u\right) \quad \text{and} \quad y = \frac{a}{2} \tag{5–56}$$

For u large and negative, $x \simeq au/2\pi$ increases to 0 as u goes to zero. As u passes zero, e^u it becomes larger than $1 + u$ and x retraces its path at $\frac{1}{2}a$ above the x axis from 0 to $-\infty$. The line in the f plane folds back on itself at $u = 0$ on being mapped to the z plane. Similarly, the line at $v = -\pi$ maps to the half-infinite line at $y = -a/2$ with negative x. If we let the lines at $\pm i\pi$ in the f plane be the constant potential plates of the infinite parallel plate capacitor of Figure 5.9, the mapping carries this potential to that of the semi-infinite capacitor on the right. We can, with this mapping, find the fringing fields and potentials near the edge of a finite parallel plate capacitor. As we will see in the next section, we will also be able find the correction to the capacitance due to the fringing fields.

Letting the plates of the infinite capacitor in the f plane be at potential $\pm V_0/2$, with the top plate positive, the potential at any point in between the plates becomes $V = V_0 v/2\pi$. The analytic function of which V is the imaginary part is easily obtained as $\Phi(f) = V_0 f/2\pi$. Again it is somewhat awkward to substitute for f in terms of x and y. Instead, we separate the equation for z into real and imaginary components to get

$$x = \frac{a}{2\pi}\left(1 + u + e^u \cos v\right) \tag{5–57}$$

and

$$y = \frac{a}{2\pi}\left(v + e^u \sin v\right) \tag{5–58}$$

which we can evaluate parametrically by varying u to obtain the constant v (consequently also the constant $V = V_0 v/2\pi$) curves. Similarly, the electric field lines are obtained in the x-y plane by varying v. The field and potential lines are plotted in Figure 5.12.

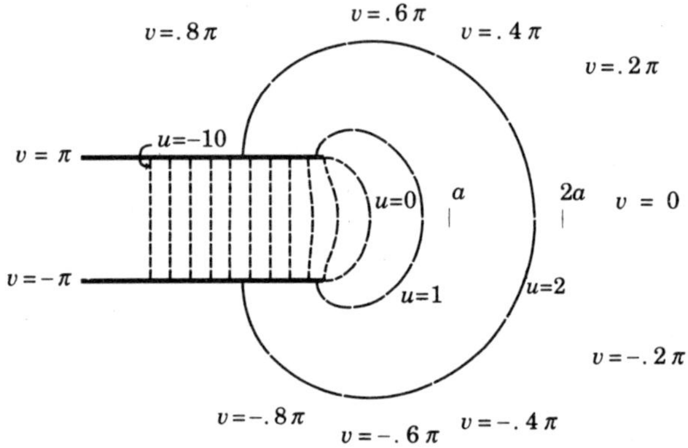

Figure 5.12: The equipotential and field lines near the edge of a semi-infinite plate capacitor.

Verifying that a given mapping does indeed map a particular surface onto another is fairly straightforward. It is not, however, obvious how mappings for given boundaries are to be obtained, other than perhaps by trial and error. For polygonal boundaries, a general method of constructing the mapping is offered by *Schwarz-Christoffel* transformations. For nonpolygonal boundaries, one must rely on "dictionaries" of mappings.[15]

5.2.5 Schwarz-Christoffel Transformations

In this segment we will briefly consider the mappings of the real axis in the f plane to a polygonal boundary not necessarily closed or continuous) in the z plane. The transformation that maps the upper half of the x plane to the "interior" of a polygon was given independently by Schwarz (also spelled Schwartz) and by Christoffel, the latter preceding Schwarz by two years. Nonetheless, the transformations are frequently referred to as simply "Schwarz transformations"

[15]For for a dictionary of *Schwarz-Christoffel* transformations, see for instance K.J. Binns and P.J. Lawrenson. (1973) *Analysis and Computation of Electric and Magnetic Field Problems, 2nd ed.* Pergamon Press, New York.

To begin, we note that a polygon is a figure made up of line segments, each of which has a constant slope, terminated by a vertex where the slope suddenly changes to a new value. We therefore consider a differential equation (an equation for the slope) relating f and z. In particular, consider the differential equation

$$\frac{dz}{df} = A\prod_{i=1}^{n}(f-u_i)^{\beta_i} \tag{5–59}$$

where the u_i are points on the real axis in the f plane. We see immediately that conformality fails at every point $f = u_i$. The real axis of the f plane will be mapped in continuous segments between these points, to distinct line segments in the z plane. Recall that a complex number may be represented in polar form, $Re^{i\theta}$. The *argument* of the number, θ, gives the inclination of a line joining the origin to z. The angle of the segments in z plane may be determined from the argument of the derivative (5–59) above. Using $\arg(f-u_i)^{\beta_i} = \arg(Re^{i\theta})^{\beta_i} = \arg(R^{\beta_i}e^{i\theta\beta_i}) = \beta_i\arg(e^{i\theta}) = \beta_i\arg(f-u_i)$, we may write

$$\arg\left(\frac{dz}{df}\right) = \arg A + \beta_1\arg(f-u_1) + \beta_2\arg(f-u_2) + \cdots + \beta_n\arg(f-u_n) \tag{5–60}$$

Since we were interested in mappings of the real axis, we restrict f to the real axis. When $u < u_1$, $f - u_i < 0$, implying each of the arguments equals π. As u increases to greater than u_1, the argument of $(f-u_1)$ vanishes. Further increases will make more and more of the terms vanish as u passes each of the roots u_i. With f on the real axis $df = du$ so that

$$\arg\left(\frac{dz}{df}\right) = \arg\left(\frac{dx+idy}{du}\right) = \arg\left(dx+idy\right) = \tan^{-1}\left(\frac{dy}{dx}\right) \tag{5–61}$$

In other words, the argument of the derivative is just the slope of the line segment at x. We abbreviate this slope in the interval $u \in (u_i, u_{i+1})$ as θ_i.

Let us evaluate this angle when u lies between u_i and u_{i+1}. u is larger than every point to the left, meaning the argument is zero and it is smaller than every point on the right, meaning the argument for all factors of the product with subscript $> i$ is π. Therefore

$$\theta_i = \arg A + \pi(\beta_{i+1} + \beta_{i+2} + \cdots + \beta_n) \tag{5–62}$$

We conclude that the image of the segment lying between u_i and u_{i+1} in the f plane is mapped to a segment in the z plane inclined at angle θ_i to the x axis. If we subtract θ_i from θ_{i+1} to find the difference between two successive segments' slopes, we obtain the bend at the point where the two segments meet,

$$\theta_{i+1} - \theta_i = -\pi\beta_{i+1} \tag{5–63}$$

We can relate this change in slope to the interior angle, α_{i+1} of the polygon at the point mapped from u_{i+1} as indicated in Figure 5.13.

$$\alpha_{i+1} = \pi + \pi\beta_{i+1} \tag{5–64}$$

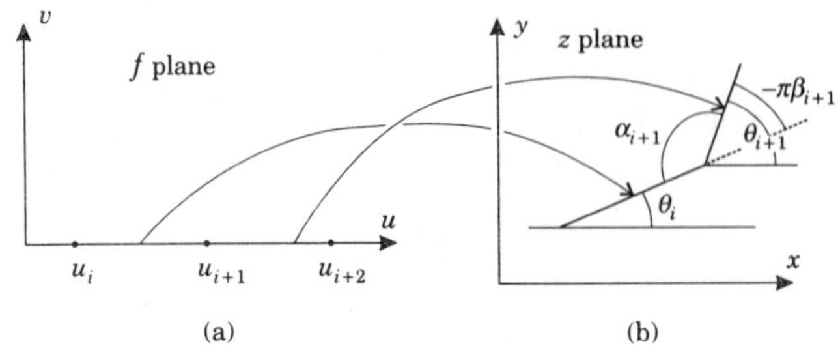

Figure 5.13: The angle α_{i+1} is complementary to $-\pi\beta_{i+1}$.

which we solve for β_{i+1} to give

$$\beta_{i+1} = \frac{\alpha_{i+1}}{\pi} - 1 \tag{5-65}$$

In terms of the interior angles, (5–59) becomes

$$\frac{dz}{df} = A\prod_{i=1}^{n}(f - u_i)^{\alpha_i/\pi-1} \tag{5-66}$$

The argument of the (complex) factor A will rotate the whole figure and the magnitude will scale it. The equation (5–66) is useful provided it can be integrated in terms of elementary functions which is frequently not possible. We will illustrate the procedure with several examples.

EXAMPLE 5.11: Use the Schwarz-Christoffel transformation to produce the 90° bent plane mapping of page 117.

Solution: We wish to produce the mapping that maps the real axis of the f plane to $+y$ axis for $u < 0$ and the $+x$ axis for $u > 0$. This means we want to produce a bend of $\pi/2$ at $u = 0$. The mapping is according to (5–66)

$$\frac{dz}{df} = A(f-0)^{1/2-1} = Af^{-1/2} \tag{Ex 5.11.1}$$

which is readily integrated to give

$$z = A'f^{1/2} + k \tag{Ex 5.11.2}$$

Choosing $k = 0$ and A' real gives the required mapping.

EXAMPLE 5.12: Find the mapping that maps the real axis onto the real axis and a line parallel to the real axis a distance a above it.

Solution: We choose as before, the point zero on the u-axis to produce the singular point, and this time as we want parallel lines we choose $\alpha = 0$. (5–66) then becomes

$$\frac{dz}{df} = A(f-0)^{0-1} = Af^{-1} \tag{Ex 5.12.1}$$

which we integrate to obtain

$$z = A \ln f + k \tag{Ex 5.12.2}$$

Again we can eliminate k without loss (it merely identifies $z = -\infty$ with $f = 0$). For positive u, we have $z = A \ln |u|$ whereas for negative u, we have $z = A(\ln |u| + i\pi)$. Thus the negative u axis maps to a line $A\pi$ above the x axis. Choosing $A = a/\pi$ gives the required mapping.

EXAMPLE 5.13: Find the mapping that carries the real axis onto the real axis with a gap of width $2a$ straddling the origin. (This would be a slotted plane in three dimensions.)

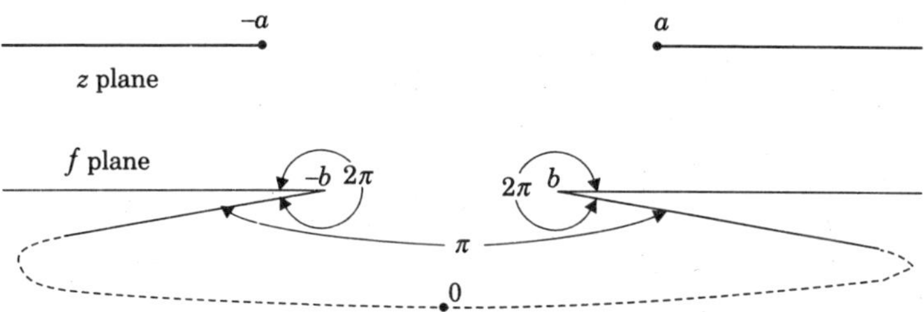

Figure 5.14: The plane with a slot can be formed by cutting the infinite plane at the center and folding the edges back at $\pm b$.

Solution: It is help to imagine how the real axis might be bent and cut to achieve this mapping. Figure 15.14 illustrates the operation. The real axis is cut at $u = 0$. At $u = \pm b$, the line is folded back, meaning the image of the segment of the u axis straddling zero has a $\pm\pi$ 'bend' while the image of $\pm b$ has a 2π bend. The prescription for the mapping then gives us

$$\frac{dz}{df} = A(f+b)(f-b)(f-0)^{-2} = A\frac{f^2 - b^2}{f^2} \tag{Ex 5.13.1}$$

Integrating (Ex 13.1) gives

$$z = A\left(f + \frac{b^2}{f}\right) + B \tag{Ex 5.13.2}$$

When $f = \pm b$ we wish $z = \pm a$ which gives $B = 0$ and $A = a/2b$ so that the required mapping is

$$z = \frac{a}{2}\left(\frac{f}{b} + \frac{b}{f}\right) \tag{Ex 5.13.3}$$

As a final example we will obtain the mapping of the top half of the semi-finite parallel plate capacitor.

EXAMPLE 5.14: Obtain a transformation to map the real axis onto the real axis and $z = id$ for $x < 0$.

Solution: For this mapping we need two vertices (the vertices at $\pm\infty$ are irrelevant), one to split the line and one to fold the upper 'electrode' back onto itself in order to have it terminate. Let the image of the vertex producing the 2π fold lie at a and the image of displacement vertex with 0 vertex angle at b on the real axis in the f plane. The mapping then obeys

$$\frac{dz}{df} = A(f - a)(f - b)^{-1} \tag{Ex 5.14.1}$$

Although we have three constants to determine, there is only one significant constant in the geometry, namely d. Therefore any two of the constants may be given convenient values with the third left to fit the correct boundary. For convenience then, we choose $b = 0$ and $a = -1$, resulting in

$$\frac{dz}{df} = A\left(\frac{f+1}{f}\right) \tag{Ex 5.14.2}$$

which, when integrated gives

$$z = A(f + \ln f) + B \tag{Ex 5.14.3}$$

For large positive real f, z is real so that we conclude B is real. For negative real f, $z = A(u + i\pi + \ln|u|) + B$ so that we conclude $A\pi = d$. Letting our fold point $u = -1$ correspond to $x = 0$ we obtain $B = d/\pi$. The required mapping is then

$$z = \frac{d}{\pi}(1 + f + \ln f) \tag{Ex 5.14.4}$$

Before leaving the example above entirely, we note that the mapping used earlier mapped the semi-infinite plate not from the real axis but from a line above the real axis. Substituting for f the results from example 5.13 will complete the transformation. In particular, letting $\ln f = w$ with $w = u + iv$ we can recast (Ex 5.14.4) as

$$z = \frac{d}{\pi}(1 + w + e^{w})$$

which maps $w = u + i\pi$ onto the top 'plate' and $w = u$ onto the x axis. Cascading maps in this fashion is frequently useful.

5.2.6 Capacitance

The mapping of two plates with known potential between them will allow us to calculate the capacitance of the mapped system. Given the potential everywhere in the mapped system, we could presumably take the normal derivative at the conductors to obtain the charge density. Integrating the charge density over the surface of one of the two conductors involved would then give the charge stored. Finally, dividing the charge by the potential difference between the conductors would give us the capacitance of the system.

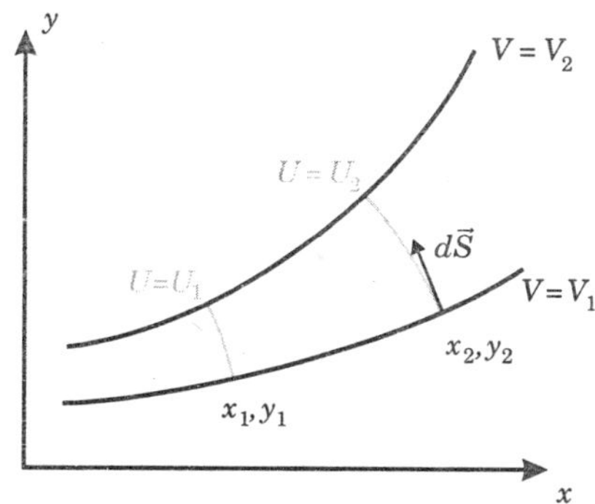

Figure 5.15: The charge on the lower plate between x_1, y_1 and x_2, y_2 is just $\varepsilon_0(U_1 - U_2)$.

The complex mapping Φ of the potential, as we have seen, has a real part U, perpendicular to the lines of constant potential, parallel to the electric field lines. The relationship between lines of constant U and $\vec{E}$ is in fact more complete. In the case of electric field lines, we interpret the density of lines as field strength. Under this interpretation, the amount of charge contained on a conductor bounded by two given field lines is just proportional to the number of lines between the boundaries. Not surprisingly the density of constant U lines, may also be interpreted as field strength. We therefore surmise that it may be possible to evaluate the charge on a conductor between $U = U_1$ and $U = U_2$ from the difference $U_2 - U_1$.

Let us consider two segments of the equipotential surfaces having $V = V_1$ and $V = V_2$ (formed perhaps by two plates of a capacitor) bounded by $U = U_1$ and $U = U_2$, as shown in Figure 5.15, extending a distance dz into the z direction (out of the page). (We abandon the use of z as a complex variable using it instead as the third spatial dimension in order that we have surfaces to deal with when we employ Gauss' law.) With the aid of Gauss' law, the charge on the lower surface is given by

$$Q = \varepsilon_0 \int_{x_1,y_1}^{x_2,y_2} \vec{E} \cdot d\vec{S} \tag{5–67}$$

Writing the electric field as $\vec{E} = -\vec{\nabla} V$, we express the charge Q as

$$Q = -\varepsilon_0 \int_1^2 \frac{\partial V}{\partial x} dS_x + \frac{\partial V}{\partial y} dS_y \tag{5–68}$$

where we have abbreviated the limits of integration (x_1, y_1) as 1 and (x_2, y_2) as 2.

We use the Cauchy-Riemann equations

$$\frac{\partial U}{\partial x} = \frac{\partial V}{\partial y} \quad \text{and} \quad \frac{\partial U}{\partial y} = -\frac{\partial V}{\partial x} \tag{5–69}$$

to recast (5–68) in terms of U:

$$Q = \varepsilon_0 \int_1^2 \frac{\partial U}{\partial y} dS_x - \frac{\partial U}{\partial x} dS_y \tag{5–70}$$

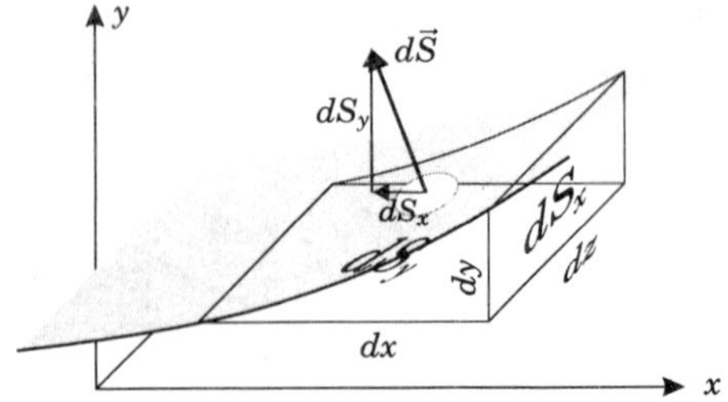

Figure 5.16: $dS_x = -dy\,dz$ and $dS_y = dx\,dz$. The decomposition of $d\vec{S}$ into components is not to scale but is intended only to show directions.

With reference to Figure 5.16, $dS_x = -dydz$ and $dS_y = dxdz$. (Had we taken an downward sloping segment, we would have had $dS_x = dydz$ and $dS_y = -dxdz$.) Substituting for dS_x and dS_y in (5–70), we have

$$\frac{dQ}{dz} = -\varepsilon_0 \int_1^2 \frac{\partial U}{\partial y} dy + \frac{\partial U}{\partial x} dx$$

$$= -\varepsilon_0 \int_1^2 dU = -\varepsilon_0 (U_2 - U_1) \qquad (5\text{–}71)$$

Recognizing that the sign is inconsequential, the capacitance per unit length is now easily obtained as

$$C_{1-2} = \varepsilon_0 \frac{U_2 - U_1}{V_2 - V_1} \qquad (5\text{–}72)$$

We use this result to find the capacitance of a large width w of the semi-infinite capacitor considered on page 119. The deviation from the infinite capacitor result will yield the contribution of the fringing field to the capacitance. We begin by writing Φ in the f plane, $\Phi = V_0 f/2\pi = V_0(u + iv)/2\pi = U + iV$. Equation (5–72) is therefore equivalent to

$$C = \varepsilon_0 \frac{u_2 - u_1}{v_2 - v_1} \qquad (5\text{–}73)$$

We note from Figure 5.12, any point x on the plate has two corresponding points u, one on the outside surface with $u > 0$, and one on the inside with $u < 0$. To obtain the capacitance of a strip of width w to the edge of the plate, we choose u_1 and u_2 on one of the two plates, say the upper plate, corresponding to $x = -w$. In other words, u_1 and u_2 are the roots of $(a/2\pi)(1 + u - e^u) = -w$. When $|w/a| \gg 1$, the approximate roots are $u_1 \cong -2\pi wa$ and $u_2 \cong \ln(2\pi w/a)$. The capacitance per unit length of a width w (including the edge) of a long parallel plate capacitor is now easily obtained:

$$C = \varepsilon_0 \left(\frac{w}{a} + \frac{1}{2\pi} \ln \frac{2\pi w}{a} \right) \qquad (5\text{–}74)$$

The term $\varepsilon_0 w/a$ is just the capacitance per unit length of a segment width w of an unbounded parallel plate capacitor. The remaining term reflects the effect of the edge on the capacitance. It is now a simple matter to deduce the capacitance per unit length of a finite width (width $\gg$ spacing) strip bounded on both sides to be

$$C = \varepsilon_0\left(\frac{w}{a} + \frac{1}{\pi}\ln\frac{2\pi w}{a}\right) \tag{5–75}$$

Moreover, the capacitance of a strip of metal above a parallel ground plane satisfies exactly the same boundary conditions so that its capacitance is also given by (5–75). Finally, we can get a good approximation of the capacitance of a pair of rectangular plates of dimensions $w \times b$ (each $\gg$ the spacing a) to be

$$C = \varepsilon_0\left[\frac{wb}{a} + \frac{1}{\pi}\left(b\ln\frac{2\pi w}{a} + w\ln\frac{2\pi b}{a}\right)\right] \tag{5–76}$$

The corners, of course, make this only an approximate solution becoming less valid as the plates get smaller.

5.2.7 Numerical Solution

When finding an analytic solution to Laplace's equation, we have seen that we frequently obtain a series solution whose evaluation requires the substitution of numerical values. If only numerical solutions are required, it is often much faster, particularly in Cartesian coordinates, simply to solve Laplace's equation numerically. In addition, the numerical solution can cope with arbitrarily complex boundaries. To see why Laplace's equation lends itself particularly well to numerical solution, we consider the Taylor expansion for the potential $V(x \pm h, y)$ and $V(x, y \pm h)$ about (x, y):

$$\begin{aligned}
V(x+h,y) &= V(x,y) + \frac{\partial V}{\partial x}h + \frac{1}{2!}\frac{\partial^2 V}{\partial x^2}h^2 + \frac{1}{3!}\frac{\partial^3 V}{\partial x^3}h^3 + \mathcal{O}(h^4)\\
V(x-h,y) &= V(x,y) - \frac{\partial V}{\partial x}h + \frac{1}{2!}\frac{\partial^2 V}{\partial x^2}h^2 - \frac{1}{3!}\frac{\partial^3 V}{\partial x^3}h^3 + \mathcal{O}(h^4)\\
V(x,y+h) &= V(x,y) + \frac{\partial V}{\partial y}h + \frac{1}{2!}\frac{\partial^2 V}{\partial y^2}h^2 + \frac{1}{3!}\frac{\partial^3 V}{\partial y^3}h^3 + \mathcal{O}(h^4)\\
V(x,y-h) &= V(x,y) - \frac{\partial V}{\partial y}h + \frac{1}{2!}\frac{\partial^2 V}{\partial y^2}h^2 - \frac{1}{3!}\frac{\partial^3 V}{\partial y^3}h^3 + \mathcal{O}(h^4)
\end{aligned} \tag{5–77}$$

where $\mathcal{O}(h^4)$ stands for terms of order h^4 or higher. Summing all four equations gives

$$V(x+h,y)+V(x-h,y)+V(x,y+h)+V(x,y-h) = 4V(x,y)+h^2\nabla^2V(x,y)+\mathcal{O}(h^4) \tag{5–78}$$

If V satisfies Laplace's equation, $\nabla^2 V = 0$ and we find that

$$V(x,y) = \frac{V(x+h,y)+V(x-h,y)+V(x,y+h)+V(x,y-h)}{4} + \mathcal{O}\left(h^4\right) \tag{5–79}$$

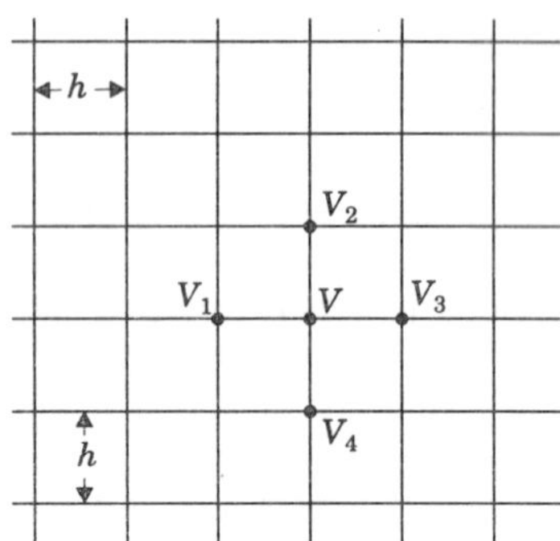

Figure 5.17: The potential at any point is to an excellent approximation the average of the potentials of its four nearest neighbors points.

If we were to draw a square grid over the region of interest, then (5–79) indicates that the potential at any point on the grid may be approximated by the average of its four nearest neighbors' potentials (Figure 5.17). We can make an initial guess of the solution inside a region whose bounding potential is given and then improve on our guess by successively replacing the potential at each point by that of the average of its four nearest neighbors. This process may be iterated until the values at the grid points stabilize. If h is sufficiently small (varying h can allow one to extrapolate to $h = 0$), the grid values "relax" to the solution of Laplace's equation with the given boundary conditions. The algorithm above is well suited to implementation in almost any computer language and yields solutions for arbitrary boundaries. Inspection of the method by which we arrived at equation (5–79) leads to the immediate generalization in three dimensions where we find the potential at any point should be well approximated by the average of that of the *six* nearest neighbor points.

5.3 $\nabla^2 V = 0$ in Three Dimensions

In this section we will explore the three-dimensional solutions to Laplace's equation. In the case of Cartesian coordinates, a straightforward generalization of the two-dimensional results applies. Likewise, the preceding numerical method immediately generalizes to three dimensions. Only in the curvilinear coordinate systems are new results obtained. For the three-dimensional examples we will obtain the general solution in cylindrical coordinates and in spherical polar coordinates. Lest the reader be led to believe that spherical polar and cylindrical systems are the only ones of interest, a final example (really a single-variable problem) is solved in ellipsoidal coordinates.

5.3.1 Cylindrical Polar Coordinates

In cylindrical polar coordinates $\nabla^2 V = 0$ takes the form

$$\frac{1}{r}\frac{\partial}{\partial r}\left(r\frac{\partial V}{\partial r}\right) + \frac{1}{r^2}\frac{\partial^2 V}{\partial \varphi^2} + \frac{\partial^2 V}{\partial z^2} = 0 \tag{5–80}$$

We assume a separable solution of the form $V(r, \varphi, z) = \mathrm{R}(r)\Phi(\varphi)\mathrm{Z}(z)$; then, dividing $\nabla^2 V$ by V, we obtain

$$\frac{\nabla^2 V}{V} = \frac{\dfrac{1}{r}\dfrac{\partial}{\partial r}\left(r\dfrac{\partial \mathrm{R}}{\partial r}\right)}{\mathrm{R}} + \frac{\dfrac{1}{r^2}\dfrac{\partial^2 \Phi}{\partial \varphi^2}}{\Phi} + \frac{\dfrac{\partial^2 \mathrm{Z}}{\partial z^2}}{\mathrm{Z}} = 0 \tag{5–81}$$

Taking $(\partial^2\mathrm{Z}/\partial z^2)/\,\mathrm{Z}$ to the righthand side of the equation we have an expression independent of z on the left, from which we conclude that either expression (on the right or on the left) must be a constant. Explicitly putting in the sign (which must still be determined from the boundary conditions) of the separation constant, we have

$$\frac{\dfrac{\partial^2 \mathrm{Z}}{\partial z^2}}{\mathrm{Z}} = \lambda^2 \quad \Rightarrow \quad \mathrm{Z}(z) = A\sinh\lambda z + B\cosh\lambda z \tag{5–82}$$

or, alternatively,

$$\frac{\dfrac{\partial^2 \mathrm{Z}}{\partial z^2}}{\mathrm{Z}} = -\lambda^2 \quad \Rightarrow \quad \mathrm{Z}(z) = A\sin\lambda z + B\cos\lambda z \tag{5–83}$$

For the moment we restrict ourselves to the choice of sign made in (5–82). The Laplace equation then reduces to

$$\frac{r\dfrac{d}{dr}\left(r\dfrac{d\mathrm{R}}{dr}\right)}{\mathrm{R}} + \lambda^2 r^2 = -\frac{\dfrac{d^2\Phi}{d\varphi^2}}{\Phi} = m^2 \tag{5–84}$$

where the choice of sign is dictated by the requirement of Φ's periodicity. Equation (5–84) is now easily solved for Φ:

$$\Phi(\varphi) = C\sin m\varphi + D\cos m\varphi \tag{5–85}$$

with m an integer. The remaining equation

$$r\frac{d}{dr}\left(r\frac{d\mathrm{R}}{dr}\right) + (\lambda^2 r^2 - m^2)\mathrm{R} = 0 \tag{5–86}$$

is Bessel's equation (Appendix E), having solutions

$$\mathrm{R}(r) = E\mathrm{J}_m(\lambda r) + F\mathrm{N}_m(\lambda r) \tag{5–87}$$

where J_m and N_m are Bessel and Neumann functions of order m. Had we picked the negative separation constant as in equation (5–83), we would have obtained for R the modified Bessel equation

$$r\frac{d}{dr}\left(r\frac{d\mathrm{R}}{dr}\right) + (-\lambda^2 r^2 - m^2)\mathrm{R} = 0 \tag{5–88}$$

having as solutions the modified Bessel functions $\mathrm{I}_m(\lambda r)$ and $\mathrm{K}_m(\lambda r)$. (K_m and N_m diverge at $r = 0$ and are therefore excluded from problems where the region of

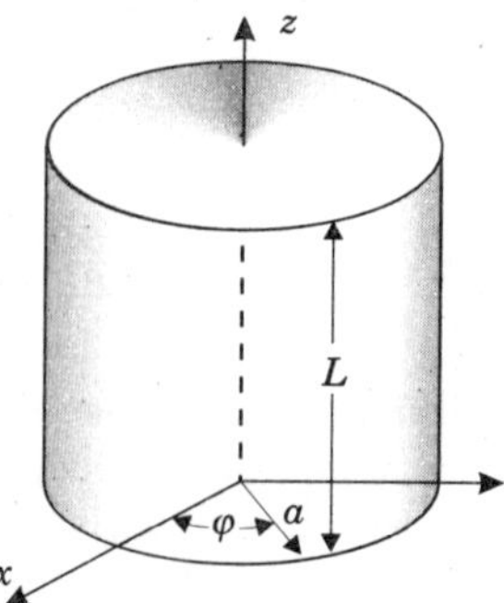

Figure 5.18: The cylinder has potential $V_L(r,\varphi)$ on its top face, 0 on the bottom, and $V_a(\varphi,z)$ on the curved face.

interest includes $r = 0$, while I_m diverges as $r \to \infty$ and will therefore be excluded from any exterior solution.) The complete solution is then of the form

$$V_{\lambda m}(r,\varphi,z) = \sum_{\lambda,m} \left\{ \begin{array}{c} \mathrm{J}_m(\lambda r) \\ \mathrm{N}_m(\lambda r) \end{array} \right\} \cdot \left\{ \begin{array}{c} \sin m\varphi \\ \cos m\varphi \end{array} \right\} \cdot \left\{ \begin{array}{c} \sinh \lambda z \\ \cosh \lambda z \end{array} \right\}$$

$$+ \sum_{\lambda,m} \left\{ \begin{array}{c} \mathrm{I}_m(\lambda r) \\ \mathrm{K}_m(\lambda r) \end{array} \right\} \cdot \left\{ \begin{array}{c} \sin m\varphi \\ \cos m\varphi \end{array} \right\} \cdot \left\{ \begin{array}{c} \sin \lambda z \\ \cos \lambda z \end{array} \right\} \qquad (5\text{–}89)$$

where the braces { } stand for the arbitrary linear combination of the two terms within.

To see how the general solution may be applied to a given problem, let us consider the problem of finding the potential inside the right circular cylinder of length L and radius a shown in Figure 5.18. For simplicity we set the potential on the bottom surface to zero while the top surface has potential $V_L(r,\varphi)$ and the curved side has potential $V_a(\varphi,z)$. As in the case of Cartesian coordinates, we divide the problem into two (three in general, when the top, bottom and sides are all non-zero) distinct problems; we solve first for the potential holding the top surface at $V = 0$, and then solve for the potential when the curved face is held at zero. The sum of the two solutions satisfies the complete boundary condition.

(a) Solution with top (and bottom) surface grounded: $V(r,\varphi,L) = V(r,\varphi,0) = 0$. To avoid an infinite solution at $r = 0$, the K_m and N_m must be absent from the solution. The requirement that V vanish at $z = 0$ and $z = L$ eliminates any terms in $\sinh \lambda z$ and $\cosh \lambda z$ (as no linear combination of these will vanish at two discrete points). Finally, this same condition eliminates the $\cos \lambda z$ term and requires $\lambda = n\pi/L$, leaving

$$V_1(r,\varphi,z) = \sum_{m,n} (A_{mn} \sin m\varphi + B_{mn} \cos m\varphi)\, \mathrm{I}_m\left(\frac{n\pi r}{L}\right) \sin \frac{n\pi z}{L} \qquad (5\text{–}90)$$

At $r = a$, this becomes

$$V_a(\varphi, z) = \sum_{n,m} \mathrm{I}_m\left(\frac{n\pi a}{L}\right)\sin\frac{n\pi z}{L}\left(A_{mn}\sin m\varphi + B_{mn}\cos m\varphi\right) \tag{5–91}$$

The right side of (5–91) is of course just the double Fourier expansion of V_a. The coefficients are then easily found from

$$\begin{pmatrix} A_{mn} \\ B_{mn} \end{pmatrix} = \frac{2}{L\pi \mathrm{I}_m\left(\frac{n\pi a}{L}\right)} \int_0^{2\pi}\int_0^L V_a(\varphi, z)\begin{pmatrix}\sin m\varphi \\ \cos m\varphi\end{pmatrix}\sin\frac{n\pi z}{L}\,d\varphi dz \tag{5–92}$$

(The stacked $\sin m\varphi$ and $\cos m\varphi$ are intended to be read as either the top line in both sides of the equation or the bottom line in both sides of the equation, much the way $\pm$ is used.)

(b) Solution with the curved surface grounded: $V(a, \varphi, z) = 0$. Again, the interior solution cannot support K_m or N_m. I_m has no zeros, and its inclusion in the solution would prevent V from vanishing at $r = a$ for all φ and z. Equation (5–89) reduces to

$$V_2 = \sum_m \mathrm{J}_m(\lambda r)\left\{\begin{array}{c}\sin m\varphi \\ \cos m\varphi\end{array}\right\}\cdot\left\{\begin{array}{c}\sinh\lambda z \\ \cosh\lambda z\end{array}\right\} \tag{5–93}$$

In order for V to vanish at $z = 0$, the $\cosh\lambda z$ term must be eliminated, and finally, the requirement that V vanish at $r = a$ translates to requiring λa to be a root of J_m. We denote the ith root of J_m by ρ_{mi} Appendix E), giving $\lambda = \rho_{mi}a$. The general solution that satisfies this half of the boundary conditions is

$$V_2(r, \varphi, z) = \sum_{m,i} \sinh\frac{\rho_{mi}z}{a}\mathrm{J}_m\left(\frac{\rho_{mi}r}{a}\right)\left(A_{mi}\sin m\varphi + B_{mi}\cos m\varphi\right) \tag{5–94}$$

At $z = L$, this reduces to

$$V_L(r, \varphi) = \sum_{m,i} \sinh\frac{\rho_{mi}L}{a}\mathrm{J}_m\left(\frac{\rho_{mi}r}{a}\right)\left(A_{mi}\sin m\varphi + B_{mi}\cos m\varphi\right) \tag{5–95}$$

The orthogonality of the trigonometric functions and weighted orthogonality of the Bessel functions (E–20) may be exploited to evaluate the coefficients of the expansion. Abbreviating $C_{mi} = A_{mi}\sinh(\rho_{mi}L/a)$ and $D_{mi} = B_{mi}\sinh(\rho_{mi}L/a)$, we multiply both sides of (5–95) by

$$\mathrm{J}_\ell\left(\frac{\rho_{\ell j}r}{a}\right)\begin{pmatrix}\cos\ell\varphi \\ \sin\ell\varphi\end{pmatrix}$$

and integrate over the entire top surface of the cylinder to obtain

$$\sum_{m,i}\int_0^a\int_0^{2\pi}\left(C_{mi}\sin m\varphi + D_{mi}\cos m\varphi\right)\mathrm{J}_m\left(\frac{\rho_{mi}r}{a}\right)\mathrm{J}_\ell\left(\frac{\rho_{\ell j}r}{a}\right)\begin{pmatrix}\cos\ell\varphi \\ \sin\ell\varphi\end{pmatrix}r\,dr\,d\varphi$$

$$= \int_0^a \int_0^{2\pi} V_L(r,\varphi) \mathrm{J}_\ell\left(\frac{\rho_{\ell j} r}{a}\right) \begin{pmatrix} \cos \ell\varphi \\ \sin \ell\varphi \end{pmatrix} r dr d\varphi \equiv \Im_{\ell j} \quad (5\text{–}96)$$

Performing the integration over φ and using the orthogonality of the trigonometric functions to eliminate all terms for which $m \neq \ell$ (for $\ell = 0$, replace π by 2π for the cosine term in all the following formulas), we find

$$\sum_i \begin{pmatrix} C_{\ell i} \\ D_{\ell i} \end{pmatrix} \pi \int_0^a \mathrm{J}_\ell\left(\frac{\rho_{\ell i} r}{a}\right) \mathrm{J}_\ell\left(\frac{\rho_{\ell j} r}{a}\right) r dr d\varphi = \Im_{\ell j} \quad (5\text{–}97)$$

and using (E–20)

$$\int_0^a \mathrm{J}_\ell\left(\frac{\rho_{\ell i} r}{a}\right) \mathrm{J}_\ell\left(\frac{\rho_{\ell j} r}{a}\right) r dr = \tfrac{1}{2} a^2 \mathrm{J}_{\ell+1}^2(\rho_{\ell j}) \delta_{ij} \quad (5\text{–}98)$$

we perform the remaining integration over r to obtain

$$\begin{pmatrix} C_{\ell j} \\ D_{\ell j} \end{pmatrix} \tfrac{1}{2} \pi a^2 \mathrm{J}_{\ell+1}^2(\rho_{\ell j}) = \int_0^a \int_0^{2\pi} V_L(r,\varphi) \mathrm{J}_\ell\left(\frac{\rho_{\ell j} r}{a}\right) \begin{pmatrix} \cos \ell\varphi \\ \sin \ell\varphi \end{pmatrix} r dr d\varphi \quad (5\text{–}99)$$

The original expansion coefficients are now easily found:

$$\begin{pmatrix} A_{\ell j} \\ B_{\ell j} \end{pmatrix} = \frac{2}{\pi a^2 \mathrm{J}_{\ell+1}^2(\rho_{\ell j}) \sinh\left(\frac{\rho_{\ell j} L}{a}\right)} \int_0^a \int_0^{2\pi} V_L(r,\varphi) \mathrm{J}_\ell\left(\frac{\rho_{\ell j} r}{a}\right) \begin{pmatrix} \sin \ell\varphi \\ \cos \ell\varphi \end{pmatrix} r dr d\varphi \quad (5\text{–}100)$$

The complete solution satisfying both boundary conditions is then the sum of (5–90) with coefficients given by (5–92) and of (5–94) with coefficients given by (5–100).

5.3.2 Spherical Polar Coordinates

When the potential (or the field) varies both in θ and φ on the surface of a sphere, it is advantageous to use spherical polar coordinates to solve the problem. We will find that several of the steps retrace our solution of the φ independent solution of Laplace's equation. Laplace's equation in spherical polars reads as follows:

$$\frac{1}{r}\frac{\partial^2}{\partial r^2}(rV) + \frac{1}{r^2 \sin\theta}\frac{\partial}{\partial\theta}\left(\sin\theta \frac{\partial V}{\partial\theta}\right) + \frac{1}{r^2 \sin^2\theta}\frac{\partial^2 V}{\partial\varphi^2} = 0 \quad (5\text{–}101)$$

We assume a separable solution of the form $V = \mathrm{R}(r)\Theta(\theta)\Phi(\varphi)$, then $r^2 \sin^2\theta \cdot \nabla^2 V/V = 0$ becomes

$$\frac{r^2 \sin^2\theta}{r\mathrm{R}}\frac{d^2}{dr^2}(r\mathrm{R}) + \frac{\sin\theta}{\Theta}\frac{d}{d\theta}\left(\sin\theta \frac{d\Theta}{d\theta}\right) = -\frac{1}{\Phi}\frac{d^2\Phi}{d\varphi^2} = m^2 \quad (5\text{–}102)$$

The separation constant has been chosen positive in anticipation of the solution $\Phi = A\cos(\mathrm{h})\, m\varphi + B\sin(\mathrm{h})\, m\varphi$. Since Φ must be periodic with period 2π, we accept only the trigonometric solution with m an integer. We eliminate Φ from (5–102). After dividing the remaining equation by $\sin^2\theta$, it is separated as

$$\frac{r^2}{r\mathrm{R}}\frac{d^2}{dr^2}(r\mathrm{R}) = -\frac{1}{\Theta \sin\theta}\frac{d}{d\theta}\left(\sin\theta \frac{d\Theta}{d\theta}\right) + \frac{m^2}{\sin^2\theta} = \ell(\ell+1) \quad (5\text{–}103)$$

with no restrictions on ℓ yet. Our rather peculiar choice of writing the separation constant anticipates a well-behaved solution. The radial equation

$$r^2 \frac{d^2}{dr^2}(r\mathrm{R}) = \ell(\ell+1)(r\mathrm{R}) \tag{5–104}$$

is easily solved to give $r\mathrm{R} = Cr^{\ell+1} + Dr^{-\ell}$, whence we conclude

$$\mathrm{R} = Cr^\ell + \frac{D}{r^{\ell+1}} \tag{5–105}$$

The remaining equation for Θ,

$$\frac{1}{\sin\theta}\frac{d}{d\theta}\left(\sin\theta\frac{d\Theta}{d\theta}\right) + \left[\ell(\ell+1) - \frac{m^2}{\sin^2\theta}\right]\Theta = 0 \tag{5–106}$$

may be converted to the associated Legendre equation by setting $x = \cos\theta \Rightarrow d/d\theta = -\sin\theta \cdot d/dx$, which results in

$$\frac{d}{dx}\left[(1-x^2)\frac{d\Theta}{dx}\right] + \left[\ell(\ell+1) - \frac{m^2}{1-x^2}\right]\Theta = 0 \tag{5–107}$$

The solutions to (5–107) are the associated Legendre functions $\mathrm{P}_\ell^m(x)$ and $\mathrm{Q}_\ell^m(x)$. $\mathrm{Q}_\ell^m(\pm 1) = \pm\infty$ and $\mathrm{P}_\ell^m(x)$ diverges as $x \to \pm 1$ unless ℓ is an integer. $\mathrm{P}_\ell^0(x) \equiv \mathrm{P}_\ell(x)$ are the Legendre Polynomials encountered in section 5.2.3 and earlier.

Recapitulating, we have found the well-behaved solutions to $\nabla^2 V = 0$ to be of the form

$$V(r,\theta,\varphi) = \sum_{\ell,m}(A_m \sin m\varphi + B_m \cos m\varphi)\left(C_\ell r^\ell + \frac{D_\ell}{r^{\ell+1}}\right)\mathrm{P}_\ell^m(\cos\theta) \tag{5–108}$$

or, more briefly, the entire solution may be written in terms of spherical harmonics,

$$V(r,\theta,\varphi) = \sum_{\ell,m}\left(A_{\ell,m} r^\ell + \frac{B_{\ell,m}}{r^{\ell+1}}\right)\mathrm{Y}_\ell^m(\theta,\varphi) \tag{5–109}$$

where $A_{\ell,m}$ and $B_{\ell,m}$ are complex constants.

Example 5.15: Find the potential inside a hollow sphere of radius a given that the potential on the surface is $V_a(\theta,\varphi)$.

Solution: Since V is finite at the origin, $B_{\ell,m}$ of equation (5–109) must vanish; hence the general solution in the region of interest is of the form

$$V(r,\theta,\varphi) = \sum_{\ell,m} A_{\ell,m} r^\ell \mathrm{Y}_\ell^m(\theta,\varphi) \tag{Ex 5.15.1}$$

At $r = a$, this specializes to

$$V_a(\theta,\varphi) = \sum_{\ell,m} A_{\ell,m} a^\ell \mathrm{Y}_\ell^m(\theta,\varphi) \tag{Ex 5.15.2}$$

Multiplying both sides by $Y_{\ell'}^{*m'}(\theta,\varphi)$, the complex conjugate of $Y_{\ell'}^{m'}(\theta,\varphi)$, and integrating over the entire solid angle, we have with the aid of the orthonormality (F–43) of the spherical harmonics

$$\int_0^{2\pi}\int_0^{\pi} V_a(\theta,\varphi)Y_{\ell'}^{*m'}(\theta,\varphi)\sin\theta d\theta d\varphi = A_{\ell',m'}a^{\ell'} \tag{Ex 5.15.3}$$

The coefficients $A_{\ell,m}$ of the expansion follow immediately.

EXAMPLE 5.16: A sphere of radius a centered on the origin has potential $V = V_0\sin^2\theta\cos 2\varphi$ on its surface. Find the potential both inside and outside the sphere.

Solution: Inside the sphere the potential must take the form

$$V(r<a,\theta,\varphi) = \sum_{\ell,m} A_{\ell,m}r^\ell Y_\ell^m(\theta,\varphi) \tag{Ex 5.16.1}$$

Whereas we could in principle use the prescription in the previous problem to determine the coefficients it is easier to write the potential on the surface in spherical harmonics and then depend on the linear independence of the spherical harmonics to find the coefficients. Therefore, using (F–42), we write the boundary condition for the problem as

$$V_a(\theta,\varphi) = \tfrac{1}{2}V_0\sin^2\theta\left(e^{2i\varphi}+e^{-2i\varphi}\right) = \sqrt{\frac{8\pi}{15}}\left(Y_2^2(\theta,\varphi)+Y_2^{-2}(\theta,\varphi)\right) \tag{Ex 5.16.2}$$

Equating the general potential (Ex 5.16.1) to (Ex 5.16.2) at $r=a$, we have

$$\sum_{\ell,m} A_{\ell,m}a^\ell Y_\ell^m(\theta,\varphi) = \sqrt{\frac{8\pi}{15}}V_0\left(Y_2^2(\theta,\varphi)+Y_2^{-2}(\theta,\varphi)\right) \tag{Ex 5.16.3}$$

We conclude immediately

$$A_{2,2} = A_{2,-2} = \frac{V_0}{a^2}\sqrt{\frac{8\pi}{15}} \tag{Ex5.16.4}$$

We can, of course express the potential in terms of trigonometric functions by substituting the trigonometric expressions for $Y_2^{\pm 2}(\theta,\varphi)$

$$V(r<a,\theta,\varphi) = \frac{V_0r^2}{a^2}\sqrt{\frac{8\pi}{15}}\sqrt{\frac{15}{32\pi}}\sin^2\theta\left(e^{2i\varphi}+e^{-2i\varphi}\right) = \frac{V_0r^2}{a^2}\sin^2\theta\cos 2\varphi \tag{Ex 5.16.5}$$

The exterior solution is obtained in much the same way. The solution must take the form:

$$V(r>a,\theta,\varphi) = \sum_{\ell,m}\frac{B_{\ell,m}}{r^{\ell+1}}Y_\ell^m(\theta,\varphi) \tag{Ex 5.16.6}$$

We again equate this potential at a to that on the sphere to obtain

$$\sum_{\ell,m} \frac{B_{\ell,m}}{a^{\ell+1}} \mathrm{Y}_\ell^m(\theta,\varphi) = \sqrt{\frac{8\pi}{15}} V_0 \left(\mathrm{Y}_2^2(\theta,\varphi) + \mathrm{Y}_2^{-2}(\theta,\varphi)\right) \tag{Ex 5.16.7}$$

leading to the conclusion that the only nonvanishing coefficients are

$$B_{2,\pm 2} = a^3 V_0 \sqrt{\frac{8\pi}{15}} \tag{Ex 5.16.8}$$

and the potential for $r > a$ is

$$V(r > a, \theta, \varphi) = \frac{a^3 V_0}{r^3} \sqrt{\frac{8\pi}{15}} \left(\mathrm{Y}_2^2(\theta,\varphi) + \mathrm{Y}_2^{-2}(\theta,\varphi)\right) = \frac{a^3 V_0}{r^3} \sin^2\theta \cos 2\varphi \tag{Ex 5.16.9}$$

It is evident in both cases that the solution satisfies the boundary conditions.

Example 5.17: Find the potential inside and outside a sphere of radius R that has surface charge with density $\sigma = \sigma_0 \sin^2\theta \cos 2\varphi$ distributed on its surface.

Solution: The presence of a surface charge means the electric field is discontinuous by σ/ε_0 across the surface of the sphere. The second boundary condition that the tangential component of $\vec{E}$ is continuous across the surface means that both $\partial V/\partial\theta$ and $\partial V/\partial\varphi$ are continuous which in turn implies that the potential is continuous across the surface. The form of the potential is

$$V(r < R, \theta, \varphi) = \sum_{\ell,m} A_{\ell,m} r^\ell \mathrm{Y}_\ell^m(\theta,\varphi) \tag{Ex 5.17.1}$$

$$V(r > R, \theta, \varphi) = \sum_{\ell,m} B_{\ell,m} r^{-(\ell+1)} \mathrm{Y}_\ell^m(\theta,\varphi) \tag{Ex 5.17.2}$$

We apply the radial boundary condition to obtain

$$\left.\frac{\partial V}{\partial r}\right|_{R_-} - \left.\frac{\partial V}{\partial r}\right|_{R_+} = \frac{\sigma}{\varepsilon_0} \tag{Ex 5.17.3}$$

or

$$\sum_{\ell,m} \left(\ell A_{\ell,m} R^{\ell-1} + \frac{(\ell+1)B_{\ell,m}}{R^{\ell+2}}\right) \mathrm{Y}_\ell^m(\theta,\varphi) = \frac{\sigma_0}{\epsilon_0} \sqrt{\frac{8\pi}{15}} \left(\mathrm{Y}_2^2(\theta,\varphi) + \mathrm{Y}_2^{-2}(\theta,\varphi)\right) \tag{Ex 5.17.4}$$

The linear independence of the spherical harmonics means that for $(\ell, m) \neq (2, \pm 2)$

$$A_{\ell,m} = -\frac{(\ell+1)B_{\ell,m}}{\ell R^{2\ell+1}} \tag{Ex 5.17.5}$$

while for $(\ell, m) = (2, \pm 2)$ we have

$$2A_{2,\pm 2} R + \frac{3B_{2,\pm 2}}{R^4} = \frac{\sigma_0}{\varepsilon_0} \sqrt{\frac{8\pi}{15}} \tag{Ex 5.17.6}$$

The continuity of the potential across the boundary gives for all (ℓ, m)

$$A_{\ell,m} = \frac{B_{\ell,m}}{R^{2\ell+1}} \tag{Ex 5.17.7}$$

(Ex 5.17.7) and (Ex 5.17.5) together yield only trivial solutions for $A_{\ell,m}$ and $B_{\ell,m}$, leaving us to solve only (Ex 5.17.7) and (Ex 5.17.6) for the $(\ell, m) = (2, \pm 2)$ coefficients. Solving these two equations simultaneously yields

$$A_{2,\pm 2} = \frac{\sigma_0}{5\varepsilon_0 R}\sqrt{\frac{8\pi}{15}} \quad \text{and} \quad B_{2,\pm 2} = \frac{\sigma_0 R^4}{5\varepsilon_0}\sqrt{\frac{8\pi}{15}} \tag{Ex 5.17.8}$$

We conclude the potentials are

$$V(r < R, \theta, \varphi) = \frac{\sigma_0 r^2}{5\varepsilon_0 R}\sqrt{\frac{8\pi}{15}}\left(Y_2^{-2}(\theta,\varphi) + Y_2^2(\theta,\varphi)\right) \tag{Ex 5.17.9}$$

and

$$V(r > R, \theta, \varphi) = \frac{\sigma_0 R^4}{5\varepsilon_0 r^3}\sqrt{\frac{8\pi}{15}}\left(Y_2^{-2}(\theta,\varphi) + Y_2^2(\theta,\varphi)\right) \tag{Ex 5.18.10}$$

⋆ 5.3.3 Oblate Ellipsoidal Coordinates

As a last example of separation of variables we consider a problem in oblate (pancake shaped) ellipsoidal coordinates. This by no means exhausts the separable coordinate systems: prolate ellipsoidal (cigar-shaped), bispherical, conical, and toroidal systems (not strictly speaking separable, see Appendix B) come to mind immediately.[16] Rather than obtain the general solution we will solve what is really a one variable problem, the potential due to a charged, constant potential oblate ellipsoid of revolution. We will specialize to find the potential around a charged, conducting, flat circular plate. We begin by constructing the Laplacian in oblate ellipsoidal coordinates ρ, α, and φ defined by

$$\left.\begin{aligned} x &= a\rho\sin\alpha\cos\varphi \\ y &= a\rho\sin\alpha\sin\varphi \end{aligned}\right\} \quad s \equiv \sqrt{x^2+y^2} = a\rho\sin\alpha$$
$$z = a\sqrt{\rho^2-1}\cos\alpha \qquad \rho \ge 1,\ 0 \le \alpha \le \pi,\ 0 \le \varphi \le 2\pi \tag{5–110}$$

where, for convenience we have defined the cylindrical radius s as ρ has already been used as one of the oblate elliptical coordinates. Surfaces of constant ρ obey the equation

$$\frac{s^2}{a^2\rho^2} + \frac{z^2}{a^2(\rho^2-1)} = 1 \tag{5–111}$$

[16] Toroidal coordinates are discussed in Appendix B. The metric tensor, curl, grad, div, and the Laplacian for a large number of orthogonal coordinate systems may be found in Parry Moon and Domina E. Spencer. (1961) *Field Theory Handbook*, Springer-Verlag, Berlin. The separation of the Laplace equation and the Helmholtz equation is effected for each of the coordinate systems in this small volume, and the general solutions are given.

demonstrating that the constant ρ surface is an ellipsoid of revolution with major radius $a\rho$ in the x-y plane and minor radius $a(\rho^2 - 1)^{1/2}$ along the z axis. The surface described by $\rho = 1$ has $z \equiv 0$ and s lies between 0 and a on this surface. It is a (two-sided) circular flat disk of radius a.

Although there are other methods, we construct the Laplacian using the methods of Appendix B, by means of the metric tensor whose nonzero elements are

$$g_{\rho\rho} = \left(\frac{\partial x}{\partial \rho}\right)^2 + \left(\frac{\partial y}{\partial \rho}\right)^2 + \left(\frac{\partial z}{\partial \rho}\right)^2 = \frac{a^2(\rho^2 - \sin^2\alpha)}{\rho^2 - 1}$$

$$g_{\alpha\alpha} = \left(\frac{\partial x}{\partial \alpha}\right)^2 + \left(\frac{\partial y}{\partial \alpha}\right)^2 + \left(\frac{\partial z}{\partial \alpha}\right)^2 = a^2(\rho^2 - \sin^2\alpha)$$

$$g_{\varphi\varphi} = \left(\frac{\partial x}{\partial \varphi}\right)^2 + \left(\frac{\partial y}{\partial \varphi}\right)^2 + \left(\frac{\partial z}{\partial \varphi}\right)^2 = a^2\rho^2\sin^2\alpha \tag{5–112}$$

The Laplacian may now be constructed:

$$\nabla^2 V = \frac{\sqrt{\rho^2 - 1}}{a^2\rho^2(\rho^2 - \sin^2\alpha)}\frac{\partial}{\partial\rho}\left(\rho\sqrt{\rho^2 - 1}\,\frac{\partial V}{\partial \rho}\right)$$
$$+ \frac{1}{a^2\sin\alpha(\rho^2 - \sin^2\alpha)}\frac{\partial}{\partial\alpha}\left(\sin\alpha\frac{\partial V}{\partial\alpha}\right) + \frac{1}{a^2\rho^2\sin^2\alpha}\frac{\partial^2 V}{\partial\varphi^2} \tag{5–113}$$

For the particular problem of the (constant potential) metal circular disk, the boundary conditions are independent of φ or α (as $\rho \to \infty$, $\alpha \to \theta$), leading us to seek a solution depending only on ρ. V must then satisfy

$$\frac{\partial}{\partial\rho}\left(\rho\sqrt{\rho^2 - 1}\,\frac{\partial V}{\partial\rho}\right) = 0 \tag{5–114}$$

Integrating twice we obtain

$$\frac{\partial V}{\partial \rho} = \frac{k}{\rho\sqrt{\rho^2 - 1}} \tag{5–115}$$

followed by

$$V(\rho) = -k\sin^{-1}\left(\frac{1}{\rho}\right) + C \tag{5–116}$$

where k and C are constants of integration. At very large values of ρ, (where $a\rho$ is indistinguishable from r) we expect the potential to vanish. The value of C follows, $V(\infty) = 0 = -k\sin^{-1}0 + C \Rightarrow C = 0$. Setting the potential on the surface of the ellipsoid defined by $\rho = \rho_0$ to V_0, we find $V_0 = -k\sin^{-1}(1/\rho_0)$. [For the flat plate defined by $\rho_0 = 1$, $V_0 = -k\sin^{-1}(1) = -\frac{1}{2}k\pi$]. The potential at any point ρ is then

$$V(\rho) = \frac{V_0}{\sin^{-1}(1/\rho_0)}\sin^{-1}\left(\frac{1}{\rho}\right) \tag{5–117}$$

To relate V_0 to the total charge on the ellipsoid, we first compute the electric field

$$\vec{E} = -\vec{\nabla} V = -\frac{1}{\sqrt{g_{\rho\rho}}}\frac{\partial V}{\partial \rho}\hat{\rho} \tag{5–118}$$

where $\hat{\rho}$ is a unit vector in the ρ direction. The electric field is then

$$\vec{E} = \frac{V_0}{\sin^{-1}(1/\rho_0)}\frac{\hat{\rho}}{a\rho\sqrt{\rho^2 - \sin^2\alpha}} \tag{5–119}$$

To obtain the charge on the ellipsoid, we integrate $\vec{E}$ over any constant $\rho = \rho_0$ surface containing the charged ellipsoid and equate the surface integral to Q/ε_0 (Gauss' law).

$$\begin{aligned}\frac{Q}{\varepsilon_0} &= \oint \vec{E}\cdot d\vec{S} = \oint E_\rho \sqrt{g_{\alpha\alpha}g_{\varphi\varphi}}\,d\alpha d\varphi \\ &= \frac{V_0}{\sin^{-1}(1/\rho_0)}\int_0^{2\pi}\int_0^{\pi}\frac{a^2\rho_0\sin\alpha\sqrt{\rho_0^2-\sin^2\alpha}}{\rho_0 a\sqrt{\rho_0^2-\sin^2\alpha}}\,d\alpha d\varphi \\ &= \frac{V_0}{\sin^{-1}(1/\rho_0)}\int_0^{2\pi}\int_0^{\pi} a\sin\alpha\,d\alpha d\varphi = \frac{4\pi V_0 a}{\sin^{-1}(1/\rho_0)}\end{aligned} \tag{5–120}$$

The potential V_0 on the conductor is then related to the charge by

$$V_0 = \frac{Q\sin^{-1}(1/\rho_0)}{4\pi\varepsilon_0 a} \tag{5–121}$$

giving

$$V(\rho) = \frac{Q}{4\pi\varepsilon_0 a}\sin^{-1}\left(\frac{1}{\rho}\right) \tag{5–122}$$

and

$$\vec{E} = \frac{Q}{4\pi\varepsilon_0 a}\frac{\hat{\rho}}{a\rho\sqrt{\rho^2-\sin^2\alpha}} \tag{5–123}$$

Specializing now to the flat circular plate of radius a corresponding to $\rho = 1$, we have

$$\vec{E}(\rho = 1) = \frac{Q\hat{\rho}}{4\pi\varepsilon_0 a\sqrt{a^2 - s^2}} \tag{5–124}$$

with $\hat{\rho} = \pm\hat{k}$.

The surface charge density on the plate is given by

$$\sigma = \varepsilon_0 E_\perp = \frac{Q}{4\pi a\sqrt{a^2-s^2}} \tag{5–125}$$

(At $s = 0$, the center of the plate, this is just half of the charge density it would have if the charge were uniformly distributed over both sides.)

Exercises and Problems

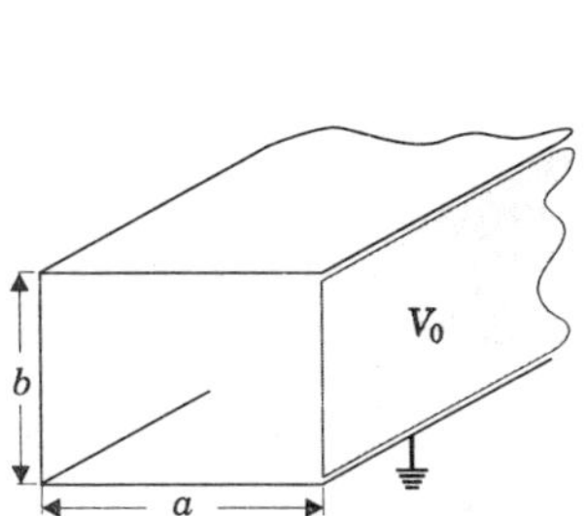

Figure 5.19: The rectangular pipe has three sides grounded, while the fourth has uniform potential V_0.

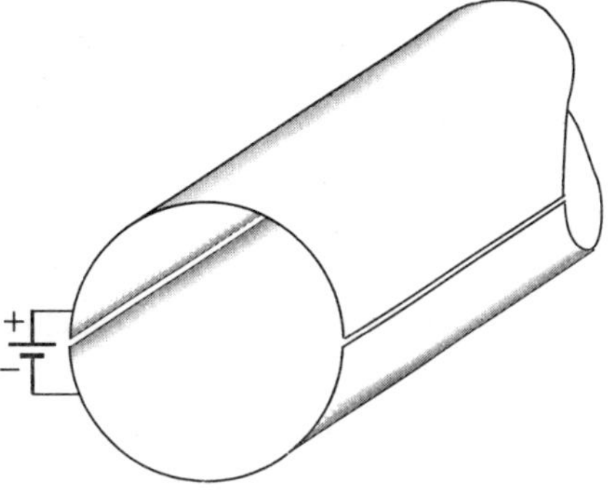

Figure 5.20: The pipe has potential difference V_0 applied between the upper and lower halves.

5-1 Show that the potential at the center of a charge-free sphere is precisely the average of the potential on the surface of the sphere.

5-2 Two large flat conducting plates are placed so that they form a wedge of angle $\alpha < \pi/2$. The plates are insulated from each other and have potentials 0 on one and $V = V_0$ on the other. Find the potential between the plates.

5-3 Two conducting coaxial cones with vertex at the origin and apex angles α_1 and α_2 are isolated from each other. The inner and outer cones have potential V_1 and V_2, respectively. Find the potential between the cones.

5-4 A rectangular pipe (Figure 5.19) of dimensions $a = 10$ cm and $b = 8$ cm has three of its sides maintained at zero potential while the remaining side is insulated and maintained at V_0. Use separation of variables to evaluate the potential along the $y = b/2$ line at $x = 1$ cm, 5 cm, and 9 cm to three significant figures.

5-5 A cylindrical pipe (Figure 5.20) of radius a is sawn lengthwise into two equal halves. A battery connected between the two halves establishes a potential difference of V_0 between the two halves. Use separation of variables to find the potential inside and outside the pipe.

5-6 A conducting sphere of radius a with its center at the origin is cut into two halves at the x-y plane. The two halves are separated slightly, and the top half is charged to V_0 while the bottom half is charged to $-V_0$. Find the potential both interior to the sphere and exterior to the sphere.

5-7 Find the scalar magnetic potential in the vicinity of the center of a solenoid of length L with N turns, each carrying current I.

5-8 Use the mapping

$$f = \ln\left(\frac{a+z}{a-z}\right)$$

to map the infinite parallel plates at $v = \pm\pi/2$ to the cross section of the split cylinder of Figure 5.20. (Hint: Express $[a+z]/[a-z]$ as $Re^{i\alpha}$, giving $u + iv =$

$\ln R + i\alpha$ or $v = \alpha$.) Express the potential in the pipe in closed form, and compare this to the result obtained by summing the first five nonzero terms of the series of exercise 5-5 for the point $y = 0.2a$ and $x = 0$.

5-9 Obtain the mapping that carries the $+u$ axis in the f plane to the positive x axis and the $-u$ axis to an axis inclined at α to the x axis.

5-10 Obtain the charge density on the constant potential right angled plate of Figure 5.9 (page 116).

5-11 Verify that the mapping

$$z = \frac{2a}{\pi} \ln \cosh f$$

carries the infinite lines at $v = \pi/2$, 0, and $-\pi/2$ to two infinite lines and one half-infinite line. Use this mapping to find the capacitance of a metal plate inserted in the middle between two larger grounded plates.

5-12 Calculate the capacitance of two parallel flat circular conducting plates with a space d between them and a radius $R \gg d$. How good an approximation is your result?

5-13 Draw the image in the z plane of the $v = 1$ and $v = 2$ (potential) lines as well as the $u = \frac{1}{4}b$ and $\frac{1}{2}b$ (field) lines under the mapping (Ex 5.13.3).

5-14 Find the image of the real axis under the mappings and determine what happens off the real axis

$$z = \frac{\imath - f}{i + f} \quad \text{and} \quad z = \frac{i + f}{i - f}$$

5-15 Find the image of the real axis under the mappings

$$z = \frac{\sqrt{f} - 1}{\sqrt{f} + 1} \quad \text{and} \quad z = \left(\frac{\sqrt{f} - 1}{\sqrt{f} + 1}\right)^{1/n}$$

5-16 Find the image of the real axis under the mapping

$$z = d(f^{1/2} - f^{-1/2})$$

5-17 Find the image of the real axis under the mapping

$$z = \{2(f+1)^{1/2} - 2\ln[(f+1)^{1/2} + 1] + \ln f\}$$

5-18 Find the image of the upper half plane under the mapping $z = a\sqrt{f^2 - 1}$.

5-19 A right circular cylinder of radius a has a potential

$$V = V_0\left(1 - \frac{r^2}{a^2}\right)$$

on its top surface and zero on all its other surfaces. Find the potential anywhere inside the cylinder. Note that

$$\frac{d}{dz}\left(z^\nu \mathrm{J}_\nu(z)\right) = z^\nu \mathrm{J}_{\nu-1}(z)$$

implies that $\int x^n \mathrm{J}_{n-1}(x)dx = x^n \mathrm{J}_n(x)$, allowing $x\mathrm{J}_0(x)$ to be integrated directly while $x^3\mathrm{J}_0(x)$ may be integrated by parts.

5-20 A hollow spherical shell has surface charge $\sigma_0 \cos\theta$ distributed on its surface. Find the electro-static potential and electric field due to this distribution both inside and outside the sphere.

5-21 A sphere of radius a has potential $\sin 2\theta \cos\varphi$ on its surface. Find the potential at all points outside the sphere.

5-22 A hollow spherical shell carries charge density $\sigma(\theta, \varphi) = \sigma_0 \sin\theta \sin\varphi$. Calculate the potential both inside and outside the sphere

5-23 Use prolate ellipsoidal coordinates

$$x = a\sqrt{\rho^2 - 1}\sin\alpha\sin\varphi$$
$$y = a\sqrt{\rho^2 - 1}\sin\alpha\cos\varphi$$
$$z = a\rho\cos\alpha$$

to compute the potential in the vicinity of a thin conducting needle carrying charge Q. Show that the charge distributes itself uniformly along the needle.

5-24 Bispherical coordinates defined by

$$x = \frac{a\sin\alpha\cos\varphi}{\cosh\rho - \cos\alpha}$$
$$y = \frac{a\sin\alpha\sin\varphi}{\cosh\rho - \cos\alpha}$$
$$z = \frac{a\sinh\rho}{\cosh\rho - \cos\alpha}$$

may be used to obtain the potential in the vicinity of two identical conducting spheres of radius 1 m with centers separated by 10 m charged to potentials V_a and V_b, respectively. The relevant coordinate surface satisfies

$$x^2 + y^2 + (z - a\coth\rho)^2 = \frac{a^2}{\sinh^2\rho}$$

Sketch the method of obtaining the solution. (Hint: Use the substitution

$$V = (\cosh\rho - \cos\alpha)^{1/2}U$$

to separate the variables.)

Chapter 6

Static Potentials with Sources—Poisson's Equation

6.1 Poisson's Equation

Frequently, the charge distribution in the region of interest does not vanish. The presence of a source term for the fields changes Laplace's equation to Poisson's equation. When sources are present, the static electric field $\vec{E}$ is still curl free, meaning that it can still be expressed as $\vec{E} = -\vec{\nabla} V$. Combining this with $\vec{\nabla} \cdot \vec{E} = \rho/\varepsilon_0$, we obtain Poisson's equation

$$\nabla^2 V = -\frac{\rho}{\varepsilon_0} \tag{6–1}$$

For the magnetic induction field, things are not quite so straightforward, since sources in the form of currents necessarily introduce a curl in the field. However, if the magnetic field arises from magnetic point dipoles, the curl vanishes everywhere (Stokes' theorem gives zero line integral about any point) and we still usefully express $\vec{B}$ as the gradient of a scalar potential. Because there is no magnetic equivalent of a conductor, we delay consideration of the magnetic field problems to the next chapter, where we will deal with fields in matter.

In this chapter we investigate the solution of (6–1), initially by the use of *image charges*, and then using *Green's functions* for a number of common geometries.

To put our problem into perspective, we might consider a charge q in the vicinity of a conductor on whose surface it induces a variable charge density $\rho(\vec{r}\,')$. The resulting potential at any point $\vec{r}$ is then given by

$$V(\vec{r}) = \frac{q}{4\pi\varepsilon_0 |\vec{r} - \vec{r}_q|} + \frac{1}{4\pi\varepsilon_0} \int \frac{\rho(\vec{r}\,')}{|\vec{r} - \vec{r}\,'|} d^3r' \tag{6–2}$$

Our problem would consist of determining the charge density induced by the introduction of known charges. Even more subtly, we might encounter a charged conductor that adjusts its charge density in the presence of another conductor. In fact, it is usually more practical to determine the potential by solving (6–1) and then to determine the induced charge density from the field at the surface of the conductor.

6.2 Image Charges

It will often prove possible to find a single charge whose field precisely mimics that of the induced charge distribution. Such a charge is known as an image of the inducing charge. We will find image charges for infinite plane conducting surfaces, for cylindrical conducting surfaces, and for spherical conducting surfaces in the vicinity of a given charge by choosing images such that the sum potential of both the source and the image is constant on the conducting surface. The uniqueness theorem will guarantee that the solution that satisfies the boundary condition is *the* solution. In each case it will be easy to extend the result to source charge distributions.

6.2.1 The Infinite Conducting Plane

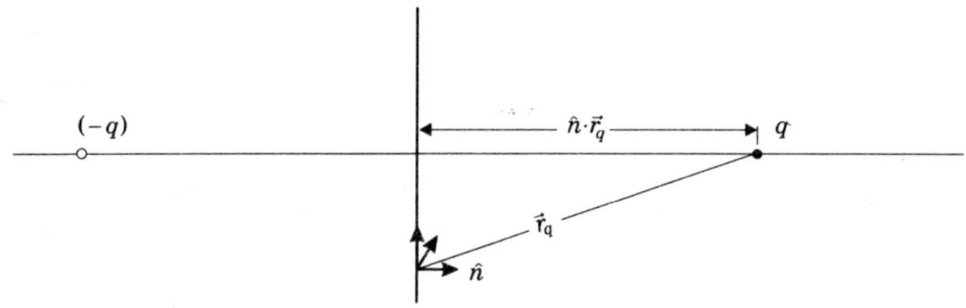

Figure 6.1: The charge q on the right of the conducting plane has an image an equal distance behind the plane on the left.

Consider the potential arising from a charge q placed a distance z to the right of the conducting x-y plane (Figure 6.1). If the conducting plane were removed, and instead an image charge $-q$ was placed a distance z behind the now missing x-y plane, then we would find a vanishing potential midway between the two charges. Reinserting the conducting surface along this plane would make no difference as the potential already satisfies the boundary condition (V = constant on a conductor). In other words, although q induces a nonzero charge distribution on the plate in order to maintain a zero potential, the potential arising from this distribution is mimicked exactly (for $z \geq 0$) by the image charge $-q$ behind the conducting plane. Thus

$$V(x, y, z \geq 0) = \frac{1}{4\pi\varepsilon_0}\left(\frac{q}{|\vec{r}-\vec{r}_q|} - \frac{q}{|\vec{r}-\vec{r}_{-q}|}\right)$$

$$= \frac{q}{4\pi\varepsilon_0}\left(\frac{1}{|\vec{r}-\vec{r}_q|} - \frac{1}{|\vec{r}-\vec{r}_q + 2(\hat{n}\cdot\vec{r}_q)\hat{n}|}\right) \tag{6–3}$$

where $\hat{n}$ is an outward-facing normal to the conducting plane. We easily generalize this result to the source charge distribution $\rho(\vec{r}\,')$ instead of q.

$$V(x, y, z \geq 0) = \frac{1}{4\pi}\int \frac{\rho(\vec{r}\,')}{\varepsilon_0}\left(\frac{1}{|\vec{r}-\vec{r}\,'|} - \frac{1}{|\vec{r}-\vec{r}\,' + 2(\hat{n}\cdot\vec{r}\,')\hat{n}|}\right) d^3r' \tag{6–4}$$

As we will see later, the term in parentheses is the *Green's function* to this problem. It should be noted that the solution above is correct only for the region of space containing the source charge. Behind the conducting plate, the field and the potential vanish.

EXAMPLE 6.1: Find the charge density induced on an infinite conducting plate by a point charge q at distance a from the plate.

Solution: To be more explicit, we take the charge to lie along the z axis at $z = a$ while the conducting plane defines the x-y plane. The electric field produced in the x-y plane by a charge $+q$ at $z = a$ and its image $-q$ at $z = -a$ is

$$\vec{E}(x, y, 0) = \frac{-2qa\hat{k}}{4\pi\varepsilon_0 \left(x^2 + y^2 + a^2\right)^{3/2}} \tag{Ex 6.1.1}$$

from which we obtain the surface charge density as $\sigma = \varepsilon_0 \vec{E} \cdot \hat{n}$. The surface charge density is therefore

$$\sigma = \frac{-qa}{2\pi \left(x^2 + y^2 + a^2\right)^{3/2}} \tag{Ex 6.1.2}$$

Not surprisingly, the total charge on the plate, assuming it was grounded, is just equal to $-q$ as is easily seen by integrating σ over the entire plate.

$$\int \sigma dA = -qa \int_0^\infty \frac{2\pi r\, dr}{2\pi(r^2 + a^2)^{3/2}} = \left.\frac{qa}{\sqrt{a^2 + r^2}}\right|_0^\infty = -q \tag{Ex 6.1.3}$$

A charge placed near the corner of two intersecting plates has images not only behind each of the conducting planes, but each of the images in turn has an image in the other planes and so on, very much like an optical image. Thus when two plates meet at right angles, a negatively charged image "appears" behind each plate, and a positive image lies diagonally through the point of intersection. It is not hard to see that this arrangement does indeed lead to equipotential surfaces along the planes defined by the conductors. When the planes intersect at 60°, five images of alternating sign will be required (the kaleidoscope is the optical analogue of this arrangement). Clearly when the angle is not commensurate with 360°, this method will not work because an infinite number of images would be required. The reader might reflect on what happens when two conducting planes intersect at 72°.

EXAMPLE 6.2: Two conducting planes along the x and y axes intersect at the origin, as shown in Figure 6.2. A charge q is placed distance b above the x axis and a distance a to the right of the y axis. Find the force on the charge.

Solution: The force is most easily found in terms of its x and y components.

$$F_x = \frac{q^2}{4\pi\varepsilon_0} \left[-\frac{1}{4a^2} + \frac{a}{4\left(a^2 + b^2\right)^{3/2}} \right] \tag{Ex 6.2.1}$$

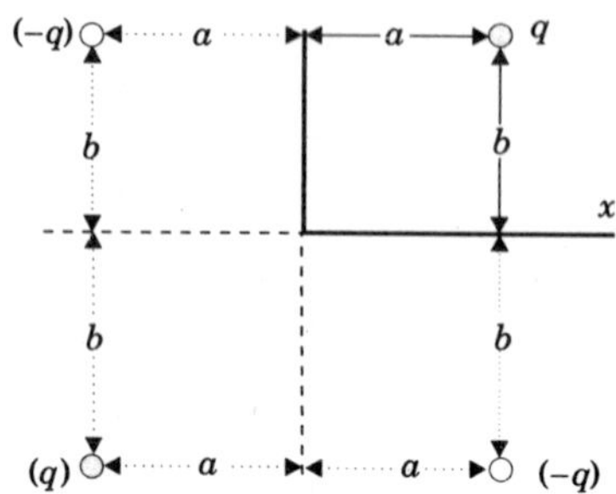

Figure 6.2: A charge placed near the corner of a large grounded conducting plate with a 90° bend requires three image charges to mimic the field.

and

$$F_y = \frac{q^2}{4\pi\varepsilon_0}\left[-\frac{1}{4b^2} + \frac{b}{4\left(a^2+b^2\right)^{3/2}}\right] \tag{Ex 6.2.2}$$

6.2.2 The Conducting Sphere

We again start by trying to find an image charge q' that will mimic the surface charge induced by q on the sphere (Figure 6.3). The potential on the surface of the sphere resulting from both charges must be constant (say zero, since it can always be changed to a different constant by the placement of a charge at the center of the sphere). By symmetry, if q lies on the z axis, then q' must also lie on the z axis. We place q' at a distance b from the center of the sphere. The potential on the surface is then given by

$$V(R,\theta) = \frac{1}{4\pi\varepsilon_0}\left(\frac{q}{r_1} + \frac{q'}{r_2}\right)$$

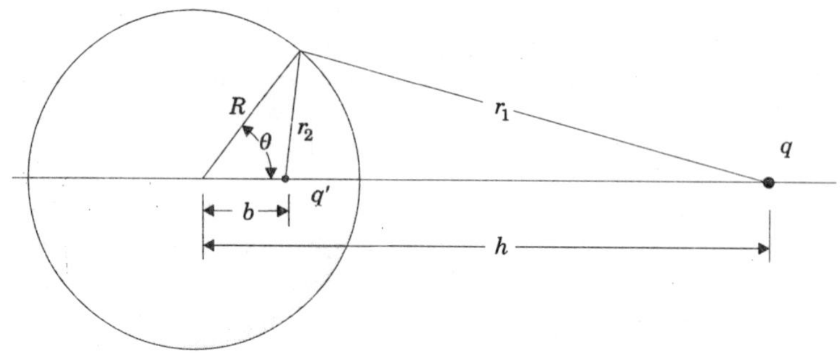

Figure 6.3: A point charge in the vicinity of a spherical conductor induces surface charge whose field is mimicked by a smaller image charge placed a distance b from the center of the sphere.

$$= \frac{1}{4\pi\varepsilon_0}\left(\frac{q}{\sqrt{h^2+R^2-2Rh\cos\theta}}+\frac{q'}{\sqrt{b^2+R^2-2Rb\cos\theta}}\right)=0 \quad (6\text{–}5)$$

We require then that b be chosen so that

$$q^2(b^2+R^2-2Rb\cos\theta) = q'^2(h^2+R^2-2Rh\cos\theta) \quad (6\text{–}6)$$

As this equality must hold independent of θ, we have

$$q^2(b^2+R^2) = q'^2(h^2+R^2) \quad (6\text{–}7)$$

and

$$q^2Rb = q'^2Rh \quad \Rightarrow \frac{q'^2}{q^2} = \frac{b}{h} \quad (6\text{–}8)$$

Substituting for q'^2 in (6–7), we get

$$b^2+R^2 = \frac{b}{h}(h^2+R^2) \quad (6\text{–}9)$$

which is solved by $b = R^2/h$ to give

$$\frac{q'^2}{q^2} = \frac{R^2}{h^2} \quad \text{or} \quad q' = -\frac{R}{h}q \quad (6\text{–}10)$$

More generally, a charge at $\vec{r}\,'$ has an image charge $q' = -(R/r\,')q$ placed at position

$$\vec{r}\,'_2 = \frac{R^2}{r'}\hat{r}' = \frac{R^2}{r'^2}\vec{r}\,' \quad (6\text{–}11)$$

leading to

$$V(\vec{r}) = \frac{q}{4\pi\varepsilon_0}\left(\frac{1}{|\vec{r}-\vec{r}\,'|} - \frac{|R/r'|}{\left|\vec{r}-\dfrac{R^2}{r'^2}\vec{r}\,'\right|}\right) \quad (6\text{–}12)$$

The latter expression is easily generalized to give the expression for the potential of a continuous charge distribution in the neighborhood of a conducting sphere

$$V(\vec{r}) = \frac{1}{4\pi}\int\frac{\rho(\vec{r}\,')}{\varepsilon_0}\left(\frac{1}{|\vec{r}-\vec{r}\,'|} - \frac{|R/r'|}{\left|\vec{r}-\dfrac{R^2}{r'^2}\vec{r}\,'\right|}\right)d^3r' \quad (6\text{–}13)$$

for $r > R$.

It is important to recognize the reciprocity of q and q'. When the original charge is interior to the sphere and the potential inside the sphere is sought, the image charge is found outside the sphere at h ($hb = R^2$). In other words, the results are identical with only (q, r) and (q', r') interchanged.

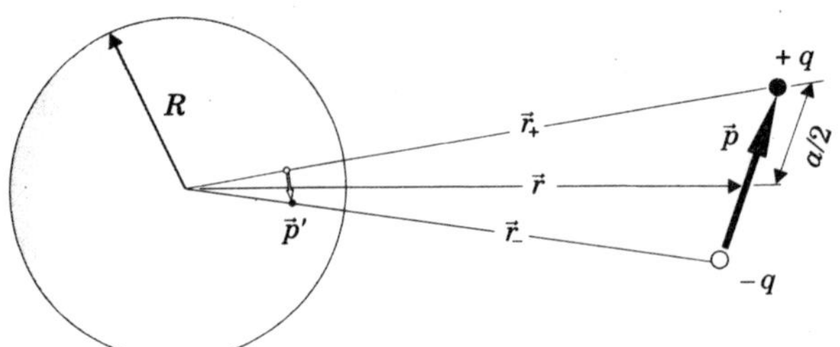

Figure 6.4: The image of the point dipole is most easily found as the limit of the finite size dipole. In this case each charge in finite dipole is separately imaged by charges inside the sphere.

EXAMPLE 6.3: An electric dipole $\vec{p}$ is placed in the vicinity of a grounded conducting sphere of radius R. Find the image dipole whose field mimics that of the induced charge on the sphere.

Solution: We usefully replace the point dipole of the problem above with a pair of charges separated by small displacement $\vec{a}$, as illustrated in Figure 6.4. We let the dipole be located at $\vec{r}$ and the positive and negative charges at $\vec{r}_+$ and $\vec{r}_-$, respectively. The dipole moment of the charges shown is $\vec{p} = q\vec{r}_+ - q\vec{r}_- = q\vec{a}$. To allow for point dipoles, we set $q = p/a$, which now gives $\vec{p} = p\hat{a}$ even when $a \to 0$. The charge $+q$ has an image charge $-qR/r_+$ located at $(R^2/r_+^2)\vec{r}_+$, while $-q$ has an image charge $+qR/r_-$ located at $(R^2/r_-^2)\vec{r}_-$. The dipole moment $\vec{p}'$ of the image dipole is now readily computed as

$$\begin{aligned}\vec{p}' &= \frac{qR}{r_-}\cdot\frac{R^2}{r_-^2}\vec{r}_- - \frac{qR}{r_+}\cdot\frac{R^2}{r_+^2}\vec{r}_+ \\ &= qR^3\left(\frac{\vec{r}_-}{r_-^3} - \frac{\vec{r}_+}{r_+^3}\right)\end{aligned} \tag{Ex 6.3.1}$$

We wish to find the limit of this expression as $|\vec{a}\,| \to 0$. We write

$$\vec{r}_+ = \vec{r} + \tfrac{1}{2}\vec{a} \qquad \text{and} \qquad \vec{r}_- = \vec{r} - \tfrac{1}{2}\vec{a} \tag{Ex 6.3.2}$$

giving $1/r_+^3 = (\vec{r}_+\cdot\vec{r}_+)^{-3/2} = (r^2 + \vec{r}\cdot\vec{a} + \frac{1}{4}a^2)^{-3/2}$ and a similar expression for $\vec{r}_-$. Expanding these expressions with the binomial theorem, discarding terms quadratic (and higher power) in a, we have

$$\frac{1}{r_+^3} \simeq \frac{1}{r^3}\left(1 + \frac{\vec{r}\cdot\vec{a}}{r^2}\right)^{-3/2} = \frac{1}{r^3}\left[1 - \frac{3}{2}\frac{\vec{r}\cdot\vec{a}}{r^2} + \mathcal{O}(a^2)\right] \tag{Ex 6.3.3}$$

and

$$\frac{1}{r_-^3} \simeq \frac{1}{r^3}\left(1 - \frac{\vec{r}\cdot\vec{a}}{r^2}\right)^{-3/2} = \frac{1}{r^3}\left[1 + \frac{3}{2}\frac{\vec{r}\cdot\vec{a}}{r^2} + \mathcal{O}(a^2)\right] \tag{Ex 6.3.4}$$

The induced dipole is then

$$\vec{p}' = \frac{qR^3}{r^3}\left[\vec{r}_-\left(1 + \frac{3}{2}\frac{\vec{r}\cdot\vec{a}}{r^2}\right) - \vec{r}_+\left(1 - \frac{3}{2}\frac{\vec{r}\cdot\vec{a}}{r^2}\right)\right]$$

$$= \frac{pR^3}{ar^3}\left[(\vec{r}_- - \vec{r}_+) + \frac{3}{2}\frac{\vec{r}\cdot\vec{a}}{r^2}(\vec{r}_- + \vec{r}_+)\right]$$

$$= \frac{R^3}{r^3}\left[-\vec{p} + \frac{3(\vec{r}\cdot\vec{p})\vec{r}}{r^2}\right] \qquad \text{(Ex 6.3.5)}$$

and is located at $\vec{r}\,' = (R^2/r^2)\vec{r}$ relative to the center of the sphere.

If the sphere in Example 6.3 had been neutral and isolated, the different charges on the two images would have left a net charge inside, detectable by Gauss' law, unless we put a compensating point charge at the center. The net charge left by the image dipole may be calculated in the limit of a point dipole as

$$q'_+ + q'_- = q\left(\frac{R}{r_-} - \frac{R}{r_+}\right) = qR\left(\frac{r_+ - r_-}{r_+ r_-}\right) \equiv \frac{qRa\cos\theta}{r^2} \qquad (6\text{–}14)$$

with θ the angle $\vec{a}$ makes with $\vec{r}$. We can, of course write the $qa\cos\theta$ as $\vec{p}\cdot\hat{r}$ so that the required neutralizing charge, q_c becomes

$$q_c = -\frac{R(\vec{p}\cdot\vec{r})}{r^3} \qquad (6\text{–}15)$$

6.2.3 Conducting Cylinder and Image Line Charges

Although the conducting cylinder does not yield image charges for point charges, it more usefully images a uniform line charge λ to an image line charge λ' inside the cylinder to give essentially the two-dimensional analogue of the spherical problem above.

Consider a line charge λ, $(dq = \lambda dz)$ at a distance h from the center of a conducting cylinder of radius R (Figure 6.5). We would like to find an image line charge λ' that would make the surface of the cylinder an equipotential surface. Suppose λ' to be at a distance b from the center. At any point $\vec{r}$ $(r \geq R)$, as indicated in Figure 6.6, the potential is given by

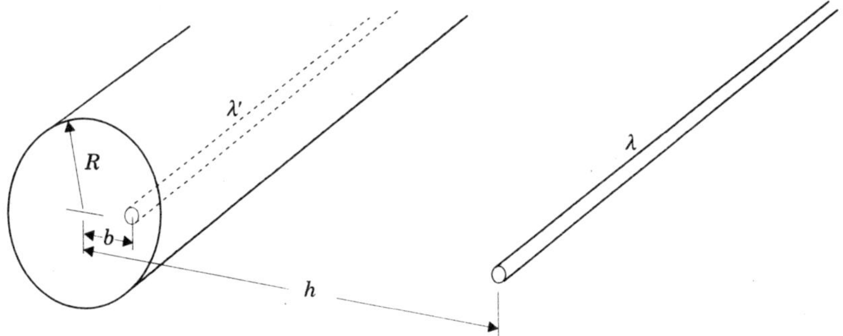

Figure 6.5: The surface charge induced on a conducting cylinder by a line charge λ is mimicked by the image line charge λ' inside the cylinder.

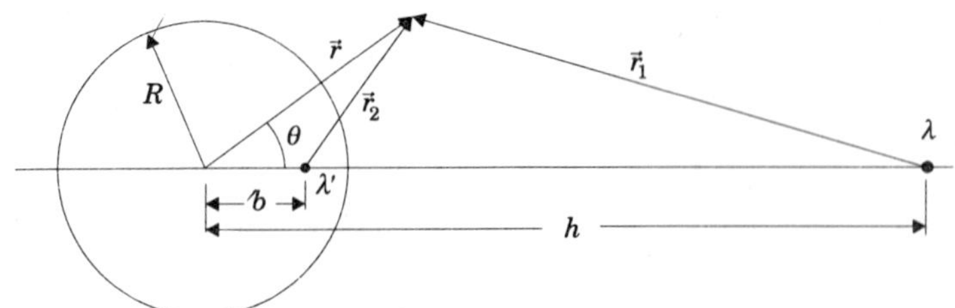

Figure 6.6: The line charge and its image seen end-on.

$$
\begin{aligned}
V(r \geq R) &= -\frac{\lambda}{2\pi\varepsilon_0}\ln r_1 - \frac{\lambda'}{2\pi\varepsilon_0}\ln r_2 \\
&= \frac{-1}{2\pi\varepsilon_0}\left(\lambda \ln\sqrt{r^2+h^2-2rh\cos\theta} + \lambda'\ln\sqrt{r^2+b^2-2rb\cos\theta}\right) \\
&= \frac{-1}{4\pi\varepsilon_0}\left[\lambda\ln(r^2+h^2-2rh\cos\theta) + \lambda'\ln(r^2+b^2-2rb\cos\theta)\right] \qquad (6\text{–}16)
\end{aligned}
$$

At $r = R$, we wish to have $\partial V/\partial\theta = 0$. Differentiating, we obtain

$$
\begin{aligned}
\left.\frac{\partial V}{\partial\theta}\right|_{r=R} &= -\frac{\partial}{\partial\theta}\left\{\frac{1}{4\pi\varepsilon_0}\left[\lambda\ln(R^2+h^2-2Rh\cos\theta) + \lambda'\ln(R^2+b^2-2Rb\cos\theta)\right]\right\} \\
&= \frac{-1}{2\pi\varepsilon_0}\left.\left(\frac{\lambda Rh\sin\theta}{r_1^2} + \frac{\lambda' Rb\sin\theta}{r_2^2}\right)\right|_{r=R} \qquad (6\text{–}17)
\end{aligned}
$$

from which we conclude

$$
\left.\left(\frac{h\lambda}{r_1^2} + \frac{b\lambda'}{r_2^2}\right)\right|_{r=R} = 0 \quad\Rightarrow\quad \left.\frac{r_1^2}{r_2^2}\right|_{r=R} = -\frac{h}{b}\frac{\lambda}{\lambda'} \qquad (6\text{–}18)
$$

Writing the ratio $\left.(r_1^2/\,r_2^2)\right|_{r=R}$ out explicitly, we find

$$
R^2 + h^2 - 2Rh\cos\theta = -\frac{h}{b}\frac{\lambda}{\lambda'}(R^2+b^2-2Rb\cos\theta) \qquad (6\text{–}19)
$$

We appeal to the linear independence of the constant term and the cosine term of (6–19) to equate the constants and the coefficients of $\cos\theta$ separately in the equation. Therefore,

$$
\begin{aligned}
2Rh\cos\theta &= -\frac{h}{b}\frac{\lambda}{\lambda'}\cdot 2Rb\cos\theta \\
&= -\frac{\lambda}{\lambda'}\cdot 2Rh\cos\theta \quad\Rightarrow\quad \lambda' = -\lambda \qquad (6\text{–}20)
\end{aligned}
$$

Inserting this into the θ-independent portion of (6–19), we have

$$
R^2 + h^2 = \frac{h}{b}(R^2+b^2) \quad\Rightarrow\quad hb = R^2 \qquad (6\text{–}21)
$$

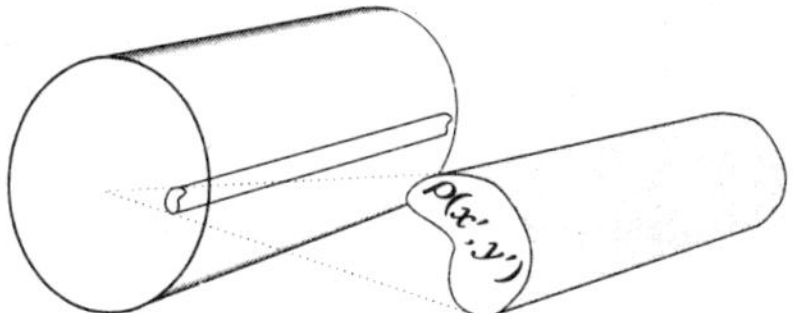

Figure 6.7: A continuous distribution invariant to displacements along the z axis has a spatially compressed image inside the cylinder.

The expression for the position of the image line charge is identical with that for the image of the point charge and the conducting sphere. By contrast, the magnitude of the image line charge is the same as the source line charge. More generally, for a line charge at $\vec{r}\,'$ we need an image charge $\lambda' = -\lambda$ at $\vec{r}\,'' = (R^2/r'^2)\vec{r}\,'$, giving potential

$$V(r \geq R) = -\frac{\lambda}{2\pi\varepsilon_0}\ln\left(\frac{r_1}{r_2}\right) = -\frac{\lambda}{2\pi\varepsilon_0}\ln\left(\frac{|\vec{r}-\vec{r}\,'|}{|\vec{r}-(R^2/r'^2)\vec{r}\,'|}\right) \tag{6–22}$$

We generalize this result to continuous distributions that are invariant with respect to displacements along the z axis (Figure 6.7).

$$V(r \geq R) = -\frac{1}{2\pi}\int_S \frac{\rho(\vec{r}\,')}{\varepsilon_0}\ln\left(\frac{|\vec{r}-\vec{r}\,'|}{|\vec{r}-(R^2/r'^2)\vec{r}\,'|}\right)d^3r' \tag{6–23}$$

It should be evident that the equipotentials of two equal uniform line charges of opposite sign are nonconcentric cylinders. At the midpoint between the two line charges lies an equipotential plane. This means that in general, parallel conducting cylinders or conducting cylinders parallel to a conducting plane can have their potentials mimicked by line charges.

Example 6.4: Find the capacitance per unit length of two parallel cylindrical wires of radius R spaced at center-to-center distance D.

Solution: The wires may be considered the equipotential surfaces of two line charges $+\lambda$ and $-\lambda$, each displaced inward from the center of its cylinder by distance b. We

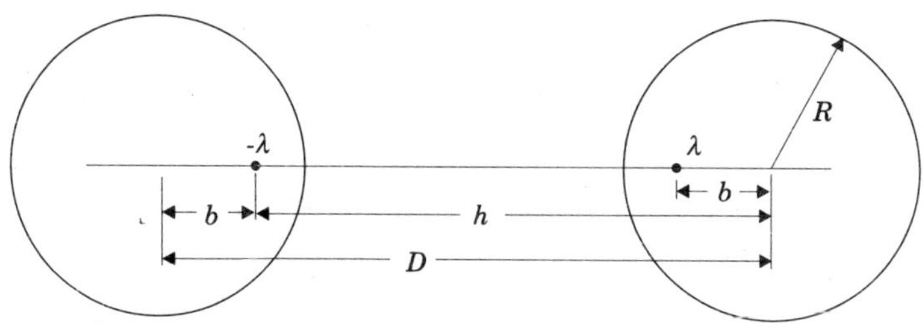

Figure 6.8: Geometry of two cylindrical wires, each of radius R, spaced at center-to-center distance D.

relate the geometry to our results with images to obtain the positions and equate the potential of these line charges to those of the wires at the surface of the wires. From Figure 6.8 we have $h = D - b$, while from our earlier results we have $bh = R^2$. Combining these relations, we have $b(D-b) = R^2$, which is easily solved by the quadratic formula to give

$$b_\pm = \frac{D \pm \sqrt{D^2 - 4R^2}}{2} \tag{Ex 6.4.1}$$

(Note that $b_+ b_- = R^2$; i.e.,$b_+ = h$.)

The potential on the surface of the wire bearing $+\lambda$ is then

$$V(r=R) = -\frac{\lambda}{2\pi\varepsilon_0}\ln\frac{r_1}{r_2} = -\frac{\lambda}{4\pi\varepsilon_0}\ln\left(\frac{r_1}{r_2}\right)^2\Bigg|_{r=R} \tag{Ex 6.4.2}$$

With the help of (6–18), the argument of the logarithm may be written as $h/b = R^2/b^2$ to give

$$V(r=R) = -\frac{\lambda}{2\pi\varepsilon_0}\ln\left(\frac{R}{b}\right) = \frac{\lambda}{2\pi\varepsilon_0}\ln\left(\frac{D-\sqrt{D^2-4R^2}}{2R}\right) \tag{Ex 6.4.3}$$

The other wire has a potential of the same magnitude but opposite sign. Therefore, using $\ln(b/R) = -\ln(h/R)$ we obtain the result

$$\Delta V = \pm\frac{\lambda}{\pi\varepsilon_0}\ln\left(\frac{D-\sqrt{D^2-4R^2}}{2R}\right) \equiv \mp\frac{\lambda}{\pi\varepsilon_0}\ln\left(\frac{D+\sqrt{D^2-4R^2}}{2R}\right) \tag{Ex 6.4.4}$$

The capacitance per unit length is now obtained as $\lambda\Delta V$.

$$\frac{C}{\ell} = \frac{\pi\varepsilon_0}{\ln\left(\dfrac{D+\sqrt{D^2-4R^2}}{2R}\right)} = \frac{\pi\varepsilon_0}{\cosh^{-1}(D/2R)} \tag{Ex 6.4.5}$$

6.3 Green's Functions

As we have seen in each of the cases considered, when we satisfied the boundary conditions of a problem using image charges, the potential was simply the integral of the source charge density multiplied by some function that varied with the boundary conditions. The multiplying function known as the *Green's* function to the problem could be deduced from the solution for a point source charge. Thus we might at the most elementary level think of a Green's function as the solution to the Poisson equation (with given boundary conditions) with only a point source. The Green's function that we will denote $G(\vec{r}, \vec{r}\,')$ is in this case a function of two variables, $\vec{r}$ and $\vec{r}\,'$, where $\vec{r}\,'$ is the position vector of the point source charge. In fact, the utility of Green's functions stretches far beyond solving Poisson's equation, and we will first give a somewhat heuristic demonstration of how Green's functions may be used to solve inhomogeneous linear partial differential equations.

Consider the function $G(x, x'; t, t')$ that solves the linear differential equation $\mathbf{D}G(x, x'; t, t') = -\delta(x - x')\delta(t - t')$ (we could of course have more variables each with its own δ function source term) where $\mathbf{D}$ is a linear differential operator acting on x and t. The function Ψ defined by $\Psi(x, t) \equiv \int f(x', t')G(x, x'; t, t')dx'dt'$ then solves $\mathbf{D}\Psi(x, t) = -f(x, t)$, as is shown below. Allowing $\mathbf{D}$ to act on the integral form of Ψ,

$$\mathbf{D}\Psi(x,t) = \mathbf{D}\int f(x',t')G(x,x';t,t')dx'dt' = \int f(x',t')\mathbf{D}G(x,x';t,t')dx'dt'$$

$$= \int -f(x',t')\delta(x-x')\delta(t-t')dx'dt' = -f(x,t) \qquad (6\text{–}24)$$

In the case that the boundary condition on G is that it vanishes when x lies on the boundary, Ψ clearly satisfies the same boundary condition. If, however, G does not vanish on the boundary, our naive approach cannot quite give Ψ with the same boundary condition. As we have noted at the end of section 6.2.2, the Green's function is symmetric in its spatial arguments x and x',

$$G(x, x') = G(x', x) \qquad (6\text{–}25)$$

but, to preserve the time ordering of causally related events

$$G(t, t') = G(-t', -t) \qquad (6\text{–}26)$$

Many authors choose to include a factor of 4π in their definition of the Green's function, so that G satisfies $\mathbf{D}G(x, x') = -4\pi\delta(x - x')$. We now turn to a somewhat more formal approach to Green's functions and their use in solving Poisson's equation.

6.3.1 Green's Theorem

Green's theorem, also known as Green's second identity, provides much of the formal framework for converting Poisson's differential equation to an integral equation and is easily derived. The divergence theorem

$$\int_\tau \vec{\nabla}\cdot\vec{F}d^3r = \oint_S \vec{F}\cdot d\vec{S} = \oint_S \vec{F}\cdot\hat{n}dS \qquad (6\text{–}27)$$

applies to any vector field $\vec{F}$. We choose in particular $\vec{F} = V\vec{\nabla}\Psi$, then $\vec{\nabla}\cdot\vec{F} = \vec{\nabla}V\cdot\vec{\nabla}\Psi + V\nabla^2\Psi$. Applying the divergence theorem to $\vec{\nabla}\cdot\vec{F}$ yields

$$\int_\tau (V\nabla^2\Psi + \vec{\nabla}V\cdot\vec{\nabla}\Psi)d^3r = \oint_S V\vec{\nabla}\Psi\cdot d\vec{S} = \oint_S V\frac{\partial\Psi}{\partial n}dS \qquad (6\text{–}28)$$

a result known as *Green's first identity.* Writing the same identity with V and Ψ interchanged gives

$$\int_\tau (\Psi\nabla^2 V + \vec{\nabla}V\cdot\vec{\nabla}\Psi)d^3r = \oint_S \Psi\vec{\nabla}V\cdot d\vec{S} = \oint_S \Psi\frac{\partial V}{\partial n}dS \qquad (6\text{–}29)$$

Subtracting this from the first form we obtain *Green's second identity* (22):

$$\int_\tau \left(V\nabla^2\Psi - \Psi\nabla^2 V\right) d^3r = \oint_S \left(V\frac{\partial\Psi}{\partial n} - \Psi\frac{\partial V}{\partial n}\right) dS \tag{6–30}$$

6.3.2 Poisson's Equation and Green's Functions

Poisson's equation, $\nabla^2 V = -\rho/\varepsilon_0$, can be converted to an integral equation by choosing Ψ in (6–30) such that Ψ satisfies. $\nabla^2\Psi = -\delta(\vec{r}-\vec{r}\,')$. Green's second identity then becomes

$$\int_\tau \left(-\delta(\vec{r}-\vec{r}\,')V(\vec{r}\,') + \Psi\,\frac{\rho(\vec{r}\,')}{\varepsilon_0}\right) d^3r' = \oint_{S'} \left(V\frac{\partial\Psi}{\partial n} - \Psi\frac{\partial V}{\partial n}\right) dS' \tag{6–31}$$

or, carrying out the integration of the δ function,

$$V(\vec{r}) = \int_\tau \frac{\Psi\rho(\vec{r}\,')}{\varepsilon_0} d^3r' + \oint_{S'} \left(\Psi\frac{\partial V}{\partial n} - V\frac{\partial\Psi}{\partial n}\right) dS' \tag{6–32}$$

To gain some insight into this expression, let us use a familiar solution to $\nabla^2\Psi = -\delta(\vec{r}-\vec{r}\,')$, namely $\Psi = 1/(4\pi|\vec{r}-\vec{r}\,'|)$. Equation (6–32) then becomes

$$V(\vec{r}) = \frac{1}{4\pi\varepsilon_0}\int_\tau \frac{\rho(\vec{r}\,')}{|\vec{r}-\vec{r}\,'|} d^3r' + \frac{1}{4\pi}\oint_{S'} \left(\frac{\partial V/\partial n}{|\vec{r}-\vec{r}\,'|} - V\frac{\partial}{\partial n}\frac{1}{|\vec{r}-\vec{r}\,'|}\right) dS' \tag{6–33}$$

The first integral is well known to us as the Coulomb integral over the source charge density. The surface integrals reflect the contribution of induced charges or fields on the boundary to the potential within the volume. The numerator of the first term of the surface integral, $\partial V/\partial n$, is just the normal component of the field at the boundary. The potential due to this term is the same as that from surface charge $\sigma = -\varepsilon_0\partial V/\partial n$. The last term is the potential corresponding to a dipole layer $\vec{D} = \varepsilon_0 V\hat{n}$ on the surface. The surface charge and dipole layer need not be real; we have shown only that the potential inside a volume can always be expressed in these terms. It is interesting that higher multipole fields from outside the boundary do not affect the potential inside the boundary. If the surface integral vanishes as it would when $S' \to \infty$, we recover the familiar Coulomb result.

The choice of $\Psi(\vec{r},\vec{r}\,')$ is not unique, as $\nabla^2\Psi = -\delta(\vec{r}-\vec{r}\,')$ is satisfied by

$$\Psi(\vec{r},\vec{r}\,') = \frac{1}{4\pi|\vec{r}-\vec{r}\,'|} + \mathrm{F}(\vec{r},\vec{r}\,') \tag{6–34}$$

where F satisfies $\nabla^2\mathrm{F}(\vec{r},\vec{r}\,') = 0$. This freedom makes it possible to choose Ψ such that the potential, $V(\vec{r})$, calculated inside the boundary depends explicitly only on either V or on the value of its normal derivative $\partial V/\partial n$ (the electric field), on S'. We handle the Dirichlet and Neumann problems separately.

If the boundary condition specifies V on S' (Dirichlet problem), we choose $\Psi = 0$ on S', in which case (6–32) reduces to

$$V(\vec{r}) = \frac{1}{\varepsilon_0}\int_\tau \rho(\vec{r}\,')\Psi(\vec{r},\vec{r}\,')d^3r' - \oint_{S'} V(\vec{r}\,')\frac{\partial\Psi(\vec{r},\vec{r}\,')}{\partial n}dS' \tag{6–35}$$

and, assuming Ψ is known, solving Poisson's equation is reduced to evaluating two definite integrals.

If instead the boundary condition specifies the normal derivative $\partial V/\partial n$ on S' (the Neumann problem), it would be tempting to choose $\partial\Psi/\partial n = 0$, but this would lead to an inconsistency with the definition of Ψ. Because the demonstration of the inconsistency holds the resolution, we produce it below. Since

$$\nabla^2\Psi(\vec{r},\vec{r}\,') = \vec{\nabla}\cdot(\vec{\nabla}\Psi) = -\delta(\vec{r}-\vec{r}\,') \tag{6–36}$$

we find, on integrating $\nabla^2\Psi$ over r', that

$$\int_\tau \nabla^2\Psi d^3r' = -1 = \oint_{S'} \vec{\nabla}\Psi\cdot d\vec{S}' = \oint_{S'} \frac{\partial\Psi}{\partial n}dS' \tag{6–37}$$

which cannot be satisfied when $\partial\Psi/\partial n = 0$ on S'. In fact, choosing

$$\left.\frac{\partial\Psi}{\partial n}\right|_{r' \text{ on } S'} = -\frac{1}{S'} \tag{6–38}$$

does satisfy the δ function normalization, and with this replacement (6–32) becomes

$$\begin{aligned} V(\vec{r}) &= \frac{1}{\varepsilon_0}\int_\tau \rho(\vec{r}\,')\Psi(\vec{r},\vec{r}\,')\,d^3r' + \oint_{S'}\left(V(\vec{r}\,')\frac{1}{S'} + \frac{\partial V(\vec{r}\,')}{\partial n}\Psi(\vec{r},\vec{r}\,')\right)dS' \\ &= \langle V\rangle + \int_\tau \frac{\rho(\vec{r}\,')}{\varepsilon_0}\Psi(\vec{r},\vec{r}\,')d^3r' + \oint_{S'}\frac{\partial V(\vec{r}\,')}{\partial n}\Psi(\vec{r},\vec{r}\,')dS' \end{aligned} \tag{6–39}$$

where $\langle V\rangle$ is the average potential on the boundary,

$$\langle V\rangle = \frac{1}{S'}\oint_{S'} V(\vec{r}\,')dS' \tag{6–40}$$

Once the Green's function is known, the solution of Poisson's equation for all problems with that boundary reduces to several integrations. We now proceed to obtain the Green's function for the special case of the Dirichlet problem with a spherical (inner or outer) boundary. We will henceforth drop the masquerade and denote the Green's function Ψ by $\mathrm{G}(\vec{r},\vec{r}\,')$.

6.3.3 Expansion of the Dirichlet Green's Function in Spherical Harmonics

The potential of a point charge q located at $\vec{r}\,'$ in the vicinity of a grounded sphere solves the equation

$$\nabla^2 V = -\frac{q\delta(\vec{r}-\vec{r}\,')}{\varepsilon_0} \tag{6–41}$$

with the boundary condition $V = 0$ at the surface of the sphere. Except for the size of the source inhomogeneity, this is precisely the same equation and boundary condition that the Green's function must satisfy. We may therefore conclude that the expression (6–13) gives us the Green's function for the problem of a charge distribution in the neighborhood of (but exterior to) a spherical boundary of radius a, namely

$$G(\vec{r},\vec{r}\,') = \frac{1}{4\pi}\left(\frac{1}{|\vec{r}-\vec{r}\,'|} - \frac{a}{r'|\vec{r}-(a^2/r'^2)\vec{r}\,'|}\right) \tag{6–42}$$

We have previously expressed the first term in spherical polar coordinates (Section 2.1.2), using as an intermediate step the generating function for Legendre polynomials (F–35)

$$\frac{1}{|\vec{r}-\vec{r}\,'|} = \frac{1}{r_>}\sum_{\ell=0}^{\infty}\left(\frac{r_<}{r_>}\right)^{\ell} P_\ell(\cos\gamma) \tag{6–43}$$

where γ is the angle between $\vec{r}$ and $\vec{r}\,'$. A similar expansion may be found for the second term.

$$\frac{a}{r'|\vec{r}-(a^2/r'^2)\vec{r}\,'|} = \frac{a}{\sqrt{r^2r'^2 + a^4 - 2a^2rr'\cos\gamma}} \tag{6–44}$$

If the charge distribution and the field point $\vec{r}$ lie outside the sphere, both $\vec{r}$ and $\vec{r}\,'$ are larger than a and we factor these from the radical in order to obtain a convergent expansion

$$\begin{aligned}\frac{a}{r'|\vec{r}-(a^2/r'^2)\vec{r}\,'|} &= \frac{a}{rr'\sqrt{1+\dfrac{a^4}{r^2r'^2} - \dfrac{2a^2}{rr'}\cos\gamma}}\\ &= \frac{a}{rr'}\sum_{\ell=0}^{\infty}\left(\frac{a^2}{rr'}\right)^{\ell} P_\ell(\cos\gamma)\end{aligned} \tag{6–45}$$

Combining the two terms and rewriting the product rr' as $r_>r_<$, we have

$$\begin{aligned}G(\vec{r},\vec{r}\,') &= \frac{1}{4\pi}\sum_{\ell=0}^{\infty}\left(\frac{r_<^{\ell}}{r_>^{\ell+1}} - \frac{a^{2\ell+1}}{r_<^{\ell+1}r_>^{\ell+1}}\right)P_\ell(\cos\gamma)\\ &= \frac{1}{4\pi}\sum_{\ell=0}^{\infty}\frac{1}{r_>^{\ell+1}}\left(r_<^{\ell} - \frac{a^{2\ell+1}}{r_<^{\ell+1}}\right)P_\ell(\cos\gamma)\end{aligned} \tag{6–46}$$

Now using (F–47)) to rewrite $P_\ell(\cos\gamma)$ in terms of the polar coordinates of $\vec{r}$ and $\vec{r}\,'$, we obtain the desired Green's function for a charge distribution outside the sphere.

$$G(\vec{r},\vec{r}\,') = \sum_{\ell=0}^{\infty}\frac{1}{2\ell+1}\sum_{m=-\ell}^{\ell}\frac{1}{r_>^{\ell+1}}\left(r_<^{\ell} - \frac{a^{2\ell+1}}{r_<^{\ell+1}}\right)Y_\ell^m(\theta,\varphi)Y_\ell^{*m}(\theta',\varphi') \tag{6–47}$$

It might be observed that G vanishes on the boundary, as of course it should.

When the source and field points are inside the sphere, the expansion above no longer converges. We distinguish the result for the source and field points interior to the sphere by renaming the radius b and regain a convergent series by factoring b^2 from the modified radical in (6–44). We proceed in the same fashion to obtain

$$\frac{b}{r'|\vec{r}-(b^2/r'^2)\vec{r}\,'|} = \frac{1}{b\sqrt{1+\dfrac{r^2r'^2}{b^4} - \dfrac{2rr'}{b^2}\cos\gamma}}$$

$$= \frac{1}{b}\sum_{\ell}\left(\frac{rr'}{b^2}\right)^{\ell} \mathrm{P}_\ell(\cos\gamma) \tag{6–48}$$

As before, we combine the two terms in G and rewrite the product rr' as $r_>r_<$ to get

$$\mathrm{G}(\vec{r},\vec{r}\,') = \sum_{\ell,m}\frac{1}{2\ell+1}\left[\frac{1}{r_>}\left(\frac{r_<}{r_>}\right)^{\ell} - \frac{1}{b}\left(\frac{r_<r_>}{b^2}\right)^{\ell}\right]\mathrm{Y}_\ell^m(\theta,\varphi)\mathrm{Y}_\ell^{*m}(\theta',\varphi')$$

$$= \sum_{\ell,m}\frac{r_<^\ell}{2\ell+1}\left(\frac{1}{r_>^{\ell+1}} - \frac{r_>^\ell}{b^{2\ell+1}}\right)\mathrm{Y}_\ell^m(\theta,\varphi)\mathrm{Y}_\ell^{*m}(\theta',\varphi') \tag{6–49}$$

Again it is clear that the Green's function vanishes at the surface of the enclosing sphere where $r = r_> = b$.

EXAMPLE 6.5: Find the potential due to a uniformly charged thin circular disk of radius a bearing charge Q placed in the center of a grounded conducting sphere of radius $b > a$.

Solution: The charge density on the disk for $r' < a$ may be written

$$\rho(\vec{r}\,') = \frac{Q\delta(\cos\theta')}{\pi a^2 r'} \tag{Ex 6.5.1}$$

as is readily verified by integrating the density over the volume of a sphere containing part of the disk. In general, the potential inside the sphere may be found from

$$V(\vec{r}) = \frac{1}{\varepsilon_0}\int_\tau \rho(\vec{r}\,')\mathrm{G}(\vec{r},\vec{r}\,')d^3r' - \oint_{S'} V(\vec{r}\,')\frac{\partial \mathrm{G}}{\partial n}dS' \tag{Ex 6.5.2}$$

As the sphere is grounded, the surface integral vanishes, reducing V to

$$V = \frac{1}{\varepsilon_0}\int_\tau \rho(\vec{r}\,')\sum_\ell\sum_m \frac{\mathrm{Y}_\ell^m(\theta,\varphi)\mathrm{Y}_\ell^{*m}(\theta',\varphi')}{2\ell+1} r_<^\ell\left(\frac{1}{r_>^{\ell+1}} - \frac{r_>^\ell}{b^{2\ell+1}}\right)d^3r' \tag{Ex 6.5.3}$$

Since V cannot depend on φ, only the $\mathrm{Y}_\ell^0(\theta,\varphi) = \sqrt{(2\ell+1)/4\pi}\,\mathrm{P}_\ell(\cos\theta)$ can enter into the sum. We therefore write

$$V = \frac{1}{4\pi\varepsilon_0}\sum_\ell\int_\tau \mathrm{P}_\ell(\cos\theta)\mathrm{P}_\ell(\cos\theta')\rho(\vec{r}\,')r_<^\ell\left(\frac{1}{r_>^{\ell+1}} - \frac{r_>^\ell}{b^{2\ell+1}}\right)d^3r'$$

$$= \frac{Q}{4\pi\varepsilon_0\pi a^2}\sum_\ell \mathrm{P}_\ell(\cos\theta)\int_\tau \frac{1}{r'}\delta(\cos\theta')\mathrm{P}_\ell(\cos\theta')r_<^\ell\left(\frac{1}{r_>^{\ell+1}} - \frac{r_>^\ell}{b^{2\ell+1}}\right)$$
$$\times r'^2 dr' d(\cos\theta')d\varphi' \tag{Ex 6.5.4}$$

We perform the integration over θ' and φ' to get

$$V = \frac{Q}{2\pi\varepsilon_0 a^2}\sum_\ell \mathrm{P}_\ell(0)\mathrm{P}_\ell(\cos\theta)\int_0^a r_<^\ell\left(\frac{1}{r_>^{\ell+1}} - \frac{r_>^\ell}{b^{2\ell+1}}\right)r'dr' \tag{Ex 6.5.5}$$

We focus our attention on the integral in (Ex 6.5.5) which we abbreviate $\Im(r)$. When $r > a$, r' is $r_<$ throughout the entire range of integration. Hence

$$\Im(r) = \left(\frac{1}{r^{\ell+1}} - \frac{r^\ell}{b^{2\ell+1}}\right)\int_0^a r'^{\ell+1}dr' = \frac{a^{\ell+2}}{\ell+2}\left(\frac{1}{r^{\ell+1}} - \frac{r^\ell}{b^{2\ell+1}}\right) \qquad \text{(Ex 6.5.6)}$$

When $r < a$, the form of the integrand changes as r' passes r. To accommodate this, we split the range of integration into two segments; one where r' is $r_<$ and one where r' is $r_>$. When $\ell \neq 1$,

$$\Im(r) = \left(\frac{1}{r^{\ell+1}} - \frac{r^\ell}{b^{2\ell+1}}\right)\int_0^r r'^{\ell+1}dr' + r^\ell \int_r^a \left(\frac{1}{r'^\ell} - \frac{r'^{\ell+1}}{b^{2\ell,+1}}\right)dr'$$

$$= \frac{1}{(\ell-1)(\ell+2)}\left[(2\ell+1)r - \frac{r^\ell}{a^{\ell-1}}\left(\ell+2 - \frac{(\ell-1)a^{2\ell+1}}{b^{2\ell+1}}\right)\right] \qquad \text{(Ex 6.5.7)}$$

The explicit values of $\mathrm{P}_\ell(0)$, $\mathrm{P}_{2\ell+1}(0) = 0$, and $\mathrm{P}_{2\ell}(0) = (-1)^\ell(2\ell-1)!!/2^\ell \ell!$ may be used to eliminate all odd terms from the series. Making this substitution we get

$$V(r > a) = \frac{Q}{2\pi\varepsilon_0}\sum_{\ell=0}^{\infty}\frac{(-1)^\ell(2\ell-1)!!a^{2\ell}}{2^\ell(2\ell+2)\ell!}\left(\frac{1}{r^{2\ell+1}} - \frac{r^{2\ell}}{b^{4\ell+1}}\right)\mathrm{P}_{2\ell}(\cos\theta) \qquad \text{(Ex 6.5.8)}$$

$$V(r < a) = \frac{Q}{2\pi\varepsilon_0 a^2}\sum_{\ell=0}^{\infty}\frac{(-1)^\ell(2\ell-3)!!}{2^\ell \ell!(2\ell+2)}\left\{(4\ell+1)r - \frac{r^2\ell}{a^{2\ell-1}}\left[(2\ell+2) - \frac{(2\ell-1)a^{4\ell+1}}{b^{4\ell+1}}\right]\right\}\mathrm{P}_{2\ell}(\cos\theta) \qquad \text{(Ex 6.5.9)}$$

EXAMPLE 6.6: Find the potential due to a uniformly charged ring of radius a and total charge Q enclosed concentrically within a grounded conducting sphere of radius b (Figure 6.9).

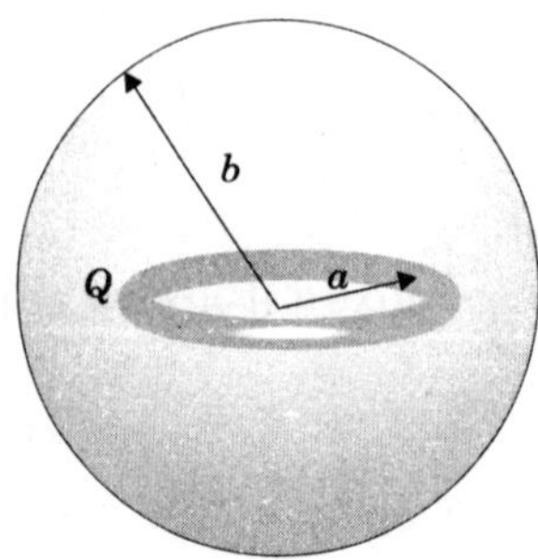

Figure 6.9: A conducting sphere of radius b has a uniformly charged concentric ring of radius a enclosed.

Solution: The charge density on the ring is

$$\rho(\vec{r}\,') = \frac{Q}{2\pi a r'}\delta(r' - a)\delta(\cos\theta') \qquad \text{(Ex 6.6.1)}$$

The sphere is grounded, therefore the surface integral in (6–30) vanishes, reducing the expression for the potential to

$$V = \frac{1}{\varepsilon_0}\int_\tau \rho(\vec{r}\,')\sum_\ell\sum_m \frac{\mathrm{Y}_\ell^m(\theta,\varphi)\mathrm{Y}_\ell^{*m}(\theta',\varphi')}{2\ell+1} r_<^\ell\left(\frac{1}{r_>^{\ell+1}} - \frac{r_>^\ell}{b^{2\ell+1}}\right)d^3r' \qquad \text{(Ex 6.6.2)}$$

and the lack of φ dependence again allows us to replace the spherical harmonics with Legendre polynomials to give

$$\begin{aligned} V &= \frac{1}{4\pi\varepsilon_0}\sum_\ell \int_\tau \mathrm{P}_\ell(\cos\theta)\mathrm{P}_\ell(\cos\theta')\rho(\vec{r}\,')r_<^\ell\left(\frac{1}{r_>^{\ell+1}} - \frac{r_>^\ell}{b^{2\ell+1}}\right)d^3r' \\ &= \frac{Q}{4\pi\varepsilon_0 2\pi a}\sum_\ell \mathrm{P}_\ell(\cos\theta)\int_\tau \delta(r'-a)\delta(\cos\theta')\mathrm{P}_\ell(\cos\theta')r_<^\ell\left(\frac{1}{r_>^{\ell+1}} - \frac{r_>^\ell}{b^{2\ell+1}}\right) \\ &\qquad\qquad \times r'dr'd(\cos\theta')'d\varphi' \qquad \text{(Ex 6.6.3)} \end{aligned}$$

The integration over θ' and φ' is easily performed to give

$$V = \frac{Q}{4\pi\varepsilon_0}\sum \mathrm{P}_\ell(0)\mathrm{P}_\ell(\cos\theta)\int \delta(r'-a)r_<^\ell\left(\frac{1}{r_>^{\ell+1}} - \frac{r_>^\ell}{b^{2\ell+1}}\right)dr' \qquad \text{(Ex 6.6.4)}$$

As in the previous example, the explicit values of $\mathrm{P}_\ell(0)$ eliminate all odd terms from the series. The δ function ensures that $r' = a$. For $r < a$, $r = r_<$, allowing us to write

$$V(r<a) = \frac{Q}{4\pi\varepsilon_0}\sum_{\ell=0}^{\infty}\frac{(-1)^\ell(2\ell-1)!!}{2^\ell \ell!}r^{2\ell}\left(\frac{1}{a^{2\ell+1}} - \frac{a^{2\ell}}{b^{4\ell+1}}\right)\mathrm{P}_{2\ell}(\cos\theta) \qquad \text{(Ex 6.6.5)}$$

while for $r > a$, $r' = r_<$, giving

$$V(r>a) = \frac{Q}{4\pi\varepsilon_0}\sum_{\ell=0}^{\infty}\frac{(-1)^\ell(2\ell-1)!!}{2^\ell \ell!}a^{2\ell}\left(\frac{1}{r^{2\ell+1}} - \frac{r^{2\ell}}{b^{4\ell+1}}\right)\mathrm{P}_{2\ell}(\cos\theta) \qquad \text{(Ex 6.6.6)}$$

In the limit as $b \to \infty$, this should reduce to the expansion for the potential of a charged ring in free space as found in Example 5.8.

This result (Ex 6.6.6) could of course also have been obtained using a ring of image charge at $a' = b^2/a$ bearing total charge $Q' = (-b/a)Q$.

⋆ 6.3.4 Dirichlet Green's function from the Differential equation

When there are both an inner and an outer spherical bounding surfaces there would be an infinite number of images as the image produced by each surface is reflected in the other. Under these conditions it proves necessary to solve the defining differential equation for G, $\nabla^2\mathrm{G} = -\delta(\vec{r}-\vec{r}\,')$. Expressing $\nabla^2\mathrm{G}$ in spherical polars using (42) we have

$$\frac{1}{r}\frac{\partial^2(r\mathrm{G})}{\partial r^2} + \frac{1}{r^2\sin\theta}\frac{\partial}{\partial\theta}\left(\sin\theta\frac{\partial\mathrm{G}}{\partial\theta}\right) + \frac{1}{r^2\sin^2\theta}\frac{\partial^2\mathrm{G}}{\partial\varphi^2} = -\delta(\vec{r}-\vec{r}\,') \qquad (6\text{–}50)$$

Considering $\mathrm{G}(\vec{r},\vec{r}\,')$ as a function of $\vec{r}$ we may generally expand it in spherical polars as

$$\mathrm{G}(\vec{r},\vec{r}\,') = \sum_{\ell,m} A_{\ell,m}(\theta',\varphi')\mathrm{R}_\ell(r,r')\mathrm{Y}_\ell^m(\theta,\varphi) \qquad (6\text{–}51)$$

Substituting this expansion into (6–50) with the help of (F–39) we obtain

$$\sum_{\ell,m}\left(\frac{1}{r}\frac{d^2(r\mathrm{R}_\ell)}{dr^2} - \frac{\ell(\ell+1)\mathrm{R}_\ell}{r^2}\right)A_{\ell,m}(\theta',\varphi')\mathrm{Y}_\ell^m(\theta,\varphi) = -\delta(\vec{r}-\vec{r}\,') \qquad (6\text{–}52)$$

We can also expand the right hand side of (6–52) first in terms of one dimensional δ functions

$$\delta(\vec{r}-\vec{r}\,') = \frac{1}{r^2}\delta(r-r')\,\delta(\cos\theta-\cos\theta')\,\delta(\varphi-\varphi') \qquad (6\text{–}53)$$

and then in terms of spherical harmonics with the aid of the completeness relation (F-46)

$$\delta(\vec{r}-\vec{r}\,') = \frac{1}{r^2}\delta(r-r')\sum_{\ell,m}\mathrm{Y}_\ell^m(\theta,\varphi)\mathrm{Y}_\ell^{*m}(\theta',\varphi') \qquad (6\text{–}54)$$

We immediately identify the coefficient $A_{\ell,m}(\theta',\varphi')$ with $\mathrm{Y}_\ell^{*m}(\theta',\varphi')$ allowing us to remove these two terms from the equation. The remaining equation is

$$\sum_{\ell,m}\left(\frac{1}{r}\frac{d^2(r\mathrm{R}_\ell)}{dr^2} - \frac{\ell(\ell+1)\mathrm{R}_\ell}{r^2}\right)\mathrm{Y}_\ell^m(\theta,\varphi) = -\frac{1}{r^2}\delta(r-r')\sum_{\ell,m}\mathrm{Y}_\ell^m(\theta,\varphi) \qquad (6\text{–}55)$$

Finally we appeal to the linear independence of the Y_ℓ^m to equate the coefficients on the left and the right, resulting in

$$\frac{1}{r}\frac{d^2(r\mathrm{R}_\ell)}{dr^2} - \frac{\ell(\ell+1)\mathrm{R}_\ell}{r^2} = -\frac{1}{r^2}\delta(r-r') \qquad (6\text{–}56)$$

So long as $r \neq r'$, the right hand side of (6–56) vanishes it may be solved for $r\mathrm{R}$ in terms of r

$$r\mathrm{R}_\ell = A_\ell r^{\ell+1} + \frac{B_\ell}{r^\ell} \qquad (6\text{–}57)$$

however, the δ function inhomogeneity at $r = r'$ means the expansion coefficients will differ for $r < r'$ and $r > r'$ so that

$$\mathrm{R}_\ell(r<r') = A_\ell r^\ell + \frac{B_\ell}{r^{\ell+1}} \quad \text{and} \quad \mathrm{R}_\ell(r>r') = C_\ell r^\ell + \frac{D_\ell}{r^{\ell+1}} \qquad (6\text{–}58)$$

The boundary condition on G, and hence on R, is that it vanish on the boundary. At the inner sphere of radius a, $r < r'$ so that

$$\mathrm{R}_\ell(a) = A_\ell a^\ell + \frac{B_\ell}{a^{\ell+1}} \quad \Rightarrow \quad B_\ell = -A_\ell a^{2\ell+1} \tag{6–59}$$

and

$$\mathrm{R}_\ell(b) = C_\ell b^\ell + \frac{D_\ell}{b^{\ell+1}} \quad \Rightarrow \quad C_\ell = -\frac{D_\ell}{b^{2\ell+1}} \tag{6–60}$$

so that, explicitly recognizing that R is a function of both r and r', we rewrite (6–58) as

$$\mathrm{R}_\ell(r < r') = A_\ell(r')\left(r^\ell - \frac{a^{2\ell+1}}{r^{\ell+1}}\right) \tag{6–61}$$

and

$$\mathrm{R}_\ell(r > r') = D_\ell(r')\left(\frac{1}{r^{\ell+1}} - \frac{r^\ell}{b^{2\ell+1}}\right) \tag{6–62}$$

recalling that the Green's function is invariant under interchange of r and r', we can also express these results (6–61) and (6–62) in terms of r and r' interchanged

$$\mathrm{R}_\ell(r < r') = D_\ell(r)\left(\frac{1}{r'^{\ell+1}} - \frac{r'^\ell}{b^{2\ell+1}}\right) \tag{6–63}$$

and

$$\mathrm{R}_\ell(r > r') = A_\ell(r)\left(r'^\ell - \frac{a^{2\ell+1}}{r'^{\ell+1}}\right) \tag{6–64}$$

(6–61) and (6–63) taken together imply that

$$\mathrm{R}_\ell(r < r') = B'_\ell\left(\frac{1}{r'^{\ell+1}} - \frac{r'^\ell}{b^{2\ell+1}}\right)\left(r^\ell - \frac{a^{2\ell+1}}{r^{\ell+1}}\right) \tag{6–65}$$

and (6–62) and (6–64) similarly lead to

$$\mathrm{R}_\ell(r > r') = C'_\ell\left(\frac{1}{r^{\ell+1}} - \frac{r^\ell}{b^{2\ell+1}}\right)\left(r'^\ell - \frac{a^{2\ell+1}}{r'^{\ell+1}}\right) \tag{6–66}$$

Finally, the continuity of G (required to make V continuous at $r = r'$ implies $B'_\ell = C'_\ell$ and we can write (6–65) and (6–66) in the more economical form

$$\mathrm{R}_\ell(r, r') = C'_\ell\left(\frac{1}{r_>^{\ell+1}} - \frac{r_>^\ell}{b^{2\ell+1}}\right)\left(r_<^\ell - \frac{a^{2\ell+1}}{r_<^{\ell+1}}\right) \tag{6–67}$$

The constant C'_ℓ is yet to be determined from the size of the source term, in this case the δ function inhomogeneity. To evaluate the effect of the δ function, we must integrate (6–56) over a vanishingly small interval about r'.

$$\int_{r'-\epsilon}^{r'+\epsilon}\left(\frac{d^2(r\mathrm{R}_\ell)}{dr^2} - \frac{\ell(\ell+1)\mathrm{R}_\ell}{r}\right)dr = -\int_{r'-\epsilon}^{r'+\epsilon}\frac{\delta(r-r')}{r}\,dr \tag{6–68}$$

or

$$\left.\frac{d(rR_\ell)}{dr}\right|_{r'-\epsilon}^{r'+\epsilon} - \left\langle\frac{\ell(\ell+1)R_\ell}{r'}\right\rangle 2\epsilon = -\frac{1}{r'} \tag{6–69}$$

The term

$$\lim_{\epsilon\to 0}\left\langle\frac{\ell(\ell+1)R_\ell}{r'}\right\rangle 2\epsilon = 0 \tag{6–70}$$

leaving

$$\left.\frac{d(rR_\ell)}{dr}\right|_{r'_+} - \left.\frac{d(r\dot{R}_\ell)}{dr}\right|_{r'_-} = -\frac{1}{r'} \tag{6–71}$$

For $r = r'_+$, $r > r'$ and we use (6–66) to compute the required derivative

$$\left.\frac{d(rR_\ell)}{dr}\right|_{r'_+} = C'_\ell\left(r'^\ell - \frac{a^{2\ell+1}}{r'^{\ell+1}}\right)\left[\frac{d}{dr}\left(\frac{1}{r^\ell} - \frac{r^{\ell+1}}{b^{2\ell+1}}\right)\right]_{r=r'}$$

$$= C'_\ell\left(r'^\ell - \frac{a^{2\ell+1}}{r'^{\ell+1}}\right)\left(\frac{-\ell}{r'^{\ell+1}} - \frac{(\ell+1)r'^\ell}{b^{2\ell+1}}\right) \tag{6–72}$$

In the same fashion we use (6-64) (with C'_ℓ substituted for B'_ℓ) to compute the derivative of rR at $r = r'_-$

$$\left.\frac{d(rR_\ell)}{dr}\right|_{r'_-} = C'_\ell\left(\frac{1}{r'^{\ell+1}} - \frac{r'^\ell}{b^{2\ell+1}}\right)\left[\frac{d}{dr}\left(r^{\ell+1} - \frac{a^{2\ell+1}}{r^\ell}\right)\right]_{r=r'}$$

$$= C'_\ell\left(\frac{1}{r'^{\ell+1}} - \frac{r'^\ell}{b^{2\ell+1}}\right)\left((\ell+1)r'^\ell + \frac{\ell a^{2\ell+1}}{r'^{\ell+1}}\right) \tag{6–73}$$

Inserting (6–72) and (6–73) into (6–71) we obtain after some algebra

$$C'_\ell = \frac{1}{(2\ell+1)\left[1 - \left(\dfrac{a}{b}\right)^{2\ell+1}\right]} \tag{6–74}$$

Collecting terms we use (6–74) in (6–67) to construct the Green's function in (6–51)

$$G(\vec{r},\vec{r}') = \sum_{\ell=0}^{\infty}\sum_{m=-\ell}^{\ell}\frac{Y_\ell^m(\theta,\varphi)Y_\ell^{*m}(\theta',\varphi')}{(2\ell+1)\left[1 - \left(\dfrac{a}{b}\right)^{2\ell+1}\right]}\left(r_<^\ell - \frac{a^{2\ell+1}}{r_<^{\ell+1}}\right)\left(\frac{1}{r_>^{\ell+1}} - \frac{r_>^\ell}{b^{2\ell+1}}\right) \tag{6–75}$$

When there is either no inner boundary or no outer bounding sphere, it suffices to let a go to zero (no inner boundary) or let b go to ∞ (no outer boundary) to recover our earlier solutions.

Exercises and Problems

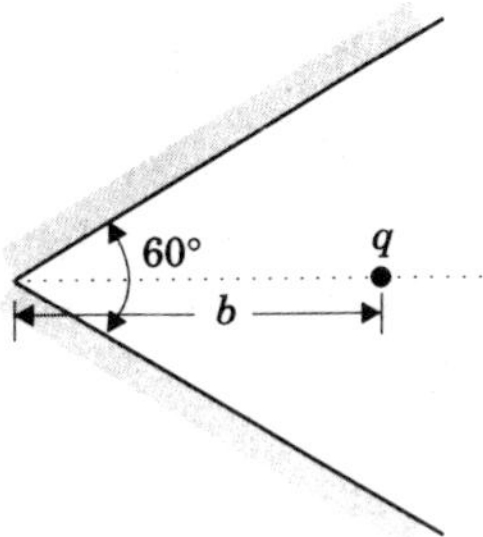

Figure 6.10: The bent, infinite, grounded conducting sheet has a charge q midway between the plates at distance b from the vertex.

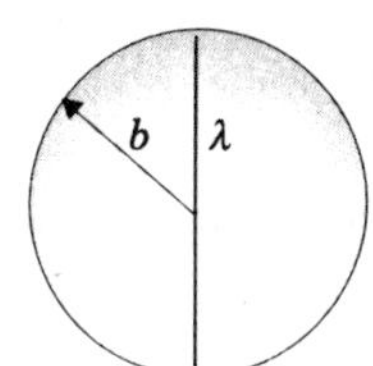

Figure 6.11: A line charge $\lambda = Q/2b$ spans the enclosing sphere along the z axis.

6-1 Find the force between an isolated conducting sphere of radius R and a point charge q in its vicinity

6-2 It is tempting to write the potential of a charge q situated a distance z above a grounded conducting plane as $V = q/(4\pi\varepsilon_0 2z)$. Show that one obtains twice the correct electric field when computing $-\vec{\nabla}V$. Explain the reason for the too-large result.

6-3 Show that when the potential at a charge located at $\vec{r}$ near a grounded conducting sphere due to its image charge is expressed in terms of the charge position $\vec{r}$, $-\vec{\nabla}V$ gives twice the correct electric field at $\vec{r}$.

6-4 Use image charges to obtain the force and the torque on a electric dipole in the vicinity of a grounded conducting plane.

6-5 A point charge is placed at distance a from the left plate between two parallel conducting grounded plates spaced distance D. Find a series expression for the force on the charge.

6-6 A point electric dipole is placed at the center of a conducting sphere. Find the resulting electric field both inside and outside the sphere.

6-7 Use image charges to find the potential along the z axis due to a uniformly charged thin ring of radius a concentric with a larger grounded sphere of radius b (illustrated in Figure 6.9). Generalize this result for $r \in (a, b)$.

6-8 A small (point) electric dipole is placed in the vicinity of a neutral conducting sphere. Find the energy of the dipole due to the induced dipole field of the sphere. This potential has the general form of the dipole-induced dipole encountered in molecular physics.

6-9 Find the force and torque on the dipole in problem 6-8.

6-10 Find the image of a uniform (straight) line charge in the vicinity of a grounded conducting sphere.

6-11 Show that the problem of an uncharged conducting sphere placed in an initially uniform field can be solved by images. (Hint: Place charges $Q = \pm 2\pi\varepsilon_0 L^2 E_0$ at sufficiently large $\pm L$ to produce a uniform electric field E_0 in the region the sphere will be placed.)

6-12 A charge is placed distance b along the bisector of the vertex of a bent, infinite plane conductor as shown in Figure 6.10. Find the force on the charge.

6-13 Generalize problem 6-9 to vertex with a small angle that divides 360° evenly.

6-14 Determine the images of a point charge q placed above the center of a hemispherical bump (radius a) in an otherwise flat plate extending to infinity. Assume all surfaces are grounded conductors.

6-15 A line charge λ is placed inside a grounded, conducting cylinder parallel to the axis of the cylinder. Find the potential inside the cylinder.

6-16 Calculate the capacitance of two nested cylinders whose radii are $a < b$ and whose parallel axes are separated by $D < b - a$.

6-17 A long wire of radius a is placed with its axis parallel to a large conducting (assume infinite) sheet at distance d from the sheet. Find the capacitance of this arrangement.

6-18 Find the capacitance of two long wires of differing radii a and b carrying equal charges of opposite sign distributed over their surfaces.

6-19 A hollow conducting sphere of radius b encloses a uniform line charge $\lambda = Q/2b$ along the z axis (Figure 6.11). Use Green's functions to find the potential interior to the sphere. Hint: $\rho(\vec{r}\,') = Q/4b\pi r'^2[\delta(\cos\theta' - 1) + \delta(\cos\theta' + 1)]$ and care should be taken integrating

$$\int_0^b r_<^\ell \left(\frac{1}{r_>}^{\ell+1} - \frac{r_>^\ell}{b^{2\ell+1}}\right) dr'$$

Break it up into intervals $0 \le r' \le r$. and $r \le r' \le b$ to obtain

$$\int_0^b \left(\cdots\right) dr' = \frac{2\ell+1}{\ell(\ell+1)}\left[1 - \left(\frac{r}{b}\right)^\ell\right]$$

This result is indeterminate for $\ell = 0$; direct integration works best in the $\ell = 0$ case.

6-20 A point charge Q is placed midway between two concentric conducting grounded spheres of radius a and b. For convenience assume the charge to be at $\theta' = 0$. Find the potential at other points between the two spheres.

6-21 Obtain the Green's function for Poisson's equation in cylindrical polar coordinates using a Bessel function expansion for the Dirac δ function.

Chapter 7

Static Electromagnetic Fields in Matter

7.1 The Electric Field Due to a Polarized Dielectric

We begin this discussion with a fairly phenomenological consideration of the electric field arising from charges in matter. We consider a *dielectric* having charges, electric dipoles, quadrupoles, etc. distributed throughout the material. The charge density will, as before, be denoted by $\rho(\vec{r}\,')$; the dipole moment density, or simply the *polarization*, by $\vec{P}(\vec{r}\,')$; and in principle, the quadrupole moment density by $\overleftrightarrow{Q}(\vec{r}\,')$; and so on. The potential at position $\vec{r}$ (with components (x_1, x_2, x_3) due to this distribution is given by

$$
\begin{aligned}
V(\vec{r}) &= \frac{1}{4\pi\varepsilon_0}\int\left[\frac{\rho(\vec{r}\,')}{|\vec{r}-\vec{r}\,'|}+\frac{(\vec{r}-\vec{r}\,')\cdot\vec{P}(\vec{r}\,')}{|\vec{r}-\vec{r}\,'|^3}+\sum_{i,j}\frac{(x_i-x_i')(x_j-x_j')Q_{ij}}{2|\vec{r}-\vec{r}\,'|^5}\cdots\right]d^3r' \\
&= \frac{1}{4\pi\varepsilon_0}\int\left[\frac{\rho(\vec{r}\,')}{|\vec{r}-\vec{r}\,'|}+\sum_i\frac{\partial}{\partial x_i'}\left(\frac{1}{|\vec{r}-\vec{r}\,'|}\right)P_i(\vec{r}\,')\right. \\
&\qquad\left.+\frac{1}{6}\sum_{i,j}\frac{\partial}{\partial x_i'}\frac{\partial}{\partial x_j'}\left(\frac{1}{|\vec{r}-\vec{r}\,'|}\right)Q_{ij}+\cdots\right]d^3r' \qquad (7\text{–}1)
\end{aligned}
$$

If the series is truncated at the dipole moment distribution, we obtain

$$
V(\vec{r}) = \frac{1}{4\pi\varepsilon_0}\int\left[\frac{\rho(\vec{r}\,')}{|\vec{r}-\vec{r}\,'|}+\vec{P}(\vec{r}\,')\cdot\vec{\nabla}'\left(\frac{1}{|\vec{r}-\vec{r}\,'|}\right)\right]d^3r' \qquad (7\text{–}2)
$$

We would like to investigate the effect of matter on Maxwell's equations. We begin with its effect on $\vec{\nabla}\cdot\vec{E}$. In particular, we would like to express the divergence of the electric field in terms of the charge density and the polarization of the material. To this end we write $\vec{E} = -\vec{\nabla}V$, and $\vec{\nabla}\cdot\vec{E} = -\nabla^2 V$. Thus

$$
\vec{\nabla}\cdot\vec{E} = -\frac{1}{4\pi\varepsilon_0}\nabla^2\int\left[\frac{\rho(\vec{r}\,')}{|\vec{r}-\vec{r}\,'|}+\vec{P}(\vec{r}\,')\cdot\vec{\nabla}'\left(\frac{1}{|\vec{r}-\vec{r}\,'|}\right)\right]d^3r'
$$

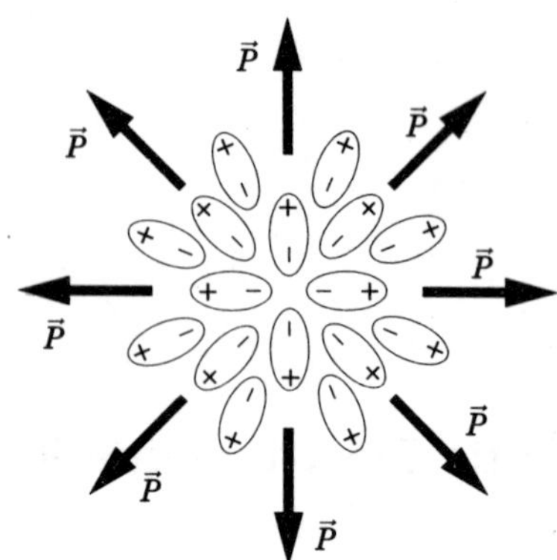

Figure 7.1: The divergent polarization leads to a concentration of negative charge that acts as a source of electric field.

$$= -\frac{1}{4\pi\varepsilon_0}\int\left[\rho(\vec{r}\,')\nabla^2\frac{1}{|\vec{r}-\vec{r}\,'|} + \vec{P}(\vec{r}\,')\cdot\vec{\nabla}'\left(\nabla^2\frac{1}{|\vec{r}-\vec{r}\,'|}\right)\right]d^3r' \qquad (7\text{–}3)$$

Using the relation (26), $\nabla^2\big(1/|\vec{r}-\vec{r}\,'|\big) = -4\pi\delta(\vec{r}-\vec{r}\,')$, we have

$$\vec{\nabla}\cdot\vec{E} = \frac{1}{\varepsilon_0}\int\left[\rho(\vec{r}\,')\delta(\vec{r}-\vec{r}\,') + \vec{P}(\vec{r}\,')\cdot\vec{\nabla}'\delta(\vec{r}-\vec{r}\,')\right]d^3r' \qquad (7\text{–}4)$$

The first term of the integral in (7–4) integrates without difficulty, and the second part can be expanded in Cartesian coordinates

$$\int\vec{P}(\vec{r}\,')\cdot\vec{\nabla}'\delta(\vec{r}-\vec{r}\,') = \int P_x(\vec{r}\,')\frac{\partial}{\partial x'}\left[\delta(x-x')\,\delta(y-y')\,\delta(z-z')\right]dx'dy'dz'$$

$$+\int P_y(\vec{r}\,')\frac{\partial}{\partial y'}\delta(\vec{r}-\vec{r}\,')d^3r' + \int P_z(\vec{r}\,')\frac{\partial}{\partial z'}\delta(\vec{r}-\vec{r}\,')d^3r' \qquad (7\text{–}5)$$

With the help of $\int_{-\infty}^{\infty}f(x)\delta'(x-a)dx = -f'(a)$ we integrate each term. Focusing our attention on the first integral on the right hand side of (7–5) we obtain

$$\int P_x(\vec{r}\,')\frac{\partial}{\partial x'}\delta(x-x')\,\delta(y-y')\,\delta(z-z')dx'dy'dz' = \int P_x(x',y,z)\frac{\partial}{\partial x'}\delta(x-x')dx'$$

$$= -\frac{\partial P_x(x,y,z)}{\partial x} = -\frac{\partial P_x(\vec{r})}{\partial x} \qquad (7\text{–}6)$$

In exactly the same fashion, the second and third integrals of (7–5) integrate respectively to

$$-\frac{\partial P_y(\vec{r})}{\partial y} \quad \text{and} \quad -\frac{\partial P_z(\vec{r})}{\partial z} \qquad (7\text{–}7)$$

to give

$$\vec{\nabla}\cdot\vec{E} = \frac{\rho}{\varepsilon_0} - \frac{\vec{\nabla}\cdot\vec{P}(\vec{r})}{\varepsilon_0} \qquad (7\text{–}8)$$

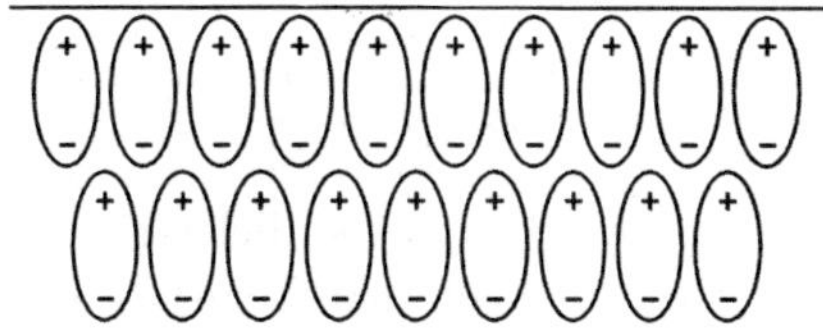

Figure 7.2: At the boundary of a polarized dielectric the exposed ends of the dipoles generate a bound surface charge equal to $\vec{P} \cdot \hat{n}$, where $\hat{n}$ is the outward facing normal.

In other words, $-\vec{\nabla} \cdot \vec{P}$ acts as a source of electric field and is often given the name *bound charge.* Its origin is physically obvious if we consider a collection of dipoles having a nonzero divergence. The divergence of $\vec{P}$ in Figure 7.1 is positive, resulting in an accumulation of negative charge in the center.

At discontinuities in $\vec{P}$, as at the boundaries of dielectrics, we get an accumulation of bound surface charge even for uniform polarizations (Figure 7.2). This observation is readily verified from equation (7–8). Setting the free charge to zero, we write

$$\vec{\nabla} \cdot \vec{E} = -\frac{\vec{\nabla} \cdot \vec{P}}{\varepsilon_0} \tag{7–9}$$

Applying Gauss' law to a small flat pillbox of Figure 7.3 with top and bottom surface parallel to the dielectric surface, we find

$$\int_\tau \vec{\nabla} \cdot \vec{E}\, d^3r = -\int_\tau \frac{\vec{\nabla} \cdot \vec{P}}{\varepsilon_0} d^3r \tag{7–10}$$

whence,

$$\oint_S \vec{E} \cdot d\vec{S} = -\frac{1}{\varepsilon_0} \oint_S \vec{P} \cdot d\vec{S} \tag{7–11}$$

Noting that the surface of the pillbox inside the dielectric is directed opposite to the outward-facing normal $\hat{n}$, we find

$$(\vec{E}_e - \vec{E}_i) \cdot \vec{S}_e = -\frac{1}{\varepsilon_0} \vec{P} \cdot \vec{S}_i = \frac{\vec{P} \cdot \vec{S}_e}{\varepsilon_0} \tag{7–12}$$

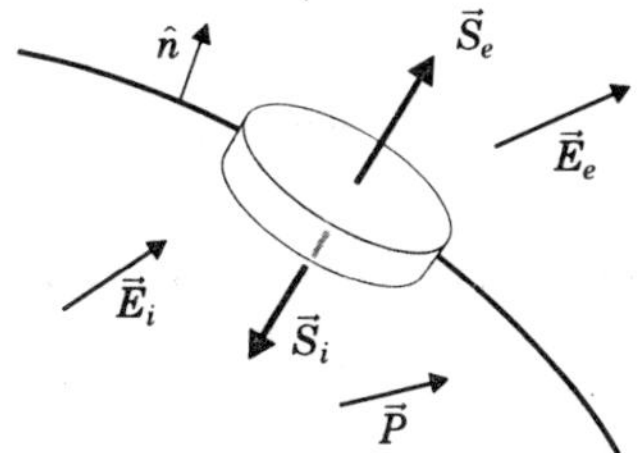

Figure 7.3: The Gaussian pillbox has 'bottom' surface $\vec{S}_i$ with normal facing into the dielectric and 'top' surface $\vec{S}_e$ with normal facing out.

requiring that

$$(\vec{E}_e - \vec{E}_i) \cdot \hat{n} = \frac{1}{\varepsilon_0} \vec{P} \cdot \hat{n} \tag{7–13}$$

where $\hat{n}$ is a normal facing outward from the dielectric (Figure 7.3). This discontinuity in the field is exactly the effect that a surface charge, $\sigma = \vec{P} \cdot \hat{n}$, at the surface of the dielectric would have. In total then, the electric field produced by a polarized dielectric medium is identical to that produced by a bound charge density $\rho_b = -\vec{\nabla} \cdot \vec{P}$ and a bound surface charge density $\sigma_b = \vec{P} \cdot \hat{n}$ on the surface. The resulting electric field may be written

$$\vec{E}(\vec{r}) = \frac{1}{4\pi\varepsilon_0} \int_\tau \frac{\left[-\vec{\nabla}' \cdot \vec{P}(\vec{r}\,')\right](\vec{r} - \vec{r}\,')}{|\vec{r} - \vec{r}\,'|^3} d^3r' + \frac{1}{4\pi\varepsilon_0} \oint_S \frac{(\vec{P} \cdot \hat{n})(\vec{r} - \vec{r}\,')}{|\vec{r} - \vec{r}\,'|^3} dS' \tag{7–14}$$

EXAMPLE 7.1: Find the electric field along the z-axis of a dielectric cylinder of length L and radius a whose axis coincides with the z axis when the cylinder is uniformly polarized along its axis.

Solution: For convenience we take one face, say the bottom, in the x-y plane. According to (7–14) the electric field in the cylinder is given by

$$\begin{aligned}
\vec{E}(z) &= \frac{1}{4\pi\varepsilon_0} \oint \frac{(\vec{P} \cdot \hat{n})(z\hat{k} - \vec{r}\,')}{|\vec{r} - \vec{r}\,'|^3} dS' \\
&= \frac{1}{4\pi\varepsilon_0} \int_0^a \int_0^{2\pi} \frac{-P(z\hat{k} - r\hat{r})r dr d\varphi}{(z^2 + r^2)^{3/2}} + \frac{1}{4\pi\varepsilon_0} \int_0^a \int_0^{2\pi} \frac{P(z\hat{k} - L\hat{k} - r\hat{r})r\, dr d\varphi}{[(z - L)^2 + r^2]^{3/2}} \\
&= \frac{P\hat{k}}{2\varepsilon_0} \left\{ \frac{z}{(z^2 + r^2)^{1/2}} - \frac{(z - L)}{[(z - L)^2 + r^2]^{1/2}} \right\} \Bigg|_0^a \\
&= \frac{P\hat{k}}{2\varepsilon_0} \left\{ \frac{z}{(z^2 + a^2)^{1/2}} - \frac{z - L}{[(z - L)^2 + a^2]^{1/2}} - 2 \right\}
\end{aligned} \tag{Ex 7.1.1}$$

Inside the cylinder, first two terms are both positive and less than one, meaning the electric field points in the negative z direction. If $a \gg L$, the electric field becomes simply $\vec{E} = P\hat{k}/\varepsilon_0$, the field between two parallel plates carrying charge density $\sigma = P$.

7.1.1 Empirical Description of Dielectrics

The molecules of a dielectric may be classed as either polar or nonpolar. We consider first the case of nonpolar molecules. When such molecules (or atoms) are placed in an electric field, the positive charges will move slightly in the direction of the field while the negative charges move slightly in the opposite direction, creating a polarization of the medium.

If, on the other hand, the molecules have intrinsic (permanent) dipole moments that in the absence of an electric field are randomly oriented, they will attempt to align with the electric field, and their non-random alignment will lead to a polarization of the medium. In either case, the resulting polarization will be a function of the local electric field (including that of neighboring dipoles). We may write the empirical relation

$$P^i = P_0^i + \chi_j^i \varepsilon_0 E^j + \chi_{jk}^{i(2)} \varepsilon_0^2 E^j E^k + \chi_{jk\ell}^{i(3)} \varepsilon_0^3 E^j E^k E^\ell + \cdots \tag{7–15}$$

(summation over repeated indices is implied). For isotropic materials, only the diagonal terms of the *dielectric susceptibility* tensor χ survive, and (7–15) becomes merely a power series expansion for the polarization $\vec{P}$.

Materials exhibiting large spontaneous polarization are known as ferroelectrics (clearly there must be polar molecules involved). In analogy to magnets, ferroelectric objects are known as electrets. The best known example of a ferroelectric crystal is $BaTiO_3$. Mechanical distortions of the crystal may result in large changes of the polarization, giving rise to *piezoelectricity.* Similarly, changes in temperature give rise to *pyroelectricity.*

A number of crystals have sufficiently large second or third order susceptibility that optical radiation traversing the crystal may excite a polarization with $\cos^2 \omega t$ or $\cos^3 \omega t$ dependence giving rise to the generation of frequency doubled or tripled light. The efficiency of such doubling or tripling would be expected to increase—linearly for doubling or quadratically for tripling—with incident field strength.

In sufficiently small electric fields, the relationship between $\vec{E}$ and $\vec{P}$ for isotropic materials becomes simply

$$\vec{P} = \chi \varepsilon_0 \vec{E} \tag{7–16}$$

The constant χ is called the *linear dielectric susceptibility* of the dielectric. The ratio between the induced molecular dipole and $\varepsilon_0 \vec{E}$, the polarizing field, is known as the *polarizability*, α. Thus

$$\vec{p} = \alpha \varepsilon_0 \vec{E} \tag{7–17}$$

7.1.2 Electric Displacement Field

The calculation of the microscopic field $\vec{E}$ arising from charges and molecular dipoles of the medium requires considerable care. It is frequently useful to think of the polarization of the medium as merely a property of the medium rather than as a source of field. To do so requires the definition of the *electric displacement* field. The differential equation (7–8),

$$\vec{\nabla} \cdot \vec{E} = \frac{\rho}{\varepsilon_0} - \frac{\vec{\nabla} \cdot \vec{P}}{\varepsilon_0}$$

is more conveniently written

$$\vec{\nabla} \cdot (\varepsilon_0 \vec{E} + \vec{P}) = \rho \tag{7–18}$$

The quantity $\vec{D} \equiv \varepsilon_0 \vec{E} + \vec{P}$ is the *electric displacement field.* In terms of $\vec{D}$, Maxwell's first equation becomes

$$\vec{\nabla} \cdot \vec{D} = \rho \tag{7–19}$$

The dipoles of the medium are not a source for $\vec{D}$; only the so-called free charges act as sources.

When $\vec{P}$ can be adequately approximated by $\vec{P} = \chi\varepsilon_0\vec{E}$ we find that

$$\vec{D} = \varepsilon_0\vec{E} + \chi\varepsilon_0\vec{E} = \varepsilon_0(1+\chi)\vec{E} = \varepsilon\vec{E} \tag{7–20}$$

The constant ε is called the *permittivity* of the dielectric. The *dielectric constant*, κ, is defined by

$$\kappa \equiv \frac{\varepsilon}{\varepsilon_0} = 1 + \chi \tag{7–21}$$

In general, because it takes time for dipoles to respond to the applied field, all three constants–χ, ε, and κ–are frequency dependent.

Anticipating the boundary condition implied by (7–19), namely that $\vec{D}_\perp$ is continuous across a dielectric interface bearing no free charge, we illustrate these ideas with a simple example.

EXAMPLE 7.2: A large parallel plate capacitor has a potential V applied across its plates. A slab of dielectric with dielectric constant κ fills 9/10 of the gap between its plates with air ($\kappa_{air} = 1$) filling the remaining space. Find the resulting electric field in the air and in the dielectric between the capacitor plates as well as the capacitance. Assume a separation t between the plates.

Solution: We take the lower capacitor plate to lie in the x-y plane and for simplicity assume the air layer is the top 10%. In terms of the electric field E_a in air and the electric field E_d in the dielectric, the potential difference V between the plates is

$$V = 0.1tE_a + 0.9tE_d \tag{Ex 7.2.1}$$

The continuity of the perpendicular components of $\vec{D}$ gives $\varepsilon_0 E_a = \varepsilon E_d$, or $E_a = \kappa E_d$. With this substitution, we solve the equation above to get

$$E_d = \frac{V}{(0.9+0.1\kappa)t} \quad \text{and} \quad E_a = \left(\frac{\kappa}{0.9+0.1\kappa}\right)\frac{V}{t} \tag{Ex 7.2.2}$$

Note that the electric field in the air space is κ times as large as that in the dielectric.

The stored charge is most easily found from the air value of the electric field since $\sigma = \varepsilon_0 E_\perp$. We have then

$$Q = \left(\frac{\varepsilon_0\kappa A}{0.9+0.1\kappa}\right)\frac{V}{t} \tag{Ex 7.2.3}$$

leading to capacitance

$$C = \frac{Q}{V} = \frac{\varepsilon_0\kappa A}{(0.9+0.1\kappa)t} \tag{Ex 7.2.4}$$

As the fraction of air layer decreases, the capacitance tends to κ times that of the air spaced capacitor. The electric field in the dielectric is evidently smaller than it is in air. The diminution may be attributed to shielding produced by the aligned dipoles at the air-dielectric interface. The thin air layer, on the other hand, has a significantly increased electric field due to the dielectric, and is a major contributor to the voltage difference between the plates.

7.2 Magnetic Induction Field Due to a Magnetized Material

Just as a material might contain within it a smooth distribution of electric dipoles $\vec{p}$, it can also contain within it a smooth distribution of magnetic dipoles $\vec{m}$. The vector potential due to a molecular dipole $\vec{m}$ at position $\vec{r}_j$ is given by a slight generalization of (2–23) as

$$\vec{A}(\vec{r}) = \frac{\mu_0}{4\pi}\frac{\vec{m}\times(\vec{r}-\vec{r}_j)}{|\vec{r}-\vec{r}_j|^3} \tag{7–22}$$

For n such molecular magnetic dipoles per unit volume, the magnetization $\vec{M}$ is defined by $\vec{M} \equiv n\langle\vec{m}\rangle$. The vector potential arising from both currents and magnetic dipoles in the material is then

$$\vec{A}(\vec{r}) = \frac{\mu_0}{4\pi}\int_\tau \frac{\vec{J}(\vec{r}\,')d^3r'}{|\vec{r}-\vec{r}\,'|} + \frac{\mu_0}{4\pi}\int_\tau \frac{\vec{M}(\vec{r}\,')\times(\vec{r}-\vec{r}\,')}{|\vec{r}-\vec{r}\,'|^3}d^3r' \tag{7–23}$$

The second integral may be written

$$\int \frac{\vec{M}(\vec{r}\,')\times(\vec{r}-\vec{r}\,')}{|\vec{r}-\vec{r}\,'|^3}d^3r' = \int \vec{M}(\vec{r}\,')\times\vec{\nabla}'\left(\frac{1}{|\vec{r}-\vec{r}\,'|}\right)d^3r' \equiv \Im \tag{7–24}$$

Using the identity (6), $\vec{\nabla}\times(\Psi\vec{M}) = \vec{\nabla}\Psi\times\vec{M} + \Psi\vec{\nabla}\times\vec{M}$,

$$\begin{aligned}\Im &= \int \frac{\vec{\nabla}'\times\vec{M}(\vec{r}\,')}{|\vec{r}-\vec{r}\,'|}d^3r' - \int \vec{\nabla}'\times\left(\frac{\vec{M}(\vec{r}\,')}{|\vec{r}-\vec{r}\,'|}\right)d^3r' \\ &= \int_\tau \frac{\vec{\nabla}'\times\vec{M}(\vec{r}\,')}{|\vec{r}-\vec{r}\,'|}d^3r' + \oint_{S'} \frac{\vec{M}(\vec{r}\,')}{|\vec{r}-\vec{r}\,'|}\times d\vec{S}'\end{aligned} \tag{7–25}$$

where we have used (21) for the last step. If $\vec{M}$ is localized to a finite region and S' lies outside this region (this must necessarily be so, as the volume of integration τ in (7–23) must include all $\vec{M}$ and $\vec{J}$), the second integral vanishes, giving

$$\vec{A}(\vec{r}) = \frac{\mu_0}{4\pi}\int \frac{\vec{J}(\vec{r}\,') + \vec{\nabla}'\times\vec{M}(\vec{r}\,')}{|\vec{r}-\vec{r}\,'|}d^3r' \tag{7–26}$$

Thus, according to (7–26), the magnetization of the medium contributes to the vector potential like an effective current $\vec{J}_m = \vec{\nabla}\times\vec{M}$.

To obtain a modified Maxwell equation for $\vec{\nabla}\times\vec{B}$, we note $\vec{\nabla}\times\vec{B} = \vec{\nabla}\times(\vec{\nabla}\times\vec{A}) = \vec{\nabla}(\vec{\nabla}\cdot\vec{A}) - \nabla^2\vec{A}$. In the Coulomb gauge (electro -/magneto- statics) we set $\vec{\nabla}\cdot\vec{A} = 0$ and use (26) to obtain,

$$\vec{\nabla}\times\vec{B}(\vec{r}) = -\nabla^2\frac{\mu_0}{4\pi}\int \frac{\vec{J}(\vec{r}\,') + \vec{\nabla}'\times\vec{M}(\vec{r}\,')}{|\vec{r}-\vec{r}\,'|}d^3r'$$

$$= \mu_0 \int \left[\vec{J}(\vec{r}\,') + \vec{\nabla}' \times \vec{M}(\vec{r}\,')\right] \delta(\vec{r} - \vec{r}\,') d^3r'$$

$$= \mu_0 \left[\vec{J}(\vec{r}) + \vec{\nabla} \times \vec{M}(\vec{r})\right] \tag{7–27}$$

Again we find that, according to equation (7–27), the curl of the magnetization behaves like a current density. Just as the polarization contributes an effective surface charge at discontinuities, a discontinuity in magnetization, as, for example, at the boundaries of magnetic materials, contributes an effective surface *current.*

In arriving at (7–26), we eliminated the surface integral by forcing the volume of integration τ to extend beyond the region containing magnetized materials. The penalty for this procedure is that the discontinuity of $\vec{M}$ at the boundary of the material makes $\vec{\nabla} \times \vec{M}$ undefined. If we had instead taken the region of integration to be the "open" region containing all the magnetization, but not the boundary, we could exclude the discontinuity from the region where the curl must be computed. We would then obtain, instead of (7-26),

$$\vec{A}(\vec{r}) = \frac{\mu_0}{4\pi} \int_{S' \notin \tau} \frac{\vec{J}(\vec{r}\,') + \vec{\nabla}' \times \vec{M}(\vec{r}\,')}{|\vec{r} - \vec{r}\,'|} d^3r' + \frac{\mu_0}{4\pi} \oint_{S'} \frac{\vec{M}(\vec{r}\,') \times d\vec{S}'}{|\vec{r} - \vec{r}\,'|} \tag{7–28}$$

The added term is just that which would have been produced by a surface current $\vec{j} = \vec{M} \times \hat{n}$. This same conclusion can of course be reached from (7–27), as is shown below. Consider a uniformly magnetized bar with magnetization $\vec{M} = M_z\hat{k}$. Inside the magnet, $\vec{\nabla} \times \vec{M} = 0$, implying that $\vec{\nabla} \times \vec{B} = \mu_0 \vec{J}$; the uniform magnetization makes no contribution to the induction field! At the boundary, since $\vec{\nabla} \times \vec{M}$ is undefined, we resort to a different stratagem. We draw a thin rectangular loop with long sides parallel to the magnetization $\vec{M}$ straddling the boundary (Figure 7.4) and integrate the curl equation over the area included in the loop:

$$\int (\vec{\nabla} \times \vec{B}) \cdot d\vec{S} = \mu_0 \int (\vec{\nabla} \times \vec{M}) \cdot d\vec{S} \tag{7–29}$$

For simplicity we have taken $\vec{J} = 0$ and take $\vec{M}$ to be directed along z.

By means of Stokes' theorem (18) both surface integrals may be recast as line integrals

$$\oint \vec{B} \cdot d\vec{\ell} = \mu_0 \oint \vec{M} \cdot d\vec{\ell} \tag{7–30}$$

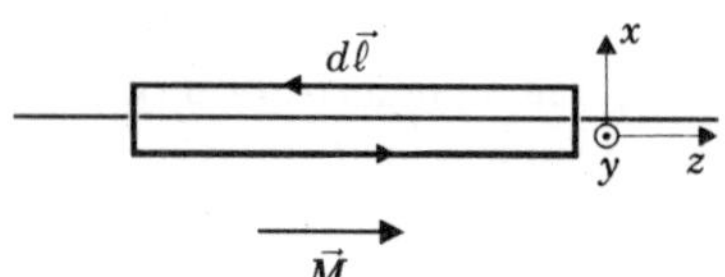

Figure 7.4: The Ampèrian loop lies in the x-z plane straddling the boundary.

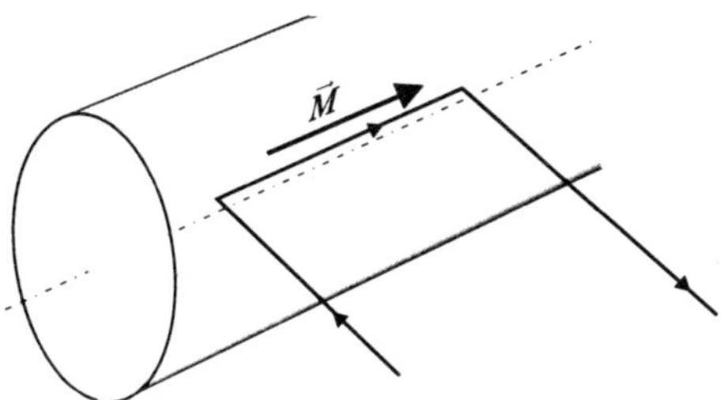

Figure 7.5: The short side of the loop, inside the bar magnet, is taken parallel to the magnetization.

which, after letting the width shrink to zero, gives $B_z^{int} - B_z^{ext} = \mu_0 M$. A surface current $\vec{j} = M\hat{\jmath}$ would produce exactly this kind of discontinuity in $\vec{B}$. Thus, at the boundary, the effect of a discontinuity in the magnetization is exactly the same as that of a surface current $\vec{j} = \vec{M} \times \hat{n}$.

EXAMPLE 7.3: Find the magnetic induction field inside a long (assume infinite) bar magnet with uniform magnetization $\vec{M}$.

Solution: We integrate the equation

$$\vec{\nabla} \times \vec{B}(\vec{r}) = \mu_0 \vec{\nabla} \times \vec{M}(\vec{r}) \tag{Ex 7.3.1}$$

over the area of a rectangular loop with one short side inside the magnet and its other short side sufficiently far removed that $\vec{B}$ vanishes, as shown in Figure 7.5:

$$\int (\vec{\nabla} \times \vec{B}) \cdot d\vec{S} = \mu_0 \int (\vec{\nabla} \times \vec{M}) \cdot d\vec{S} \tag{Ex 7.3.2}$$

Applying Stokes' theorem to both integrals we obtain

$$\oint \vec{B} \cdot d\vec{\ell} = \mu_0 \oint \vec{M} \cdot d\vec{\ell} \tag{Ex 7.3.3}$$

The translational invariance of the problem makes the integral of $\vec{B}$ along the long (perpendicular to the bar) side cancel (they would each vanish when they're perpendicular to $\vec{B}$ in any case). The integral along the distant short side also vanishes since $\vec{B}$ vanishes, reducing the integral for $\vec{B}$ to $B_z L$. Similarly, the integral of the magnetization reduces to $M_z L$. We conclude therefore that the magnetic induction field inside the magnet is $\vec{B} = \mu_0 \vec{M}$. Having found $\vec{B}$ inside the magnet, we can now repeat the argument with the distant end of the loop brought close to the magnet to obtain $\vec{B} = 0$ outside the magnet.

EXAMPLE 7.4: Find the magnetic induction field along the center line of a uniformly magnetized cylindrical bar magnet of length L.

Solution: Recognizing that we may replace the magnetization by a surface current, we note that this problem is identical to that of a solenoid of length L with azimuthal current $I = ML/N$. The equivalent current density along the periphery of the magnet is $j = M$, giving a current $dI = j'dz' = M'dz'$ for a current "loop" of width dz' centered at z'. According to example 1.8, the field due to a circular plane loop of radius R, carrying current dI centered at z', is

$$d\vec{B}(0,0,z) = \frac{\mu_0 R^2 dI\hat{k}}{2\left[R^2 + (z-z')^2\right]^{3/2}} = \frac{\mu_0 R^2 \vec{M}'dz'}{2\left[R^2 + (z-z')^2\right]^{3/2}} \tag{Ex 7.4.1}$$

Integrating this expression from $-\frac{1}{2}L$ to $\frac{1}{2}L$ to obtain the contribution from all parts of the magnet, we find that

$$\vec{B}(0,0,z) = \frac{\mu_0}{2}\int_{-\frac{1}{2}L}^{\frac{1}{2}L} \frac{\vec{M}R^2 dz'}{\left[R^2 + (z-z')^2\right]^{3/2}} = \frac{\mu_0 R^2 \vec{M}}{2} \cdot \frac{-(z-z')}{R^2\sqrt{R^2 + (z-z')^2}}\Bigg|_{-\frac{1}{2}L}^{\frac{1}{2}L}$$

$$= \frac{\mu_0 \vec{M}}{2}\left[\frac{\frac{1}{2}L + z}{\sqrt{(\frac{1}{2}L + z)^2 + R^2}} + \frac{\frac{1}{2}L - z}{\sqrt{(\frac{1}{2}L - z)^2 + R^2}}\right] \tag{Ex 7.4.2}$$

Note the resemblance of this result to the equivalent result (Ex 7.1.1) for the polarized dielectric cylinder if this result is restated for $\vec{D} = \varepsilon_0\vec{E} + \vec{P}$ and the appropriate translation of the origin is made.

7.2.1 Magnetic Field Intensity

As was true in the case of electric polarization, it is frequently preferable to ascribe the magnetization to the medium as an attribute rather than having it act as a source of the appropriate field. We write equation (7-16), $\vec{\nabla} \times \vec{B} = \mu_0(\vec{J} + \vec{\nabla} \times \vec{M})$, in more convenient form. Gathering all the curl terms on the left, we obtain

$$\vec{\nabla} \times \left(\frac{\vec{B}}{\mu_0} - \vec{M}\right) = \vec{J} \tag{7–31}$$

The quantity $\vec{H} \equiv \vec{B}/\mu_0 - \vec{M}$ is called the *magnetic field intensity.* $\vec{H}$ satisfies

$$\vec{\nabla} \times \vec{H} = \vec{J} \tag{7–32}$$

Because $\vec{H}$ is directly proportional to the controllable variable $\vec{J}$, the relation between $\vec{B}$ and $\vec{H}$ is conventionally regarded as $\vec{B}$ being a function of $\vec{H}$ with $\vec{H}$ the independent variable. This perspective leads us to write $\vec{B} = \mu_0[\vec{H} + \vec{M}(\vec{H})]$. In the linear isotropic approximation, $\vec{M}(\vec{H}) = \chi_m\vec{H}$, where χ_m is known as the *magnetic susceptibility.* Under this approximation,

$$\vec{B} \equiv \mu_0(\vec{H} + \vec{M}) = \mu_0(1 + \chi_m)\vec{H} \equiv \mu\vec{H} \tag{7–33}$$

where μ is known as the *magnetic permeability.* Materials with $\chi_m > 1$ are *paramagnetic* whereas those with $\chi_m < 0$ are *diamagnetic.* For most materials, both paramagnetic and diamagnetic, $|\chi_m| \ll 1$. When $\chi_m \gg 1$, the material is ferromagnetic. In this latter case the relation $\vec{B} = \mu_0(\vec{H} + \vec{M})$ still holds, but $\vec{M}$ is usually a very complicated, nonlinear, multivalued function of $\vec{H}$.

It is worth pointing out that to this point $\vec{H}$ and $\vec{D}$ appear to be nothing but mathematical constructs derivable from the fields $\vec{B}$ and $\vec{E}$. The Lorentz force on a charged particle is $\vec{F} = q(\vec{E} + \vec{v} \times \vec{B})$. This means the fields $\vec{E}$ and $\vec{B}$ are detectable through forces they exercise on charges. The magnetic flux through a loop remains $\Phi = \int \vec{B} \cdot d\vec{S}$. However, as we are gradually converting to the view that fields, possessing momentum and energy have an existence independent of interacting particles, there is no reason to suggest that $\vec{D}$ and $\vec{H}$ should have less reality than $\vec{E}$ and $\vec{B}$. As we recast Maxwell's equations into their eventual form, $\vec{H}$ and $\vec{D}$ will assume an equal footing with $\vec{E}$ and $\vec{B}$. It is worthwhile to shift our perspective on the fields, regarding $\vec{E}$ and $\vec{B}$ as the fields responsible for forces on charged particles, whereas $\vec{H}$ and $\vec{D}$ are the fields generated by the sources. In gravitational theory, by contrast, the gravitational mass and inertial mass are identical meaning that the source and force fields are the same.

⋆ 7.3 Microscopic Properties of Matter

In the following sections we will briefly discuss the microscopic behavior of materials responsible for polarization and magnetization. We also touch briefly on conduction in metals for the sake of completeness.

7.3.1 Polar Molecules (Langevin-Debye Formula)

A polar molecule has a permanent dipole moment. If the nearest neighbor interaction energies are small, a material made of such molecules will normally have the dipoles oriented randomly (to maximize entropy) in the absence of an electric field. In an exceedingly strong field all the dipoles will align with the electric field, giving a maximum polarization $\vec{P} = n\vec{p}$ (n is the number density of molecules and $\vec{p}$ is the dipole moment of each). At field strengths normally encountered, thermal randomizing will oppose the alignment to some extent. The average polarization may be found from thermodynamics.

According to Boltzmann statistics, the probability of finding a molecule in a state of energy W is proportional to $e^{-W/kT}$. We consider only the energy of the dipole in the electric field, $W = -\vec{p} \cdot \vec{E} = -pE\cos\theta$. The mean value of $\vec{p}$ must just be the component along $\vec{E}$, the perpendicular components averaging to zero. Hence the mean polarization is

$$\langle p\cos\theta \rangle = \frac{\displaystyle\int p\cos\theta e^{+pE\cos\theta/kT} d\Omega}{\displaystyle\int e^{+pE\cos\theta/kT} d\Omega} = p\left(\coth\frac{pE}{kT} - \frac{kT}{pE}\right) \tag{7–34}$$

a result known as the Langevin formula.

The low field limit of the polarization is readily found. We abbreviate $x = pE/kT$ to write

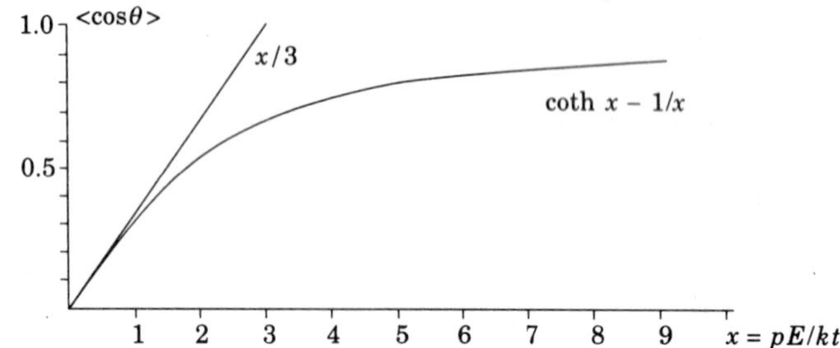

Figure 7.6: The mean dipole moment of a molecule in thermal equilibrium in an electric field. The low field susceptibility is $\frac{1}{3}p^2E/kT$.

$$\langle p\cos\theta \rangle = p\left(\coth x - \frac{1}{x}\right) = p\left(\frac{e^x + e^{-x}}{e^x - e^{-x}} - \frac{1}{x}\right) \tag{7–35}$$

This expression may be expanded for small x as

$$\begin{aligned} p\left[\frac{2+x^2}{2(x+x^3/3!)} - \frac{1}{x}\right] &\simeq p\left[\frac{(2+x^2)(1-x^2/3!)}{2x} - \frac{1}{x}\right] \\ &\simeq p\left[\frac{1}{x} + \frac{(x^2-x^2/3)}{2x} - \frac{1}{x}\right] \\ &= \frac{px}{3} = \frac{p^2E}{3kT} \end{aligned} \tag{7–36}$$

The Langevin-Debye results as well as their small x limit are plotted in Figure 7.6. We obtain the polarization by multiplying the average dipole moment by n, the number of dipoles per unit volume:

$$\vec{P} = \frac{np^2\vec{E}}{3kT} = \chi\varepsilon_0\vec{E} \tag{7–37}$$

The susceptibility is read directly from (7–37).

7.3.2 Nonpolar Molecules

Taking a simple classical harmonic oscillator model for an atom or molecule with 'spring' constant $m\omega_0^2$, we find that the displacement of charge at frequencies well below the resonant (angular) frequency ω_0 is given by

$$\Delta\vec{r} = \frac{q\vec{E}}{m\omega_0^2} \tag{7–38}$$

where q and m are, respectively, the charge and the reduced mass of the electron. The induced molecular dipole moment is then

$$\vec{p} = q\Delta\vec{r} = \frac{q^2\vec{E}}{m\omega_0^2} \tag{7–39}$$

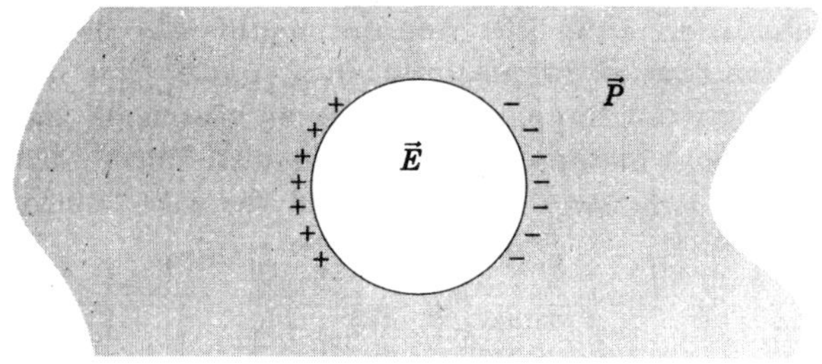

Figure 7.7: The electric field in the cavity is equal to the field in dielectric augmented by the field resulting from the exposed ends of the molecular dipoles.

We deduce that the polarizability is $\alpha = q^2/\varepsilon_0 m\omega_0^2$ and the susceptibility becomes $\chi = nq^2/\varepsilon_0 m\omega_0^2$. Thus for molecular hydrogen with its lowest electronic resonance near $\omega_0 \simeq 1.8 \times 10^{16}$ sec^{-1}($\lambda \simeq 100$nm) at STP (standard temperature and pressure), $n = 2.69 \times 10^{25}$m^{-3}) we obtain $\chi \simeq 2.64 \times 10^{-4}$. The experimental value is also (somewhat fortuitously) 2.64×10^{-4}. Such good agreement should not be expected for substances other than hydrogen and helium; generally a sum over all resonant frequencies is required to obtain reasonable agreement. It is worth noting that this value should be fairly good up to and above optical frequencies. By contrast, the orientation of polar molecules fails for frequencies approaching rotational frequencies of the molecule, typically a few GHz. Thus water has $\chi \simeq 80$ (it has a strong dependence on temperature, varying from 87 at 0° C to 55 at 100° C) at low frequencies, decreasing to $\chi \simeq .8$ at optical frequencies. It will be recognized that an exact calculation of the molecular dipole moment will require the quantum mechanical evaluation of the expectation value of the dipole moment $\langle\psi|\sum e_i\vec{r}_i|\psi\rangle$, where $|\psi\rangle$ is the ground state of the atom or molecule involved.

7.3.3 Dense Media—The Clausius-Mosotti Equation

In the foregoing treatment, we have tacitly assumed that the electric field experienced by a molecule is in fact the average macroscopic field in the dielectric. In gases, where the molecular distances are large, there is little difference between the macroscopic field and the field acting on any molecule. In dense media, however, the closely spaced neighboring dipoles give rise to an internal field $\vec{E}_i$ at any given molecule that must be added to the externally applied field $\vec{E}$. A useful dodge is to exclude the field arising from molecules within some small sphere of radius R about the chosen molecule (small on the scale of inhomogeneities of $\vec{P}$ but still containing many molecules) and then to add the near fields of the molecules contained in the sphere. As we will show in example 7.8, the electric field in the spherical cavity formed by the removal of all the near neighbors is given by

$$\vec{E}_{cav} = \vec{E}_0 + \frac{\vec{P}}{3\varepsilon_0} \qquad (7\text{–}40)$$

The physical origin of the polarization contribution to the field is evident from

Figure 7.7. The calculation of the field from the nearby molecules is more difficult, depending on the structure of the medium. In a simple cubic lattice of dipoles this field vanishes at any lattice point, and it seems reasonable that the field will also vanish for amorphous materials including liquids. Under this condition the polarizing field for the molecule of interest is just the electric field in the cavity (7–40). Therefore, we find

$$\vec{p} = \varepsilon_0 \alpha \left(\vec{E}_0 + \frac{\vec{P}}{3\varepsilon_0} \right) \tag{7–41}$$

The polarization due to n such induced dipoles per unit volume is

$$\vec{P} = n\varepsilon_0 \alpha \left(\vec{E}_0 + \frac{\vec{P}}{3\varepsilon_0} \right) \tag{7–42}$$

which, solved for $\vec{P}$, gives

$$\vec{P} = \frac{n\alpha}{1 - n\alpha/3} \varepsilon_0 \vec{E} \tag{7–43}$$

The electric susceptibility may now be read from (7–43)

$$\chi = \frac{n\alpha}{1 - n\alpha/3} \tag{7–44}$$

The relationship (7–44) is known as the Clausius-Mosotti equation. When $n\alpha$ is small, as is the case for a dilute gas, the $n\alpha/3$ in the denominator is inconsequential. For denser liquids, $n\alpha$ is of order unity and is not negligible.

7.3.4 Crystalline Solids

The near fields on a molecule within a crystal will not vanish for all crystal structures; nonetheless, the net result is generally not large. For the purpose of this discussion let us assume that the we can replace $n\alpha/3$ in (7–43) by $n\alpha/\eta$ with η not very different from 3 to account for the field from nearby molecules.

A number of materials, when cooled in an electric field, freeze in an electric polarization. A piece of such a material is called an electret. Electrets are much less noticeable than magnets because the surfaces very quickly attract neutralizing charges. When the polarization of the electret is changed, however, a net charge will appear on the surface. This change of polarization may be brought about by exceedingly small changes in the physical parameters when $n\alpha$ is near η. Thus, heating a crystal decreases the density, n, giving rise to the pyroelectric effect. Compressing the crystal increases n sometimes producing very large voltages. This and the inverse effect are known as the piezoelectric effect.

At first sight it would appear that there is nothing to prevent $n\alpha$ from exceeding η, resulting in a negative susceptibility, χ. Physically, however, as $n\alpha$ is increased from less than η, the polarization becomes greater, in turn giving an increased local field. If, in small field, $n\alpha$ is larger than η, then the extra field produced by the polarization is larger than the original field producing it. The polarization grows spontaneously until nonlinearities prevent further growth. A material with

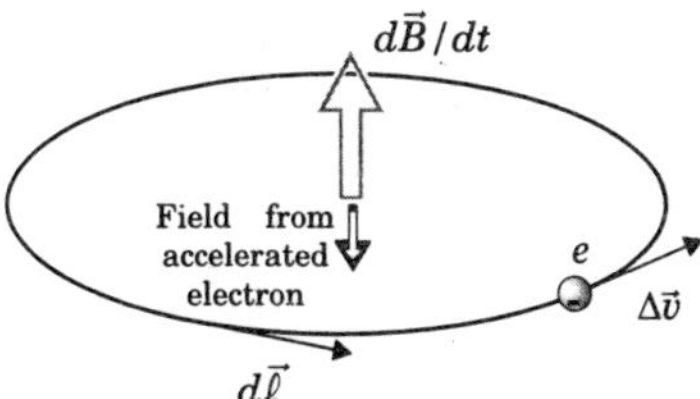

Figure 7.8: As the flux through the loop is increased, the electron is tangentially accelerated. The change in field resulting from this acceleration must oppose the externally imposed $d\vec{B}/dt$.

this property is ferroelectric. On heating the material it is possible to decrease the density until at the Curie point $n\alpha$ no longer exceeds η and the material ceases to have a spontaneous polarization. For $BaTiO_3$ the Curie point is 118° C. Slightly above this temperature χ may be as large as 50,000.

7.3.5 Simple Model of Paramagnetics and Diamagnetics

All materials exhibit diamagnetism. To better understand its origin, let us consider the atoms and molecules of matter as Bohr atoms with electrons in plane orbit about the nucleus. The orbiting electrons have magnetic moments, but because the moments are randomly oriented, no net magnetization results. When a magnetic field is introduced, Lenz' law predicts that the electron orbits ought to change in such a manner that field from their change in magnetic moment opposes the applied field, yielding a negative magnetic susceptibility. Let us make this observation somewhat more quantitative by considering an electron in a circular orbit (Figure 7.8).

The electromotive force around the loop of the electron's orbit is given by

$$\mathcal{E} = \frac{1}{e}\oint \vec{F}\cdot d\vec{\ell} = -\int \frac{d\vec{B}}{dt}\cdot d\vec{S} \tag{7-45}$$

Replacing the force by the rate of change of momentum, we evaluate the two integrals when the loop is perpendicular to the magnetic field to obtain

$$\frac{d|\vec{p}|}{dt}2\pi r = e\pi r^2\frac{dB}{dt} \tag{7-46}$$

Integrating this expression over time, we have

$$\Delta p = \frac{er\Delta B}{2} \tag{7-47}$$

We would like to relate the change of momentum to the change of the (orbital) magnetic moment of the electron. The magnetic moment of an orbiting electron with mass m_e is given by

$$\vec{m} = \frac{e\vec{v}\times\vec{r}}{2} = \frac{e\vec{p}\times\vec{r}}{2m_e} \tag{7-48}$$

leading to change in magnetic moment in response to the introduction of the magnetic field

$$\Delta|\vec{m}| = \frac{e\,|\Delta\vec{p}\times\vec{r}\,|}{2m_e} = \frac{e^2r^2\Delta B}{4m_e} \tag{7-49}$$

For several electrons inside an atom, the planes of the orbit will clearly not all be perpendicular to the field and r^2 should be replaced by $\langle r^2\cos^2\theta\rangle$, where θ is the inclination of the orbit (the component of H perpendicular to the orbit and the component of $\Delta\vec{m}$ along $\vec{H}$ are each decreased by a factor $\cos\theta$). For an isotropic distribution of orbits,

$$\begin{aligned}\langle r^2\cos^2\theta\rangle &= \langle r^2\rangle\int\frac{\cos^2\theta d\Omega}{4\pi}\\ &= \frac{\langle r^2\rangle}{2}\int_0^\pi\cos^2\theta\sin\theta d\theta = \tfrac{1}{3}\langle r^2\rangle\end{aligned} \tag{7-50}$$

Since the currents inside atoms flow without resistance, the dipoles created by the imposition of the field will persist until the magnetic induction field is turned off again. The resulting magnetic susceptibility is $\chi_m = -\frac{1}{12}ne^2\langle r^2\rangle\mu_0/m_e$.

Paramagnetism arises when the molecules' nuclei have a nonzero magnetic moment that attempts to align with the local $\vec{B}$ in much the same fashion that polar molecules align with $\vec{E}$. This interaction tends to be very similar in size to the diamagnetic interaction so that it is hard to predict whether any particular substance will have net positive or negative susceptibility. Because the nuclear magnetic dipole–field interaction is so much smaller than that for the electric field, large alignments are attainable only at very low temperatures. A few molecules with unpaired electrons such as O_2, NO, and $GdCl_3$ have a paramagnetic susceptibility several hundred times larger due to the much larger (spin) magnetic moment of the electron (compared to that of the nucleus).

Although it is tempting to ascribe ferromagnetism to a mechanism similar to that of ferroelectricity, the magnetic dipole–field interaction is so much weaker than the electric dipole–field interaction that thermal agitation would easily overwhelm the aligning tendencies. A much stronger quantum mechanical spin–spin exchange interaction must be invoked to obtain sufficiently large aligning forces. With the exchange force responsible for the microscopic spin–spin interaction, the treatment of ferromagnetism parallels that of ferroelectricity.

In metallic ferromagnets, magnetic moments over large distances (magnetic domains) spontaneously align. An applied field will reorient or expand entire domains, resulting in a very large magnetization. Materials exhibiting ferromagnetism are usually very nonlinear, and the magnetization depends on the history of the material.

7.3.6 Conduction

Qualitatively, a conductor is a material that contains (sub-) microscopic charged particles or charge carriers that are free to move macroscopic distances through the medium. In the absence of an electric field these charges move erratically through the conductor in a random fashion. When an electric field is present, the charges accelerate briefly in the direction of (or opposite to) the field before being scattered by other relatively immobile components. The random component of the carriers' velocity yields no net current, but the short, directed segments yield a drift velocity along the field (for isotropic conductors). This leads us to postulate that

$$\vec{J} = g(E)\vec{E} \tag{7–51}$$

For many materials, $g(E)$ is almost independent of E, in which case the material is labelled *ohmic* with constitutive relation $\vec{J} = g\vec{E}$. The constant g is called the conductivity of the material and is generally a function of temperature as well as dislocations in the material. (Many authors use σ to denote the conductivity.) The resistivity $\eta \equiv 1/g$ is also frequently employed.

A rough microscopic description can be given in terms of the carriers' mean time between collisions, τ, since $\langle\vec{v}\rangle \simeq \frac{1}{2}\vec{a}\tau = \frac{1}{2}q\vec{E}\tau/m$. The quantity $q\tau/m$ is commonly called the carrier mobility. Computing the net current density as $\vec{J} = nq\langle\vec{v}\rangle = \frac{1}{2}(nq^2\tau/m)\vec{E}$ where n is the carrier number density, leads us to write the conductivity as

$$g = \frac{nq^2\tau}{2m} \tag{7–52}$$

When a magnetic field is present, we expect the current to be influenced by the magnetic force on the charge carriers. The modified law of conduction should read

$$\vec{J} = g(\vec{E} + \langle\vec{v}\rangle \times \vec{B}) \tag{7–53}$$

This form of the conduction law governs the decay of magnetic fields in conductors. Substituting (7-53) into Ampère's law and assuming, that inside the conductor $\partial\vec{E}/\partial t$ is sufficiently small to ignore, we have $\vec{\nabla} \times \vec{B} = \mu\vec{J}$. Taking the curl once more we obtain

$$\nabla^2\vec{B} = g\mu\left[\frac{\partial\vec{B}}{\partial t} - \vec{\nabla} \times (\langle\vec{v}\rangle \times \vec{B})\right] \tag{7–54}$$

If the conductor is stationary, the equation above reduces to a diffusion equation.

7.4 Boundary Conditions for the Static Fields

The time-independent simplified Maxwell equations (3–27) are modified in the presence of matter to read

$$\begin{aligned} \vec{\nabla} \cdot \vec{D} &= \rho \qquad & \vec{\nabla} \times \vec{E} &= 0 \\ \vec{\nabla} \cdot \vec{B} &= 0 \qquad & \vec{\nabla} \times \vec{H} &= \vec{J} \end{aligned} \tag{7–55}$$

where $\vec{D} = \varepsilon_0\vec{E} + \vec{P}$ and $\vec{B} = \mu_0(\vec{H} + \vec{M})$. The first and the last of these equations may be integrated to give, respectively, Gauss' law

$$\oint_S \vec{D} \cdot d\vec{S} = \int_\tau \rho d^3r \tag{7–56}$$

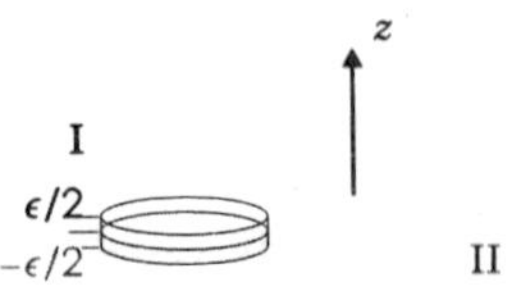

Figure 7.9: The Gaussian pillbox straddles the interface between two dielectrics.

and Ampère's law

$$\oint_\Gamma \vec{H} \cdot d\vec{\ell} = \int_S \vec{J} \cdot d\vec{S} \tag{7-57}$$

in the presence of matter. To determine the behavior of each of the fields at the interface between differing materials, we integrate each of the equations (7–55). For ease of referring to various direction, we take, without loss of generality, the x-y plane to be tangential to the dielectric interface.

$$\boxed{\vec{\nabla} \cdot \vec{D} = \rho}$$

Consider $\vec{\nabla} \cdot \vec{D}$ inside a thin pillbox of width ϵ whose flat sides lie on opposite sides of a dielectric interface, as shown in Figure 7.9. We let the charge density be described by $\rho = \rho_v(x, y, z) + \sigma(x, y)\delta(z)$, where ρ_v is a volume charge density and σ a surface charge density confined to the interface. Integrating $\vec{\nabla} \cdot \vec{D} = \rho$ over the volume of the pillbox, we get

$$\int_\tau \vec{\nabla} \cdot \vec{D}\, d^3r = \int_\tau \rho\, d^3r = \int_\tau \rho_v(x, y, z)\, d^3r + \int_S \sigma(x, y)\, dxdy \tag{7-58}$$

which, with the aid of the mean value theorem and the divergence theorem (20), becomes

$$\oint_S \vec{D} \cdot d\vec{S} = \bar{\rho}_v \cdot \tau + \int_S \sigma(x, y) dxdy \tag{7-59}$$

We break the surface integral of the electric displacement into integrals over each of the three surfaces of the pillbox to write

$$\int_S D_z^I(x, y, \epsilon/2) dxdy - \int_S D_z^{II}(x, y, -\epsilon/2) dxdy + \int_{\substack{\text{curved}\\ \text{side}}} \vec{D} \cdot d\vec{S} = \bar{\rho}_v S\epsilon + \int_S \sigma(x, y) dxdy \tag{7-60}$$

In the limit of vanishing ϵ this becomes

$$\int_S (D_z^I - D_z^{II}) dxdy = \int_S \sigma(x, y) dxdy \tag{7-61}$$

which can hold true for arbitrary S only if $D_z^I - D_z^{II} = \sigma$, or to make our conclusion coordinate independent,

$$\left(\vec{D}^I - \vec{D}^{II}\right) \cdot \hat{n} = \sigma \tag{7-62}$$

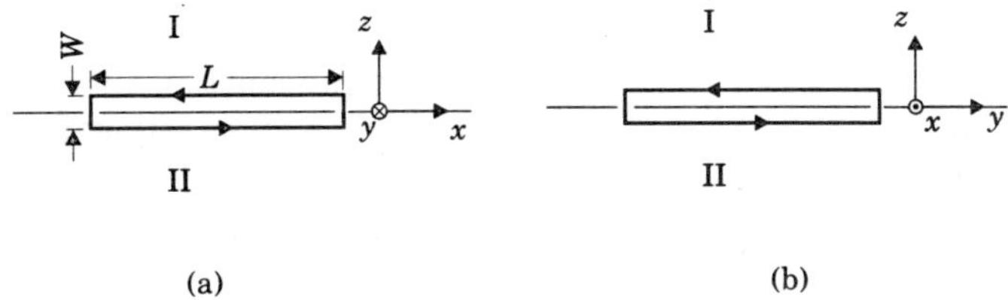

Figure 7.10: The rectangular loop in (a) lies in the x-z plane, and y points into the page. In (b), the loop is placed in the y-z plane leaving x to point out of the page.

where $\hat{n}$ is a unit normal to the interface pointing from region II to region I. In conclusion, *the perpendicular component of $\vec{D}$ is discontinuous by σ.*

$$\boxed{\vec{\nabla} \cdot \vec{B} = 0}$$

The same argument as above, may be applied to $\vec{\nabla} \cdot \vec{B} = 0$ and results in

$$\left(\vec{B}^I - \vec{B}^{II}\right) \cdot \hat{n} = 0 \tag{7–63}$$

In other words, *the perpendicular component of $\vec{B}$ is continuous* across the interface.

$$\boxed{\vec{\nabla} \times \vec{E} = 0}$$

We obtain the boundary condition on the tangential component of $\vec{E}$ by integrating $\vec{\nabla} \times \vec{E}$ over the area of the thin loop whose two long sides lie on opposite sides of the interface illustrated in Figure 7.10a.

$$0 = \int_S (\vec{\nabla} \times \vec{E}) \cdot d\vec{S} = \oint \vec{E} \cdot d\vec{\ell}$$

$$= \int_{x_0}^{x_0+L} \left(E_x^{II} - E_x^I\right) dx + W\left[\bar{E}_z(x_0 + L) - \bar{E}_z(x_0)\right] \tag{7–64}$$

Taking the limit as $W \to 0$, we require

$$\int_{x_0}^{x_0+L} \left(E_x^{II} - E_x^I\right) dx = 0 \tag{7–65}$$

which can hold for all L only if $E_x^{II} = E_x^I$. Generally, then, *the tangential component of $\vec{E}$ is continuous across the interface.*

$$\boxed{\vec{\nabla} \times \vec{H} = \vec{J}}$$

Consider $\vec{\nabla} \times \vec{H}$ inside the loop of Figure 7.10a, and let the current density consist of a body current density $\vec{J}(x, y, z)$ and a surface current density $\vec{j}(x, y)\delta(z)$ confined to the interface. (Surface currents are, strictly speaking, possible only with perfect conductors, but they often constitute a good approximation to the currents on metallic interfaces.) As $\vec{j}$ lies in the x-y plane, $j_z = 0$. Integrating $\vec{\nabla} \times \vec{H}$ over the surface included by the loop of figure 7.10a gives

$$\int (\vec{\nabla} \times \vec{H}) \cdot d\vec{S} = \int -J_y dS + \int -j_y \delta(z) dx dz$$

$$\oint \vec{H} \cdot d\vec{\ell} = -J_y S - \int_{x_0}^{x_0+L} j_y dx \tag{7-66}$$

Again, taking the limit as $W \to 0$, we get

$$\int_{x_0}^{x_0+L} \left(H_x^{II} - H_x^{I}\right) dx = -\int_{x_0}^{x_0+L} j_y dx \tag{7-67}$$

independent of x_0 or L. Equating the integrands, gives $H_x^I - H_x^{II} = j_y$. Next, placing the loop in the z-y plane as illustrated in Figure 7.10b, we obtain a similar result for H_y: $H_y^I - H_y^{II} = -j_x$. (The change in sign occurs because the x axis points out of the page, meaning that $d\vec{S} = +\hat{\imath} dx\, dy$.) We can generalize these results by writing

$$\hat{n} \times \left(\vec{H}^I - \vec{H}^{II}\right) = \vec{j} \tag{7-68}$$

In words, the *parallel component of $\vec{H}$ is discontinuous by j.*

We summarize these boundary conditions below, labelling the components perpendicular and parallel to the surface by $\perp$ and $\|$ respectively.

$(\vec{D})_\perp$ is discontinuous by σ
$(\vec{E})_\|$ is continuous
$(\vec{B})_\perp$ is continuous
$(\vec{H})_\|$ is discontinuous by $\vec{j}$

(7-69)

EXAMPLE 7.5: A large slab of uniform dielectric is placed in a uniform electric field $\vec{E}$ with its parallel faces making angle θ with the field. Determine the angle that the internal electric field makes with the faces. (Note that this is not Snell's law, which arises from interference of waves.)

Solution: The following are the boundary conditions for the electric field: $D_\perp$ is continuous and $E_\|$ is continuous. Labelling the vacuum fields with the subscript v and those in the dielectric with subscript d, we have in vacuum $D_{v\perp} = \varepsilon_0 E_v \cos\theta_v$ and $E_\| = E_v \sin\theta_v$. The boundary conditions then translate to

$$\varepsilon_0 E_v \cos\theta_v = \varepsilon_d E_d \cos\theta_d \quad \text{and} \quad E_v \sin\theta_v = E_d \sin\theta_d \tag{Ex 7.5.1}$$

Dividing one equation by the other, we have $\tan\theta_d = (\varepsilon_d/\varepsilon_0)\tan\theta_v$. In other words, the electric field bends *away* from the normal on entering the dielectric.

A magnetic induction field, in exactly the same fashion, also bends away from the normal when entering a medium of high permeability.

As a simple illustration of using the boundary conditions at the interface with an anisotropic medium, we reconsider the preceding example with a hypothetical dielectric having two different permittivities.

EXAMPLE 7.6: A crystal whose two equivalent principal axes lie in the x-y plane while the third lies along the z axis has two distinct permittivities, $\varepsilon_0(1+\chi_{xx}) = \varepsilon_0(1+\chi_{yy}) = \varepsilon_{xx} = 2\varepsilon_0$ and $\varepsilon_0(1+\chi_{zz}) = \varepsilon_{zz} = 3\varepsilon_0$. A large slab of this material is placed in a uniform electric field in vacuum, making angle θ with the (normal) z axis of the slab. Determine the directions of $\vec{E}$ and $\vec{D}$ in the dielectric.

Solution: In the medium, we have, denoting the dielectric values with subscript d

$$\begin{pmatrix} D_{d,x} \\ D_{d,z} \end{pmatrix} = \begin{pmatrix} 2\varepsilon_0 & 0 \\ 0 & 3\varepsilon_0 \end{pmatrix} \begin{pmatrix} E_{d,x} \\ E_{d,z} \end{pmatrix} = \begin{pmatrix} 2\varepsilon_0 E_d \sin\theta_{E,d} \\ 3\varepsilon_0 E_d \cos\theta_{E,d} \end{pmatrix} \tag{Ex 7.6.1}$$

which relates the displacement field to the electric field. We denote the vacuum field and angles with a subscript v. The boundary conditions

$$E_{\parallel} \text{ is continuous} \quad \Rightarrow \quad E_v \sin\theta_v = E_d \sin\theta_{E,d} \tag{Ex 7.6.2}$$

$$\text{and } D_{\perp} \text{ is continuous} \quad \Rightarrow \quad \varepsilon_0 E_v \cos\theta_v = 3\varepsilon_0 E_d \cos\theta_{E,d} \tag{Ex 7.6.3}$$

Dividing (Ex 7.6.2) by (Ex 7.6.3) gives $\tan\theta_{E,d} = 3\tan\theta_v$. Further, from (Ex 7.6.1) $\tan\theta_{D,d} = \frac{2}{3}\tan\theta_{E,d}$. The angle of the electric displacement field is now obtained from $\tan\theta_{D,d} = \frac{2}{3}\tan\theta_{E,d} = 2\tan\theta_v$.

7.5 Electrostatics and Magnetostatics in Linear Media

The electrostatic field $\vec{E}$ satisfies $\vec{\nabla}\times\vec{E} = 0$, implying that $\vec{E}$ may be written $\vec{E} = -\vec{\nabla}V$. Furthermore, $\vec{\nabla}\cdot\vec{D} = \rho$ becomes for a linear, isotropic dielectric, $\vec{\nabla}\cdot(\varepsilon\vec{E}) = \vec{\nabla}\cdot(-\varepsilon\vec{\nabla}V) = \rho$. If ε is piecewise constant, then the potential, V, must satisfy Poisson's equation, $\nabla^2 V = -\rho/\varepsilon$. In a region of space devoid of free charges, $\nabla^2 V = 0$. The same methods as those used for boundary condition problems in vacuum may now be applied when dielectrics are involved, as illustrated below.

EXAMPLE 7.7: Find the potential in the neighborhood of a sphere of radius R, made of a linear isotropic dielectric placed in an initially uniform field $\vec{E}_0$.

Solution: Solving the Laplace equation in spherical polar coordinates with the z axis chosen to lie along $\vec{E}_0$ and assuming no azimuthal angle dependence, we have

$$V(r<R) = \sum_{\ell=0} A_\ell r^\ell \mathrm{P}_\ell(\cos\theta) \tag{Ex 7.7.1}$$

We eliminate all terms that grow faster than r^1 and separate the $\ell = 1$ term from the rest of the sum

$$V(r>R) = \sum_{\ell=0}^{\infty} \frac{B_\ell}{r^{\ell+1}} \mathrm{P}_\ell(\cos\theta) - E_0 r\cos\theta$$

$$= \frac{B_0}{r} + \left(\frac{B_1}{r^2} - E_0 r\right)\cos\theta + \sum_{\ell=2}^{\infty} \frac{B_\ell}{r^{\ell+1}} \mathrm{P}_\ell(\cos\theta) \qquad \text{(Ex 7.7.2)}$$

The term B_0/r leads to an electric field $\vec{E} = -B_0/r^2\hat{r}$ and could result only from a net charge $Q = -4\pi\varepsilon_0 B_0$ on the sphere. We will neglect this term. The boundary conditions relating the interior and exterior solutions are

$$D_\perp(R_+) = D_\perp(R_-) \quad \text{and} \quad E_\parallel(R_+) = E_\parallel(R_-) \qquad \text{(Ex 7.7.3)}$$

In terms of the potentials, these two boundary conditions become

$$\varepsilon_1 \frac{\partial V}{\partial r}\bigg|_{R_-} = \varepsilon_0 \frac{\partial V}{\partial r}\bigg|_{R_+} \qquad \text{(Ex 7.7.4)}$$

and

$$\frac{1}{R}\frac{\partial V}{\partial \theta}\bigg|_{R_-} = \frac{1}{R}\frac{\partial V}{\partial \theta}\bigg|_{R_+} \qquad \text{(Ex 7.7.5)}$$

The first of these equations, (Ex 7.7.4), leads to

$$\varepsilon_1 A_1 \cos\theta + \varepsilon_1 \sum_{\ell=2}^{\infty} \ell A_\ell R^{\ell-1} \mathrm{P}_\ell(\cos\theta)$$
$$= -\varepsilon_0\left(\frac{2B_1}{R^3} + E_0\right)\cos\theta - \varepsilon_0 \sum_{\ell=2}^{\infty} \frac{(\ell+1)B_\ell}{R^{\ell+2}} \mathrm{P}_\ell(\cos\theta) \qquad \text{(Ex 7.7.6)}$$

giving for $\ell = 1$

$$\varepsilon_1 A_1 = -\varepsilon_0\left(\frac{2B_1}{R^3} + E_0\right) \qquad \text{(Ex 7.7.7)}$$

and for $\ell \neq 1$

$$\varepsilon_1 \ell A_\ell R^{\ell-1} = -\frac{\varepsilon_0(\ell+1)B_\ell}{R^{\ell+2}} \qquad \text{(Ex 7.7.8)}$$

The second equation, (Ex 7.7.5), yields

$$-A_1 \sin\theta + \sum_{\ell=2}^{\infty} A_\ell R^{\ell-1} \frac{\partial}{\partial\theta}\left[\mathrm{P}_\ell(\cos\theta)\right]$$
$$= -\left(\frac{B_1}{R^3} - E_0\right)\sin\theta + \sum_{\ell=2}^{\infty} \frac{B_\ell}{R^{\ell+2}} \frac{\partial}{\partial\theta}\left[\mathrm{P}_\ell(\cos\theta)\right] \qquad \text{(Ex 7.7.9)}$$

Since the derivatives of distinct Legendre polynomials are linearly independent, we equate the coefficients of each order to obtain for $\ell = 1$

$$A_1 = \frac{B_1}{R^3} - E_0 \qquad \text{(Ex 7.7.10)}$$

and for $\ell \neq 1$

$$A_\ell R^{\ell-1} = \frac{B_\ell}{R^{\ell+2}} \qquad \text{(Ex 7.7.11)}$$

We now solve the equations for A_1 and B_1 simultaneously to find

$$B_1 = \left(\frac{\varepsilon_1 - \varepsilon_0}{\varepsilon_1 + 2\varepsilon_0}\right) R^3 E_0 \quad \text{and} \quad A_1 = \frac{B_1}{R^3} - E_0 = -\frac{3\varepsilon_0}{\varepsilon_1 + 2\varepsilon_0} E_0 \qquad \text{(Ex 7.7.12)}$$

Attempting to solve for the remaining coefficients B_ℓ, $\ell \neq 1$, of the expansions we obtain $[\varepsilon_1 \ell + \varepsilon_0(\ell+1)]B_\ell = 0$, which implies, except for peculiar combinations of ε_0 and ε_1, that $B_\ell = 0$. The solution is then

$$V(r \le R, \theta) = A_1 r \cos\theta = -\frac{3\varepsilon_0}{\varepsilon_1 + 2\varepsilon_0} E_0 r \cos\theta \qquad \text{(Ex 7.7.13)}$$

and

$$V(r > R, \theta) = -E_0 r \cos\theta + \frac{B_1}{r^2}\cos\theta = \left[E_0 \left(\frac{\varepsilon_1 - \varepsilon_0}{\varepsilon_1 + 2\varepsilon_0}\right)\frac{R^3}{r^3} - E_0\right] r\cos\theta \qquad \text{(Ex 7.7.14)}$$

Inside the sphere, the electric field is unform and outside the field is the superposition of the external field and that resulting from a dipole $\vec{p} = 4\pi\varepsilon_0 R^3(\varepsilon_1 - \varepsilon_0)\vec{E}_0/(\varepsilon_1 + 2\varepsilon_0)$. If the two permittivities differ little this result may be approximated as $\vec{p} = \frac{4\pi}{3} R^3 \Delta\varepsilon \vec{E}_0$.

EXAMPLE 7.8: Find the electric field inside a spherical cavity in an otherwise uniformly polarized dielectric when the electric field far from the cavity is $\vec{E}_0$ and the polarization is $\vec{P}$ (parallel to $\vec{E}_0$).

Solution: We choose the z axis to be directed along $\vec{E}_0$, and, as in the last problem, express the solution to Laplace's equation in spherical polar coordinates:

$$V(r \le R) = \sum A_\ell r^\ell \mathrm{P}_\ell(\cos\theta) \qquad \text{(Ex 7.8.1)}$$

$$V(r > R) = -E_0 r \cos\theta + \sum \frac{B_\ell}{r^{\ell+1}} \mathrm{P}_\ell(\cos\theta) \qquad \text{(Ex 7.8.2)}$$

The boundary conditions are, as before, $D_\perp(R_+) = D_\perp(R_-)$ and $E_\|(R_+) = E_\|(R_-)$. We cannot, this time, write $\vec{D} = \varepsilon\vec{E}$ because there is no reason to assume that a linear relationship exists between the polarization and the electric field. Instead, we write $\vec{D} = \varepsilon_0\vec{E} + \vec{P}$, whence, outside the cavity, $D_r = -\varepsilon_0(\vec{\nabla}V)_r + P\cos\theta$. The boundary conditions then become,

$$\sum -\varepsilon_0 \ell A_\ell R^{\ell-1} \mathrm{P}_\ell(\cos\theta) = P\cos\theta + \varepsilon_0 E_0 \cos\theta + \sum \varepsilon_0 \frac{(\ell+1)B_\ell}{R}^{\ell+2} \mathrm{P}_\ell(\cos\theta) \qquad \text{(Ex 7.8.3)}$$

and

$$\sum -\frac{\partial}{\partial\theta}\left[\mathrm{P}_\ell(\cos\theta)\right] A_\ell R^{\ell-1} = \frac{\partial}{\partial\theta} E_0 \cos\theta - \frac{\partial}{\partial\theta} \sum \frac{B_\ell}{R^{\ell+2}} \mathrm{P}_\ell(\cos\theta) \qquad \text{(Ex 7.8.4)}$$

Separating the $\ell = 1$ term from the others and using the linear independence of the Legendre polynomials, we equate their coefficients, to obtain

$$A_1 = -\frac{P}{\varepsilon_0} - E_0 - \frac{2B_1}{R^3} \tag{Ex 7.8.5}$$

and

$$A_1 = -E_0 + \frac{B_1}{R^3} \tag{Ex 7.8.6}$$

which, when solved for α_1, give $\alpha_1 = -E_0 - \frac{1}{3}P/\varepsilon_0$.

The $\ell \neq 1$ terms vanish as in the previous example. The potential inside the spherical cavity is then

$$V(r \leq R) = -\left(E_0 + \tfrac{1}{3}P/\varepsilon_0\right) r\cos\theta = -\left(E_0 + \tfrac{1}{3}P/\varepsilon_0\right) z \tag{Ex 7.8.7}$$

leading to an electric field $\vec{E} = \vec{E}_0 + \frac{1}{3}\vec{P}/\varepsilon_0$, a result we quoted earlier in Section 7.3.3.

EXAMPLE 7.9: Magnetic shielding. A spherical shell of permeable material having inner radius a and outer radius b is placed in an initially uniform magnetic induction field $\vec{B}_0$ (Figure 7.11). Find the magnetic induction field inside the spherical shell.

Solution: In the absence of any free currents, $\vec{\nabla} \times \vec{H} = 0$, implying that $\vec{H}$ may be written as the gradient of a potential, $\vec{H} = -\vec{\nabla}V_m$. Furthermore,

$$\vec{\nabla} \cdot \vec{B} = \vec{\nabla} \cdot (\mu\vec{H}) = \vec{\nabla} \cdot (-\mu\vec{\nabla}V_m) = 0 \tag{Ex 7.9.1}$$

which, for piecewise constant μ, may be written

$$\mu\nabla^2 V_m = 0 \tag{Ex 7.9.2}$$

Taking the z axis along the initial field $\vec{B}_0$, we have $\vec{B} = -\mu\vec{\nabla}V_m$. Thus at sufficiently large distance from the shell, $V_m(r \to \infty) = -B_0 z/\mu_0 = -(B_0/\mu_0) r\cos\theta$. Expanding $V_m(\vec{r})$ in spherical polar coordinates in each of the three regions, explicitly putting in the asymptotic form of V_m for large r, we get

$$V_m(r > b) = -\frac{B_0}{\mu_0} r\cos\theta + \sum \frac{A_\ell}{r^{\ell+1}} \mathrm{P}_\ell(\cos\theta) \tag{Ex 7.9.3}$$

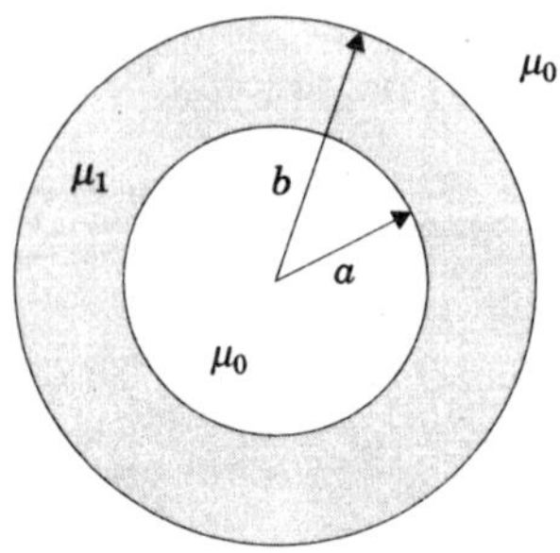

Figure 7.11: The permeable shell has permeability μ_1.

$$V_m(a < r < b) = \sum \left(C_\ell r^\ell + \frac{D_\ell}{r^{\ell+1}} \right) \mathrm{P}_\ell(\cos\theta) \tag{Ex 7.9.4}$$

$$V_m(r < a) = \sum F_\ell r^\ell \mathrm{P}_\ell(\cos\theta) \tag{Ex 7.9.5}$$

The boundary conditions at $r = a$ and $r = b$ are

$$\begin{aligned} B_r(a_+) &= B_r(a_-) \qquad & B_r(b_+) &= B_r(b_-) \\ H_\theta(a_+) &= H_\theta(a_-) \qquad & H_\theta(b_+) &= H_\theta(b_-) \end{aligned} \tag{Ex 7.9.6}$$

or, in terms of the magnetic scalar potential,

$$\mu_1 \left.\frac{\partial V_m}{\partial r}\right|_{a_+} = \mu_0 \left.\frac{\partial V_m}{\partial r}\right|_{a_-} \qquad \mu_0 \left.\frac{\partial V_m}{\partial r}\right|_{b_+} = \mu_1 \left.\frac{\partial V_m}{\partial r}\right|_{b_-} \tag{Ex 7.9.7}$$

and

$$\left.\frac{\partial V_m}{\partial \theta}\right|_{a_+} = \left.\frac{\partial V_m}{\partial \theta}\right|_{a_-} \qquad \left.\frac{\partial V_m}{\partial \theta}\right|_{b_+} = \left.\frac{\partial V_m}{\partial \theta}\right|_{b_-} \tag{Ex 7.9.8}$$

Using arguments akin to those used in the electrostatic examples, but more laboriously, we can show that all the coefficients with $\ell \neq 1$ vanish. The equations for the $\ell = 1$ coefficients are explicitly

$$\mu_1 \left(C_1 - \frac{2D_1}{a^3} \right) = \mu_0 F_1 \qquad \mu_0 \left(-\frac{B_0}{\mu_0} - \frac{2A_1}{b^3} \right) = \mu_1 \left(C_1 - \frac{2D_1}{b^3} \right) \tag{Ex 7.9.9}$$

$$C_1 a + \frac{D_1}{a^2} = F_1 a \qquad -\frac{B_0}{\mu_0} b + \frac{A_1}{b^2} = C_1 b + \frac{D_1}{b^2} \tag{Ex 7.9.10}$$

Eliminating C and D, we solve for A_1 and F_1 to obtain

$$A_1 = \left[\frac{(2\mu_1 + \mu_0)(\mu_1 - \mu_0)}{(2\mu_1 + \mu_0)(\mu_1 + 2\mu_0) - 2(a/b)^3(\mu_1 - \mu_0)^2} \right] (b^3 - a^3) \frac{B_0}{\mu_0} \tag{Ex 7.9.11}$$

$$F_1 = \left[\frac{-9\mu_1}{(2\mu_1 + \mu_0)(\mu_1 + 2\mu_0) - 2(a/b)^3(\mu_1 - \mu_0)^2} \right] B_0 \tag{Ex 7.9.12}$$

The exterior potential

$$V_m = -\frac{B_0}{\mu_0} r\cos\theta + \frac{A_1}{r^2}\cos\theta \tag{Ex 7.9.13}$$

consists of that for a uniform field $\vec{H} = \vec{B}_0/\mu_0$ plus the field of a dipole with magnetic dipole moment $4\pi A_1$, oriented parallel to $\vec{B}_0$. Inside the cavity, there is a uniform magnetic induction $\vec{B} = -\mu_0 F_1 \vec{B}_0$. When the permeability of the shell μ_1 is much greater than that of vacuum, the coefficient $A_1 \simeq b^3 B_0/\mu_0$ and $F_1 \simeq -9B_0/2\mu_1(1 - a^3/b^3)$. For shields of high permeability, μ_1 ranges from $10^3\mu_0$ to $10^9\mu_0$; even relatively thin shells cause a great reduction of B in the interior of the shell. For example, taking $\mu_1 = 10^5\mu_0$ and $a/b = .95$, we find

$$\vec{B}(r < a) = \frac{9\vec{B}_0}{2 \times 10^5[1 - (.95)^3]} = 3.16 \times 10^{-4}\vec{B}_0 \tag{Ex 7.9.13}$$

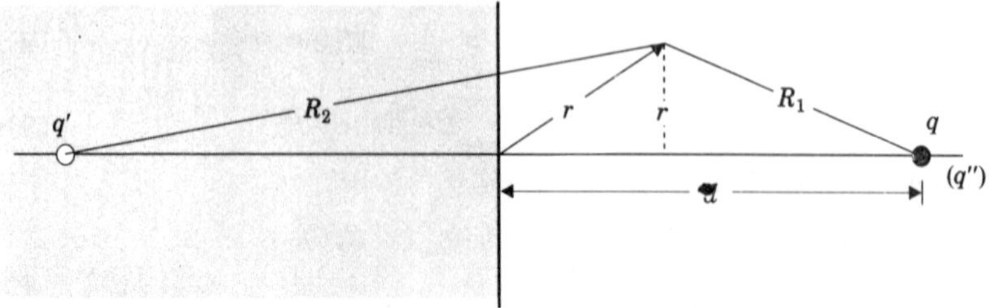

Figure 7.12: The electric field outside the dielectric arising from the induced polarization of the dielectric may be accounted for by the image charge q'. Inside the dielectric, q' does not exist, instead we see q screened by the induced polarization. The screened charge q'' takes the place of q and the induced polarization.

7.5.1 Electrostatics with Dielectrics Using Image Charges

Consider a charge q placed at distance d from the plane interface of a semi-infinite dielectric (Figure 7.12). We expect the electric field of the charge to induce a polarization of the medium near the surface, the field of which an observer unaware of the dielectric (not in the dielectric) might well suppose to be due to a charge q' behind the interface. Thus the observer (in vacuum) might suppose that the potential outside the dielectric would be given by the expression

$$V(\vec{r}) = \frac{1}{4\pi\varepsilon_0}\left(\frac{q}{R_1} + \frac{q'}{R_2}\right) = \frac{1}{4\pi\varepsilon_0}\left[\frac{q}{\sqrt{\rho^2+(d-z)^2}} + \frac{q'}{\sqrt{\rho^2+(d+z)^2}}\right] \tag{7–70}$$

where ρ is the cylindrical radial coordinate. From a point inside the dielectric, we cannot maintain the fiction of the image charge, since we must have $\vec{\nabla}\cdot\vec{D} = 0$ inside the medium. Instead we expect to see q partially screened by the dipoles in the medium (or by the effective surface charge induced), thereby being effectively reduced to q''.

The potential inside the medium would under these conditions become

$$V(\vec{r}) = \frac{1}{4\pi\varepsilon_1}\frac{q''}{R_1} = \frac{1}{4\pi\varepsilon_1}\frac{q''}{\sqrt{\rho^2+(d-z)^2}} \tag{7–71}$$

where $z \le 0$. Applying the boundary conditions

$$E_\rho(z=0_+) = E_\rho(z=0_-) \qquad D_z(z=0_+) = D_z(z=0_-) \tag{7–72}$$

that must be satisfied, we solve for q' and q'' (thereby verifying that these image charges do indeed give solutions that satisfy the boundary conditions). In terms of the potential, the boundary conditions become

$$\left.\frac{\partial V}{\partial \rho}\right|_{0_+} = \left.\frac{\partial V}{\partial \rho}\right|_{0_-} \tag{7–73}$$

and

$$\varepsilon_0 \frac{\partial V}{\partial z}\bigg|_{0_+} = \varepsilon_1 \frac{\partial V}{\partial z}\bigg|_{0_-} \tag{7–74}$$

After we perform the differentiation, (7–73) yields

$$\frac{1}{4\pi\varepsilon_0}\left[\frac{-q\rho}{(\rho^2+d^2)^{3/2}} - \frac{q'\rho}{(\rho^2+d^2)^{3/2}}\right] = \frac{1}{4\pi\varepsilon_1}\left[\frac{-q''\rho}{(\rho^2+d^2)^{3/2}}\right] \tag{7–75}$$

which clearly implies

$$\frac{q+q'}{\varepsilon_0} = \frac{q''}{\varepsilon_1} \tag{7–76}$$

while (7–74) yields

$$\frac{qd}{(\rho^2+d^2)^{3/2}} - \frac{q'd}{(\rho^2+d^2)^{3/2}} = \frac{q''d}{(\rho^2+d^2)^{3/2}} \tag{7–77}$$

giving

$$q - q' = q'' \tag{7–78}$$

Finally, we solve (7-76) and (7-78) for q' and q'':

$$q' = -\frac{\varepsilon_1 - \varepsilon_0}{\varepsilon_1 + \varepsilon_0} q \quad \text{and} \quad q'' = \frac{2\varepsilon_1}{\varepsilon_1 + \varepsilon_0} q \tag{7–79}$$

The potential inside the dielectric may therefore be written

$$V(\vec{r}) = \frac{1}{4\pi\varepsilon_1} \cdot \frac{2\varepsilon_1}{\varepsilon_1 + \varepsilon_0} \frac{q(\vec{r}\,')}{|\vec{r} - \vec{r}\,'|} = \frac{2}{4\pi(\varepsilon_1 + \varepsilon_0)} \frac{q(\vec{r}\,')}{|\vec{r} - \vec{r}\,'|} \tag{7–80}$$

which we immediately generalize to obtain the potential inside the dielectric arising from an arbitrary charge distribution $\rho(\vec{r}\,')$ outside the dielectric:

$$V(\vec{r}) = \frac{2}{4\pi(\varepsilon_1 + \varepsilon_0)} \int \frac{\rho(\vec{r}\,')}{|\vec{r} - \vec{r}\,'|} d^3r' \tag{7–81}$$

Similarly, outside the dielectric, the potential is given by

$$V(\vec{r}) = \frac{1}{4\pi\varepsilon_0} \int \rho(\vec{r}\,') \left[\frac{1}{|\vec{r} - \vec{r}\,'|} - \frac{(\varepsilon_1 - \varepsilon_0)/(\varepsilon_1 + \varepsilon_0)}{|\,\vec{r} - [\vec{r}\,' - 2(\hat{n}\cdot\vec{r}\,')\hat{n}]\,|}\right] d^3r' \tag{7–82}$$

7.5.2 Image Line Charges for the Dielectric Cylinder

The polarization of the medium in a cylindrical dielectric caused by an exterior line charge λ can also be mimicked by image line charges as seen outside the dielectric and by a partial screening of the charge as seen from inside the dielectric (Figure 7.13). Because we may in principle measure the field from the dielectric on a Gaussian cylinder surrounding the dielectric to find zero net charge inside, any image

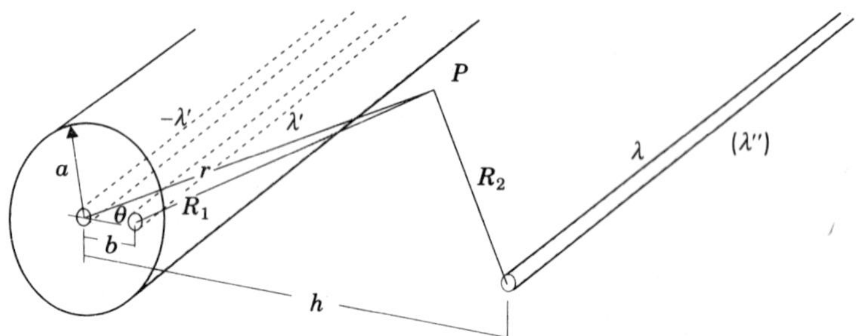

Figure 7.13: The dielectric cylinder must remain neutral and has two line charges, λ' and $-\lambda'$ to accomplish this requirement. Inside the cylinder, only the screened line charge λ'' is detected.

charge λ' inside the dielectric that we choose to satisfy the boundary conditions must be neutralized by a second line charge $-\lambda'$ at the center of the cylinder (at this position the line charge does not disturb the boundary conditions because the field arising from it is purely perpendicular to the surface). From inside the cylinder, we see again only the screened charge λ''. To be explicit, we consider the line charge λ placed at distance h from the center of the cylinder of radius a, inducing image charge λ' at distance b from the center line of the cylinder. The potential as seen from a point P outside the dielectric is then

$$V(\vec{r}) = -\frac{1}{2\pi\varepsilon_0}(\lambda \ln R_2 + \lambda' \ln R_1 - \lambda' \ln r) \tag{7-83}$$

while for a point r inside the cylinder it is

$$V(\vec{r}) = -\frac{1}{2\pi\varepsilon_1}\lambda'' \ln R_2 \tag{7-84}$$

Expressing R_1 and R_2 in terms of r and θ (the point's polar angle with respect to the center of the cylinder), we have

$$R_1 = \sqrt{r^2 + b^2 - 2rb\cos\theta} \qquad \text{and} \qquad R_2 = \sqrt{r^2 + h^2 - 2rh\cos\theta} \tag{7-85}$$

The boundary conditions that must be satisfied are

$$\varepsilon_0 \frac{\partial V}{\partial r}\bigg|_{a_+} = \varepsilon_1 \frac{\partial V}{\partial r}\bigg|_{a_-} \tag{7-86}$$

and

$$\frac{\partial V}{\partial \theta}\bigg|_{a_+} = \frac{\partial V}{\partial \theta}\bigg|_{a_-} \tag{7-87}$$

The derivatives are explicitly

$$\frac{\partial V}{\partial r}\bigg|_{a} = -\frac{\lambda''}{2\pi\varepsilon_1}\frac{1}{R_2(a)}\frac{\partial R_2}{\partial r}\bigg|_{a} = -\frac{\lambda''}{2\pi\varepsilon_1}\frac{a - h\cos\theta}{R_2^2(a)} \tag{7-88}$$

$$\left.\frac{\partial V}{\partial r}\right|_{a_+} = -\frac{1}{2\pi\varepsilon_0}\left[\frac{\lambda(a-h\cos\theta)}{R_2^2(a)} + \frac{\lambda'(a-b\cos\theta)}{R_1^2(a)} - \frac{\lambda'}{a}\right] \tag{7-89}$$

$$\left.\frac{\partial V}{\partial \theta}\right|_{a_-} = -\frac{\lambda''}{2\pi\varepsilon_1}\frac{1}{R_2(a)}\left.\frac{\partial R_2}{\partial\theta}\right|_{a_-} = -\frac{\lambda''}{2\pi\varepsilon_1}\frac{ah\sin\theta}{R_2^2(a)} \tag{7-90}$$

$$\left.\frac{\partial V}{\partial \theta}\right|_{a_+} = -\frac{a}{2\pi\varepsilon_0}\left[\frac{\lambda h\sin\theta}{R_2^2(a)} + \frac{\lambda' b\sin\theta}{R_1^2(a)}\right] \tag{7-91}$$

Substituting (7–90) and (7–91) into (7–87) we have

$$\frac{1}{\varepsilon_1}\frac{\lambda'' h}{R_2^2(a)} = \frac{1}{\varepsilon_0}\left[\frac{\lambda h}{R_2^2(a)} + \frac{\lambda' b}{R_1^2(a)}\right] \tag{7-92}$$

whereas inserting (7–88) and (7–89) into (7–86) we obtain

$$\frac{\lambda''(a-h\cos\theta)}{R_2^2(a)} = \frac{\lambda(a-h\cos\theta)}{R_2^2(a)} + \frac{\lambda'(a-b\cos\theta)}{R_1^2(a)} - \frac{\lambda'}{a} \tag{7-93}$$

The solution of equations (7-92) and (7-93) is straightforward although slightly laborious. Equation (7-92) may be rewritten

$$\frac{\varepsilon_0}{\lambda' b}\left(\frac{\lambda''}{\varepsilon_1} - \frac{\lambda}{\varepsilon_0}\right)h = \frac{R_2^2(a)}{R_1^2(a)} = \frac{a^2+h^2-2ah\cos\theta}{a^2+b^2-2ab\cos\theta} \tag{7-94}$$

We abbreviate the (constant) left side of this equation as K^2 and find

$$\left(a^2+h^2-2ah\cos\theta\right) = K^2\left(a^2+b^2-2ab\cos\theta\right) \tag{7-95}$$

which can be satisfied for all θ only if $2ah = 2K^2ab$, or $K^2 = h/b$. When we substitute this value of K^2 into (7–95), the remaining terms give

$$\left(a^2+h^2\right) = K^2\left(a^2+b^2\right) = \frac{a^2h}{b} + bh \tag{7-96}$$

implying that $a^2 = bh$, the same result as for the conducting cylinder and the conducting sphere. Returning once more to (7–94), we substitute $K^2 = h/b = h^2/a^2$ for $[R_2(a)/R_1(a)]^2$, and obtain

$$\left(\frac{\lambda''}{\varepsilon_1} - \frac{\lambda}{\varepsilon_0}\right)h = \frac{\lambda' b}{\varepsilon_0}\frac{h^2}{a^2} = \frac{\lambda'(bh)h}{\varepsilon_0 a^2} = \frac{\lambda' h}{\varepsilon_0} \tag{7-97}$$

or

$$\frac{\lambda''}{\varepsilon_1} - \frac{\lambda}{\varepsilon_0} = \frac{\lambda'}{\varepsilon_0} \tag{7-98}$$

To obtain a second equation in λ' and λ'', we rewrite (7–93) as

$$(\lambda''-\lambda)(a-h\cos\theta) = \lambda'\frac{R_2^2(a)}{R_1^2(a)}(a-b\cos\theta) - \frac{\lambda' R_2^2(a)}{a}$$

$$= \lambda' \left[\frac{h^2}{a^2}(a - b\cos\theta) - \frac{a^2 + h^2 - 2ah\cos\theta}{a}\right]$$

$$= -\lambda'(a - h\cos\theta) \tag{7–99}$$

Equating the terms in $\cos\theta$ (or alternatively the θ independent parts) gives

$$\lambda'' - \lambda = -\lambda' \tag{7–100}$$

The solutions for the image charges are then obtained from (7–98) and (7–100) as

$$\lambda' = \frac{\varepsilon_0 - \varepsilon_1}{\varepsilon_0 + \varepsilon_1}\lambda \quad \text{and} \quad \lambda'' = \frac{2\varepsilon_1}{\varepsilon_0 + \varepsilon_1}\lambda \tag{7–101}$$

Finally, for $r > a$, the potential becomes

$$V(r > a) = -\frac{\lambda}{2\pi\varepsilon_0}\left(\ln R_2 + \frac{\varepsilon_0 - \varepsilon_1}{\varepsilon_0 + \varepsilon_1}\ln\frac{R_1}{r}\right) \tag{7–102}$$

while for $r < a$, it is

$$V(r < a) = -\frac{\lambda}{\pi(\varepsilon_0 + \varepsilon_1)}\ln R_2 \tag{7–103}$$

It is worth pointing out that no image charge can be found for the analogous point charge in the vicinity of a dielectric sphere.

7.5.3 Magnetostatics and Magnetic Poles

Although $\vec{\nabla}\cdot\vec{B} = 0$ implies that no magnetic monopoles exist, they are nevertheless a convenient fiction, very useful for visualizing the fields and for constructing image dipoles. To begin, we note that from the definition of $\vec{H}$, $\vec{B} \equiv \mu_0(\vec{H} + \vec{M})$, $\vec{\nabla}\cdot\vec{B} = 0$ implies that

$$\vec{\nabla}\cdot\vec{H} = -\vec{\nabla}\cdot\vec{M} \tag{7–104}$$

Furthermore, in regions of space where $\vec{J} = 0$, $\vec{\nabla}\times\vec{H} = 0$, implying that $\vec{H} = -\vec{\nabla}V_m$, which combined with (7–104) yields

$$\nabla^2 V_m = \vec{\nabla}\cdot\vec{M} \tag{7–105}$$

Clearly, $\vec{\nabla}\cdot\vec{M}$ acts as a source for the magnetic scalar potential V_m just as $\vec{\nabla}\cdot\vec{P}$ is a source for the electric potential V. We also note that just as $\vec{\nabla}\cdot\vec{E} = \rho/\varepsilon_0$ leads to $\vec{E} = -\vec{\nabla}V$, it is properly $\vec{H}$ that is derived from V_m, as $\vec{H} = -\vec{\nabla}V_m$, **not** $\vec{B}$ as we may have been led to believe in section 1.2.6.

If the potential vanishes at some appropriately distant boundary, Poisson's equation is easily solved using Green's functions (Section 6.3.2) or by analogy with the electrostatic equations to give the solution

$$V_m(\vec{r}) = -\frac{1}{4\pi}\int_\tau \frac{\vec{\nabla}'\cdot\vec{M}(\vec{r}\,')}{|\vec{r} - \vec{r}\,'|}d^3r' \tag{7–106}$$

where the volume of integration, τ, must contain all the magnetization.

Expression (7–106) may be cast into a number of more useful forms. Using the identity (7),

$$\vec{\nabla}' \cdot \frac{\vec{M}}{|\vec{r}-\vec{r}\,'|} = \frac{\vec{\nabla}' \cdot \vec{M}}{|\vec{r}-\vec{r}\,'|} + \vec{M} \cdot \vec{\nabla}'\left(\frac{1}{|\vec{r}-\vec{r}\,'|}\right) \tag{7–107}$$

we rewrite V_m as

$$V_m(\vec{r}) = -\frac{1}{4\pi}\int_\tau \vec{\nabla}' \cdot \frac{\vec{M}}{|\vec{r}-\vec{r}\,'|} d^3r' + \frac{1}{4\pi}\int_\tau \frac{\vec{M}\cdot(\vec{r}-\vec{r}\,')}{|\vec{r}-\vec{r}\,'|^3} d^3r' \tag{7–108}$$

Converting the first integral to a vanishing surface integral (as $\vec{M}$ vanishes on the boundary of the volume of integration) with the divergence theorem, the potential becomes

$$V_m(\vec{r}) = \frac{1}{4\pi}\int_\tau \frac{\vec{M}\cdot(\vec{r}-\vec{r}\,')}{|\vec{r}-\vec{r}\,'|^3} d^3r' \tag{7–109}$$

as might well have been anticipated from the form of the scalar potential of a magnetic dipole.

Integral (7–109) can also be written as

$$\begin{aligned} V_m(\vec{r}) &= -\frac{1}{4\pi}\int_\tau \vec{M} \cdot \vec{\nabla}\frac{1}{|\vec{r}-\vec{r}\,'|} d^3r' \\ &= -\frac{1}{4\pi}\int_\tau \vec{\nabla} \cdot \frac{\vec{M}}{|\vec{r}-\vec{r}\,'|} d^3r' \\ &= -\frac{1}{4\pi}\vec{\nabla} \cdot \int_\tau \frac{\vec{M}}{|\vec{r}-\vec{r}\,'|} d^3r' \end{aligned} \tag{7–110}$$

a form that is frequently useful.

EXAMPLE 7.10: Find the magnetic scalar potential of a magnetized sphere of radius a having magnetization $\vec{M}(\vec{r}\,') = M_0 z' \hat{\imath}$.

Solution: Using the form (7-110) for the scalar potential,

$$\begin{aligned} V_m(\vec{r}) &= -\frac{1}{4\pi}\vec{\nabla} \cdot \int_\tau \frac{M_0 z' \hat{\imath}}{|\vec{r}-\vec{r}\,'|} d^3r' \\ &= -\frac{M_0}{4\pi}\frac{\partial}{\partial x}\int_\tau \frac{z'}{|\vec{r}-\vec{r}\,'|} d^3r' \end{aligned} \tag{Ex 7.10.1}$$

The integrals is most easily evaluated by expanding $|\vec{r}-\vec{r}\,'|^{-1}$ in terms of spherical harmonics

$$\frac{1}{|\vec{r}-\vec{r}\,'|} = \sum_{\ell,m} \frac{4\pi}{2\ell+1} Y_\ell^m(\theta,\varphi) Y_\ell^{*m}(\theta',\varphi') \frac{r_<^\ell}{r_>^{\ell+1}} \tag{Ex 7.10.2}$$

and noting that $z' = \sqrt{4\pi/3}\, r' Y_1^0(\theta',\varphi')$. Making these substitutions, we get

$$\begin{aligned} V_m(\vec{r}) = -\frac{M_0}{4\pi}\frac{\partial}{\partial x}\sum_{\ell,m} \frac{4\pi}{2\ell+1} Y_\ell^m(\theta,\varphi) \\ \times \int_{4\pi} \sqrt{\frac{4\pi}{3}} Y_1^0(\theta'\varphi') Y_\ell^{*m}(\theta',\varphi') d\Omega' \int_0^a \frac{r_<^\ell}{r_>^{\ell+1}} r'^3 dr' \end{aligned} \tag{Ex 7.10.3}$$

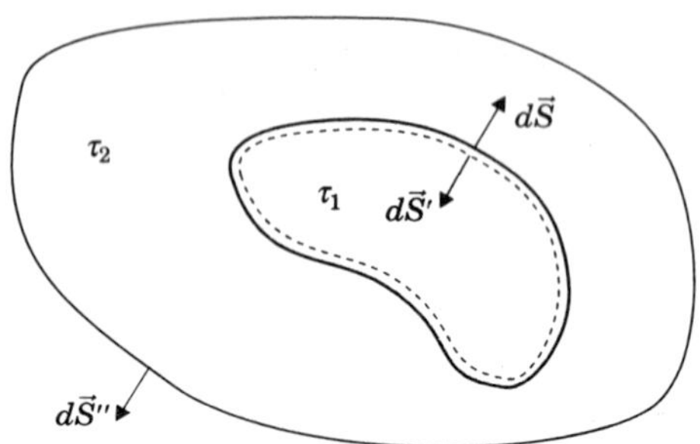

Figure 7.14: The volume of integration is subdivided into a region of uniform magnetization, τ_1 and an exterior region τ_2, containing only the edge of the magnetized body.

Only the $\ell = 1, m = 0$, term survives the integration over Ω', giving

$$V_m(\vec{r}) = -\frac{M_0}{3}\frac{\partial}{\partial x}\left[\sqrt{\frac{4\pi}{3}}\mathrm{Y}_1^0(\theta,\varphi)\int_0^a \frac{r_<}{r_>^2} r'^3 dr'\right]$$

$$= -\frac{M_0}{3}\frac{\partial}{\partial x}\int_0^a \frac{z}{r}\frac{r_<}{r_>^2} r'^3 dr' \qquad \text{(Ex 7.10.4)}$$

When r is outside the sphere, r' is $r_<$, so that the integral becomes

$$\int_0^a \frac{zr'^4}{r^3} dr' = \frac{za^5}{5r^3} \qquad \text{(Ex 7.10.5)}$$

giving

$$V_m(r > a) = -\frac{M_0 z}{3}\frac{\partial}{\partial x}\frac{a^5}{5r^3} = \frac{M_0 z x a^5}{5r^5} \qquad \text{(Ex 7.10.6)}$$

For regions inside the sphere we must break the integral over r' into two intervals: 0 to r, for which $r' = r_<$, and r to a, where $r' = r_>$. Thus, for $r < a$:

$$V_m(r < a) = -\frac{M_0 z}{3}\frac{\partial}{\partial x}\left(\int_0^r \frac{r'^4}{r^3} dr' + \int_r^a r' dr'\right)$$

$$= -\frac{M_0 z}{3}\frac{\partial}{\partial x}\left(-\frac{3r^2}{10} + \frac{a^2}{2}\right)$$

$$= \frac{M_0 z x}{5} \qquad \text{(Ex 7.10.7)}$$

If $\vec{M}$ is uniform, except for discontinuities at the edge of the material, a slightly different approach is often useful. Intuitively we expect the exposed ends of the magnetic dipoles to appear like magnetic surface charges (poles).

At the edges, where $\vec{M}$ is discontinuous, the divergence is singular and (7–106) cannot be integrated as it stands. To deal with the discontinuity of $\vec{M}$ at the boundary of the material, we divide the volume over which we integrate into two regions: τ_1, ending a small distance ϵ inside the body, and a second volume τ_2, beginning a distance ϵ inside the material and extending past the outer edge of the original volume of integration, τ (Figure 7.14).

Thus, formally

$$V_m(\vec{r}) = \int_{\tau_1} \frac{-\vec{\nabla}' \cdot \vec{M}}{4\pi|\vec{r} - \vec{r}'|} d^3r' + \int_{\tau_2} \frac{-\vec{\nabla}' \cdot \vec{M}}{4\pi|\vec{r} - \vec{r}'|} d^3r' \tag{7–111}$$

Using the vector identity (7), $\vec{\nabla} \cdot (f\vec{M}) = \vec{\nabla} f \cdot \vec{M} + f\vec{\nabla} \cdot \vec{M}$, we recast the second integral as

$$\int_{\tau_2} \frac{\vec{\nabla}' \cdot \vec{M}}{4\pi|\vec{r} - \vec{r}'|} d^3r' = \int_{\tau_2} \vec{\nabla}' \cdot \left(\frac{\vec{M}}{4\pi|\vec{r} - \vec{r}'|}\right) d^3r' - \int_{\tau_2} \frac{\vec{M} \cdot (\vec{r} - \vec{r}')}{4\pi|\vec{r} - \vec{r}'|^3} d^3r' \tag{7–112}$$

The divergence theorem allows us to rewrite the first of the two integrals on the right as a surface integral. Note that there is both an exterior surface and an interior surface to the volume τ_2. We have then

$$\int_{\tau_2} \frac{\vec{\nabla}' \cdot \vec{M} d^3r'}{4\pi|\vec{r} - \vec{r}'|} = \oint_{S'} \frac{\vec{M} \cdot d\vec{S}'}{4\pi|\vec{r} - \vec{r}'|} + \oint_{S''} \frac{\vec{M} \cdot d\vec{S}''}{4\pi|\vec{r} - \vec{r}'|} - \int_{\tau_2} \frac{\vec{M} \cdot (\vec{r} - \vec{r}')}{4\pi|\vec{r} - \vec{r}'|^3} d^3r' \tag{7–113}$$

The integral over S'' vanishes as $\vec{M}$ is zero at the exterior boundary. If we let ϵ tend to zero, the remaining volume integral over τ_2 will also vanish, while S' will tend to the boundary S of the magnetized material. We note, however, that $d\vec{S}$, the outward-pointing normal to the magnetized body, is $-d\vec{S}'$ (the "outward-pointing" normal to τ_2). Then, replacing the integral over S' by one over S, we obtain

$$V_m(\vec{r}) = \frac{1}{4\pi} \int_{\tau_1} \frac{-\vec{\nabla}' \cdot \vec{M}}{|\vec{r} - \vec{r}'|} d^3r' + \frac{1}{4\pi} \oint_S \frac{\vec{M} \cdot d\vec{S}}{|\vec{r} - \vec{r}'|} \tag{7–114}$$

where the surface integral extends over all free surfaces and the volume integral excludes all those surfaces. The surface integral is exactly the contribution to the scalar potential that we would have expected from a magnetic surface charge $\vec{M} \cdot \hat{n}$ on the surface of the object.

EXAMPLE 7.11: Find the magnetic scalar potential along the center line of a uniformly magnetized cylindrical magnet of length L magnetized along its length.

Solution: The uniform magnetization in the volume of the magnet has no divergence and therefore makes no contribution to the scalar magnetic potential. The only contribution comes from the surface integrals at the ends of the magnet. For convenience we place the origin at the center of the magnet with the z axis along the center line. Then using (7–114)

$$V_m(0,0,z) = \frac{1}{4\pi} \int_{\substack{\text{right}\\\text{face}}} \frac{M dS}{\sqrt{(z - \frac{1}{2}L)^2 + \rho^2}} + \frac{1}{4\pi} \int_{\substack{\text{left}\\\text{face}}} \frac{-M dS}{\sqrt{(z + \frac{1}{2}L)^2 + \rho^2}}$$

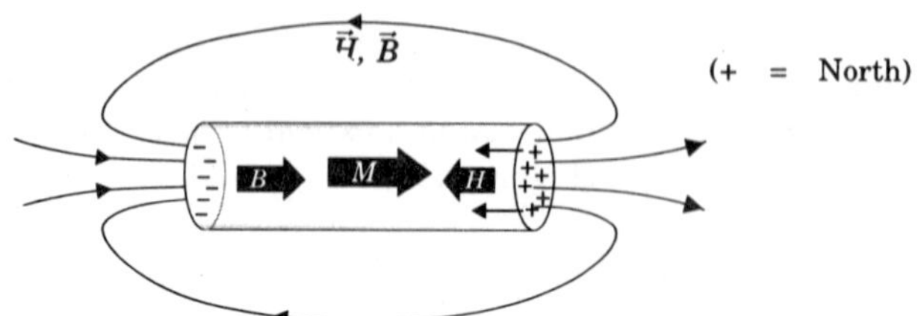

Figure 7.15: The magnetic field intensity $\vec{H}$ both inside and outside the magnet points from North pole to South pole. Inside the magnet that means that $\vec{H}$ opposes the magnetization.

$$
\begin{aligned}
&= \frac{M}{2}\int_0^a \frac{\rho d\rho}{\sqrt{(z-\frac{1}{2}L)^2+\rho^2}} - \frac{M}{2}\int_0^a \frac{\rho d\rho}{\sqrt{(z+\frac{1}{2}L)^2+\rho^2}} \\
&= -\frac{M}{2}\Big[\sqrt{(z+\tfrac{1}{2}L)^2+a^2} - \sqrt{(z+\tfrac{1}{2}L)^2} \\
&\qquad\qquad - \sqrt{(z-\tfrac{1}{2}L)^2+a^2} + \sqrt{(z-\tfrac{1}{2}L)^2}\Big] \\
&= -\frac{M}{2}\Big[\sqrt{(\tfrac{1}{2}L+z)^2+a^2} - \sqrt{(\tfrac{1}{2}L-z)^2+a^2} - 2z\Big] \qquad \text{(Ex 7.11.1)}
\end{aligned}
$$

We might continue the calculation to obtain the z-component of the magnetic induction field $\vec{B}$:

$$
\begin{aligned}
\vec{B} &= \mu_0\vec{H} + \mu_0\vec{M} \\
&= \mu_0(-\vec{\nabla}V_m + \vec{M}) \\
&= \frac{\mu_0\vec{M}}{2}\left[\frac{\frac{1}{2}L+z}{\sqrt{(\frac{1}{2}L+z)^2+a^2}} + \frac{\frac{1}{2}L-z}{\sqrt{(\frac{1}{2}L-z)^2+a^2}}\right] \qquad (7\text{–}115)
\end{aligned}
$$

in agreement with the results obtained in example 7.4.

Note that $\vec{H}$ points in a direction *opposite* to $\vec{M}$ inside the magnet, a result easily envisaged (Figure 7.14) in terms of the bound magnetic surface charges. We see then, that it is the magnetic field intensity $\vec{H}$ that points from north pole to south poles, not the magnetic induction field.

EXAMPLE 7.12: Find the magnetic scalar potential of the magnetized sphere with magnetization $\vec{M}(\vec{r}\,') = M_0 z'\hat{\imath}$ of Example 7.10 using (7–114).

Solution: Using the form (7-114) for the scalar potential, the divergence of magnetization within the sphere vanishes and the term $\hat{\imath}\cdot d\vec{S}$ in the second integral may be written $\sin\theta'\cos\varphi' a^2 d\Omega$ allowing us to write

$$
V_m(\vec{r}) = \frac{1}{4\pi}\int_\tau \frac{-\vec{\nabla}'\cdot M_0 z'\hat{\imath}}{|\vec{r}-\vec{r}\,'|}d^3r' + \frac{1}{4\pi}\oint \frac{M_0 z'\hat{\imath}\cdot d\vec{S}}{|\vec{r}-\vec{r}\,'|}
$$

$$= \frac{M_0}{4\pi}\oint \frac{z'\sin\theta'\cos\varphi' a^2 d\Omega'}{|\vec{r}-\vec{r}\,'|} \tag{Ex 7.12.1}$$

We again use the expansion of $\left|\vec{r}-\vec{r}\,'\right|^{-1}$ in terms of spherical harmonics to evaluate the integral. We substitute $z' = a\cos\theta'$ and use the expansion (Ex 7.10.2) to write

$$\frac{M_0}{4\pi}\oint \frac{z'\sin\theta'\cos\varphi' a^2 d\Omega'}{|\vec{r}-\vec{r}\,'|}$$
$$= M_0 a^3 \sum_{\ell,m} \frac{\mathrm{Y}_\ell^m(\theta,\varphi)}{2\ell+1}\oint \frac{\mathrm{Y}_\ell^{*m}(\theta',\varphi')\cos\theta'\sin\theta'\cos\varphi' r_<^\ell d\Omega'}{r_>^{\ell+1}} \tag{Ex 7.12.2}$$

The trigonometric terms may be written as spherical harmonics using (F-42)

$$\cos\theta'\sin\theta'\cos\varphi' = \sqrt{\frac{2\pi}{15}}\Big(\mathrm{Y}_2^{-1}(\theta',\varphi') - \mathrm{Y}_2^{+1}(\theta',\varphi')\Big) \tag{Ex 7.12.3}$$

so that the contribution from the second integral, using the orthogonality of spherical harmonics, may be written

$$\frac{1}{4\pi}\oint \frac{\vec{M}\cdot d\vec{S}}{|\vec{r}-\vec{r}\,'|} = \frac{M_0 a^3}{5}\sqrt{\frac{2\pi}{15}}\Big(\mathrm{Y}_2^{-1}(\theta,\varphi) - \mathrm{Y}_2^{+1}(\theta,\varphi)\Big)\frac{r_<^2}{r_>^3}$$
$$= \frac{M_0 a^3}{5}\frac{r_<^2}{r_>^3}\cos\theta\sin\theta\cos\varphi = \frac{M_0 a^3}{5}\frac{r_<^2}{r_>^3}\frac{zx}{r^2} \tag{Ex 7.12.4}$$

Thus for $r < a$, $r_> = a$ and the magnetic scalar potential is

$$V_m(r<a) = \frac{M_0 zx}{5} \tag{Ex 7.12.5}$$

Outside the sphere, when $r = r_>$ and $r_< = a$ we have

$$V_m(r>a) = \frac{M_0 zxa^5}{5r^5} \tag{Ex 7.12.6}$$

It is readily verified that these results are identical to those found in example 7.10.

7.5.4 Magnetic Image Poles

Despite the fact that magnetic monopoles appear not to exist, they are nonetheless a convenient fiction in dealing with induced magnetizations. Supposing for the moment that magnetic charges q_m do exist and produce magnetic scalar potential

$$V_m(\vec{r}) = \frac{q_m}{4\pi|\vec{r}-\vec{r}\,'|} \tag{7–116}$$

We postulate that the induced dipoles at the interface of a permeable substance can be mimicked by an image charge q_m' inside the material if seen outside the material, while from inside the material, the shielding they produce reduces the

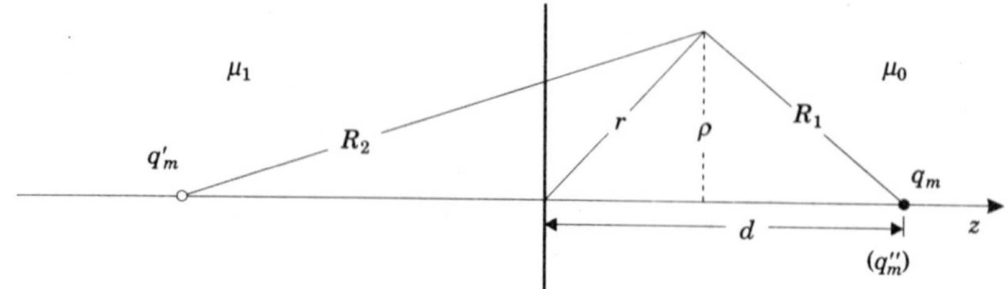

Figure 7.16: A hypothetical magnetic charge induces magnetic dipoles that screen the 'charge' and reduce the apparent charge to q''_m. Outside the material, an image charge q'_m imitates the magnetization.

effective charge to q''_m (Figure 7.16). The situation is entirely analogous to the electrical equivalent. For convenience we place the x-y plane on the interface with the magnetic charge q_m lying at distance d from the origin along the z axis.

In the material the potential is then given by

$$V_m(z<0) = \frac{q''_m}{4\pi R_1} = \frac{q''_m}{4\pi\sqrt{(d-z)^2+\rho^2}} \tag{7–117}$$

while outside the material it is given by

$$V_m(z>0) = \frac{q_m}{4\pi R_1} + \frac{q'_m}{4\pi R_2} = \frac{q_m}{4\pi\sqrt{(d-z)^2+\rho^2}} + \frac{q'_m}{4\pi\sqrt{(d+z)^2+\rho^2}} \tag{7–118}$$

At the interface we must have $H_\| = -\partial V_m/\partial\rho$ continuous and $B_\perp = -\mu\partial V_m/\partial z$ continuous. At the surface, $R_1 = R_2 = (d^2+\rho^2)^{1/2}$, which for brevity we abbreviate as R. Performing the required differentiations we have

$$\frac{\rho q_m}{R^3} + \frac{\rho q'_m}{R^3} = \frac{\rho q''_m}{R^3} \quad \text{or} \quad q_m + q'_m = q''_m \tag{7–119}$$

and

$$-\mu_0\frac{q_m d}{R^3} + \mu_0\frac{q'_m d}{R^3} = -\mu_1\frac{q''_m}{R^3} \quad \text{or} \quad -\mu_0 q_m + \mu_0 q'_m = -\mu_1 q''_m \tag{7–120}$$

Solving (7–119) and (7–120) simultaneously, we obtain

$$q'_m = \frac{\mu_0-\mu_1}{\mu_0+\mu_1}q_m \quad \text{and} \quad q''_m = \frac{2\mu_0}{\mu_0+\mu_1}q_m \tag{7–121}$$

As there are no real magnetic monopoles, we immediately construct dipoles from two monopoles.

Let q_m be situated at $\vec{r}_+$ and $-q_m$ at $\vec{r}_-$. Then $\vec{m} = q_m(\vec{r}_+ - \vec{r}_-)$. The pole q_m has an image q'_m located at $\vec{r}_+ - 2(\vec{r}_+\cdot\hat{n})\hat{n}$ while $-q_m$ has an image $-q'_m$ located at $\vec{r}_- - 2(\vec{r}_-\cdot\hat{n})\hat{n}$. The image dipole is therefore

$$\vec{m}' = q'_m\left[(\vec{r}_+ - \vec{r}_-) - 2\hat{n}(\vec{r}_+ - \vec{r}_-)\cdot\hat{n}\right] = \frac{\mu_0-\mu_1}{\mu_0+\mu_1}\left[\vec{m} - 2(\vec{m}\cdot\hat{n})\hat{n}\right] \tag{7–122}$$

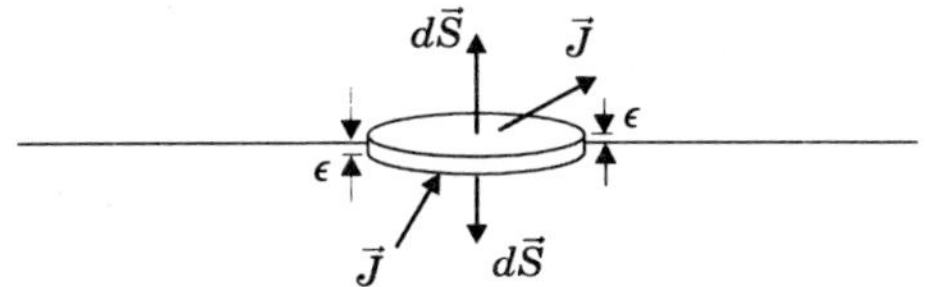

Figure 7.17: The difference between the current flowing into and out of a gaussian pillbox is the rate that charge accumulates within the box.

Similarly, the screened dipole seen from inside the material is

$$\vec{m}'' = \frac{2\mu_0}{\mu_0 + \mu_1}\vec{m} \tag{7–123}$$

Torques and forces on magnetic dipoles in vicinity of a permeable material are then found simply as the interaction of a dipole with its image. These results could of course have been obtained, albeit much more laboriously, without invoking monopoles.

7.6 Conduction in Homogeneous Matter

In ohmic materials, the current density $\vec{J}$ is given by $\vec{J} = g\vec{E}$. In the absence of charge imbalances, $\vec{J}$ must then satisfy $\vec{J} = -g\vec{\nabla}V$ with $\nabla^2 V = 0$. The boundary conditions on $\vec{J}$ are easily obtained from the continuity equation and from $\vec{\nabla}\times\vec{E} = 0$. Thus $\vec{\nabla}\cdot\vec{J} = -\partial\rho/\partial t$ is easily integrated over the volume of a Gaussian pillbox (Figure 7.17) of thickness 2ϵ spanning the boundary taken on the x-y plane between two media to give

$$\int_\tau \vec{\nabla}\cdot\vec{J}d^3r = -\int_\tau \frac{\partial\rho}{\partial t}d^3r \tag{7–124}$$

or

$$\oint \vec{J}\cdot d\vec{S} = -\int \frac{\partial\rho}{\partial t}d^3r \tag{7–125}$$

For a charge density made up of a volume charge density ρ_v and a surface charge density $\sigma\delta(z)$, the integrals become

$$\int \left(J_z\Big|_\epsilon - J_z\Big|_{-\epsilon} \right) dS + \int_{\substack{\text{curved}\\\text{side}}} \vec{J}\cdot d\vec{S} \simeq -2\epsilon\int \frac{\partial\rho}{\partial t}dS - \int \frac{\partial\sigma}{\partial t}dS \tag{7–126}$$

Shrinking the thickness of the pillbox to zero yields

$$J_z(0_+) - J_z(0_-) = -\frac{\partial\sigma}{\partial t} \tag{7–127}$$

This result is, of course, fully consistent with our intuitive view that if more charge arrives at the surface than leaves, charge must pile up at the surface. If no charge accumulates on the interface, then the component of $\vec{J}$ perpendicular to the interface is continuous, leading us to conclude that $gE_\perp$ is continuous and $E_\parallel$ is continuous.

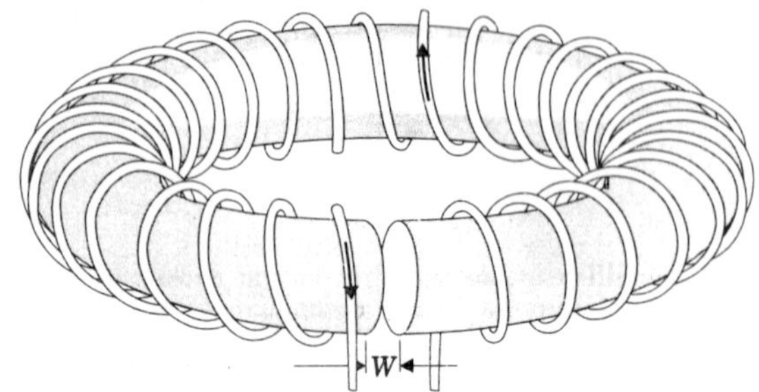

Figure 7.18: A permeable toroidal core wrapped with N turns of wire has a thin gap of width W perpendicular to the core axis.

Current flow problems can be solved using exactly the same techniques as electrostatics problems. When current sources are present they behave like charges, and image source techniques can be used.

7.7 Magnetic Circuits

Although of no great fundamental importance, the notion of magnetic circuits is of considerable practical interest in the design of transformers, electromagnets, and electric motors. In such circuits, current windings constitute a source of flux, while permeable materials provide the conductors.

Consider the field of a torus of N turns uniformly wound around a core of permeability μ (Figure 7.18). The magnetic field strength in the core a distance r from center is easily found using Ampère's law. Setting $2\pi r$ equal to ℓ, we have $H_\varphi(r) \cdot \ell = NI$. The corresponding magnetic induction field is then

$$B_\varphi = \frac{\mu NI}{\ell} \tag{7–128}$$

If we were to cut a small gap of width W in the core, the line integral of the field intensity H is still given by NI, but there is no reason to assume it will be constant around the torus. Assuming that H is still largely contained in the same cross section, we can find H by setting $H_{gap} = B_{gap}/\mu_0$ and using the continuity of $B_\perp$ boundary condition at the interface, $B_{gap} = B_{core}$. Thus the line integral of the field intensity around the loop becomes

$$\oint \vec{H} \cdot d\vec{\ell} = H_{core}(\ell - W) + \frac{\mu H_{core}}{\mu_0} W \tag{7–129}$$

and Ampère's law gives

$$H_{core}\left[\ell + \left(\frac{\mu}{\mu_0} - 1\right) W\right] = NI \tag{7–130}$$

Clearly H_{core} and $B_{core} = \mu H_{core}$ are smaller with the gap than without, indicating that the gap offers greater *reluctance* to the field than the permeable core does. The observation that when μ is $\gg \mu_0$, the flux is fairly well constrained to the permeable material justifies the notion of magnetic circuits.

Let us consider a more general series "circuit" of several different core materials surrounded by a toroidal winding of N turns carrying current I. From the application of Ampère's law,

$$\oint \vec{H} \cdot d\vec{\ell} = NI \tag{7–131}$$

It is convenient to express H at each point in terms of the magnetic flux $\Phi = BA = \mu HA$ where A is the cross-sectional area of the circuit at the point under consideration. Then

$$\oint \frac{\Phi d\ell}{\mu A} = NI \tag{7–132}$$

and, since Φ is essentially constant around the circuit,

$$\Phi \oint \frac{d\ell}{\mu A} = NI \tag{7–133}$$

This is the basic magnetic circuit equation that enables us to solve for the flux in terms of the circuit parameters. The equation is reminiscent of the elementary series circuit equation $I(\Sigma R) = \mathcal{E}$. (The resistance of a similar electric circuit is $R = \oint d\ell/gA$.) By analogy, the *reluctance*, $\Re$, is defined as

$$\Re = \oint \frac{d\ell}{\mu A} \tag{7–134}$$

and NI is referred to as the *magnetomotive force* (mmf). With these definitions, $\Phi\Re$ = (mmf), and the conserved magnetic flux, Φ, plays the role of 'magnetic current'. If the circuit is made of several homogeneous pieces, each of uniform cross section, the reluctance may be approximated as

$$\Re = \sum_j \frac{\ell_j}{\mu_j A_j} = \sum_j \Re_j \tag{7–135}$$

Clearly reluctances in series and parallel add like resistors in series and parallel. Although the theory is not exact (because of flux leakage), it is useful in engineering applications. As the flux fringing is particularly severe for air gaps, an empirical adjustment of increasing the effective cross-section dimensions (diameter or width and height) by the length of the gap, is conventionally made by engineers.

EXAMPLE 7.13: The magnetic circuit sketched in Figure 7.19 is wound with 100 turns of wire, carrying a current of 1 A. The winding is located on the extreme left-hand leg of the circuit. All legs have the same 6 cm^2 cross section and permeability of 5000 μ_0.

Solution: The analogous electric circuit has $3R$ in parallel with R giving $.75R$ in series with $3R$ for a total of $3.75R$. Transposing to the magnetic circuit, $\Re = 3.75\,\Re_j$ with $\Re_j = (10\text{ cm})/(\mu \cdot 6\,\text{cm}^2) = 10^3/(6\mu)$ (SI units). The flux through the coils is then

$$\Phi = \frac{NI}{3.75\,\Re_j} = \frac{100 \times 6\mu}{3.75 \times 10^3} = .16\mu \text{ (SI)} \tag{Ex 7.13.1}$$

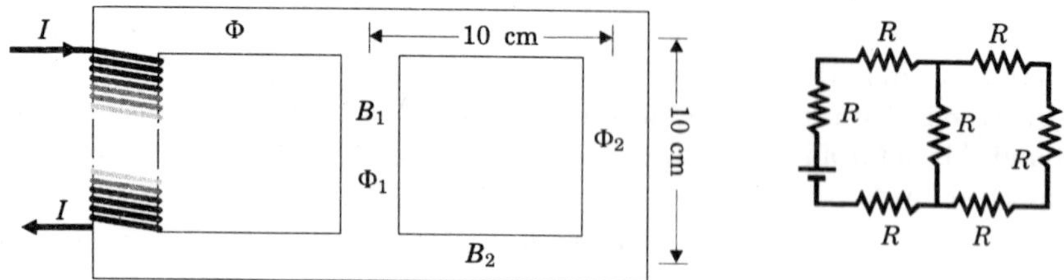

Figure 7.19: The double-holed yoke has 100 turns on its leftmost leg. The analogous electric circuit is shown on the right.

Of this flux, $\frac{3}{4}$ passes through the central leg, and $\frac{1}{4}$ passes though the right-hand leg:

$$\begin{aligned} \Phi_1 &= \tfrac{3}{4} \times .16\mu = (0.12 \times 5000 \times 4\pi \times 10^{-7}) = 7.54 \times 10^{-4}\,\text{Wb} \\ \Phi_2 &= \tfrac{1}{4} \times .16\mu = 2.51 \times 10^{-4}\,\text{Wb} \end{aligned} \tag{Ex 7.13.2}$$

giving magnetic induction fields

$$\begin{aligned} B_1 &= \frac{\Phi_1}{A} = \frac{7.54 \times 10^{-4}}{6 \times 10^{-4}} = 1.256\,\text{T} \\ B_2 &= \frac{\Phi_2}{A} = 0.418\,\text{T} \end{aligned} \tag{Ex 7.13.3}$$

7.7.1 Magnetic Circuits Containing Permanent Magnets

Let us consider a magnetic circuit devoid of currents, but having instead a permanent magnet of length L_{PM} as the source of fields, as shown in Figure 7.20. Ampère's law gives for a loop around the magnet and the yoke

$$\oint \vec{H} \cdot d\vec{\ell} = 0 \tag{7-136}$$

or, integrating around the loop in clockwise direction from a to b

$$\int_{a\ \text{(yoke)}}^{b} H_Y\, d\ell = -\int_{b\ \text{(PM)}}^{a} H_{PM}\, d\ell \tag{7-137}$$

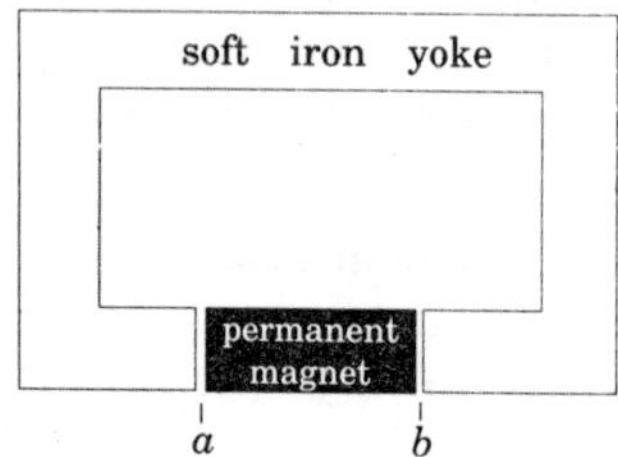

Figure 7.20: A soft iron yoke connects the two end of a permanent magnet.

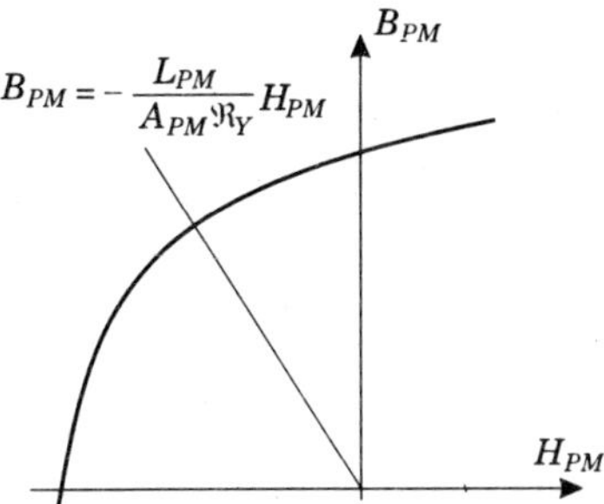

Figure 7.21: A quadrant of a typical hysteresis curve for a permanent magnet.

In addition, we use the constancy of flux around the loop to find $B_{PM}A_{PM} = B_Y A_Y = \mu A_Y H_Y$. Substituting for H_Y in the integrals (7–137) above, we obtain

$$\frac{A_{PM}B_{PM}L_Y}{\mu A_Y} = -H_{PM}L_{PM} \tag{7–138}$$

which we invert to obtain

$$B_{PM} = -\frac{L_{PM}}{A_{PM}}\frac{\mu A_Y}{L_Y}H_{PM} \tag{7–139}$$

an equation linking B and H in the magnet. The resulting fields may then be found from the hysteresis curve for the permanent magnet (figure 7.21). The ends of the magnet produce a demagnetizing field H to reduce B, as we had noted in passing earlier.

The result above may be generalized to more complicated yokes by recognizing $L_Y/\mu A_Y$ as the reluctance of the yoke. More generally, then

$$B_M = -\frac{L_M}{A_M\Re_Y}H_{PM} \tag{7–140}$$

7.7.2 The Hysteresis Curve of a Ferromagnet

For a ferromagnetic material, $\vec{B} = \mu(H)\vec{H}$, where μ is usually a very strong function of H. A typical hysteresis curve is sketched in Figure 7.21. This curve will in general depend on the history of the material. The induction field that remains even when there is no field intensity H to induce it is known as the *remanence*. The magnetic field intensity required to reduce the flux density to zero is the *coercive force*.

It is interesting to calculate the amount of work that must be done to move the material from one point on the hysteresis curve to another. We consider a closely wound torus with a core of the material of interest (Rowland's ring). The EMF, $\mathcal{E}$, induced on the N turns of the winding as we change $\vec{B}$ from B_1 to B_2 is given by $\mathcal{E} = -Nd\Phi/dt$. The current required to produce the field strength H to cause the change in B does work against this EMF at the rate $dW/dt = NId\Phi/dt$. Assuming

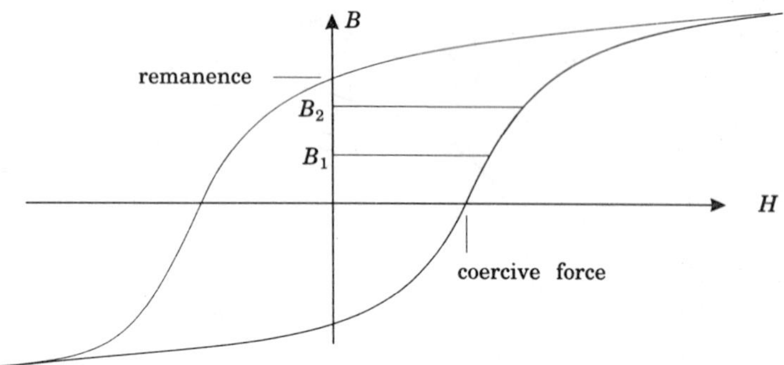

Figure 7.22: A typical hysteresis curve for a permanent ferromagnet. The lower trace is produced by $\vec{H}$ increasing from large negative values.

that the ring is large enough that B is constant over the cross section S of the core, we find that $\Phi = BS$, and write NI in terms of H, $NI = H\ell$ so that

$$\frac{dW}{dt} = \ell H \frac{d}{dt}(BS) = VH\frac{dB}{dt} \tag{7–141}$$

where $\ell S = V$ is the volume of the core. Consequently, the work required to change B to $B + dB$ in a unit volume is

$$d\left(\frac{W}{V}\right) = HdB \tag{7–142}$$

leading to the conclusion that

$$\frac{W}{V} = \int_{B_1}^{B_2} H(B)dB \tag{7–143}$$

We interpret the integral (7–143) geometrically as the shaded area of Figure 7.22. If the material is cycled around its hysteresis loop, the energy input per unit volume is given by the area of the loop. A transformer core would cycle around

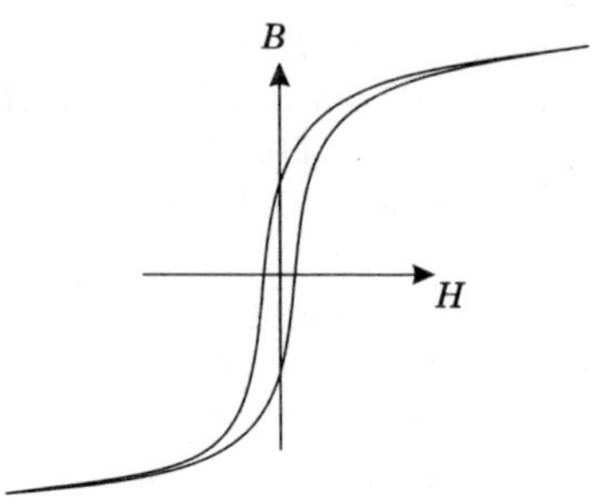

Figure 7.23: A good transformer core should have a thin hysteresis curve.

the loop many times. To minimize flux losses, the permeability should be large. In order to minimize hysteresis losses one would use a ferromagnetic core with a very thin loop. As B must remain large, this means that the curve should be very thin in H, as illustrated in Figure 7.23.

Exercises and Problems

7-1 A large parallel plate capacitor with plates separated by d has a dielectric slab with dielectric constant κ and thickness t partially filling the space between the plates. A potential ΔV is applied across the plates. Find $\vec{E}$ in the intervening space and in the dielectric.

7-2 A large block of dielectric with polarization $\vec{P}$ has a long thin slot cut in it parallel to the polarization. Find the $\vec{E}$ and $\vec{D}$ in the slot.

7-3 A large block of dielectric with polarization $\vec{P}$ has a thin disk-shaped cavity perpendicular to the polarization. Find $\vec{E}$ and $\vec{D}$ in the cavity.

7-4 Find the electric field at points along the z axis outside a dielectric (right circular) cylinder of length L and radius a whose axis coincides with the z axis and is uniformly polarized along its axis.

7-5 A block of uniformly magnetized material has the slot and disk of the two problems above respectively parallel and perpendicular to the magnetization. Find $\vec{B}$ and $\vec{H}$ in each case.

7-6 Find the force between two identical charges embedded in a dielectric of permittivity ε.

7-7 Find the force between two identical charges embedded in different dielectrics with permittivities ε_1 and ε_2. Assume that the charges lie at equal distances from the interface along a normal to the interface. (Beware of images.)

7-8 A dielectric sphere of radius R contains a uniform distribution of free charge. Find the potential at the center.

7-9 A thick hemispherical shell of inner radius a and outer radius b is uniformly magnetized along its symmetry axis. Show that the magnetic induction field at the center of curvature of the shell vanishes.

7-10 An electric point dipole is placed in the neighborhood of a thick large sheet of dielectric of permeability ε. Find the force and torque on the dipole resulting from the induced polarization of the dielectric.

7-11 A parallel plate capacitor with plates of dimension $a \times b$ spaced at d has the corner of a large slab of dielectric of thickness $d/2$ and permittivity ε partly inserted to depth of Δx and Δy in the space between the plates. Find the force on the slab, ignoring fringing fields.

7-12 A capacitor constituted of two vertical concentric cylinders of radii a and b has one end immersed in oil of dielectric constant κ. To what height does the oil rise in the space between the two cylinders when the potential difference V is applied between them? Ignore capillarity and fringing fields.

7-13 A charged, conducting cylinder of radius a carrying line charge density λ in vacuum is placed near and parallel to the plane interface with a dielectric medium. Find the resulting charge distribution on the surface of the cylinder.

7-14 A magnetron magnet has two opposing pole faces spaced at 1 cm and an induction field in the gap of 2T. A screwdriver ($\mu = 10{,}000\ \mu_0$, assume it linear although it will clearly not be) of cross section 0.5 cm^2 is inadvertently inserted partway between the poles. Find approximately the force on the screw driver. Neglect fringing fields and assume that inserting the screwdriver does not appreciably affect the reluctance of the magnet.

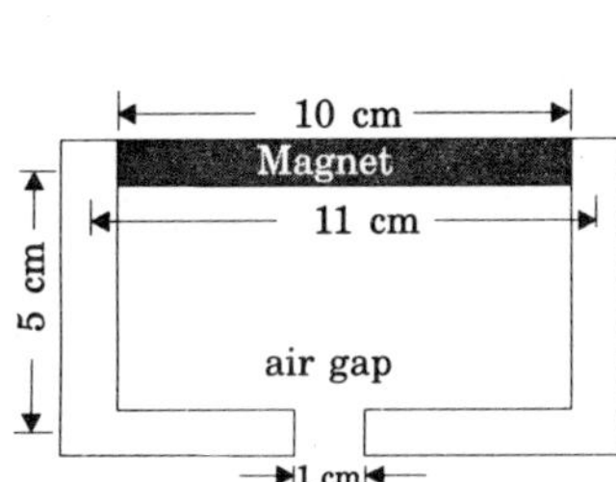

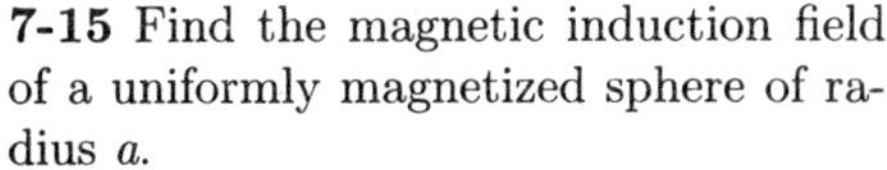
Figure 7.24: Permanent magnet and yoke.

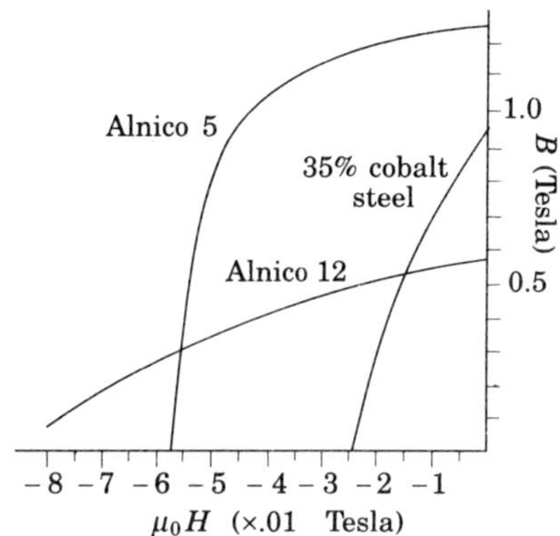

Figure 7.25: Quadrant of hysteresis curve for several permanent magnetic materials.

7-15 Find the magnetic induction field of a uniformly magnetized sphere of radius a.

7-16 Find $\vec{B}$ in a spherical cavity in a large, uniformly magnetized block of material when $\vec{H}$ far from the cavity is $H_0\hat{k}$. What does this become when H_0 is zero but $\vec{M}$ is not?

7-17 Find $\vec{B}$ and $\vec{H}$ in a thin circular disk of material uniformly magnetized perpendicular to the flat face of the material.

7-18 Use equation (7–114) to find the magnetic scalar potential on the symmetry axis in the gap be-tween the circular pole pieces of a permanent magnet having uniform magnetization M. Assume that the radius of the pole pieces is a and the width of the gap is b.

7-19 A long cylinder of radius a and permeability μ is placed in an initially uniform field $\vec{B}_0$ so that the cylinder axis is perpendicular to $\vec{B}_0$. Calculate the resulting field $\vec{B}$.

7-20 Find the dipole moment of a thin cylindrical magnet with uniform magnetization M along its axis.

7-21 Find the force on a permanent magnet of volume V and magnetization M held in the neighborhood of a large sheet of soft iron of permeability $10^4\mu_0$ (a) when the magnetization (of the magnet) is parallel to the interface and (b) when the magnetization is perpendicular to the interface.

7-22 An Alnico-5 magnet of length 10 cm and cross section 1 cm^2 is placed in a magnetic circuit completed by permalloy yoke of permeability $\mu = 25,000\mu_0$ and 1 cm^2 cross section with a 1 cm air gap, as illustrated in Figure 7.24. Use the appropriate hysteresis curve from Figure 7.25 to find B and H in the magnet, the permalloy yoke, and the air gap.

7-23 Find the electric displacement field inside the cylinder of example 7.1. Translate the origin so that the end faces lie at $z = \pm\frac{1}{2}L$ and compare this result with (Ex 7.4.2).

7-24 Find the magnetic induction field both inside and outside a magnetized sphere with magnetization $\vec{M} = M_0 x\hat{\imath}$.

7-25 Show that the bulk permittivity of an isotropic mixture of dielectrics with permittivities ε_i is given by

$$\varepsilon_{mix} = \left\langle \varepsilon_i^{1/3} \right\rangle^3$$

where $\langle\ \rangle$ means the average over volume. (Hint: write

$$\begin{aligned}\langle \vec{D} \rangle &= \varepsilon_{mix} \langle \vec{E} \rangle \\ &= \Big\langle (\langle \varepsilon \rangle + \delta\varepsilon)(\langle \vec{E} \rangle + \delta \vec{E}) \Big\rangle\end{aligned}$$

and show that

$$\langle \delta\varepsilon \delta \vec{E} \rangle = \tfrac{1}{3} \langle \vec{E} \rangle \langle (\delta\varepsilon)^2 \rangle / \langle \varepsilon \rangle.)$$

7-26 Use the diffusion equation (7–54) that governs the temporal behavior of a magnetic induction field in a conductor to estimate the time required for an initial field $\vec{B}_0$ in a stationary copper sphere of radius 1 m to decay. Repeat the exercise using a nonrotating model of the earth having a 3000 km core of iron with conductivity $g = 10^6/\Omega\text{m}$ (600° C) (assume $\mu = \mu_0$). What does this imply for the timescale of reversals of the earth's magnetic polarity?

Chapter 8

Time-Dependent Electromagnetic Fields in Matter

8.1 Maxwell's Equations

We begin this chapter with a consideration of how Maxwell's equations ought to be amended when matter is present. To start, we restate Maxwell's equations in vacuum (equations 3–27):

$$\vec{\nabla}\cdot\vec{E} = \frac{\rho}{\varepsilon_0} \qquad \vec{\nabla}\times\vec{E} = -\frac{\partial\vec{B}}{\partial t}$$
$$\vec{\nabla}\cdot\vec{B} = 0 \qquad \vec{\nabla}\times\vec{B} = \mu_0\left(\vec{J} + \varepsilon_0\frac{\partial\vec{E}}{\partial t}\right)$$

As we saw in section 7.1, the source term for $\vec{E}$, $\rho(t)$, is augmented in the presence of matter by the bound charge density $\rho_b(t) = -\vec{\nabla}\cdot\vec{P}(t)$, leading to

$$\vec{\nabla}\cdot\vec{D} = \rho \tag{8–1}$$

as in the static case. Because there was no time dependence in the vacuum equations, there is no reason to expect one when matter is present.

The $\vec{\nabla}\times\vec{E}$ equation

$$\vec{\nabla}\times\vec{E} = -\frac{\partial\vec{B}}{\partial t} \tag{8–2}$$

remains unchanged also because this is a relationship between the force fields that does not involve the sources. (If magnetic monopoles existed, the magnetic current would depend on the sources, and this term would change in the presence of matter.)

The $\vec{\nabla}\cdot\vec{B}$ equation

$$\vec{\nabla}\cdot\vec{B} = 0 \tag{8–3}$$

also remains unchanged, as there appear to be no magnetic monopoles (or, if there are, they do not occur in any experiment we might consider).

The last of the four equations, in contrast to the others, must be modified. As we have already seen in the static case, magnetizations are a source of magnetic field taken into account by augmenting the current density $\vec{J}$ by $\vec{\nabla} \times \vec{M}$. In the nonstatic case an additional term arises.

When a molecular dipole changes its orientation or size, a current flows on a microscopic scale. When the change in polarization is uniform, the transient current corresponds to the bound surface charge being transported from one side of the medium to the other. When the change of polarization is not uniform, charge concentrations in the form of bound charge accumulate. The accumulation of bound charge corresponds to a current, which constitutes a source of field not yet considered. To quantify this observation, we consider the bound charge in some fixed volume τ, and use the divergence theorem (20) to obtain

$$Q_b = \int_\tau \rho_b d^3r = -\int_\tau \vec{\nabla} \cdot \vec{P} d^3r = -\oint \vec{P} \cdot d\vec{S} \tag{8–4}$$

Taking the time derivative of both sides of the equation, we have

$$\frac{dQ_b}{dt} = -\oint \vec{J}_b \cdot d\vec{S} = -\oint \frac{\partial \vec{P}}{\partial t} \cdot d\vec{S} \tag{8–5}$$

from which we conclude that

$$\vec{J}_b = \frac{\partial \vec{P}}{\partial t} \tag{8–6}$$

With this addition the last of Maxwell's equations becomes

$$\vec{\nabla} \times \vec{B} = \mu_0 \left(\vec{J} + \vec{\nabla} \times \vec{M} + \varepsilon_0 \frac{\partial \vec{E}}{\partial t} + \frac{\partial \vec{P}}{\partial t} \right) \tag{8–7}$$

Casting this equation in terms of $\vec{H}$ and $\vec{D}$, we have

$$\vec{\nabla} \times \left(\frac{\vec{B}}{\mu_0} - \vec{M} \right) = \vec{J} + \frac{\partial}{\partial t} \left(\varepsilon_0 \vec{E} + \vec{P} \right) \tag{8–8}$$

or

$$\vec{\nabla} \times \vec{H} = \vec{J} + \frac{\partial \vec{D}}{\partial t} \tag{8–9}$$

Summarizing Maxwell's equations in their final form, we have

$$\boxed{\begin{array}{ll} \vec{\nabla} \cdot \vec{D} = \rho & \vec{\nabla} \cdot \vec{B} = 0 \\ \vec{\nabla} \times \vec{E} = -\dfrac{\partial \vec{B}}{\partial t} & \vec{\nabla} \times \vec{H} = \vec{J} + \dfrac{\partial \vec{D}}{\partial t} \end{array}} \tag{8–10}$$

with $\vec{D} = \varepsilon_0 \vec{E} + \vec{P}$ and $\vec{B} = \mu_0(\vec{H} + \vec{M})$.

For linear isotropic materials we have, in addition, the constitutive relations $\vec{P} = \chi\varepsilon_0 \vec{E}$, $\vec{M} = \chi_m \vec{H}$, and $\vec{J} = g\vec{E}$. The interaction of matter with the electromagnetic fields is entirely through the Lorentz force, ($\vec{f}$ is the force per unit volume)

$$\vec{f} = \rho \vec{E} + \vec{J} \times \vec{B} \tag{8–11}$$

The equation of continuity, expressing charge conservation, is easily seen to be a consequence of Maxwell's equations. To demonstrate this, we simply take the divergence of $\vec{\nabla} \times \vec{H}$:

$$\begin{aligned} 0 = \vec{\nabla} \cdot (\vec{\nabla} \times \vec{H}) &= \vec{\nabla} \cdot \vec{J} + \vec{\nabla} \cdot \frac{\partial \vec{D}}{\partial t} \\ &= \vec{\nabla} \cdot \vec{J} + \frac{\partial}{\partial t}\left(\vec{\nabla} \cdot \vec{D}\right) \\ &= \vec{\nabla} \cdot \vec{J} + \frac{\partial \rho}{\partial t} \end{aligned} \tag{8–12}$$

It is a testament to the genius of James Clerk Maxwell that he deduced these equations (8–10) from the "lines of force" that Faraday used to interpret the results of his experiments in electromagnetism.

EXAMPLE 8.1 A circular dielectric disk of thickness d and radius a, possessing both linear and quadratic dielectric susceptibilities χ and $\chi^{(2)}$, is placed between the plates of a parallel plate capacitor also of radius a, spaced by d, that have an alternating voltage $V = V_0 e^{-i\omega t}$ difference impressed on them. Find the magnetic field intensity $\vec{H}$ near the dielectric.

Solution: We draw a loop of radius r encircling the dielectric in a plane parallel to the capacitor plates and integrate $\vec{\nabla} \times \vec{H}$ over the surface of the loop. With the aid of Stokes' theorem we have

$$\int_\Gamma \vec{H} \cdot d\vec{\ell} = \int_S \frac{\partial}{\partial t}\left[\varepsilon_0(1+\chi)Ee^{-i\omega t} + \chi^{(2)}\varepsilon_0^2 E^2 e^{-2i\omega t}\right]dS \tag{Ex 8.1.1}$$

with $E = V/d$. The rotational symmetry guarantees that the φ component of H is invariant around the loop allowing us to write

$$2\pi r H_\varphi = \pi a^2\left[\varepsilon_0(1+\chi)\frac{-i\omega V e^{-i\omega t}}{d} + \varepsilon_0^2\chi^{(2)}\frac{-2i\omega V^2 e^{-2i\omega t}}{d^2}\right] \tag{Ex 8.1.2}$$

The radial component of $\vec{H}$ must vanish as a nonzero cylindrically symmetric H_r would have a non-vanishing divergence in the absence of a magnetization.

8.1.1 Boundary Conditions for Oscillating Fields

We have, in dealing with refraction of waves in dielectrics, tacitly assumed that the boundary conditions appropriate to a static field would serve. Nonstatic conditions do alter at least some of the boundary conditions. We might anticipate, for instance, that a time-dependent polarization of the medium would require a discontinuity in current crossing the interface between two media in order to account for the arrival or departure of bound charge. In this section we investigate the boundary conditions in more detail. Our point of departure, as always, is Maxwell's equations (8–10)

$$\vec{\nabla} \cdot \vec{D} = \rho \qquad \vec{\nabla} \cdot \vec{B} = 0$$
$$\vec{\nabla} \times \vec{E} = -\frac{\partial \vec{B}}{\partial t} \qquad \vec{\nabla} \times \vec{H} = \vec{J} + \frac{\partial \vec{D}}{\partial t}$$

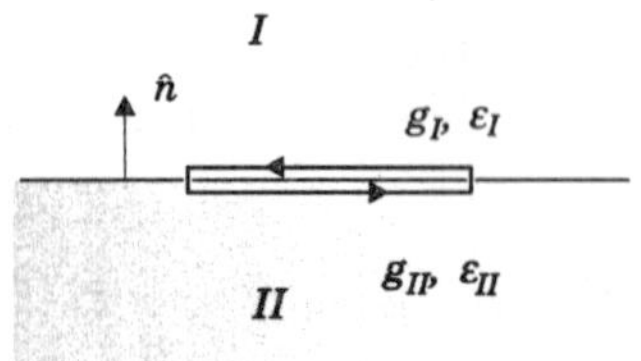

Figure 8.1: The thin loop has its long side parallel to the interface.

To obtain the boundary conditions we integrate each of the equations over a volume or surface spanning the boundary between two media.

The first two (divergence) equations are unaltered from their static form (7–53), and the same arguments lead to the same conclusions, namely that $B_\perp$ is continuous and $D_\perp$ is discontinuous by the (time-dependent) surface charge σ. Only the curl equations could be altered, and we consider them in detail.

Integrating the equation for $\vec{\nabla} \times \vec{E}$ over the area of the thin loop spanning the interface shown in Figure 8.1, we have

$$\int (\vec{\nabla} \times \vec{E}) \cdot d\vec{S} = -\int \frac{\partial \vec{B}}{\partial t} \cdot d\vec{S} \tag{8–13}$$

Application of Stokes' theorem to the left side gives

$$\oint \vec{E} \cdot d\vec{\ell} = \int -\frac{\partial \vec{B}}{\partial t} \cdot d\vec{S} \tag{8–14}$$

and as width of the loop shrinks to zero, the surface integral on the right vanishes. We conclude, then, that as before, $\vec{E}_\parallel^{II} = \vec{E}_\parallel^{I}$.

Integrating the equation for $\vec{\nabla} \times \vec{H}$ over the same thin loop of Figure 8.1, we have

$$\int (\vec{\nabla} \times \vec{H}) \cdot d\vec{S} = \int \vec{J} \cdot d\vec{S} + \int \frac{\partial \vec{D}}{\partial t} \cdot d\vec{S} \tag{8–15}$$

With the aid of Stokes' theorem the left-hand side is converted to a line integral. Letting the interface lie in the x-y plane, we write the current density as the sum of a body current and surface current density, $\vec{J} = \vec{J}_b + \vec{\jmath}\delta(z)$. Shrinking the width of the loop to 0 eliminates all but the surface current contribution to the right-hand side. The remaining terms give $H_\parallel^{II} - H_\parallel^{I} = j$, where j is the surface current that threads the loop. On generalizing we recover our earlier result, (7–68)

$$\hat{n} \times \left(\vec{H}^I - \vec{H}^{II}\right) = \vec{\jmath}$$

At this point we might well wonder where, if anywhere, the claimed distinction between the static and nonstatic boundary conditions might occur. The difference lies in the possibility of maintaining a zero surface charge density.

When fields oscillate, it is not generally possible to maintain a zero surface charge density σ, as is easily seen from the continuity equation (1–24):

$$\vec{\nabla} \cdot \vec{J} = -\frac{\partial \rho}{\partial t}$$

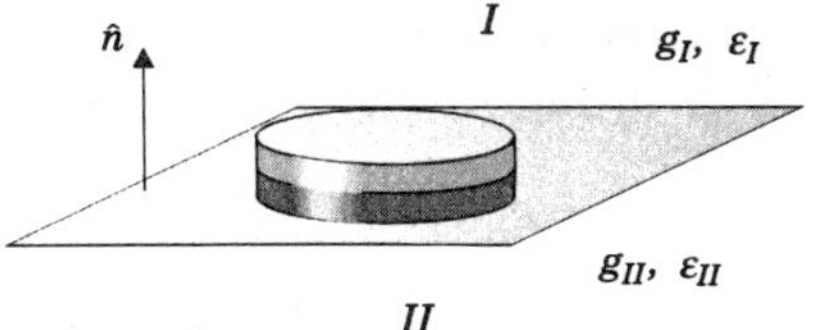

Figure 8.2: A thin Gaussian pillbox spans the interface between medium I and II.

Integrating (1–24) over the volume of the thin pillbox of Figure 8.2 when $\rho = \rho_v(x, y, z) + \sigma(x, y)\delta(z)$ (we have again taken the interface in the x-y plane), we obtain

$$\int \vec{\nabla} \cdot \vec{J} d^3r = -\frac{\partial}{\partial t}\left[\iint \rho_v(x, y, z) d^3r + \int \sigma(x, y) dS\right] \tag{8–16}$$

or

$$\oint \vec{J} \cdot d\vec{S} = -\frac{\partial}{\partial t}\left[\bar{\rho}_v \tau + \int \sigma(x, y)\, dS\right] \tag{8–17}$$

where $\bar{\rho}$ is the mean charge density in the volume of integration τ.

As the thickness of the pillbox shrinks to zero,

$$\int (\vec{J}^I \cdot \hat{n} - \vec{J}^{II} \cdot \hat{n}) dS = -\frac{\partial}{\partial t}\int \sigma(x, y)\, dS \tag{8–18}$$

which leads to

$$\left(\vec{J}^I - \vec{J}^{II}\right) \cdot \hat{n} = -\frac{\partial \sigma}{\partial t} \tag{8–19}$$

Now, for $\sigma = \sigma_0 e^{-i\omega t}$ and $\vec{J} = g\vec{E}$, this becomes $g_1 E_n^I - g_2 E_n^{II} = i\omega\sigma$. On the other hand, from $\vec{\nabla} \cdot \vec{D} = \rho$ we have $D_n^I - D_n^{II} = \sigma$. These two equations are compatible with $\sigma = 0$ only when $g_1/\varepsilon_1 = g_2/\varepsilon_2$, a condition not generally satisfied. Eliminating σ from the two equations, we obtain the following from the equation in $\vec{D}$:

$$\left(\varepsilon_1 + \frac{ig_1}{\omega}\right) E_n^I = \left(\varepsilon_2 + \frac{ig_2}{\omega}\right) E_n^{II} \tag{8–20}$$

8.1.2 Special Cases

Three special cases may now be distinguished. In all cases we assume $\vec{J} = g\vec{E}$, $\vec{D} = \varepsilon\vec{E}$, $\vec{B} = \mu\vec{H}$, $\vec{H} = \vec{H}_0 e^{-i\omega t}$ and $\vec{E} = \vec{E}_0 e^{-i\omega t}$.

— Both materials are nonconductors: $E_\parallel$, $H_\parallel$, $D_\perp$, and $B_\perp$ are all continuous across the interface.

— Both materials have nonzero, finite conductivity: Surface currents cannot be sustained in a less-than-perfect conductor; $\vec{j} = 0$. Therefore

$$E_\parallel,\ \left(\varepsilon + \frac{ig}{\omega}\right) E_\perp,\ H_\parallel, \text{and } B_\perp \text{ are all continuous.} \tag{8–21}$$

— One of the media is a perfect conductor: say $g_2 = \infty$. This is frequently a good approximation for a metal boundary. Using

$$\vec{\nabla} \times \vec{H}^{II} = \vec{J}^{II} + \frac{\partial \vec{D}^{II}}{\partial t} = (g_2 - i\varepsilon_2\omega)\vec{E}^{II} \tag{8–22}$$

we obtain for $|\vec{\nabla} \times \vec{H}^{II}| < \infty$, $\vec{E}^{II} = 0$. Next, using $\vec{\nabla} \times \vec{E}^{II} = -\partial \vec{B}^{II}/\partial t = i\mu\omega\vec{H}^{II} = 0$, we find $\vec{H}^{II} = 0$. The boundary conditions are therefore,

$$\begin{gathered} \vec{E}^{II} = \vec{H}^{II} = 0, \quad B_\perp^I = B_\perp^{II} = 0, \quad E_\parallel^I = 0, \\ \varepsilon_1 E_\perp^I = \sigma, \quad \text{and} \quad \hat{n} \times \vec{H}^I = \vec{j} \end{gathered} \tag{8–23}$$

where $\hat{n}$ is the normal pointing outward from the perfect conductor.

8.2 Energy and Momentum in the Fields

We saw in Chapter 3 that the potential energy of particles could be associated with their fields. In material media it seems fairly clear that indeed, the potential energy will be stored in the polarization and magnetization of the medium. Before finding the general result, we briefly consider the work, δW, required to add a small quantity of charge δq in constructing a charge whose potential at the time of adding δq is V. Let us suppose the charge increment is distributed over some volume τ small enough that the potential is constant. We write then $\delta q = \int \delta\rho\, d^3r$. The work performed in adding the charge is

$$\delta W = V\delta q = \int V(\delta\rho)d^3r = \int V\delta(\vec{\nabla}\cdot\vec{D})d^3r = \int V\vec{\nabla}\cdot(\delta\vec{D})d^3r \tag{8–24}$$

The integral may be integrated by parts using $\vec{\nabla}\cdot(V\delta\vec{D}) = \vec{\nabla}V\cdot\delta\vec{D} + V(\vec{\nabla}\cdot\delta\vec{D})$:

$$\begin{aligned} \delta W &= \int \vec{\nabla}\cdot(V\delta\vec{D})\, d^3r - \int \vec{\nabla}V\cdot\delta\vec{D}d^3r \\ &= \oint V(\delta\vec{D})\cdot d\vec{S} + \int \vec{E}\cdot\delta\vec{D}d^3r \end{aligned} \tag{8–25}$$

If the field is produced by a finite charge occupying a finite region of space, the potential at large r diminishes as r^{-1} (or faster) and $\vec{D}$ diminishes as r^{-2} (or faster). Therefore, at sufficiently large $\vec{r}$, $V(\delta\vec{D})$ decreases more rapidly than S increases, leading to a vanishing first integral. Dropping the first integral, we obtain

$$\delta W = \int_{\substack{all\\space}} \vec{E}\cdot\delta\vec{D}\, d^3r \tag{8–26}$$

where the integration extends over all space. The increment of work δW cannot usually be integrated unless $\vec{D}$ is given as a function of $\vec{E}$. For $\vec{D} = \varepsilon\vec{E}$, the work required to assemble the entire charge distribution

$$W = \int_0^D \delta W = \int_0^D \int_\tau \vec{E}\cdot\delta\vec{D}d^3r$$

$$= \int_\tau \int_0^D \tfrac{1}{2}\varepsilon \vec{E} \cdot \delta(\vec{E}) d^3r = \tfrac{1}{2}\int_\tau \varepsilon E^2 d^3r = \tfrac{1}{2}\int_\tau \vec{E} \cdot \vec{D} d^3r \tag{8–27}$$

In the case that the material is linear but not isotropic, we set $\varepsilon^i_j = \varepsilon_0 \chi^i_j$ and $D^i = \varepsilon^i_j E^j$, leading to

$$\delta W = \int_0^D \vec{E} \cdot \delta\vec{D} d^3r = \int_0^D E_i(\delta D^i) d^3r = \int_0^D E_i \varepsilon^i_j (\delta E^j) d^3r \tag{8–28}$$

which integrates to

$$W = \int \frac{\varepsilon^i_j E_i E^j}{2} d^3r = \int \frac{D_j E^j d^3r}{2} = \int \frac{\vec{E} \cdot \vec{D}}{2} d^3r \tag{8–29}$$

as before.

The expression for the energy density of a linear anisotropic material

$$U = \tfrac{1}{2}\varepsilon^i_j E_i E^j \equiv \tfrac{1}{2}\varepsilon^j_i E_j E^i = \tfrac{1}{2}\varepsilon_{ij} E^i E^j \tag{8–30}$$

leads incidentally to the useful fact that the permittivity tensor is symmetric under interchange of columns and rows, $\varepsilon^i_j = \varepsilon^j_i$.

8.2.1 Energy of Electric and Magnetic Fields

The more general case when both electric and magnetic fields are present is easily obtained by manipulation of Maxwell's equations. Consider the following application of identity (8):

$$\begin{aligned} \vec{\nabla} \cdot (\vec{E} \times \vec{H}) &= \vec{H} \cdot (\vec{\nabla} \times \vec{E}) - \vec{E} \cdot (\vec{\nabla} \times \vec{H}) \\ &= -\vec{H} \cdot \frac{\partial \vec{B}}{\partial t} - \vec{E} \cdot \vec{J} - \vec{E} \cdot \frac{\partial \vec{D}}{\partial t} \end{aligned} \tag{8–31}$$

Integrating this expression over some arbitrary volume τ, gives

$$\int_\tau \vec{\nabla} \cdot (\vec{E} \times \vec{H}) d^3r = -\int_\tau \left(\vec{E} \cdot \frac{\partial \vec{D}}{\partial t} + \vec{H} \cdot \frac{\partial \vec{B}}{\partial t} \right) d^3r - \int_\tau \vec{E} \cdot \vec{J} d^3r \tag{8–32}$$

The last integral of (8–32) is simply the rate that work is done by the field on currents in the region of interest ($\vec{E} \cdot \vec{J} = \rho \vec{E} \cdot \vec{v} = \vec{f} \cdot \vec{v}$). If matter is linear and isotropic, the terms within the parentheses of the second integral may be written

$$\vec{E} \cdot \frac{\partial \vec{D}}{\partial t} = \frac{1}{2}\frac{\partial}{\partial t}(\varepsilon E^2) = \frac{\partial}{\partial t}\left(\frac{\vec{E} \cdot \vec{D}}{2} \right) \tag{8–33}$$

and

$$\vec{H} \cdot \frac{\partial \vec{B}}{\partial t} = \frac{\partial}{\partial t}\left(\frac{\vec{H} \cdot \vec{B}}{2} \right) \tag{8–34}$$

Converting the left side of (8–32) to a surface integral, we write

$$\oint(\vec{E}\times\vec{H})\cdot d\vec{S} = -\frac{\partial}{\partial t}\int\left(\frac{\vec{E}\cdot\vec{D}}{2}+\frac{\vec{B}\cdot\vec{H}}{2}\right)d^3r - \int\vec{E}\cdot\vec{J}d^3r \tag{8–35}$$

We identify

$$U = \tfrac{1}{2}(\vec{E}\cdot\vec{D}+\vec{B}\cdot\vec{H}) \tag{8–36}$$

as the energy density of the fields, since whenever the left-hand side is zero, energy conservation requires that any work done on the currents be compensated by a decrease of the field energy.

The term on the left now clearly represents the rate that energy leaves the volume by flowing out, across the surface. Thus

$$\vec{S} = \vec{E}\times\vec{H} \tag{8–37}$$

the Poynting vector, is identified with the energy flux. As we remarked earlier, it is, strictly speaking, an energy flux only when it has zero curl. Conservation of energy is now expressed as

$$\oint\vec{S}\cdot d\vec{S} = -\frac{\partial}{\partial t}\int U d^3r - \frac{\partial W'}{\partial t} \tag{8–38}$$

where $\partial W'/\partial t = \int\vec{E}\cdot\vec{J}d^3r$ is the rate energy is imparted to motion of the charges constituting the current.

8.2.2 Momentum and the Maxwell Stress Tensor

We anticipate that an energy flux will have a momentum flux associated with it. It is therefore no surprise that when there is a net energy flow into or out of a given volume of space, there is a change in the momentum contained within that volume; there must be a force acting on the volume to precipitate this momentum change. While we have no difficulty conceiving of forces acting on charges within the volume, a different view emerges when we express the forces in terms of the fields created by the charges. We proceed in a fashion very similar to that of Section 4.3.2.

We consider the force on a system of charges:

$$\begin{aligned}\vec{F} &= \frac{d}{dt}\int\vec{\mathcal{P}}d^3r = \int(\rho\vec{E}+\vec{J}\times\vec{B})d^3r\\ &= \int\left[(\vec{\nabla}\cdot\vec{D})\vec{E} + \left(\vec{\nabla}\times\vec{H}-\frac{\partial\vec{D}}{\partial t}\right)\times\vec{B}\right]d^3r\\ &= \int\left[(\vec{\nabla}\cdot\vec{D})\vec{E} + (\vec{\nabla}\times\vec{H})\times\vec{B} - \frac{\partial}{\partial t}\left(\vec{D}\times\vec{B}\right) + \left(\vec{D}\times\frac{\partial\vec{B}}{\partial t}\right)\right]d^3r\end{aligned} \tag{8–39}$$

Replacing $\partial\vec{B}/\partial t$ with $-\vec{\nabla}\times\vec{E}$ and gathering all time derivatives on the left, we get

$$\frac{d}{dt}\int\left(\vec{\mathcal{P}}+\vec{D}\times\vec{B}\right)d^3r = \int\left[(\vec{\nabla}\cdot\vec{D})\vec{E} - \vec{B}\times(\vec{\nabla}\times\vec{H}) - \vec{D}\times(\vec{\nabla}\times\vec{E})\right]d^3r \tag{8–40}$$

If the fields are uniform, all the spatial derivatives of the right-hand side vanish, and conservation of momentum requires that $\vec{\mathcal{P}}_{em} \equiv \vec{D} \times \vec{B}$ be the momentum density of the fields. As before, we would like to express the right-hand side of (8–24) as the divergence of a second rank tensor. Denoting the argument of the right hand side of (8–40) by $\vec{f}$,

$$\vec{f} = (\vec{\nabla} \quad \vec{D})\vec{E} - \vec{B} \times (\vec{\nabla} \times \vec{H}) - \vec{D} \times (\vec{\nabla} \times \vec{E}) \tag{8–41}$$

and using tensor notation to facilitate the manipulations gives

$$\begin{aligned} f^j &= \partial_k D^k E^j - \epsilon^{jk\ell} B_k \epsilon_{\ell mn} \partial^m H^n - \epsilon^{jk\ell} D_k \epsilon_{\ell mn} \partial^m E^n \\ &= \partial_k D^k E^j - \epsilon^{\ell jk} \epsilon_{\ell mn} B_k \partial^m H^n - \epsilon^{\ell jk} \epsilon_{\ell mn} D_k \partial^m E^n \\ &= \partial_k D^k E^j - \left(\delta^j_m \delta^k_n - \delta^j_n \delta^k_m\right) B_k \partial^m H^n - \left(\delta^j_m \delta^k_n - \delta^j_n \delta^k_m\right) D_k \partial^m E^n \\ &= \partial_k D^k E^j - B_k \partial^j H^k + B_k \partial^k H^j - D_k \partial^j E^k + D_k \partial^k E^j \end{aligned} \tag{8–42}$$

This expression (8–42), may be symmetrized by adding the null term $(\vec{\nabla} \cdot \vec{B})H^j = (\partial_k B^k)H^j$ to the right side:

$$\begin{aligned} f^j &= \partial_k D^k E^j + D_k \partial^k E^j + B_k \partial^k H^j + \partial_k B^k H^j - \left(B_k \partial^j H^k + D_k \partial^j E^k\right) \\ &= \partial_k \left(D^k E^j\right) + \partial_k \left(B^k H^j\right) - \left(B_k \partial^j H^k + D_k \partial^j E^k\right) \end{aligned} \tag{8–43}$$

For linear materials (though not necessarily isotropic), the last term may be written

$$B_k \partial^j H^k + D_k \partial^j E^k = \tfrac{1}{2}\partial^j \left(\vec{B} \cdot \vec{H} + \vec{D} \cdot \vec{E}\right) = \tfrac{1}{2}\partial_k \delta^{kj} \left(\vec{B} \cdot \vec{H} + \vec{D} \cdot \vec{E}\right)$$

resulting in

$$f^j = \partial_k \left[D^k E^j + B^k H^j - \tfrac{1}{2}\delta^{jk}\left(\vec{B} \cdot \vec{H} + \vec{D} \cdot \vec{E}\right)\right] = -\partial_k T^{kj} \tag{8–44}$$

where we have made the obvious generalization of the Maxwell stress tensor:

$$T^{ij} = -D^i E^j - B^i H^j + \tfrac{1}{2}\delta^{ij}(\vec{B} \cdot \vec{H} + \vec{D} \cdot \vec{E}) \tag{8–45}$$

⋆ 8.2.3 Blackbody Radiation Pressure

As an application of the stress tensor let us use it to compute the pressure offered by the isotropic radiation in a cavity with absorbing walls. We will compute the force $d\vec{F}$ acting on a small segment $d\vec{S}$ of the wall. From our earlier discussion, $d\vec{F} = -\overleftrightarrow{T} \cdot d\vec{S}$. To be explicit, we take $d\vec{S}$ to be a segment of the right-hand wall with normal $-\hat{\imath}$. By symmetry, the only nonvanishing component of force must be normal to the walls. We have then

$$dF^x = -T^{xj} dS_j \tag{8–46}$$

or, in Cartesians $dF_x = +T_{xx}dS_x$. (Note that although we assume summation for repeated i, j, etc., no summation is implied when specific indices such as x, y, or z are used.)

Substituting for T_{xx}, we obtain

$$dF_x = -\left[E_xD_x + B_xH_x - \tfrac{1}{2}(\vec{E}\cdot\vec{D} + \vec{B}\cdot\vec{H})\right]dS_x \tag{8–47}$$

For isotropic radiation, $E_xD_x = E_yD_y = E_zD_z = \frac{1}{3}\vec{E}\cdot\vec{D}$ and $B_xH_x = \frac{1}{3}\vec{B}\cdot\vec{H}$. Inserting these expressions into (8–44), we obtain

$$\begin{aligned} dF_x &= -\left[\tfrac{1}{3}(\vec{E}\cdot\vec{D} + \vec{B}\cdot\vec{H}) - \tfrac{1}{2}(\vec{E}\cdot\vec{D} + \vec{B}\cdot\vec{H})\right]dS_x \\ &= \frac{1}{3}\left(\frac{\vec{E}\cdot\vec{D} + \vec{B}\cdot\vec{H}}{2}\right)dS_x = \tfrac{1}{3}U\,dS_x \end{aligned} \tag{8–48}$$

We conclude that the pressure $\wp$ $(= dF_x/dS_x)$ equals $\frac{1}{3}$ of the energy density U.

This result forms the basis for the thermodynamic derivations of the Stefan-Boltzmann law for blackbody radiation. We digress briefly to demonstrate the realization of this law. We temporarily adopt for this discussion the notations Q for the heat supplied to a system, u for the internal energy of the system, S for the entropy, and V for the volume of the system.

The adiabatic law of compression for a blackbody follows immediately from the pressure–energy density relation. For an adiabatic change the fundamental relation reads

$$dQ = 0 = du + \wp dV \tag{8–49}$$

Setting $u = 3\wp V$, we have $du = 3\wp dV + 3V d\wp$. Substituted into (8–49), this yields

$$0 = 3V d\wp + 4\wp dV \tag{8–50}$$

which, on integration, gives $\wp V^{4/3} = \text{constant}$. In other words, blackbody radiation behaves like a perfect gas with $\gamma = \frac{4}{3}$. We find the energy density u in terms of the temperature by substituting $\wp = \frac{1}{3}(\partial u/\partial V)_T$ into the thermodynamic relation[17]

$$\left(\frac{\partial u}{\partial V}\right)_T = T\left(\frac{\partial \wp}{\partial T}\right)_V - \wp \tag{8–51}$$

[17]This result is most easily obtained from $du = TdS - \wp dV$ with substitutions for dS as follows. Considering the entropy S as a function of T and V, we write

$$dS = \left(\frac{\partial S}{\partial T}\right)_V dT + \left(\frac{\partial S}{\partial V}\right)_T dV = \frac{c_V}{T}dT + \left(\frac{\partial \wp}{dT}\right)_V dV$$

With this substitution, the equation for du becomes

$$du = c_V dT + \left(T\frac{\partial \wp}{\partial T} - \wp\right)_V dV$$

Comparing this to the general expression, $du = \left(\frac{\partial u}{\partial T}\right)_V dT + \left(\frac{\partial u}{\partial V}\right)_T dV$, we see that the second term gives the required result.

to obtain

$$U = \frac{T}{3}\frac{\partial U}{\partial T} - \frac{U}{3} \tag{8–52}$$

On rearrangement this becomes

$$\frac{\partial U}{U} = 4\frac{\partial T}{T} \tag{8–53}$$

Integration of (8–53) leads to the Stefan-Boltzmann law.

$$U = aT^4 \tag{8–54}$$

Thus, the energy density of radiation contained in a perfectly absorbing cavity is seen to be proportional the fourth power of temperature. The radiation pressure, as a consequence is given by $\wp = aT^4/3$.

Radiation emitted from a small aperture of cross sectional area A in such a black-walled cavity is readily computed. The rate of radiation exiting within $d\Omega$ of angle θ to the normal derives from a column of volume $Ac\cos\theta$ and has probability $d\Omega/4\pi$ of having this direction. The rate of radiation from the aperture is then

$$\frac{du}{dt} = \int_{2\pi} \frac{UAc\cos\theta d\Omega}{4\pi} = -\frac{UAc\cos^2\theta}{4}\bigg|_0^{\frac{1}{2}\pi} = \frac{UAc}{4} = \sigma AT^4 \tag{8–55}$$

The Stefan-Boltzmann constant, σ, is found experimentally to be $5.6697 \times 10^{-8}\,\mathrm{W\,m^{-2}\,{}^\circ K^{-4}}$.

8.3 The Electromagnetic Potentials

The introduction of matter has no effect on the field equations $\vec{\nabla}\cdot\vec{B} = 0$ and $\vec{\nabla}\times\vec{E} = -\partial\vec{B}/\partial t$, or on the former's implication that $\vec{B} = \vec{\nabla}\times\vec{A}$. Combining the two curl equations we have, as before

$$\vec{\nabla}\times\left(\vec{E} + \frac{\partial\vec{A}}{\partial t}\right) = 0 \quad\Rightarrow\quad \vec{E} + \frac{\partial\vec{A}}{\partial t} = -\vec{\nabla}V \tag{8–56}$$

so that $\vec{E} = -\vec{\nabla}V - \partial\vec{A}/\partial t$. As in the vacuum case, gauge transformations allow sufficient freedom in $\vec{A}$ to pick $\vec{\nabla}\cdot\vec{A}$ at our convenience.

The remaining Maxwell's equations may be used to produce the time-dependent equivalent of Poisson's equation for linear, homogeneous matter:

$$\vec{\nabla}\times\vec{H} = \vec{J} + \varepsilon\frac{\partial\vec{E}}{\partial t} \tag{8–57}$$

Expressed in terms of the potentials this becomes

$$\vec{\nabla}\times\left(\frac{\vec{\nabla}\times\vec{A}}{\mu}\right) = \frac{1}{\mu}\left[\vec{\nabla}(\vec{\nabla}\cdot\vec{A}) - \nabla^2\vec{A}\right] = \vec{J} + \varepsilon\frac{\partial}{\partial t}\left(-\vec{\nabla}V - \frac{\partial\vec{A}}{\partial t}\right) \tag{8–58}$$

Rearranging the terms, we get

$$\nabla^2\vec{A} - \mu\varepsilon\frac{\partial^2\vec{A}}{\partial t^2} = -\mu\vec{J} + \vec{\nabla}\left[\vec{\nabla}\cdot\vec{A} + \mu\varepsilon\frac{\partial V}{\partial t}\right] \tag{8–59}$$

Similarly,

$$\vec{\nabla}\cdot\vec{D} = \vec{\nabla}\cdot\left(\varepsilon\vec{E}\right) = \varepsilon\vec{\nabla}\cdot\left(-\vec{\nabla}V - \frac{\partial\vec{A}}{\partial t}\right) = -\varepsilon\nabla^2 V - \varepsilon\frac{\partial}{\partial t}(\vec{\nabla}\cdot\vec{A}) = \rho \tag{8–60}$$

becomes

$$\nabla^2 V - \mu\varepsilon\frac{\partial^2 V}{\partial t^2} = -\frac{\rho}{\varepsilon} - \frac{\partial}{\partial t}\left(\vec{\nabla}\cdot\vec{A} + \mu\varepsilon\frac{\partial V}{\partial t}\right) \tag{8–61}$$

upon rearrangement. Both (8–59) and (8–61) are simplified considerably by choosing $\vec{\nabla}\cdot\vec{A} = -\mu\varepsilon\partial V/\partial t$, the Lorenz gauge. We then obtain

$$\nabla^2\vec{A} - \mu\varepsilon\frac{\partial^2\vec{A}}{\partial t^2} = -\mu\vec{J}(t) \qquad \text{and} \qquad \nabla^2 V - \mu\varepsilon\frac{\partial^2 V}{\partial t^2} = -\frac{\rho(t)}{\varepsilon} \tag{8–62}$$

In words, the vector potential $\vec{A}$ and scalar potential V satisfy the inhomogeneous wave equation with source terms $-\mu\vec{J}(t)$ and $-\rho(t)/\varepsilon$, respectively. We postpone consideration of how oscillating sources generate waves to chapter 10 and deal only with the homogeneous equation in this chapter.

8.4 Plane Waves in Material Media

To isolate fields $\vec{E}$ and $\vec{B}$ from Maxwell's equations, we use the usual stratagem of decoupling the equations by taking the curl of the curl equations,

$$\vec{\nabla}\times(\vec{\nabla}\times\vec{E}) = -\frac{\partial}{\partial t}(\vec{\nabla}\times\vec{B}) = -\mu\frac{\partial}{\partial t}\left(\vec{J} + \frac{\partial\vec{D}}{\partial t}\right) \tag{8–63}$$

Expanding the left side and using the constitutive relations for linear, ohmic materials to replace $\vec{J}$ with $g\vec{E}$ and $\partial\vec{D}/\partial t$ with $\varepsilon\partial\vec{E}/\partial t$, we find

$$\vec{\nabla}(\vec{\nabla}\cdot\vec{E}) - \nabla^2\vec{E} = -\frac{\partial}{\partial t}\left(\mu g\vec{E} + \mu\varepsilon\frac{\partial\vec{E}}{\partial t}\right) \tag{8–64}$$

In the absence of free charges, $\vec{\nabla}\cdot\vec{E} = 0$,[18] and, assuming harmonically varying fields $\vec{E}(\vec{r},t) = \vec{E}_0(\vec{r})e^{-i\omega t}$, (picking harmonically varying fields is in reality no restriction as we could always frequency analyze the temporal variation and apply the following to each Fourier component) we may replace $\partial/\partial t$ with $-i\omega$ to obtain

$$\nabla^2\vec{E}_0 + \mu\varepsilon\omega^2\left(1 + \frac{ig}{\omega\varepsilon}\right)\vec{E}_0 = 0 \tag{8–65}$$

[18] Actually $\vec{\nabla}\cdot\vec{D} = 0$, which implies that $\vec{\nabla}\cdot\vec{E} = 0$ only if $\vec{D}$ can be written as $\vec{D} = \varepsilon\vec{E}$ with ε independent of the coordinates.

Similarly for $\vec{B}$

$$\vec{\nabla}\times(\vec{\nabla}\times\vec{H})=\vec{\nabla}\times\left(\vec{J}+\frac{\partial\vec{D}}{\partial t}\right)=\vec{\nabla}\times\left(g\vec{E}+\varepsilon\frac{\partial\vec{E}}{\partial t}\right) \tag{8–66}$$

and, following the same steps as above,

$$\frac{1}{\mu}\left[\vec{\nabla}(\vec{\nabla}\cdot\vec{B})-\nabla^2\vec{B}\right]=-\frac{\partial}{\partial t}\left(g\vec{B}+\varepsilon\frac{\partial\vec{B}}{\partial t}\right) \tag{8–67}$$

leads to

$$\nabla^2\vec{B}_0+\mu\varepsilon\omega^2\left(1+\frac{ig}{\omega\varepsilon}\right)\vec{B}_0=0 \tag{8–68}$$

Another useful form of the wave equation is obtained by not incorporating the polarization into the displacement field. If the material is not linear, this is in fact the only legitimate procedure.

Again taking the curl of the curl equations, we obtain

$$\begin{aligned}\vec{\nabla}\times(\vec{\nabla}\times\vec{E})&=-\mu_0\frac{\partial}{\partial t}(\vec{\nabla}\times\vec{H})-\mu_0\frac{\partial}{\partial t}(\vec{\nabla}\times\vec{M})\\&=-\mu_0\varepsilon_0\frac{\partial^2\vec{E}}{\partial t^2}-\mu_0\frac{\partial^2\vec{P}}{\partial t^2}-\mu_0\frac{\partial}{\partial t}\left(\vec{J}+\vec{\nabla}\times\vec{M}\right)\end{aligned} \tag{8–69}$$

Gathering terms in $\vec{E}$, we find

$$\vec{\nabla}\times(\vec{\nabla}\times\vec{E})+\mu_0\varepsilon_0\frac{\partial^2\vec{E}}{\partial t^2}=-\frac{\mu_0\partial^2\vec{P}}{\partial t^2}-\mu_0\frac{\partial}{\partial t}\left(\vec{J}+\vec{\nabla}\times\vec{M}\right) \tag{8–70}$$

The terms on the right are source terms for electromagnetic waves. We note that time-varying currents and polarizations constitute sources of electromagnetic waves.

The magnetic field term is handled in identical fashion to give

$$\vec{\nabla}\times(\vec{\nabla}\times\vec{H})+\mu_0\varepsilon_0\frac{\partial^2\vec{H}}{\partial t^2}=-\mu_0\varepsilon_0\frac{\partial^2\vec{M}}{dt^2}+\vec{\nabla}\times\left(\vec{J}+\frac{\partial\vec{P}}{\partial t}\right) \tag{8–71}$$

In most cases, the magnetization term can be ignored.

We will investigate the plane wave solutions to (8–65) and (8–68) initially for nonconductors ($g=0$), and then for conductors.

8.4.1 Plane Waves in Linear, Isotropic Dielectrics

In nonconducting dielectrics, the fields satisfy the homogeneous wave equations

$$\nabla^2\vec{E}-\mu\varepsilon\frac{\partial^2\vec{E}}{\partial t^2}=0 \quad \text{and} \quad \nabla^2\vec{H}-\mu\varepsilon\frac{\partial^2\vec{H}}{\partial t^2}=0 \tag{8–72}$$

(We use $\vec{H}$ instead of $\vec{B}$ because of the increased symmetry in the constitutive relations, $\vec{D}=\varepsilon\vec{E}$ and $\vec{B}=\mu H$, and because of the more parallel boundary conditions:

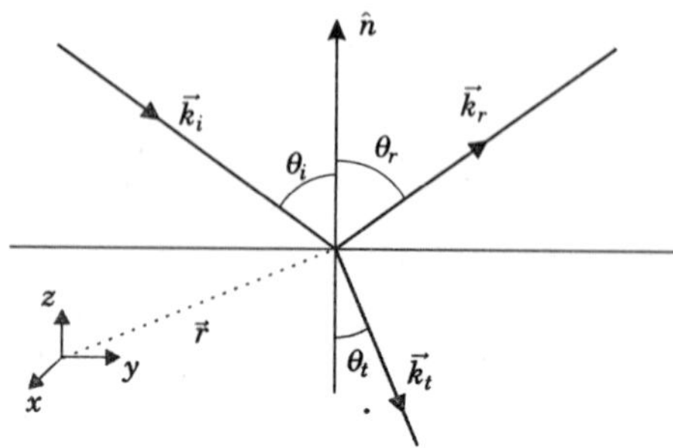

Figure 8.3: The wave vectors $\vec{k}_i$, $\vec{k}_r$ and $\vec{k}_t$ all lie in the plane of incidence.

$D_\perp$ and $B_\perp$ are continuous and $E_\parallel$ and $H_\parallel$ are continuous.) This pair of equations is satisfied (these are not the only possible solutions) by the plane wave solutions

$$\vec{E}(\vec{r},t) = \vec{E}_0 e^{i(\vec{k}\cdot\vec{r}-\omega t)} \qquad \text{and} \qquad \vec{H}(\vec{r},t) = \vec{H}_0 e^{i(\vec{k}\cdot\vec{r}-\omega t)} \tag{8–73}$$

The two solutions (8–73) are not independent but are linked by Maxwell's equations. The divergence equations give for the plane waves

$$\vec{k}\cdot\vec{D}_0 = 0 \qquad \text{and} \qquad \vec{k}\cdot\vec{B}_0 = 0 \tag{8–74}$$

implying that the wave vector $\vec{k}$ is perpendicular to each of $\vec{D}$ and $\vec{B}$. The curl equations give

$$\vec{k}\times\vec{E}_0 = \omega\mu\vec{H}_0 \qquad \text{and} \qquad \vec{k}\times\vec{H}_0 = -\omega\varepsilon\vec{E}_0 \tag{8–75}$$

implying that $\vec{E}$ and $\vec{H}$ are perpendicular to each other. For isotropic media $\vec{E}$ is parallel to $\vec{D}$, and $\vec{B}$ is parallel to $\vec{H}$. We may therefore conclude that the wave vector $\vec{k}$, the normal to the surfaces of constant phase, is parallel to the Poynting vector $\vec{\mathcal{S}} = \vec{E}\times\vec{H}$. In other words, for isotropic media the direction of energy propagation is along $\vec{k}$. We point out that this conclusion does not hold for anisotropic media. in which $\vec{\mathcal{S}}$ is perpendicular to $\vec{E}$ whereas $\vec{k}$ is perpendicular to $\vec{D}$.

Substituting the plane wave solutions (8–73) into the wave equation (8–72) immediately gives the *dispersion relation*, $-k^2 + \mu\varepsilon\omega^2 = 0$, from which we conclude that the phase velocity ω/k of the wave is given by $1/\sqrt{\mu\varepsilon}$. The group velocity, $\partial\omega/\partial k$ for isotropic media, will generally be different from the phase velocity.

Since most optical materials have $\mu \simeq \mu_0$, the velocity of light through the medium is $\simeq c/\sqrt{\varepsilon/\varepsilon_0}$. The laws of reflection and refraction are evidently merely a matter of satisfying the boundary conditions at the interface of two dielectrics.

8.4.2 Reflection and Refraction—Snell's law

Let us consider a plane wave with wave vector $\vec{k}_i$ incident on a plane interface, giving rise to a reflected wave with wave vector $\vec{k}_r$ and a transmitted wave with wave vector $\vec{k}_t$, as shown in Figure 8.3.

At a point $\vec{r}$ on the interface, the parallel component of the electric field must be the same on both sides of the interface. Therefore

$$E^{\|}_{0,i}e^{i(\vec{k}_i\cdot\vec{r}-\omega t)} + E^{\|}_{0,r}e^{i(\vec{k}_r\cdot\vec{r}-\omega t+\varphi_r)} = E^{\|}_{0,t}e^{i(\vec{k}_t\cdot\vec{r}-\omega t+\varphi_t)} \tag{8-76}$$

(a phase change on transmission or reflection is at least a theoretical possibility). If this is to be satisfied at all $\vec{r}$ on the interface, the arguments of the exponentials must all have the same functional dependence on $\vec{r}$ and t; that is,

$$\vec{k}_i\cdot\vec{r}-\omega t = \vec{k}_r\cdot\vec{r}-\omega t+\varphi_r = \vec{k}_t\cdot\vec{r}-\omega t+\varphi_t \tag{8-77}$$

From the first equality of (8–77), we obtain $(\vec{k}_i-\vec{k}_r)\cdot\vec{r}=\varphi_r$. The surface defined by this equation, the interface, is perpendicular to $(\vec{k}_i-\vec{k}_r)$. Further, as $(\vec{k}_i-\vec{k}_r)$ is perpendicular to the plane, the cross product of $(\vec{k}_i-\vec{k}_r)$ with a normal $\hat{n}$ is zero: $(\vec{k}_i-\vec{k}_r)\times\hat{n}=0$, implying that

$$k_i\sin\theta_i = k_r\sin\theta_r \tag{8-78}$$

or, since the incident and reflected waves are both in the same medium, the magnitudes k_i and k_r are equal,

$$\sin\theta_i=\sin\theta_r \quad\Rightarrow\quad \theta_i=\theta_r \qquad \text{when } \theta_i,\theta_r\in(0,\pi/2) \tag{8-79}$$

The angle of incidence is equal to the angle of reflection.

Using the second half of the equality (8–77) of arguments of the exponentials gives

$$(\vec{k}_i-\vec{k}_t)\cdot\vec{r}=\varphi_t \tag{8-80}$$

The same argument now yields $(\vec{k}_i-\vec{k}_t)\times\hat{n}=0$ leading to

$$k_i\sin\theta_i = k_t\sin\theta_t \tag{8-81}$$

after we multiply both sides of (8–81) by c/ω, this expression becomes

$$n_i\sin\theta_i = n_t\sin\theta_t \tag{8-82}$$

where the *refractive index*, n, is defined by $n=c/v_{phase}$. It is apparent that Snell's law is a consequence only of the plane wave nature of the disturbance and the requirement of continuity. It will therefore find application well outside optics.

8.4.3 Fresnel's Equations

To obtain the amplitudes $E_{0,i}$, $E_{0,r}$, and $E_{0,t}$ of the incident, reflected, and transmitted wave, we must use the boundary conditions in more detail. We must first resolve $\vec{E}$ into two components, one labelled E_p, parallel to the plane of incidence (perpendicular to the interface), and the other labelled E_s, perpendicular to the plane of incidence (parallel to the interface). (The plane of incidence is the plane containing the incident and reflected rays; the s subscript derives from *senkrecht*,

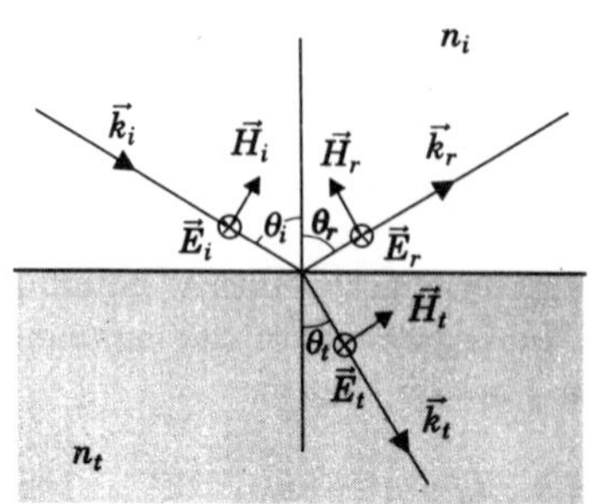

Figure 8.4: The electric vector of s-polarized light is perpendicular to the plane of incidence. The vectors illustrated refer to the fields at the vertex of the rays.

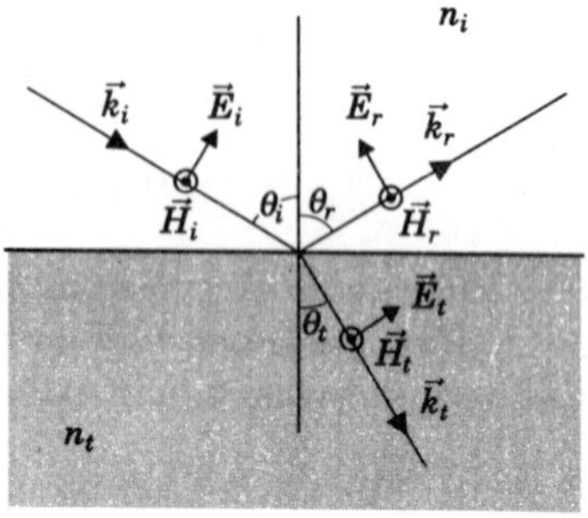

Figure 8.5: The electric vector of p-polarized light lies in the plane of incidence.

German for *perpendicular*, and the p stands for *parallel*.) The two components must be handled separately.

s-polarization: An s-polarized wave has its electric field $\vec{E}$ perpendicular to the plane of incidence, meaning that $\vec{B}$ lies in the plane of incidence. Assuming that $\vec{E}_i$, $\vec{E}_r$, and $\vec{E}_t$ all point in the same direction, we obtain for points on the interface

$$E_i + E_r = E_t \tag{8–83}$$

Applying the requirement that $H_\parallel$ be continuous, we have also from the diagram in Figure 8.4

$$H_i \cos\theta_i - H_r \cos\theta_r = H_t \cos\theta_t \tag{8–84}$$

In the medium, the magnitudes of H and E are related by $H = \sqrt{\varepsilon/\mu}\,E = (1/v\mu)E$. Multiplying (8–84) by c and substituting for H, we find

$$\frac{n_i}{\mu_i}(E_i - E_r)\cos\theta_i = \frac{n_t}{\mu_t}E_t \cos\theta_t \tag{8–85}$$

Expressions (8–83) and (8–85) may be solved for E_r and E_t to obtain

$$\left(\frac{E_r}{E_i}\right)_s = \frac{(n_i/\mu_i)\cos\theta_i - (n_t/\mu_t)\cos\theta_t}{(n_i/\mu_i)\cos\theta_i + (n_t/\mu_t)\cos\theta_t} \equiv r_s \tag{8–86}$$

and

$$\left(\frac{E_t}{E_i}\right)_s = \frac{2(n_i/\mu_i)\cos\theta_i}{(n_i/\mu_i)\cos\theta_i + (n_t/\mu_t)\cos\theta_t} \equiv t_s \tag{8–87}$$

p-polarization: When the wave is p-polarized, the electric field is parallel to the plane of incidence, while $\vec{B}$ (and therefore $\vec{H}$) is perpendicular. Assuming that $\vec{H}$ has the same direction for the incident, transmitted, and reflected waves, we construct the diagram of Figure 8.5. (Note that under this assumption, E_i and E_r point in opposite directions for a wave at normal incidence.)

Again using the continuity of $H_\parallel$ and $E_\parallel$, we have

$$H_i + H_r = H_t \tag{8–88}$$

and

$$E_i \cos\theta - E_r \cos\theta_r = E_t \cos\theta_t \tag{8–89}$$

We again replace H with $\sqrt{\varepsilon/\mu}\, E = (1/v\mu)E$ to rewrite the first of these equations as

$$\frac{n_i}{\mu_i}(E_i + E_r) = \frac{n_t}{\mu_t}E_t \tag{8–90}$$

Solving (8–89) and (8–90) for E_r and E_t, we find

$$\left(\frac{E_r}{E_i}\right)_p = \frac{(n_t/\mu_t)\cos\theta_i - (n_i/\mu_i)\cos\theta_t}{(n_t/\mu_t)\cos\theta_i + (n_i/\mu_i)\cos\theta_t} \equiv r_p \tag{8–91}$$

and

$$\left(\frac{E_t}{E_i}\right)_p = \frac{2(n_i/\mu_i)\cos\theta_i}{(n_t/\mu_t)\cos\theta_i + (n_i/\mu_i)\cos\theta_t} \equiv t_p \tag{8–92}$$

Notice the distinction between (8–86, 87) and (8–91, 92). For virtually all dielectrics of optical interest, $\mu = \mu_0$, meaning that the permeabilities μ may be cancelled from (8–85) and subsequent equations. (In microwave systems ferrites are frequently used in isolators; μ is not negligible in that case.)

Using Snell's law to eliminate the ratios of refractive indices (assuming $\mu_i = \mu_t$), we can simplify (8–86, 87) and (8–91, 92) to obtain the Fresnel equations:

$$\begin{aligned} r_s &= -\frac{\sin(\theta_i - \theta_t)}{\sin(\theta_i + \theta_t)} & r_p &= \frac{\tan(\theta_i - \theta_t)}{\tan(\theta_i + \theta_t)} \\ t_s &= \frac{2\sin\theta_t\cos\theta_i}{\sin(\theta_i + \theta_t)} & t_p &= \frac{2\sin\theta_t\cos\theta_i}{\sin(\theta_i + \theta_t)\cos(\theta_i - \theta_t)} \end{aligned} \tag{8–93}$$

An alternative way of expressing (8–86, 87) and (8-91, 92) is found by eliminating the angle of transmission θ_t from the equations. Letting $n = (n_t/\mu_t)/(n_i/\mu_i)$ and $\cos\theta_t = \sqrt{1 - \sin^2\theta_t} = \frac{1}{n}\sqrt{n^2 - \sin^2\theta_i}$, we obtain

$$r_s = \frac{\cos\theta_i - n\cos\theta_t}{\cos\theta_i + n\cos\theta_t} = \frac{\cos\theta_i - \sqrt{n^2 - \sin^2\theta_i}}{\cos\theta_i + \sqrt{n^2 - \sin^2\theta_i}} \tag{8–94}$$

$$t_s = \frac{2\cos\theta_i}{\cos\theta_i + \sqrt{n^2 - \sin^2\theta_i}} \tag{8–95}$$

$$r_p = \frac{n\cos\theta_i - \cos\theta_t}{n\cos\theta_i + \cos\theta_t} = \frac{n^2\cos\theta_i - \sqrt{n^2 - \sin^2\theta_i}}{n^2\cos\theta_i + \sqrt{n^2 - \sin^2\theta_i}} \tag{8–96}$$

$$t_p = \frac{2n\cos\theta_i}{n^2\cos\theta_i + \sqrt{n^2 - \sin^2\theta_i}} \tag{8–97}$$

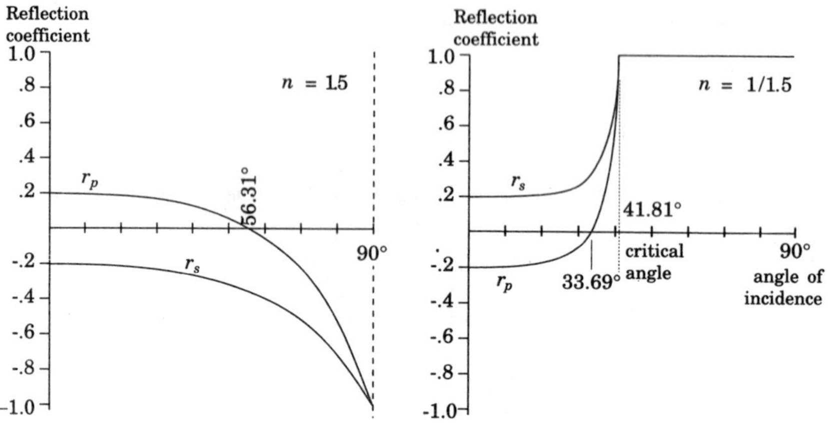

Figure 8.6: The external and internal amplitude reflection coefficients for an air-glass interface.

At normal incidence the amplitude reflection coefficients reduce to $r_s = (1-n)/(1+n)$ and $r_p = -(1-n)/(1+n)$. As there is no distinction between an s and p wave at normal incidence, at first sight, the difference in sign appears to be an error. However, for $n > 1$, both correspond to a change in sign of the electric field when a wave is reflected. The apparent contradiction is resolved by noting that E_i and E_r have been defined (Figure 8.4) so that $E_{i,s}$ and $E_{t,s}$ both point in the same direction when positive. On the other hand, we have taken $H_{i,p}$ and $H_{r,p}$ to be parallel for p waves, leading to $E_{i,p}$ and $E_{r,p}$ being antiparallel at normal incidence when both are positive (Figure 8.5). The sign of the reflection coefficient varies from author to author according to the choice for positive $E_{r,p}$.

The amplitude reflection coefficients for an air-glass interface ($n = 1.5$ for external reflection and $n = 1/1.5$ for internal reflection) are sketched in Figure 8.6. The angle where r_p vanishes ($\tan\theta_i = n$) is known as Brewster's angle.

When n is less than one, there is a critical angle beyond which the sine of θ_i exceeds n and both r_s and r_p become complex numbers of unit magnitude. The coefficients now give the phase change of the wave on reflection. To evaluate this phase change, we write

$$r_s = \frac{\cos\theta_i - i\sqrt{\sin^2\theta_i - n^2}}{\cos\theta_i + i\sqrt{\sin^2\theta_i - n^2}} \tag{8–98}$$

and

$$r_p = \frac{n^2\cos\theta_i - i\sqrt{\sin^2\theta_i - n^2}}{n^2\cos\theta_i + i\sqrt{\sin^2\theta_i - n^2}} \tag{8–99}$$

Since r_s has unit magnitude, it may be written as a complex exponential

$$r_s = e^{-i\varphi_s} = \frac{ae^{-i\alpha}}{ae^{+i\alpha}} \tag{8–100}$$

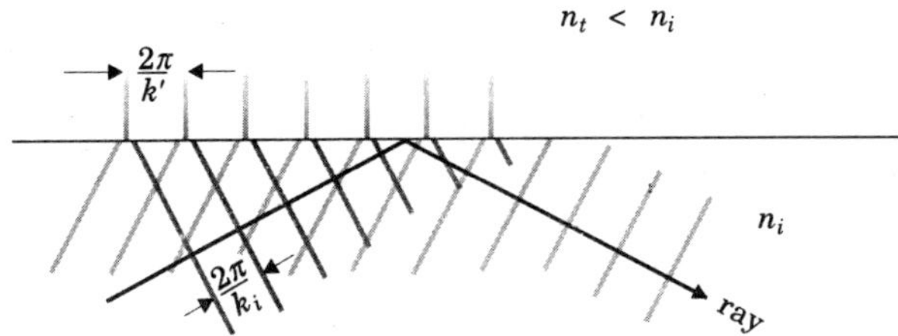

Figure 8.7: Wave fronts for a totally internally reflected wave. The evanescent wave carries half the phase shift of the reflected wave and travels along the interface.

in which we have used the fact that the numerator and denominator of (equation 8–98) are complex conjugates. We conclude then that $\varphi_s = 2\alpha$, where $\tan\alpha$ $(= \tan\frac{1}{2}\varphi_s)$ is given by

$$\tan\alpha = \tan\tfrac{1}{2}\varphi_s = \frac{\sqrt{\sin^2\theta_i - n^2}}{\cos\theta_i} \tag{8–101}$$

In the same fashion we write $r_p = e^{-\varphi_p} = be^{-i\beta}/be^{+i\beta}$ resulting in

$$\tan\tfrac{1}{2}\varphi_p = \frac{\sqrt{\sin^2\theta_i - n^2}}{n^2\cos\theta_i} \tag{8–102}$$

Each of the waves is totally internally reflected but suffers a phase shift that differs for the two (s and p) polarizations.

8.4.4 The Evanescent Wave

In spite of the fact that the incident energy is totally reflected beyond the critical angle, the electric field does not abruptly vanish at the boundary. The phase difference between the incident and reflected waves prevents the complete destructive interference required to eliminate the transmitted wave. The "transmitted" field that extends beyond the boundary of the dielectric when a wave is totally internally reflected is known as the *evanescent wave.*

To investigate the evanescent wave, we postulate a transmitted wave of the form

$$\vec{E}_t = \vec{E}_{t,0}e^{i(\vec{k}_t\cdot\vec{r}-\omega t)} \tag{8–103}$$

We choose a coordinate system so that the boundary lies in the x-y plane and $\vec{k}$ lies in the x-z plane. Then, for points $\vec{r}$ in the plane of incidence,

$$\begin{aligned}\vec{k}_t\cdot\vec{r} &= k_t x\sin\theta_t + k_t z\cos\theta_t \\ &= k_t\left(\frac{x\sin\theta_i}{n} + iz\sqrt{\frac{\sin^2\theta_i}{n^2} - 1}\right)\end{aligned} \tag{8–104}$$

which leads to the form (recall $\sin\theta_i > n$)

$$\vec{E}_t(\vec{r},t) = \vec{E}_{t,0}e^{-\alpha z}e^{i(k'x-\omega t)} \tag{8–105}$$

for the transmitted field. The wave propagates along the boundary (in the x direction) with propagation constant $k' = k_t \sin\theta_t = k_i \sin\theta_i$, diminishing exponentially with distance from the boundary (in the z direction) with decay constant α. A representation of the incident, reflected, and evanescent wave fronts is shown in Figure 8.7.

8.4.5 Plane Waves in a Tenuous Plasma

A gaseous medium with a considerable number of free (unbound) charges distributed throughout is generally known as a plasma. The constitution of the sun is mostly dense plasma with matter almost 100% ionized. Closer to home, the ionosphere surrounding the earth is a much less dense plasma whose electromagnetic properties have a significant impact on communications. The interstellar medium is also largely plasma, a fact of considerable interest to astronomers.

By a tenuous plasma, we mean a medium with free charges that do not significantly interact with one another. Thus we ignore any cooperative behavior such as one might expect from electrons in metals.

The gas we will consider will be a homogeneous mix of free electrons and a corresponding number of positive ions to maintain neutrality. Because low-mass particles move much more than heavier ones in response to an alternating electromagnetic field, we consider the motion of electrons embedded in an overall neutral gas and ignore any motion of the positive ions.

A free electron of charge e subjected to the electric field $\vec{E} = \vec{E}_0 e^{-i\omega t}$ of a passing electromagnetic wave responds with an acceleration

$$\ddot{\vec{r}} = \frac{e\vec{E}_0}{m} e^{-i\omega t} \tag{8–106}$$

This expression is easily integrated to give the displacement of a typical electron:

$$\vec{r} = -\frac{e}{m\omega^2}\vec{E} \tag{8–107}$$

leading to a dipole moment

$$\vec{p} = e\vec{r} = -\frac{e^2}{m\omega^2}\vec{E} \tag{8–108}$$

For n such electrons per unit volume, the resulting polarization is

$$\vec{P} = -\frac{ne^2}{m\omega^2}\vec{E} \equiv -\frac{\omega_p^2}{\omega^2}\varepsilon_0\vec{E} \tag{8–109}$$

where the *plasma frequency* $\omega_p/2\pi$ is defined by $\omega_p^2 = ne^2/m\varepsilon_0$. The permittivity of the plasma is therefore $\varepsilon_0(1-\omega_p^2/\omega^2)$. Substituting this form of the permittivity into the homogeneous wave equation (8–72) and inserting a plane wave solution (8–73), we obtain the dispersion relation

$$k^2 = \omega^2\mu\varepsilon = \frac{\omega^2}{c^2}\left(1 - \frac{\omega_p^2}{\omega^2}\right) \tag{8–110}$$

It is evident that k^2 tends to zero as ω tends to ω_p. When ω is smaller than ω_p, k^2 is negative, meaning that k is imaginary. Substitution of an imaginary k into the plane wave exponential results in an electric field $\vec{E} = \vec{E}_0 e^{i(\vec{k}\cdot\vec{r}-\omega t)} = \vec{E}_0 e^{-|\vec{k}\cdot\vec{r}|}e^{-i\omega t}$. The wave dies off exponentially as it enters the medium. We have not provided any energy dissipation mechanisms, leading us to deduce that the wave must be reflected when the frequency is less than the plasma frequency. For a charged particle density $n = 10^{15}$ electrons/m^3, typical of the ionosphere, $f_p = 9 \times 10^6$ Hz. Thus radio waves below several megahertz are totally reflected by the ionosphere. The higher frequency television and UHF waves (> 80 MHz.) are transmitted by the ionosphere; in consequence they will not travel over the horizon. During solar flares, the electron density fluctuates, leading to disruptions in radio communication.

⋆ 8.4.6 Plane Waves in Linear Anisotropic Dielectrics

Although an exhaustive study of wave propagation in anisotropic dielectrics is beyond the scope of this book,[19] a rudimentary discussion of double refraction will form a reasonable introduction to the electro-magnetic aspects.

The introduction of anisotropy considerably complicates the relations between the fields. Now rather than simply writing $\vec{P} = \chi\varepsilon_0\vec{E}$, we must write $\vec{P}$ as the tensor product (in Cartesian coordinates) as $P_i = \chi_{ij}\varepsilon_0 E^j$, where the susceptibility tensor may be written in matrix form:

$$\overleftrightarrow{\chi} = \begin{pmatrix} \chi_{11} & \chi_{12} & \chi_{13} \\ \chi_{21} & \chi_{22} & \chi_{23} \\ \chi_{31} & \chi_{32} & \chi_{33} \end{pmatrix} \tag{8–111}$$

For nondissipative media, this tensor is symmetric, meaning that we can transform to a principal axis system where χ has only diagonal elements χ_{11}, χ_{22}, and χ_{33}, known as the principal susceptibilities. (These elements are of course not the same as those in the matrix above.)

The wave equation for the electric field is, as always, obtained by taking the curl of $\vec{\nabla} \times \vec{E}$ in Maxwell's equations to obtain

$$\begin{aligned} \vec{\nabla} \times (\vec{\nabla} \times \vec{E}) &= \vec{\nabla} \times \left(-\frac{\partial \vec{B}}{\partial t}\right) \\ &= -\frac{\partial}{\partial t}\left(\mu_0 \vec{J} + \mu_0 \frac{\partial \vec{D}}{\partial t}\right) \end{aligned} \tag{8–112}$$

In the anisotropic dielectric, ε is not independent of the coordinates so we cannot eliminate the divergence of $\vec{E}$ from the curl curl to reduce it to a (negative) Laplacian. Taking $\vec{J} = 0$ for the nonconductor and inserting $\vec{D} = \varepsilon_0\vec{E} + \vec{P} = \varepsilon_0(\vec{E} + \overleftrightarrow{\chi} \cdot \vec{E})$, we write

$$\vec{\nabla}(\vec{\nabla} \cdot \vec{E}) - \nabla^2 \vec{E} = -\frac{1}{c^2}\frac{\partial^2 \vec{E}}{\partial t^2} - \frac{\overleftrightarrow{\chi}}{c^2} \cdot \frac{\partial^2 \vec{E}}{\partial t^2} \tag{8–113}$$

[19] The excellent optics book *Principles of Optics*, 4th edition by M. Born and E. Wolf (Pergamon, New York, 1964) has a sound and complete discussion of double refraction and related topics.

With the assumption of a plane wave solution of the form

$$\vec{E} = \vec{E}_0 e^{i(\vec{k}\cdot\vec{r}-\omega t)} \tag{8–114}$$

the cartesian i component of the wave equation becomes

$$k_i(\vec{k}\cdot\vec{E}) - k^2 E_i + \frac{\omega^2}{c^2}(\delta_{ij} + \chi_{ij})E^j = 0 \tag{8–115}$$

Writing the three components out in full in the principal axis system where the three principal susceptibilities will be denoted χ_{xx}, χ_{yy}, and χ_{zz}, we have

$$\begin{aligned}
\left[-k_y^2 - k_z^2 + \frac{\omega^2}{c^2}(1+\chi_{xx})\right] E_x \quad + \quad k_x k_y E_y \quad + \quad k_x k_z E_z &= 0 \\
k_x k_y E_x \quad + \quad \left[-k_x^2 - k_z^2 + \frac{\omega^2}{c^2}(1+\chi_{yy})\right] E_y \quad + \quad k_y k_z E_z &= 0 \\
k_x k_z E_x \quad + \quad k_y k_z E_y \quad + \quad \left[-k_x^2 - k_y^2 + \frac{\omega^2}{c^2}(1+\chi_{zz})\right] E_z &= 0
\end{aligned} \tag{8–116}$$

In order that the system have a solution, the determinant of the coefficients must vanish. With the definition of the *principal refractive indices*— $n_x = \sqrt{1+\chi_{xx}}$ $n_y = \sqrt{1+\chi_{yy}}$ and $n_z = \sqrt{1+\chi_{zz}}$—we write this condition as

$$\begin{vmatrix}
(n_x\omega/c)^2 - k_y^2 - k_z^2 & k_x k_y & k_x k_z \\
k_x k_y & (n_y\omega/c)^2 - k_x^2 - k_z^2 & k_y k_z \\
k_x k_z & k_y k_z & (n_z\omega/c)^2 - k_x^2 - k_y^2
\end{vmatrix} = 0 \tag{8–117}$$

The solution for any constant ω generates a surface in k space. To gain some insight, we restrict $\vec{k}$ to lie in one of the coordinate planes, say the y-z plane. Then $k_x = 0$ and the characteristic equation reduces to

$$\left[(n_x\omega/c)^2 - k_y^2 - k_z^2\right]\left\{\left[(n_y\omega/c)^2 - k_z^2\right]\left[(n_z\omega/c)^2 - k_y^2\right] - k_y^2 k_z^2\right\} = 0 \tag{8–118}$$

In order for the product to vanish, one or both of the factors must vanish. We obtain two distinct roots, one denoted $\vec{k}_o$ (o for *ordinary*) satisfying

$$k_{o,y}^2 + k_{o,z}^2 = \left(\frac{n_x\omega}{c}\right)^2 \tag{8–119}$$

and a second denoted $\vec{k}_e$ (e for *extraordinary*) satisfying

$$\frac{k_{e,y}^2}{(n_z\omega/c)^2} + \frac{k_{e,z}^2}{(n_y\omega/c)^2} = 1 \tag{8–120}$$

We deduce that when $\vec{k}$ lies in the y-z plane, there are two possible values for k^2; one simply $k_o = n_x\omega/c$, the other, k_e, varying with direction as a radius of the ellipse described by (8–120).

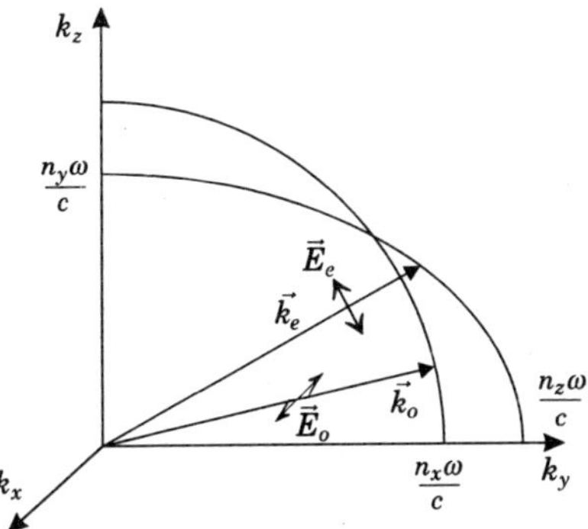

Figure 8.8: When $\vec{k}$ is confined to the y-z plane, x polarized waves travel with the same speed in all directions whereas waves with polarization parallel y-z plane travel with speed that varies with direction.

It is fruitful to continue consideration of this special case before generalizing. In particular, it is useful to determine the electric field alignment (polarization) associated with these distinct values of k. Rewriting the set of equations for E_x, E_y, and E_z when $\vec{k}$ is restricted to the y-z plane, such that

$$\begin{pmatrix} (n_x\omega/c)^2 - k^2 & 0 & 0 \\ 0 & (n_y\omega/c)^2 - k_z^2 & k_y k_z \\ 0 & k_y k_z & (n_z\omega/c)^2 - k_y^2 \end{pmatrix} \begin{pmatrix} E_x \\ E_y \\ E_z \end{pmatrix} = 0 \qquad (8\text{–}121)$$

we find that the first eigenvalue k_o, admits only $E_x \neq 0$; the other two components must vanish. In other words, the wave vector of an x-polarized wave with $\vec{k}$ in the y-z plane satisfies (8–119) travelling with phase speed $v = c/n_x$ in all directions. The second (distinct) eigenvalue k_e has an eigenvector $\vec{E}_e$ orthogonal to the first. This field therefore lies in the y-z plane. We conclude that a wave with polarization in the y-z plane has $\vec{k}_e$ satisfying (8–120); hence the phase speed v_{phase} varies with the direction of $\vec{k}$.

The circle described in the k_y-k_z plane by the first (ordinary) root has k_y and k_z intercepts of $n_x\omega/c$ while the ellipse described by the second root has intercepts $n_z\omega/c$ along the k_z axis and $n_y\omega/c$ along the z axis as illustrated in Figure 8.8. The polarization of the associated wave is indicated by the double-ended arrows placed on the k vectors.

These results are easily transported to the other coordinate planes. The locus of k on each coordinate plane consists of one circle and one ellipse. The fact that the surfaces are quadratic allows us to extrapolate to points between the coordinate planes.

If two of the principal indices are equal, the k surface consists of a sphere and an ellipsoid of revolution that touch at two points. In the direction defined by these two points, the two roots k_o and k_e are equal. A line through these points is known

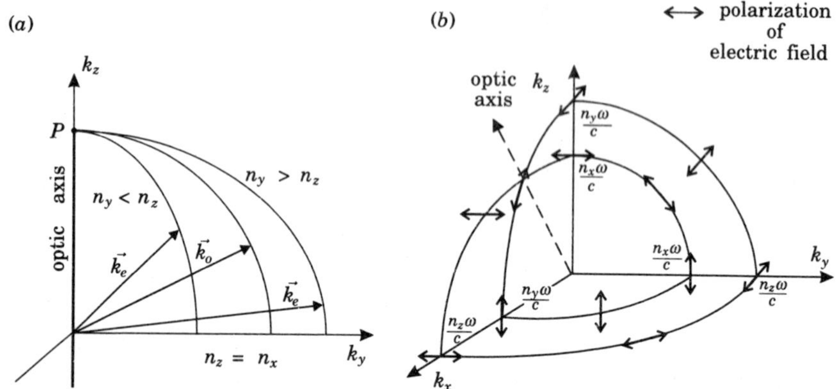

Figure 8.9: (*a*) A uniaxial crystal has its optic axis along the direction of the distinct index. (*b*) When all three indices are distinct, a self-intersecting surface results.

as the optic axis, and a crystal having a single optic axis, is said to be uniaxial.

If all three indices are distinct, the surface defined by $\vec{k}$ at constant ω is a self-intersecting surface sketched in Figure 8.9 (*b*). There will be two optic axes, and the crystal is said to be biaxial.

The character of the susceptibility tensor is related to the crystal symmetry. Cubic crystals such as NaCl can only be isotropic. Trigonal, tetragonal, and hexagonal crystals are uniaxial. The best known uniaxial crystal is probably calcite, with $n_o = 1.658$ and $n_e = 1.486$. Triclinic, monoclinic, and orthorhombic crystals are biaxial with mica, feldspar, and gypsum as the best known examples.

While $\vec{k}$ defines the normal to the constant phase planes (wave fronts) of the wave, it is not the direction of the energy flux because $\vec{k}$ and $\vec{E}$ are not generally perpendicular to each other. Instead, $\vec{k}$ and $\vec{D}$ are perpendicular. The energy travels in the direction of the Poynting vector and, as evident from Figure 8.10, the phase speed of the wave along $\vec{S}$ is $v_{phase}/\cos\theta$, where θ is the angle between $\vec{D}$

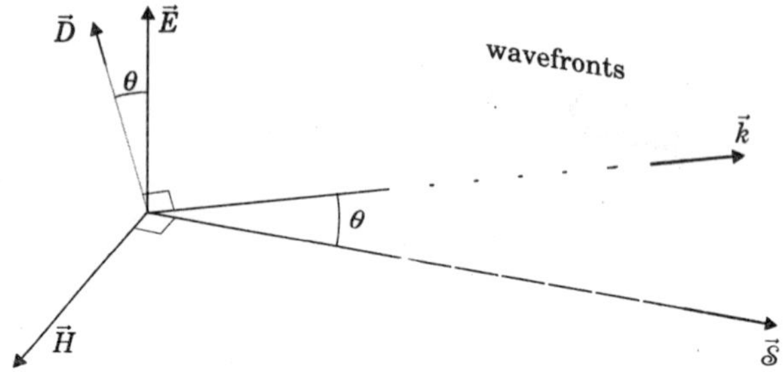

Figure 8.10: The energy flux is perpendicular to $\vec{E}$ and $\vec{H}$ whereas $\vec{k}$ is perpendicular to $\vec{D}$ and $\vec{H}$.

and $\vec{E}$. The extraordinary ray does not obey Snell's law at the interface between dielectrics.

8.4.7 Plane Waves in Isotropic, Linear Conducting Matter

In linear, isotropic, homogeneous conducting media, the fields satisfy (8–65) and (8–68)

$$\nabla^2\vec{E} + \mu\varepsilon\omega^2\left(1+\frac{ig}{\omega\varepsilon}\right)\vec{E} = 0 \qquad \text{and} \qquad \nabla^2\vec{H} + \mu\varepsilon\omega^2\left(1+\frac{ig}{\omega\varepsilon}\right)\vec{H} = 0$$

together with the constitutive relations, $\vec{D} = \varepsilon\vec{E}$, $\vec{B} = \mu\vec{H}$, and $\vec{J} = g\vec{E}$. Anticipating a damped wave solution, we consider solutions of the form

$$\vec{E}(\vec{r},t) = \vec{E}_o e^{i(\vec{K}\cdot\vec{r}-\omega t)} \qquad \text{and} \qquad \vec{H}(\vec{r},t) = \vec{H}_0 e^{i(\vec{K}\cdot\vec{r}-\omega t)} \tag{8–122}$$

where the complex wave vector, $\vec{K}$ is a vector whose components are complex numbers. Maxwell's equations applied to these expressions give the relations between the fields

$$\begin{aligned} &\vec{K}\cdot\vec{E}_0 = 0 \qquad && \vec{K}\times\vec{E}_0 = \omega\mu\vec{H}_0 \\ &\vec{K}\cdot\vec{H}_0 = 0 \qquad && \vec{K}\times\vec{H}_0 = -\omega\left(\varepsilon+\frac{ig}{\omega}\right)\vec{E}_0 \end{aligned} \tag{8–123}$$

and the wave equations (8–65) and (8–68) may be used to obtain the dispersion relation

$$K^2 = \mu\omega^2\left(\varepsilon+\frac{ig}{\omega}\right) \tag{8–124}$$

As a prelude to the general case we consider briefly the special case of the good conductor, for which $i\mu g\omega \gg 1$. In this case $K = \sqrt{i\mu g\omega} = \sqrt{\mu g\omega/2} + i\sqrt{\mu g\omega/2}$ $\equiv k + i\alpha$. Substituting this form into the solution, we find

$$\begin{pmatrix} \vec{E}(\vec{r},t) \\ \vec{H}(\vec{r},t) \end{pmatrix} = \begin{pmatrix} \vec{E}_0 \\ \vec{H}_0 \end{pmatrix} e^{-\vec{\alpha}\cdot\vec{r}} e^{i(\vec{k}\cdot\vec{r}-\omega t)} \tag{8–125}$$

The wave decays exponentially along its path. The penetration length $\delta = 1/\alpha$ is known as the skin depth. At microwave frequencies (10 GHz), the skin depth in silver ($g = 3\times 10^7\Omega^{-1}/\text{m}$) is 9.2×10^{-7}m. Very little would therefore be gained by constructing microwave components of silver rather than thinly electroplating them with silver to a thickness of several skin depths.

We return now to the more general conductor. The electric and magnetic fields are not independent but are related by equations (8–123). Thus, if we take $\vec{E}$ as above, then H must take the form

$$\vec{H}(\vec{r},t) = \frac{\vec{K}\times\vec{E}_0}{\mu\omega} e^{i(\vec{K}\cdot\vec{r}-\omega t)} \tag{8–126}$$

The time averaged Poynting vector $\langle\vec{S}\rangle$ is obtained from the complex fields as

$$\begin{aligned}
\langle\vec{S}\rangle &= \text{Re}\left(\frac{\vec{E}\times\vec{H}^*}{2}\right) = \frac{\vec{E}\times\vec{H}^* + \vec{E}^*\times\vec{H}}{4} \\
&= \frac{1}{4}\left[\frac{\vec{E}_0\times(\vec{K}^*\times\vec{E}_0^*)e^{i\vec{K}\cdot\vec{r}}\cdot\left(e^{i\vec{K}\cdot\vec{r}}\right)^*}{\omega\mu} + \frac{\vec{E}_0^*\times(\vec{K}\times\vec{E}_0)\left(e^{i\vec{K}\cdot\vec{r}}\right)^* e^{i\vec{K}\cdot\vec{r}}}{\omega\mu}\right] \\
&= \frac{1}{4}|E_0|^2\left(\frac{\vec{K}+\vec{K}^*}{\omega\mu}\right)e^{i\vec{K}\cdot\vec{r}}e^{-i\vec{K}^*\cdot\vec{r}} = \frac{1}{4}|E_0|^2\left(\frac{\vec{K}+\vec{K}^*}{\omega\mu}\right)e^{i(\vec{K}-\vec{K}^*)\cdot\vec{r}} \\
&= \frac{1}{2}|E_0|^2\frac{\text{Re}(\vec{K})}{\omega\mu}e^{-2\text{Im}(\vec{K})\cdot\vec{r}}
\end{aligned} \tag{8–127}$$

Similarly, the energy density U is calculated as

$$\begin{aligned}
\langle U\rangle &= \tfrac{1}{8}\left(\vec{E}\cdot\vec{D}^* + \vec{E}^*\cdot\vec{D} + \vec{B}\cdot\vec{H}^* + \vec{B}^*\cdot\vec{H}\right) \\
&= \tfrac{1}{2}|E_0|^2\left(\varepsilon + \frac{K^2}{\omega^2\mu}\right)e^{-2\text{Im}(\vec{K})\cdot\vec{r}}
\end{aligned} \tag{8–128}$$

To investigate more fully the effect of a complex wave vector $\vec{K}$, we again set $\vec{K} = \vec{k} + i\vec{\alpha}$, with $\vec{k}$ and $\vec{\alpha}$ real (and collinear). We relate k and α to the physical properties of the medium as follows. The dispersion relation (8–124) gives

$$K^2 = k^2 - \alpha^2 + 2i\alpha k = \mu\omega^2\left(\varepsilon + \frac{ig}{\omega}\right) \tag{8–129}$$

Assuming that g and ε are real (in the next section we will see that ε is frequently not real), we equate real and imaginary parts to obtain

$$k^2 - \alpha^2 = \mu\varepsilon\omega^2 \quad \text{and} \quad 2\alpha k = \mu\omega g \tag{8–130}$$

This pair of relations (8–130), may be solved for k and α to give

$$k^2 = \frac{\mu\varepsilon\omega^2}{2}\left[1+\sqrt{1+(g/\varepsilon\omega)^2}\right] \equiv \frac{\mu\varepsilon\omega^2}{2}\beta \quad \text{and} \quad \alpha^2 = \frac{\mu^2\omega^2 g^2}{4k^2} = \frac{\mu g^2}{2\varepsilon\beta} \tag{8–131}$$

with $\beta \equiv 1 + \sqrt{1+(g/\varepsilon\omega)^2}$.

The wave properties for waves in a homogeneous conductor can now be expressed in terms of the medium's properties as:

$$\lambda = \frac{2\pi}{k} = \frac{2\pi}{\omega\sqrt{\mu\varepsilon}}\sqrt{\frac{2}{\beta}} \simeq 2\pi\sqrt{\frac{2}{\mu g\omega}} \tag{8–132}$$

$$v_{phase} = \frac{\omega}{k} = \frac{1}{\sqrt{\mu\varepsilon}}\sqrt{\frac{2}{\beta}} \simeq \sqrt{\frac{2\omega}{\mu g}} \tag{8–133}$$

$$\delta = \frac{1}{\alpha} = \frac{2}{g}\sqrt{\frac{\varepsilon}{\mu}}\sqrt{\frac{\beta}{2}} \simeq \sqrt{\frac{2}{g\mu\omega}} \tag{8–134}$$

$$n = \frac{c}{v_{phase}} = c\sqrt{\mu\varepsilon}\sqrt{\frac{\beta}{2}} \simeq c\sqrt{\frac{\mu g}{2\omega}} \tag{8–135}$$

where the near equality in each of the four expressions above holds for good conductors only ($g \gg \varepsilon\omega$).

EXAMPLE 8.2: Find the wavelength λ, skin depth δ, and refractive index for an electromagnetic wave of angular frequency $\omega = 2\pi \times 10^{10}\ \text{s}^{-1}$ propagating through aluminum ($g = 3.53 \times 10^7\ \Omega^{-1}/\text{m}$).

Solution: We use the good conductor approximation and assume the permeability $\mu \simeq \mu_0 = 4\pi \times 10^{-7}$. While the vacuum wavelength is 3 cm, it is reduced to

$$\lambda = 2\pi\sqrt{\frac{2}{4\pi \times 10^{-7} \times 3.53 \times 10^7 \times 2\pi \times 10^{10}}} = 5.32 \times 10^{-6}\ \text{m} = 5.32\ \mu\text{m} \tag{Ex 8.2.1}$$

in aluminum. The skin depth is now easily obtained as $\lambda/2\pi = 0.85\ \mu$m. The refractive index may be found from $n = c/v_{phase} = 2\pi c/\lambda\omega = 5.64 \times 10^3$. We conclude that in a good conductor, electromagnetic waves have a very large refractive index, very small wavelength, and very small skin depth. The electric field E is cancelled in roughly one skin depth by free charges, while H is cancelled in the same distance by eddy currents induced close to the surface of the conductor. Semiconductors such a germanium are used as lens material for infrared radiation at $\lambda = 10.4\,\mu$m produced by carbon-dioxide lasers. The much lower conductivity results in a refractive index near 4.

8.4.8 Simple Model for the Frequency Dependence of Dielectric Susceptibility

As we have seen, the refractive index of a dielectric is normally given by $\sqrt{\varepsilon/\varepsilon_0}$. We now briefly investigate a simple model of the dielectric to give us some insight into the frequency dependence of the permittivity.

Consider a dielectric composed of heavy positive ion cores surrounded by electrons bound to the ions by a harmonic potential, $V = \frac{1}{2}kr^2$. The equation of motion for such an electron is

$$m\frac{d^2\vec{r}}{dt^2} + m\gamma\frac{d\vec{r}}{dt} + k\vec{r} = q\vec{E}(\vec{r}, t) \tag{8–136}$$

The viscous damping term has been introduced to allow for energy loss to the lattice and radiation damping. If we set $\vec{E} = \vec{E}_0 e^{-i\omega t}$ and $\vec{r} = \vec{r}_0 e^{-i\omega t}$, the equation of motion becomes, with $\omega_0^2 \equiv k/m$,

$$\left(-\omega^2 - i\omega\gamma + \omega_0^2\right)\vec{r} = \frac{q}{m}\vec{E} \tag{8–137}$$

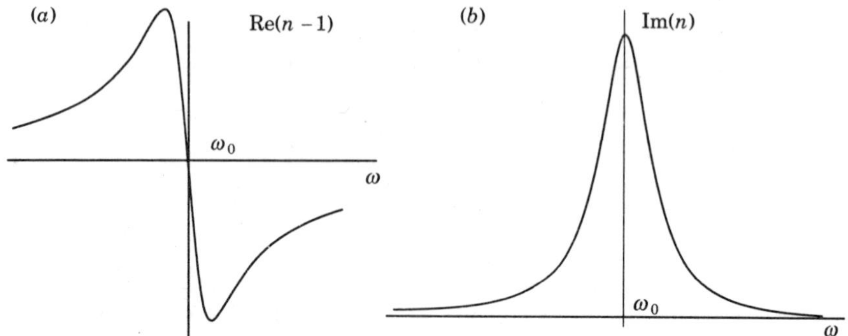

Figure 8.11: The real part of n gives the dispersion curve on the left, while the imaginary part gives the absorption curve on the right.

The polarization of a medium having n such oscillators per unit volume is then $\vec{P} = nq\langle\vec{r}\rangle = \chi\varepsilon_0\vec{E}$, implying

$$\chi = \frac{nq^2}{m\varepsilon_0\,(\omega_0^2 - \omega^2 - i\omega\gamma)} \tag{8–138}$$

Rationalizing, we write χ as

$$\chi = \frac{nq^2\left[(\omega_0^2 - \omega^2) + i\omega\gamma\right]}{m\varepsilon_0\left[(\omega_0^2 - \omega^2)^2 + \omega^2\gamma^2\right]} \tag{8–139}$$

and finally obtain

$$\varepsilon = \varepsilon_0(1+\chi) = \varepsilon_0 + \frac{nq^2(\omega_0^2 - \omega^2)}{m\left[(\omega_0^2 - \omega^2)^2 + \omega^2\gamma^2\right]} + \frac{inq^2\omega\gamma}{m\left[(\omega_0^2 - \omega^2)^2 + \omega^2\gamma^2\right]} \tag{8–140}$$

Most materials have γ in the range 10^6 to $10^{10}\ \mathrm{s}^{-1}$ at optical frequencies ($\omega \geq 10^{14}\ \mathrm{s}^{-1}$) varying approximately as ω^3. Except when ω is very close to ω_0, the imaginary term will be much smaller than the real term, leading us to conclude that ε is real except very near resonance. When not too close to resonance, we find the deviation of the refractive index from unity to be

$$n - 1 \simeq \frac{nq^2}{2m\varepsilon_0(\omega_0^2 - \omega^2)}\left(1 + \frac{i\omega\gamma}{\omega_0^2 - \omega^2}\right) \tag{8–141}$$

The imaginary part of n gives rise to absorption, while the real part is responsible for dispersion. The real and imaginary parts of the refractive index are plotted in Figure 8.11(a) and 8.11(b), respectively.

8.4.9 Simple Model of a Conductor in an Oscillating Field

A static electric field cannot penetrate into a conductor because the force on free charges in the conductor resulting from the field would make those charges move until the force vanishes. When the charges have adjusted themselves so that no further force remains, the field vanishes. The flow of charges into their shielding positions is not, however, instantaneous, so that a time-dependent electric field would be expected to penetrate into the conductor and generate transient currents.

Consider a metal having n free electrons per unit volume. Under the influence of an oscillating electric field $\vec{E} = \vec{E}_0 e^{-i\omega t}$ and a "viscous" drag force $-m\gamma\vec{v}$, we may write down the equation of motion for a free electron:

$$m\frac{d\vec{v}}{dt} = -m\gamma\vec{v} + q\vec{E}_0 e^{-i\omega t} \tag{8–142}$$

We assume an oscillating solution $\vec{v} = \vec{v}_0 e^{-i\omega t}$, which we insert into the equation above to obtain

$$\vec{v}(t) = \frac{q\vec{E}(t)}{m(\gamma - i\,\omega)} \tag{8–143}$$

The current density in the metal is $\vec{J} = nq\vec{v} = g\vec{E}$, giving the conductivity $g(\omega) = nq^2/m(\gamma - i\omega)$. To estimate an order of magnitude for γ, we consider the conductivity of copper at zero frequency. For copper $n \simeq 8 \times 10^{28}/\text{m}^3$ and $g(0) = 5.9 \times 10^7\ \Omega^{-1}/\text{m} = nq^2/m\gamma$; we deduce that $\gamma = 3.8 \times 10^{13}\ \text{s}^{-1}$. At frequencies $\omega \ll \gamma$ (including microwave frequencies), g is essentially real and independent of ω for a good conductor.

EXAMPLE 8.3: Obtain the ratio of reflected to incident power of an electromagnetic wave reflecting from a conducting surface at normal incidence.

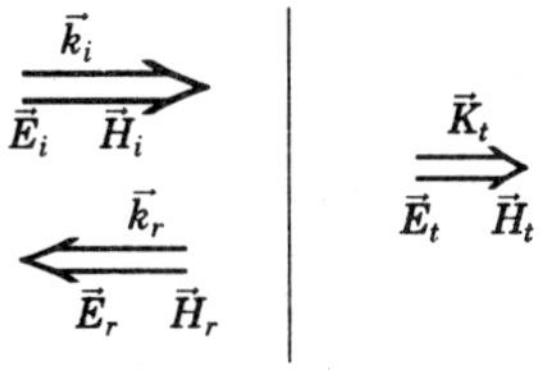

Figure 8.12: Normally incident waves at the boundary of a conductor.

Solution: We denote the incident, reflected, and transmitted fields by subscripts i, r, and t respectively. Since we expect a complex wave vector inside the conductor, we denote it with K_t in contrast to the incident k_i of the nonconducting side of incidence. Assuming transverse waves,

$$\begin{aligned}
\vec{E}_i &= \vec{E}_{0,i} e^{i(k_i z - \omega t)} \\
\vec{E}_r &= \vec{E}_{0,r} e^{i(-k_i z - \omega t)} \\
\vec{E}_t &= \vec{E}_{0,t} e^{i(K_t z - \omega t)}
\end{aligned} \tag{Ex 8.3.1}$$

and taking the interface at $z = 0$ for convenience, we apply the boundary conditions

$$\begin{aligned}
E_\parallel \text{ is continuous} \quad &\Rightarrow \quad E_{0,i} + E_{0,r} = E_{0,t} \\
H_\parallel \text{ is continuous} \quad &\Rightarrow \quad H_{0,i} + H_{0,r} = H_{0,t}
\end{aligned}$$

Further, $\vec{H}_0$ is related to $\vec{E}_0$ by $\vec{H}_0 = (\vec{K} \times \vec{E}_0)/\mu\omega$. For the nonconductor (on the side of incidence), $k/\omega = \sqrt{\mu_i \varepsilon_i}$. Substituting this into the E–H relation gives

$$H_{0,i} = \pm\sqrt{\frac{\varepsilon_i}{\mu_i}}E_{0.i} \qquad \text{and} \qquad H_{0.r} = \mp\sqrt{\frac{\varepsilon_i}{\mu_i}}E_{0.r} \tag{Ex 8.3.2}$$

while on the conducting side, the dispersion relation gives

$$K_t^2 = \mu_t\omega^2\left(\varepsilon_t + \frac{ig_t}{\omega}\right) \quad \text{leading to} \quad H_{0.t} = \sqrt{\frac{\varepsilon_t}{\mu_t} + \frac{ig_t}{\mu_t\omega}}E_{0.t} \tag{Ex 8.3.3}$$

The continuity of $H_{\|}$ then gives

$$\sqrt{\frac{\varepsilon_i}{\mu_i}}\left(E_{0.i} - E_{0.r}\right) = \sqrt{\frac{\varepsilon_t}{\mu_t} + \frac{ig_t}{\mu_t\omega}}E_{0.t} \tag{Ex 8.3.4}$$

which may be solved simultaneously with $E_{0.i} + E_{0.r} = E_{0.t}$ to give

$$E_{0.t} = \frac{2E_{0.i}}{1+\eta} \quad \text{and} \quad E_{0.r} = \frac{1-\eta}{1+\eta}E_{0.i} \quad \text{with } \eta \equiv \sqrt{\frac{\mu_i}{\varepsilon_i}\left(\frac{\varepsilon_t}{\mu_t} + \frac{ig_t}{\mu_t\omega}\right)} \tag{Ex 8.3.5}$$

When the side of incidence is air and the conductor is a non-ferrous metal with good conductivity, $\varepsilon_i \simeq \varepsilon_t \simeq \varepsilon_0$, $\mu_i \simeq \mu_t \simeq \mu_0$, and $g_t \gg \varepsilon_t\omega$. Under these conditions we may approximate η by

$$\eta = \sqrt{\frac{ig}{\omega\varepsilon_0}} = \sqrt{\frac{g}{\omega\varepsilon_0}}\left(e^{i\pi/2}\right)^{1/2} = \sqrt{\frac{g}{2\omega\varepsilon_0}}(1+i) \tag{Ex 8.3.6}$$

The ratio of the reflected to the incident power is $|E_{0.r}/E_{0.i}|^2 = |(1-\eta)/(1+\eta)|^2$, which, for $|\eta| \gg 1$, is best expressed as

$$\begin{aligned}\left|\frac{1-1/\eta}{1+1/\eta}\right|^2 &\simeq \left|1-\frac{1}{\eta}\right|^4 = \left[\left(1-\frac{1}{\eta}\right)\cdot\left(1-\frac{1}{\eta^*}\right)\right]^2 \\ &\simeq 1 - 2\left(\frac{1}{\eta} + \frac{1}{\eta^*}\right) = 1 - 2\sqrt{\frac{2\omega\varepsilon_0}{g}}\left(\frac{1}{1+i} + \frac{1}{1-i}\right) \\ &= 1 - 2\sqrt{\frac{2\omega\varepsilon_0}{g}}\end{aligned} \tag{Ex 8.3.7}$$

to first order in $1/\eta$.

Exercises and Problems

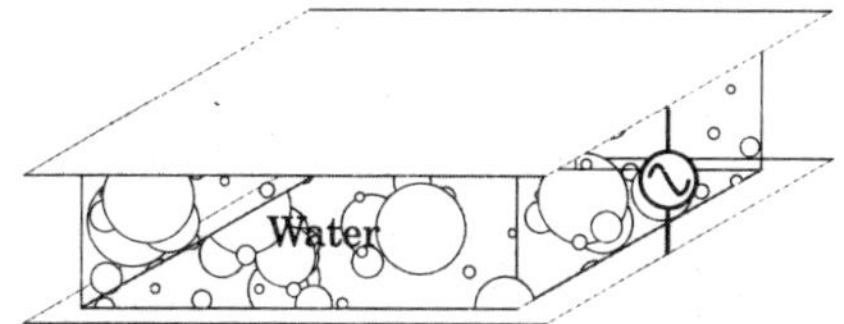

Figure 8.13: The volume between the plates is filled with salt water.

8-1 Show that in a good conductor, $\vec{E}$ and $\vec{H}$ oscillate 45° out of phase.

8-2 The conductivity of sea water is about 4.3 Ω^{-1}/m. Assume that $\mu = \mu_0$ and $\varepsilon \simeq 80\varepsilon_0$. Calculate the penetration depth in sea water of a typical ELF (extremely low frequency) wave at 100 Hz. Comment on the suitability of low-frequency radio waves as a means of communication with or between submarines.

8-3 How large could a soap bubble (surface tension σ) be before blackbody radiation pressure would destroy it? (Assume a temperature of 300°K.)

8-4 Find the relationship between the Poynting vector and the momentum density of an electromagnetic wave in a linear homogeneous dielectric.

8-5 Determine the magnitude and direction of force on a dielectric of refractive index n when a beam carrying power dW/dt impinges normally on it.

8-6 Light is incident internally on a dielectric/air interface at 60° to the normal. Find the decay distance of the evanescent wave in air when the refractive index of the dielectric is 1.7.

8-7 Find the phase difference between the reflected s and p waves in problem 8-6, assuming they were in phase before internally reflecting.

8-8 In the text, the dispersion relation for a plasma was calculated on the basis of displaced electrons producing a polarization. Take the alternative view, that the moving electrons constitute a current, and recalculate the dispersion relation on this basis. (Note that one may take one view or the other but not both.)

8-9 Obtain the dispersion relation for an electromagnetic wave travelling through a tenuous plasma with a weak axial (along $\vec{k}$) magnetic field.

8-10 Two large, flat conducting plates are separated by a 1-cm thickness of salt water ($g = 4.3\Omega^{-1}$/m, $\varepsilon \simeq 80\varepsilon_0$) (Figure 8.13). An oscillating voltage $V_0 \cos \omega t$ is applied across the plates. Find the current in the water as well as the charge accumulation on the plates when $\omega = \pi \times 10^9$ s^{-1}.

8-11 An anisotropic dielectric with principal indices $n_x = 1.3$, $n_y = 1.5$, and $n_z = 1.7$ has its principal axes aligned with the x, y, and z axes. An electromagnetic wave with electric field amplitude $E_0(\hat{\imath} + \hat{\jmath} + \hat{k})/\sqrt{3}$ and magnetic field strength amplitude $H_0(\hat{\imath} - \hat{\jmath})/\sqrt{2}$ propagates in the medium. Find the direction of $\vec{k}$ and $\vec{S}$ in the medium.

8-12 The squares of the amplitude transmission coefficients (8–87) and (8–92) do not give the ratio of transmitted to incident energy because the energy density is proportional to εE^2, not E^2, and because the cross section of any light beam is changed on refraction. Obtain the correct expression for the energy transmission coefficient.

8-13 When, early in the age of the universe the temperature had fallen to 1000°K, sufficient recombination occurred for the universe to become transparent Radiation has expanded essentially adiabatically since that time. Given that the current black-body temperature is currently about 2.73°K, but what factor has the universe increased in size since decoupling from matter?

8-14 Obtain the power reflection coefficient for visible light reflecting normally off silver ($g = 6.8 \times 10^7\, \Omega^{-1}/\text{m}$) at an angular frequency of $\omega = 10^{14}\,\text{s}^{-1}$. Obtain also the phase change of the reflected and transmitted waves.

Chapter 9

Waveguide Propagation—Bounded Waves

9.1 Bounded Waves

The plane waves we have considered in Chapter 8 have wave fronts stretching infinitely in both directions. Such waves are clearly unphysical. In Chapter 3 we briefly encountered spherical waves and saw that it was no longer possible to maintain the electric field and the magnetic field each perpendicular to the direction of propagation; instead, waves were classified as transverse electric (TE) and transverse magnetic (TM). In this chapter we consider travelling waves that are bounded either by being enclosed within metallic waveguides or guided by dielectric rods. We touch also on standing waves confined to a metal cavity. In the case that the dimensions of the guide are much larger than the wavelength, one can think of the propagation as simply a plane wave reflecting successively from the walls either by metallic reflection for metal waveguides, or total internal reflection for dielectric waveguides. For smaller guides, this picture cannot be supported. It will be shown that no TEM solutions (both the electric and magnetic fields perpendicular to the direction of propagation) exist for guided waves inside hollow waveguides.

9.1.1 TE Modes in a Rectangular Waveguide

As a preliminary to our investigation we consider the simplest (also the most commonly used) mode of propagation in a rectangular cross section waveguide. Until Section 9.2.3, where we consider losses in the guide, we will take the walls to be perfect conductors. For simplicity, we assume that the space inside the waveguide is empty and that the permeability is everywhere μ_0.

Consider the x-polarized wave having no x dependence of the form

$$\vec{E}(\vec{r},t) = E_0(y)\hat{\imath}\, e^{i(k_z z-\omega t)} \tag{9–1}$$

propagating in the z direction inside the rectangular cross section waveguide illustrated in Figure 9.1. Substituting this into the $\vec{\nabla}\times\vec{E}$ equation of Maxwell's

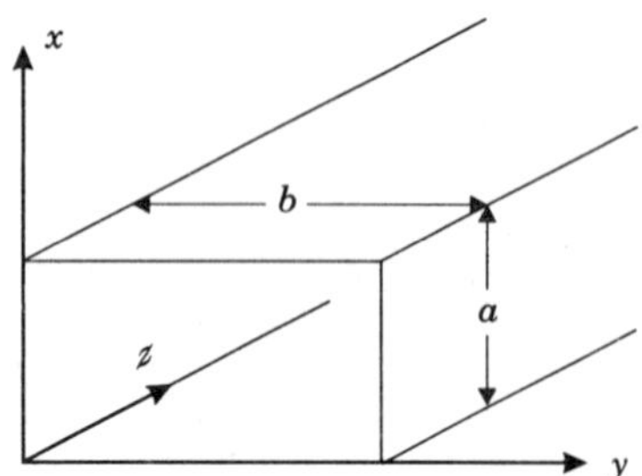

Figure 9.1: The wave propagates in the z direction inside the metal-walled rectangular cross section waveguide.

equations (8–10), we find the associated magnetic field intensity $\vec{H}$ is

$$\vec{H}(\vec{r},t) = \frac{\vec{\nabla} \times \vec{E}(\vec{r},t)}{i\mu_0\omega} = \frac{1}{i\mu_0\omega}\left[ik_z E_0(y)\hat{\jmath} - \frac{\partial E_0(y)}{\partial y}\hat{k}\right] e^{i(k_z z - \omega t)} \tag{9–2}$$

In the interior of the waveguide, where the conductivity $g = 0$ and $\mu\varepsilon = \mu_0\varepsilon_0 = 1/c^2$, the field E_0 satisfies

$$\left(\nabla^2 + \mu_0\varepsilon_0\omega^2\right) E_0(y) e^{i(k_z z - \omega t)} = 0 \tag{9–3}$$

or

$$\frac{\partial^2 E_0(y)}{\partial y^2} + \left(- k_z^2 + \frac{\omega^2}{c^2}\right) E_0(y) = 0 \tag{9–4}$$

Setting $k_y^2 = \omega^2/c^2 - k_z^2$, we abbreviate this as

$$\frac{\partial^2 E_0(y)}{\partial y^2} + k_y^2 E_0(y) = 0 \tag{9–5}$$

The form of $E_0(y)$ must clearly be $E_0(y) = A \sin k_y y + B \cos k_y y$. At a perfectly conducting surface, $E_\| = 0$, implying that $E_0 = 0$ at $y = 0$ and at $y = b$. We conclude that $B = 0$ and $k_y = n\pi/b$ with n an integer.

This particular wave, a TE wave ($\vec{H}$ has a component along z and is not transverse), has the form

$$\vec{E}_{0,n}^{\,\rm TE} = \hat{\imath} A \sin\frac{n\pi y}{b} e^{i(k_z z - \omega t)} \tag{9–6}$$

and from (9–2)

$$\vec{H}_{0,n}^{\,\rm TE} = \frac{A}{\omega\mu_0}\left[k_z \sin\frac{n\pi y}{b}\hat{\jmath} + \frac{in\pi}{b}\cos\frac{n\pi y}{b}\hat{k}\right] \tag{9–7}$$

The fields in the walls of the perfectly conducting waveguide are zero, and are therefore zero outside as well. The surface currents are given by $\vec{j} = \hat{n} \times \vec{H}$. On the top surface, for instance, $\hat{n} = -\hat{\imath}$, implying that

$$\vec{j} = \frac{A}{\omega\mu_0}\left[-k_z \sin\frac{n\pi y}{b}\hat{k} + \frac{in\pi}{b}\cos\frac{n\pi y}{b}\hat{\jmath}\right] e^{i(k_z z - \omega t)} \tag{9–8}$$

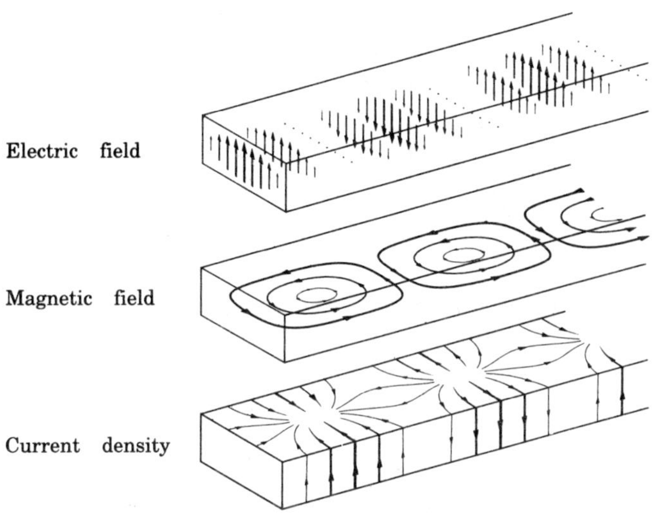

Figure 9.2: Field and current patterns of the TE_{01} mode of a rectangular conducting wall waveguide.

The field and current patterns of the TE_{01} mode are plotted in Figure 9.2. The dispersion relation is obtained by substituting $\partial^2 E_0/\partial y^2 = -(n\pi/b)^2 E_0$ into the wave equation (9–4). The substitution gives

$$k_z^2 = \frac{\omega^2}{c^2} - \left(\frac{n\pi}{b}\right)^2 \tag{9–9}$$

When $\omega < \omega_c \equiv n\pi c/b$, k_z becomes pure imaginary and the wave no longer propagates. Equivalently, when the vacuum wavelength $\lambda = 2\pi c/\omega$ exceeds the cutoff wavelength $\lambda_c = 2b/n$, the wave no longer propagates. The phase velocity, $v_{phase} = \omega/k_z$, is given by

$$v_{phase} = \frac{\omega}{k_z} = c\sqrt{1 + \frac{n^2\pi^2}{b^2k_z^2}} = c\sqrt{1 + \left(\frac{\lambda}{\lambda_c}\right)^2} > c \tag{9–10}$$

while the group velocity, $v_{group} = d\omega/dk_z$, is

$$\frac{d\omega}{dk_z} = c\frac{d}{dk_z}\sqrt{k_z^2 + \frac{n^2\pi^2}{b^2}} = \frac{k_z c^2}{c\sqrt{k_z^2 + \dfrac{n^2\pi^2}{b^2}}} = \frac{k_z c^2}{\omega} = \frac{c^2}{v_{phase}} \tag{9–11}$$

less than the velocity of light, c, as of course it must be.

Another often useful way of looking at the propagation is to write (9–6) as

$$\vec{E}_{0,n}^{\,\mathrm{TE}} = \frac{A}{2i}\hat{\imath}\left[e^{i(n\pi y/b + k_z z - \omega t)} - e^{i(-n\pi y/b + k_z z - \omega t)}\right]$$

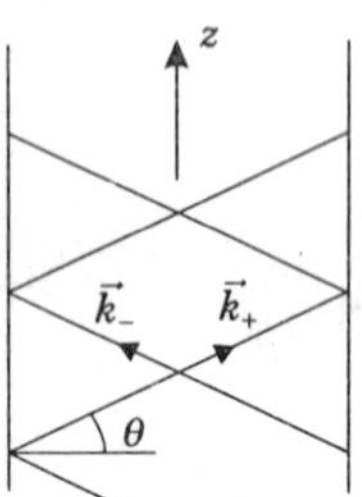

Figure 9.3: The mode may be visualized as two plane waves zigzagging along the guide.

$$= \frac{A}{2i}\hat{\imath}\left[e^{i(\vec{k}_+\cdot\vec{r}-\omega t)} - e^{i(\vec{k}_-\cdot\vec{r}-\omega t)}\right] \tag{9–12}$$

where $\vec{k}_\pm = \pm(n\pi/b)\hat{\jmath} + k_z\hat{k}$. The wave is a superposition of two, 180° out-of-phase waves of equal amplitude, zigzagging down the waveguide (Figure 9.3). Each wave's propagation vector $\vec{k}_\pm$ makes an angle $\theta = \tan^{-1}(bk_z/n\pi)$ with the normal of the waveguide wall. As n increases, θ gets smaller, implying that higher modes suffer more reflections per unit length, leading to higher losses. At cutoff, k_z vanishes, meaning that $\theta = 0$. The waves merely bounce back and forth between the walls. Measured along $\vec{k}_\pm$, the wave is characterized by a wavelength

$$\lambda_\pm = \frac{2\pi}{\sqrt{k_y^2 + k_z^2}} = \frac{2\pi}{\omega/c} = \frac{c}{\nu} = \lambda_0 \tag{9–13}$$

where λ_0 is the vacuum wavelength, and the wave travels with speed c.

Modes with $\vec{E}$ polarized in the y direction are also possible, giving TE$_{m,0}$ modes, and modes polarized in the x-y plane designated as TE$_{m,n}$ modes having

$$k_z^2 = \frac{\omega^2}{c^2} - \frac{m^2\pi^2}{a^2} - \frac{n^2\pi^2}{b^2} \tag{9–14}$$

are also possible.

In addition to the TE modes, there are others designated TM$_{m,n}$ modes, having $\vec{H}$ perpendicular to the direction of propagation while $\vec{E}$ has a longitudinal component.

9.2 Cylindrical Waveguides

In the preceding discussion, we assumed a very simple solution and verified that it satisfied the boundary conditions, and explored some of the properties of waves inside a metal tube. In this section we will obtain a tractable set of equations governing the behavior of the fields inside a metal-wall waveguide of constant cross section (the term cylindrical is not restricted to circular cylinders). In general, solving

$$(\nabla^2 + \mu\varepsilon\omega^2)\begin{pmatrix}\vec{E}\\ \vec{H}\end{pmatrix} = 0 \tag{9–15}$$

is not as simple as it seems because $\vec{E}$ and $\vec{H}$ are coupled by Maxwell's (curl) equations. In fact, the curl equations may be used to show that the electromagnetic field is entirely determined by the longitudinal (z) components of $\vec{E}$ and $\vec{H}$. These longitudinal components satisfy a simple scalar wave equation and appropriate boundary conditions. The other components of $\vec{E}$ and $\vec{H}$ can be derived from the z components.

We start by assuming that waves propagate in the z direction, and that we may write the fields in Cartesian coordinates as

$$\begin{pmatrix} \vec{E}(\vec{r},t) \\ \vec{H}(\vec{r},t) \end{pmatrix} = \begin{pmatrix} \vec{E}_0(x,y) \\ \vec{H}_0(x,y) \end{pmatrix} e^{i(kz-\omega t)} \tag{9–16}$$

Maxwell's equations then give

$$\vec{\nabla}\times\vec{E} = i\omega\vec{B} \quad \text{or} \quad \begin{cases} \dfrac{\partial E_z}{\partial y} - ikE_y = i\mu\omega H_x \\ -\dfrac{\partial E_z}{\partial x} + ikE_x = i\mu\omega H_y \end{cases} \tag{9–17}$$

$$\vec{\nabla}\times\vec{H} = -i\varepsilon\omega\vec{E} \quad \text{or} \quad \begin{cases} \dfrac{\partial H_z}{\partial y} - ikH_y = -i\omega\varepsilon E_x \\ -\dfrac{\partial H_z}{\partial x} + ikH_x = -i\omega\varepsilon E_y \end{cases} \tag{9–18}$$

The electric field component E_y may be eliminated from the top line of (9–17) and the bottom line of (9–18) to obtain

$$H_x = \frac{1}{\mu\varepsilon\omega^2 - k^2}\left(ik\frac{\partial H_z}{\partial x} - i\varepsilon\omega\frac{\partial E_z}{\partial y}\right) \tag{9–19}$$

Similarly, eliminating E_x from bottom line (9–17) and the top line (9–18), we have

$$H_y = \frac{1}{\mu\varepsilon\omega^2 - k^2}\left(ik\frac{\partial H_z}{\partial y} + i\varepsilon\omega\frac{\partial E_z}{\partial x}\right) \tag{9–20}$$

Alternatively, H_x may be eliminated from (9–17) and (9–18) to give

$$E_y = \frac{1}{\mu\varepsilon\omega^2 - k^2}\left(ik\frac{\partial E_z}{\partial y} - i\mu\omega\frac{\partial H_z}{\partial x}\right) \tag{9–21}$$

while eliminating H_y from (9–17) and (9–18) gives

$$E_x = \frac{1}{\mu\varepsilon\omega^2 - k^2}\left(ik\frac{\partial E_z}{\partial x} + i\mu\omega\frac{\partial H_z}{\partial y}\right) \tag{9–22}$$

The four equations above are conveniently combined to obtain expressions for the transverse fields in terms of only the z components. Defining $\vec{H}_t \equiv \hat{\imath}H_x + \hat{\jmath}H_y$, we get

$$\vec{H}_t = \frac{1}{\mu\varepsilon\omega^2 - k^2}\left(ik\vec{\nabla}_t H_z + i\varepsilon\omega\hat{k}\times\vec{\nabla}_t E_z\right) \tag{9–23}$$

and similarly for $\vec{E}_t \equiv \hat{\imath}E_x + \hat{\jmath}E_y$

$$\vec{E}_t = \frac{1}{\mu\varepsilon\omega^2 - k^2}\left(ik\vec{\nabla}_t E_z - i\mu\omega\hat{k} \times \vec{\nabla}_t H_z\right) \tag{9-24}$$

where $\vec{\nabla}_t \equiv \vec{\nabla} - \hat{k}(\partial/\partial z)$ is the transverse gradient operator.

It is clear now that once E_z and H_z are known, the remaining components of the fields are completely specified.

Since we have already made explicit use of the form

$$\begin{pmatrix} \vec{E}(\vec{r},t) \\ \vec{H}(\vec{r},t) \end{pmatrix} = \begin{pmatrix} \vec{E}_0(x,y) \\ \vec{H}_0(x,y) \end{pmatrix} e^{i(kz-\omega t)} \tag{9-25}$$

we can recast the wave equation (9–15) for the z component of the fields as

$$\left[\nabla_t^2 + \left(\mu\varepsilon\omega^2 - k^2\right)\right] \begin{pmatrix} E_{0,z}(x,y) \\ H_{0,z}(x,y) \end{pmatrix} = 0 \tag{9-26}$$

with $\nabla_t^2 \equiv \nabla^2 - \partial^2/\partial z^2$. The solution of this equation will be made subject to the boundary conditions corresponding to a perfect conductor–dielectric interface, namely $B_\perp = 0$ and $E_\parallel = 0$.

$E_\parallel = E_z = 0$ at the walls is easy to implement. The consequences of the other condition, $B_\perp = B_t = 0$, on the z components of the fields is not so easily fathomed. To see what the boundary condition on $B_\perp$ implies for H_z, we consider the expression for H_t in (9–23). $B_\perp = 0$ is equivalent to $\vec{H} \cdot \hat{n} = 0$, where $\hat{n}$ is a normal to the waveguide wall. Thus,

$$\begin{aligned} \vec{H} \cdot \hat{n} &= \frac{1}{\mu\varepsilon\omega^2 - k^2}\left[ik\left(\vec{\nabla}_t H_z\right) \cdot \hat{n} + i\varepsilon\omega\hat{n} \cdot \left(\hat{k} \times \vec{\nabla}_t E_z\right)\right] \\ &= \frac{1}{\mu\varepsilon\omega^2 - k^2}\left[ik\frac{\partial H_z}{\partial n} + i\varepsilon\omega(\hat{n} \times \hat{k}) \cdot \vec{\nabla}_t\, E_z\right] \end{aligned} \tag{9-27}$$

Now, $(\hat{n} \times \hat{k})$ defines a direction tangential (along the circumference) to the conducting wall of a waveguide, say the $\hat{t}$ direction. For a point $\vec{r}_0$ on the wall, an adjacent point $\vec{r}_0 + \epsilon\hat{t}$ is also a point on the wall. We can in general expand $E_z(\vec{r}_0 + \epsilon\hat{t}\,) = E_z(\vec{r}_0) + \epsilon\hat{t} \cdot \vec{\nabla} E_z(\vec{r}_0)$. Since both $E_z(\vec{r}_0)$ and $E_z(\vec{r}_0 + \epsilon\hat{t}\,)$ vanish, $\hat{t} \cdot \vec{\nabla}_t E_z(\vec{r}_0)$ must vanish for $\vec{r}_0$ on the wall.

We therefore conclude that

$$\vec{H} \cdot \hat{n}\Big|_{wall} = \frac{ik}{\mu\varepsilon\omega^2 - k^2}\frac{\partial H_z}{\partial n} \tag{9-28}$$

and that $B_\perp = 0$ on the wall implies $\partial H_z/\partial n = 0$.

The fields in the waveguide now divide naturally into two distinct categories, transverse magnetic ($H_z = 0$) and transverse electric ($E_z = 0$). The most general travelling wave will of course be a linear combination of the two types.

In principle, there is also a degenerate mode designated TEM for which both E_z and H_z vanish. In order to obtain nonvanishing transverse fields we must set

$k^2 - \mu\varepsilon\omega^2 = 0$, leading to $k/\omega = \sqrt{\mu\varepsilon}$. The wave travels as if it were in an infinite medium without bounding surfaces. The two-dimensional wave equation, (9–26). reduces to

$$\nabla_t^2 \begin{pmatrix} E_{0,z}^{\text{TEM}} \\ H_{0,z}^{\text{TEM}} \end{pmatrix} = 0 \tag{9–29}$$

The uniqueness theorem applied to $E_{0,z}$ or $H_{0,z}$ now requires that $E_{0,z}$ and $H_{0,z}$ take on the value they have on the boundary everywhere inside the boundary. A TEM mode cannot exist in a waveguide with only an exterior conductor. TEM modes are, however, the dominant modes in coaxial cables and parallel wire transmission lines. We dispose of the TEM mode with the example below before continuing with hollow waveguides.

EXAMPLE 9.1 Find the TEM mode for waves travelling between coaxial inner conducting circular cylinder of radius a and outer grounded conducting cylinder of radius b.

Solution: As the problem insists there is no z component to $\vec{H}$ or $\vec{E}$, the expression of the transverse fields in terms of the z components is not very useful. Instead of using (9–23) and (9–24), we return to Maxwell's equations to obtain for the wave of form $\vec{E}_0(x, y)e^{i(kz-\omega t)}$:

$$\frac{\partial E_x}{\partial x} + \frac{\partial E_y}{\partial y} + ikE_z = \frac{\partial E_x}{\partial x} + \frac{\partial E_y}{\partial y} = 0 \tag{Ex 9.1.1}$$

and

$$\frac{\partial E_x}{\partial y} - \frac{\partial E_y}{\partial x} = i\omega\mu H_z = 0 \tag{Ex 9.1.2}$$

Since the curl of $\vec{E}_0$ vanishes, we may derive it from a potential ϕ as $\vec{E}_0 = -\vec{\nabla}\phi$ which together with (Ex 9.1.1) means $\nabla^2\phi = 0$. We've already solved this equation in cylindrical coordinates in chapter 5, (5–24). The boundary conditions require 'no φ dependence' allowing us to eliminate the $m \geq 1$ terms of (5–24) and retain only

$$\phi(r, \varphi) = C_0 \ln r + D_0 \tag{Ex 9.1.3}$$

so that $\vec{E}_0 = -\vec{\nabla}\phi$ becomes

$$\vec{E}_0 = -\frac{C_0}{r}\hat{r} \quad \text{and} \quad \vec{H}_0 = \frac{i\vec{k} \times \vec{E}_0}{i\omega\mu} = \frac{\hat{k} \times \vec{E}_0}{v\mu} = -\frac{C_0}{v\mu r}\hat{\varphi} \tag{Ex 9.1.4}$$

so that the wave is given by

$$\vec{E}(\vec{r}, t) = -\frac{C_0}{r}\hat{r}\, e^{i(kz-\omega t)} \quad \text{and} \quad \vec{H}(\vec{r}, t) = -\frac{C_0}{v\mu r}\hat{\varphi}\, e^{i(kz-\omega t)} \tag{Ex 9.1.5}$$

The constant C_0 may be evaluated from the potential on the central conductor assuming the outer conductor is grounded,

$$V_a = -\int_b^a \vec{E} \cdot d\vec{r} = -C_0 \ln\left(\frac{b}{a}\right) e^{i(kz-\omega t)} \tag{Ex 9.1.6}$$

so that finally

$$\vec{E}(\vec{r},t) = \frac{V_a\hat{r}\,e^{i(kz-\omega t)}}{r\ln(b/a)} \quad \text{and} \quad \vec{H}(\vec{r},t) = \frac{V_a\hat{\varphi}\,e^{i(kz-\omega t)}}{v\mu r\ln(b/a)} \tag{Ex 9.1.7}$$

The surface current on the inner conductor is given by $\vec{j} = \hat{r}\times\vec{H}\big|_{r=a}$ leading to

$$I = 2\pi a\vec{j} = \frac{2\pi V_a\hat{k}\,e^{i(kz-\omega t)}}{v\mu\ln(b/a)} \tag{Ex 9.1.8}$$

It is interesting to note that the TEM mode fields correspond to the electrostatic fields in the two transverse dimensions. Thus, the TEM mode electric field for the transmission line (two parallel cylindrical wires) is just the negative gradient of the potential we found in Example 6.4.

We briefly summarize the boundary conditions on E_z and H_z and the differential equations for TM and TE modes. The reader is reminded that in general the wave will propagate as a linear combination of TE and TM modes

Transverse magnetic modes: The longitudinal electric field E_z of TM modes is a solution of

$$\left[\nabla_t^2 + \left(\mu\varepsilon\omega^2 - k^2\right)\right]E_{0,z} = 0 \tag{9–30}$$

subject to the boundary condition, $E_z = 0$ at the walls, while the longitudinal magnetic field H_z vanishes everywhere (this trivially satisfies the wave equation for H_z as well as the boundary condition $\partial H_z/\partial n = 0$ at the walls). $\vec{H}$ has only transverse components.

Transverse electric modes: The longitudinal magnetic field H_z of TE modes is a solution of

$$\left[\nabla_t^2 + \left(\mu\varepsilon\omega^2 - k^2\right)\right]H_{0,z} = 0 \tag{9–31}$$

subject to the boundary condition $\partial H_{0,z}/\partial n = 0$ on the walls while $E_{0,z}$ vanishes everywhere.

EXAMPLE 9.2: Find the dispersion relation and cutoff frequencies for the TM modes of a rectangular cross section waveguide of sides a and b.

Solution: For a TM mode, $B_{0,z} = 0$ and $E_{0,z}$ satisfies (9–30)

$$\left(\nabla_t^2 + \gamma^2\right)E_{0,z} = 0 \tag{Ex 9.2.1}$$

where we have set $\gamma^2 = \mu\varepsilon\omega^2 - k^2$ for brevity. We solve for $E_{0,z}$ by separation of variables to obtain

$$E_{0,z} = \left\{\begin{matrix}\cos\alpha x\\ \sin\alpha x\end{matrix}\right\}\left\{\begin{matrix}\cos\beta y\\ \sin\beta y\end{matrix}\right\} \tag{Ex 9.2.2}$$

with $\alpha^2 + \beta^2 = \gamma^2$ and the braces are again used to indicate the arbitrary linear combination of terms enclosed. Applying the boundary conditions $E_z(x=0) = E_z(x=a) = 0$ and $E_z(y=0) = E_z(y=b) = 0$, we reduce the solution to

$$E_{0,z} = \sum_{m,n} A_{m,n}\sin\frac{n\pi x}{a}\sin\frac{m\pi y}{b} \tag{Ex 9.2.3}$$

The dispersion relation for the m, n mode then becomes

$$\omega^2 = c^2\left(\gamma^2 + k^2\right) = c^2\left(\frac{n^2\pi^2}{a^2} + \frac{m^2\pi^2}{b^2} + k^2\right) \tag{Ex 9.2.4}$$

When $\omega^2/c^2 < (n\pi/a)^2 + (m\pi/b)^2$, k becomes imaginary, leading to an exponentially damped wave. The cutoff frequency for the waveguide carrying the m, n mode is

$$\omega_c = c\sqrt{\frac{n^2\pi^2}{a^2} + \frac{m^2\pi^2}{b^2}} \tag{Ex 9.2.5}$$

Given $E_{0,z}$ above, and $H_{0,z} = 0$, the remaining components of the field are easily obtained from (9–23) and (9–24) as

$$\vec{E}_{0,t} = \frac{ik}{\gamma^2}\vec{\nabla}_t E_{0,z} \quad \text{and} \quad \vec{H}_{0,t} = \frac{i\omega\varepsilon}{\gamma^2}\hat{k}\times\vec{\nabla}_t E_{0,z} \tag{Ex 9.2.6}$$

9.2.1 Circular Cylindrical Waveguides

For circular cross section waveguides we express the z components of the fields in terms of the polar coordinates, r and φ. As before, $E_{0,z}(r,\varphi)$ and $H_{0,z}(r,\varphi)$ satisfy the equation

$$\left(\nabla_t^2 + \gamma^2\right)\begin{pmatrix} E_{0,z} \\ H_{0,z} \end{pmatrix} = 0 \tag{9–32}$$

with $\gamma^2 = \mu\varepsilon\omega^2 - k^2$. Letting ψ represent $E_{0,z}$ for TM modes or $H_{0,z}$ for TE modes, we write in cylindrical (polar) coordinates

$$\left[\frac{1}{r}\frac{\partial}{\partial r}\left(r\frac{\partial}{\partial r}\right) + \frac{1}{r^2}\frac{\partial^2}{\partial\varphi^2} + \gamma^2\right]\psi = 0 \tag{9–33}$$

Separating variables and applying a periodic boundary condition to the azimuthal component, we find $\psi = \sum_m \mathrm{R}_m(r)e^{\pm im\varphi}$, with m an integer. The radial function R_m satisfies Bessel's equation (E–1)

$$r^2\frac{d^2\mathrm{R}_m}{dr^2} + r\frac{d\mathrm{R}_m}{dr} + \left(\gamma^2 r^2 - m^2\right)\mathrm{R}_m = 0 \tag{9–34}$$

whose general solution is of the form

$$\mathrm{R}_m(r) = A_m\mathrm{J}_m(\gamma r) + B_m\mathrm{N}_m(\gamma r) \tag{9–35}$$

Since R_m must be well behaved at $r = 0$, we set the expansion constants B_m to zero, giving

$$\psi = \sum_m A_m\mathrm{J}_m(\gamma r)e^{\pm im\varphi} \tag{9–36}$$

TM modes: For TM modes, $B_{0,z} = 0$ and $E_{0,z} = \psi$. At $r = a$, the waveguide wall, $E_{0,z}$ must vanish, implying that $\mathrm{J}_m(\gamma a) = 0$. The argument, γa, must therefore be a root of the Bessel function J_m, say the ith root we designate by $\rho_{m,i}$. Substituting for γ, we obtain the z component of the electric field for the (m, i) mode

$$E_{0,z}(r,\varphi) = A J_m\left(\frac{\rho_{m,i}}{a} r\right) e^{\pm im\varphi} \tag{9-37}$$

The transverse fields $\vec{E}_{0,t}$ and $\vec{H}_{0,t}$ may be obtained from (9–24) and (9–23):

$$\vec{E}_{0,t} = \frac{ik}{\varepsilon\mu\omega^2 - k^2}\vec{\nabla}_t E_{0,z} = \frac{ik}{\gamma^2}\vec{\nabla}_t E_{0,z} = \frac{ika^2}{\rho_{m,i}^2}\vec{\nabla}_t E_{0,z} \tag{9-38}$$

and

$$\vec{H}_{0,t} = \frac{i\varepsilon\omega}{\varepsilon\mu\omega^2 - k^2}(\hat{k} \times \vec{\nabla}_t E_{0,z}) = \frac{i\varepsilon\omega a^2}{\rho_{m,i}^2}(\vec{k} \times \vec{\nabla}_t E_{0,z}) \tag{9-39}$$

The dispersion relation reads $k^2 = \varepsilon\mu\omega^2 - \gamma^2 = \varepsilon\mu\omega^2 - \rho_{m,i}^2/a^2$. When ω^2 becomes less than $\rho_{m,i}^2/\varepsilon\mu a^2 \equiv \omega_c^2$, k becomes imaginary and the wave no longer propagates.

The $TM_{0,1}$ mode in particular (see Appendix E, Table E.1 for the first few roots of J_0 to J_3.) has $\rho_{0,1} \simeq 2.405$. The fields are then, using $J_0' = -J_1$.

$$\begin{aligned} E_{0,z} &= AJ_0\left(\frac{\rho_{0,1}r}{a}\right) & H_{0,z} &= 0 \\ \vec{E}_{0,t} &= -\frac{ikaA}{\rho_{0,1}}J_1\left(\frac{\rho_{0,1}r}{a}\right)\hat{r} & \vec{H}_{0,t} &= \frac{i\omega\varepsilon aA}{\rho_{0,1}}J_1\left(\frac{\rho_{0,1}r}{a}\right)\hat{\varphi} \end{aligned} \tag{9-40}$$

With this root, $\omega_c = \rho_{0,1}/a\sqrt{\mu\varepsilon}$. For $\mu = \mu_0$ and $\varepsilon = \varepsilon_0$, the cutoff wave-length is $\lambda_c = 2\pi c/\omega_c = 2\pi a/\rho_{0,1} = 2.61a$.

TE modes: TE modes have $E_{0,z} = 0$ and $H_{0,z} = AJ_m(\gamma r)e^{\pm im\varphi}$. The boundary condition is $\partial H_{0,z}/\partial n = -\partial H_{0,z}/\partial r = 0$ at $r = a$, which implies that $J'(\gamma a) = 0$. Therefore, γa must be a root of J_m', which we denote by $\rho_{m,i}'$. The z component of the magnetic field intensity is then

$$H_{0,z} = AJ_m\left(\frac{\rho_{m,i}' r}{a}\right)e^{\pm im\varphi} \tag{9-41}$$

and $E_{0,z} = 0$, leading to the transverse fields

$$\vec{H}_{0,t} = \frac{ika^2}{\rho_{m,i}'^2}\vec{\nabla}_t H_{0,z} \quad \text{and} \quad \vec{E}_{0,t} = -\frac{i\mu\omega a^2}{\rho_{m,i}'^2}\left(\hat{k} \times \vec{\nabla}_t H_{0,z}\right) \tag{9-42}$$

The smallest root of all the Bessel function derivatives J_m', is the first root of J_1', $\rho_{1,1}' = 1.84118\ldots$ (see Table E.2). The mode with the lowest cut-off frequency is therefore the $TE_{1,1}$ mode. Specializing to the $TE_{1,1}$ mode, we get

$$H_{0,z} = AJ_1\left(\frac{\rho_{1,1}' r}{a}\right) \tag{9-43}$$

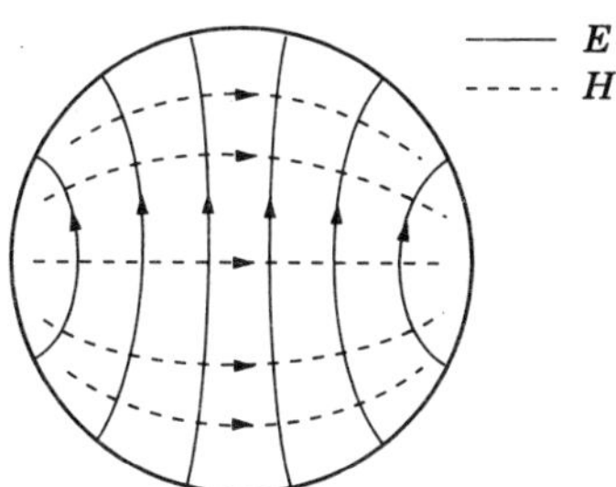

Figure 9.4: Field pattern of the $\mathrm{TE}_{1,1}$ mode in a circular cylindrical waveguide.

$$\vec{H}_{0,t} = \frac{ika^2 A}{\rho'^2_{1,1}} \left[\pm \frac{i\hat{\varphi}}{r} \mathrm{J}_1 \left(\frac{\rho'_{1,1} r}{a} \right) e^{\pm i\varphi} + \frac{1.841 \hat{r}}{a} \mathrm{J}'_1 \left(\frac{\rho'_{1,1} r}{a} \right) e^{\pm i\varphi} \right] \tag{9–44}$$

and

$$E_{0,z} = 0 \tag{9–45}$$

$$\vec{E}_{0,t} = -\frac{i\mu\omega a^2 A}{\rho'^2_{1,1}} \left[\mp \frac{i\hat{r}}{r} \mathrm{J}_1 \left(\frac{\rho'_{1,1} r}{a} \right) e^{\pm i\varphi} + \frac{\rho'_{1,1} \hat{\varphi}}{a} \mathrm{J}'_1 \left(\frac{\rho'_{1,1} r}{a} \right) e^{\pm i\varphi} \right] \tag{9–46}$$

Taking the real part of $E_{0,t}$ and $B_{0,t}$, we obtain the field pattern sketched in Figure 9.4. The cutoff frequency and cutoff wavelength of this, the lowest mode, are

$$\omega_c = \frac{1}{\sqrt{\mu\varepsilon}} \frac{\rho'_{1,1}}{a}, \qquad \lambda_c = \frac{2\pi a}{\rho'_{1,1}} = 3.41a \tag{9–47}$$

Because $\mathrm{J}'_0 = -\mathrm{J}_1$, the roots of J'_0 coincide with those of J_1, leading to a degeneracy of the $\mathrm{TM}_{1,i}$ modes and $\mathrm{TE}_{0,i}$ modes.

9.2.2 Resonant Cavities

The modes of resonant cavities are easily obtained as linear combinations of waves travelling in the $\pm z$ direction. For perfectly conducting plates at $z = 0$ and $z = d$ we find now additional boundary conditions on $\vec{H}$ and $\vec{E}$, namely $\vec{E}_{0,t} = 0$ and $H_z = 0$. The latter implies for TE modes that

$$H_{0,z} = \psi(x,y) \sin \frac{n\pi z}{d} \tag{9–48}$$

For TM modes, $\vec{E}_{0,t} = 0$ on the end walls give $\partial E_z/\partial n = \pm \partial E_z/\partial z = 0$ using an argument similar to that used in Section 9.2 for treating $B_\perp = 0$ on the side walls. For a TM mode, then, we have

$$E_{0,z} = \psi(x,y) \cos \frac{n\pi z}{d} \tag{9–49}$$

Formulas (9–23) and (9–24) made explicit use of the form $\vec{E} = \vec{E}_0(x,y)e^{i(kz-\omega t)}$, whereas now we have a standing wave of the form $(e^{ikz} \pm e^{-ikz})\, e^{-i\omega t}$. Noting that the ik arose from differentiating e^{ikz} with respect to z, we may replace ik by $\partial/\partial z$ to make the result valid for $\pm k$. The fields in a cavity are thus

$$\vec{H}_{0,\,t} = \frac{1}{\varepsilon\mu\omega^2 - k^2}\left(\vec{\nabla}_t \frac{\partial H_{0,z}}{\partial z} + i\omega\varepsilon\hat{k} \times \vec{\nabla}_t E_{0,z}\right) \tag{9–50}$$

$$\vec{E}_{0,t} = \frac{1}{\varepsilon\mu\omega^2 - k^2}\left(\vec{\nabla}_t \frac{\partial E_{0,z}}{\partial z} - i\mu\omega\hat{k} \times \vec{\nabla}_t H_{0,z}\right) \tag{9–51}$$

The transverse fields in the cavity become:

TM:

$$\vec{E}_{0,t} = -\frac{n\pi}{d\gamma^2}\sin\frac{n\pi z}{d}\vec{\nabla}_t\psi(x,y) \tag{9–52}$$

$$\vec{H}_{0,t} = \frac{i\omega\varepsilon}{\gamma^2}\cos\frac{n\pi z}{d}\hat{k} \times \vec{\nabla}_t\psi(x,y) \tag{9–53}$$

TE:

$$\vec{E}_{0,t} = -\frac{i\mu\omega}{\gamma^2}\sin\frac{n\pi z}{d}\hat{k} \times \vec{\nabla}_t\psi(x,y) \tag{9–54}$$

$$\vec{H}_{0,t} = \frac{n\pi}{d\gamma^2}\cos\frac{n\pi z}{d}\vec{\nabla}_t\;\psi(x,y) \tag{9–55}$$

with $\gamma^2 = \mu\varepsilon\omega^2 - (n\pi/d)^2$.

The circular cylindrical cavity $\text{TM}_{\ell,i,n}$ mode has $\gamma = \rho_{\ell,i}/a$ and a resonant frequency

$$\omega_{\ell,i,n} = \frac{1}{\sqrt{\mu\varepsilon}}\sqrt{\left(\frac{\rho_{\ell,i}}{a}\right)^2 + \left(\frac{n\pi}{d}\right)^2} \tag{9–56}$$

while a $\text{TE}_{\ell,i,n}$ mode has

$$\omega_{\ell,i,n} = \frac{1}{\sqrt{\mu\varepsilon}}\sqrt{\left(\frac{\rho'_{\ell,i}}{a}\right)^2 + \left(\frac{n\pi}{d}\right)^2} \tag{9–57}$$

The resonant frequency of the $\text{TM}_{\ell,i,0}$ mode is independent of the cavity length, while all others depend on d. For tunable cavities, the $\text{TE}_{1,0,n}$ mode is usually chosen. For this mode,

$$\begin{aligned}\vec{H}_{0,t} &= \frac{n\pi}{d}\frac{a^2}{\rho'^2_{1,0}}\cos\frac{n\pi z}{d}\vec{\nabla}_t \mathrm{J}_0\left(\frac{\rho'_{1,0}r}{a}\right)\\ &= -\frac{n\pi a\hat{r}}{\rho'_{1,0}d}\mathrm{J}_1\left(\frac{\rho'_{1,0}r}{a}\right)\cos\frac{n\pi z}{d}\end{aligned} \tag{9–58}$$

The surface current density on the walls is $\hat{n} \times \vec{H}$. On the end faces, $\hat{n} = \pm\hat{k}$ so that the surface current is

$$\vec{j} = \mp\frac{n\pi}{d}\begin{pmatrix}\cos 0\\ \cos n\pi\end{pmatrix}\frac{a}{\rho'_{1,0}}\mathrm{J}_1\left(\frac{\rho'_{1,0}r}{a}\right)\hat{\varphi} \tag{9–59}$$

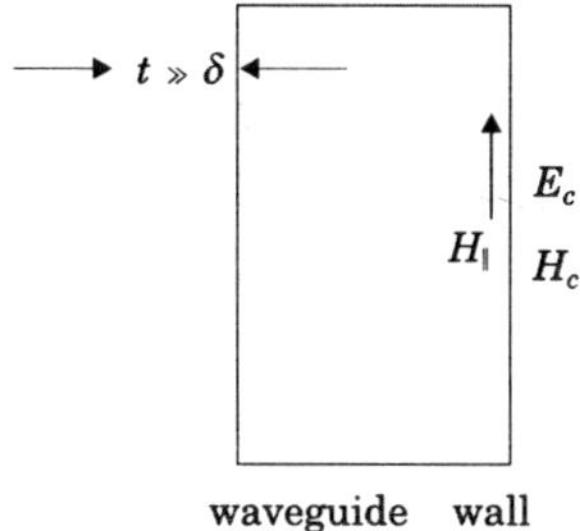

Figure 9.5: The fields penetrating into the wall generate eddy currents leading to resistive energy loss.

We see that there are no radial currents on the end plates, which means that the cavity may be tuned by moving the end plate without a need for a good electrical contact between the movable plates and the wall.

The modes of a resonant cavity may, of course, also be obtained by solving the wave equation in the appropriate coordinate system with the boundary conditions $\vec{E}_{\|} = 0$ and $\vec{H}_{\perp} = 0$ as well as the coupling equations between $\vec{E}$ and $\vec{H}$.

9.2.3 Dissipation by Eddy Currents

In real conductors, the field does not decrease to zero instantaneously at the boundary with the conductor. Instead, it decays over several skin depths δ. Solving the exact boundary condition problem would be onerous and almost always unnecessary since the mode pattern would hardly be disturbed by the finite conductivity. (The apparent dimensions of the waveguide would increase slightly, by a distance of order δ.) The only significant change brought by the finite conductivity is the resistive dissipation of energy by the eddy currents induced by H near the wall. To calculate the power loss, we assume that the fields are adequately described by the perfectly conducting wall modes in the interior of the guide, but decay exponentially in the metal walls. The subscript c will be used to denote the fields in the conductor.

Near the surface, the electric field in the wall is related to the eddy current density $\vec{J}$ by $\vec{J} = g\vec{E}_c$, leading to a power loss in the volume τ:

$$\mathbf{P} = \tfrac{1}{2}\int_\tau \vec{J}\cdot\vec{E}_c^{*}d^3r = \tfrac{1}{2}\int g|E_c|^2 dxdydz \tag{9–60}$$

The rate at which power, $\mathbf{P}$, is lost by the wave as it moves in the z direction is then

$$\frac{d\mathbf{P}}{dz} = \tfrac{1}{2}\int_S g|E_c|^2 dxdy \tag{9–61}$$

where S is a cross section of the waveguide wall shown darkened in figure 9.5. In the conductor, the electric field is related to the magnetic field intensity $\vec{H}_c$ by

$\vec{H}_c = \sqrt{g/\mu\omega}(\hat{n} \times \vec{E}_c)e^{i\pi/4}$. Then $g|E_c|^2 = \mu\omega|H_c|^2$, and we can also write the power loss integral as

$$\frac{d\mathbf{P}}{dz} = \frac{\mu\omega}{2}\int_S |H_c|^2 dx dy \tag{9–62}$$

As at the wall $\vec{E}$ has vanishing parallel component, the perpendicular component of $\vec{H}$ is nearly zero, while the parallel component of $\vec{H}$ diminishes exponentially as it enters the metal.

We specialize, for convenience, to the rectangular guide with walls of thickness t shown in Figure 9.5. We assume $t \gg \delta$. The surface integral can be split into four separate integrals, one for each of the sides. Providing that the skin depth is much smaller than the dimensions of the guide, we may neglect the contribution from the corners. Let us look in detail at the contribution to the loss from one of those sides, say the wall at $x = a$. Inside this wall, the magnetic field H_c takes the form $H_c = H_{\|}e^{-x/\delta}$, where we have used the continuity of $H_{\|}$ at the interface. The power loss at this ($x = a$) wall for $t \gg \delta$ is then

$$\begin{aligned}\left.\frac{d\mathbf{P}}{dz}\right|_{\substack{x=a\\ \text{wall}}} &= \frac{\mu\omega}{2}\int_{y=0}^{b}\int_{a}^{a+t} |H_{\|}|^2 e^{-2x/\delta} dx dy\\ &= \frac{\mu\omega\delta}{4}\int |H_{\|}|^2 dy\end{aligned} \tag{9–63}$$

Each of the four sides makes a similar contribution, whence we conclude

$$\frac{d\mathbf{P}}{dz} = \frac{\mu\omega\delta}{4}\oint |H_{\|}|^2 d\ell \tag{9–64}$$

where the integration is carried out around the periphery of the waveguide. $H_{\|}$ in the integral may of course be replaced by the fictitious perfect-conductor surface current j, using $|\vec{j}\,|^2 = |\hat{n} \times \vec{H}_{\|}|^2 = |H_{\|}|^2$. Making this substitution, we obtain

$$\frac{d\mathbf{P}}{dz} = \frac{\mu\omega\delta}{4}\oint |j|^2 d\ell \tag{9–65}$$

Further, for $g \gg \varepsilon\omega$ we have from (8–134), $\delta^2 = 2/g\mu\omega$. The power dissipation rate then becomes

$$\frac{d\mathbf{P}}{dz} = \frac{1}{2g\delta}\oint |j|^2 d\ell \tag{9–66}$$

There is a simple interpretation to this result. Pretending the current j is spread evenly over a strip of width δ, and zero beyond it, we have $J_{equiv.} = j/\delta$. The power loss rate would then have been calculated as

$$\frac{d\mathbf{P}}{dz} = \frac{1}{2}\oint \vec{J}\cdot\vec{E}\,\delta\, d\ell = \frac{1}{2g}\oint |J|^2\delta\, d\ell = \frac{1}{2g\delta}\oint |\,j\,|^2\, d\ell \tag{9–67}$$

just the result (9–66) above.

The results above are of course not limited to waveguides of rectangular cross section but apply equally to the circular guides already discussed.

EXAMPLE 9.3: Calculate the power transported by a $\text{TE}_{n,m}$ wave in a rectangular waveguide of dimensions $a \times b$. Estimate the power loss in the walls, assuming ohmic losses. Use these results to find the attenuation (relative power loss per unit length) of the waveguide.

Solution: The $\text{TE}_{n,m}$ mode has $E_z = 0$ and

$$H_z = H_0 \cos\frac{n\pi x}{a}\cos\frac{m\pi y}{b} \tag{Ex 9.3.1}$$

We begin by finding the remaining components of the fields in order that we may compute $\vec{S} = \frac{1}{2}\vec{E}\times\vec{H}^*$. Using (9–23) and (9–24) we obtain

$$\begin{aligned}\vec{H}_t &= \frac{ik}{\gamma^2}\vec{\nabla}_t H_z \\ &= \left[-\frac{ik}{\gamma^2}\left(\frac{n\pi}{a}\right)\sin\frac{n\pi x}{a}\cos\frac{m\pi y}{b}\right]\hat{\imath} + \left[-\frac{ik}{\gamma^2}\left(\frac{m\pi}{b}\right)\cos\frac{n\pi x}{a}\sin\frac{m\pi y}{b}\right]\hat{\jmath}\end{aligned} \tag{Ex 9.3.2}$$

and

$$\vec{E}_t = \frac{i\omega\mu}{\gamma^2}\hat{k}\times\vec{\nabla}_t H_z \tag{Ex 9.3.3}$$

Rather than substitute these forms immediately, we first simplify

$$\begin{aligned}\vec{E}\times\vec{H}^* &= -\frac{\mu\omega k}{\gamma^4}\left(\hat{k}\times\vec{\nabla}_t H_z\right)\times\vec{\nabla}_t H_z^* \\ &= -\frac{\mu\omega k}{\gamma^4}\left[\left(\hat{k}\cdot\vec{\nabla}_t H_z^*\right)\vec{\nabla}_t H_z - \hat{k}\left(\vec{\nabla}_t H_z^*\cdot\vec{\nabla}_t H_z\right)\right] \\ &= \frac{\mu\omega k}{\gamma^4}\hat{k}|\vec{\nabla}_t H_z|^2\end{aligned} \tag{Ex 9.3.4}$$

The Poynting vector of the fields in the guide,

$$\vec{S} = \frac{\mu\omega k H_0^2}{2\gamma^4}\hat{k}\left[\left(\frac{n\pi}{a}\right)^2\sin^2\frac{n\pi x}{a}\cos^2\frac{m\pi y}{b} + \left(\frac{m\pi}{b}\right)^2\cos^2\frac{n\pi x}{a}\sin^2\frac{m\pi y}{b}\right] \tag{Ex 9.3.5}$$

is readily integrated over the cross section of the waveguide to give the power transported:

$$\begin{aligned}\mathbf{P} &= \frac{\mu\omega k H_0^2}{2\gamma^4}\int_{x=0}^{a}\int_{y=0}^{b} dx dy\left[\left(\frac{n\pi}{a}\right)^2\sin^2\frac{n\pi x}{a}\cos^2\frac{m\pi y}{b}\right. \\ &\qquad\qquad \left. + \left(\frac{m\pi}{b}\right)^2\cos^2\frac{n\pi x}{a}\sin^2\frac{m\pi y}{b}\right] \\ &= \frac{\mu\omega k H_0^2}{2\gamma^4}\frac{ab}{4}\left[\left(\frac{n\pi}{a}\right)^2 + \left(\frac{m\pi}{b}\right)^2\right]\end{aligned}$$

$$= \frac{\mu\omega k H_0^2 ab}{8\gamma^2} \tag{Ex 9.3.6}$$

We now proceed to find the power dissipated in travelling along the waveguide. To do so, we require $H_\|$ at each of the walls. Along the bottom wall, which we take to be the x axis, $y = 0$, making

$$\vec{H}_\| = H_0 \left[\hat{k}\cos\frac{n\pi x}{a} - \hat{\imath}\frac{ik}{\gamma^2}\left(\frac{n\pi}{a}\right)\sin\frac{n\pi x}{a}\right] \tag{Ex 9.3.7}$$

The power dissipation along this ($y = 0$) wall is

$$\left.\frac{d\mathbf{P}}{dz}\right|_{y=0} = \frac{1}{2g\delta}\int |H_\||^2 dx$$

$$= \frac{H_0^2}{2g\delta}\int_0^a dx \left[\cos^2\frac{n\pi x}{a} + \frac{k^2}{\gamma^4}\left(\frac{n\pi}{a}\right)^2 \sin^2\frac{n\pi x}{a}\right]$$

$$= \frac{H_0^2}{2g\delta}\left\{\frac{a}{2}\left[1 + \frac{k^2}{\gamma^4}\left(\frac{n\pi}{a}\right)^2\right]\right\} \tag{Ex 9.3.8}$$

and the top wall gives an identical result.

On the side wall at $x = 0$, we have

$$\vec{H}_\| = H_0 \left[\hat{k}\cos\frac{m\pi y}{b} - \hat{\jmath}\frac{ik}{\gamma^2}\left(\frac{m\pi}{b}\right)\sin\frac{m\pi y}{b}\right] \tag{Ex 9.3.9}$$

and the power dissipated on the side wall is

$$\left.\frac{d\mathbf{P}}{dz}\right|_{x=0} = \frac{H_0^2}{2g\delta}\int_0^b dy \left[\cos^2\frac{m\pi y}{b} + \frac{k^2}{\gamma^4}\left(\frac{m\pi}{b}\right)^2 \sin^2\frac{m\pi y}{b}\right]$$

$$= \frac{H_0^2}{2g\delta}\left\{\frac{b}{2}\left[1 + \frac{k^2}{\gamma^4}\left(\frac{m\pi}{b}\right)^2\right]\right\} \tag{Ex 9.3.10}$$

Again, the right hand wall at $x = a$ gives the same result. Adding all four terms, we have

$$\frac{d\mathbf{P}}{dz} = \frac{H_0^2}{2g\delta}\left[a + b + \frac{\pi^2 k^2}{\gamma^4}\left(\frac{n^2}{a} + \frac{m^2}{b}\right)\right] \tag{Ex 9.3.11}$$

Finally, we compute the relative power loss as the wave proceeds down the guide:

$$\frac{\frac{d\mathbf{P}}{dz}}{\mathbf{P}} = \frac{8\gamma^2}{\mu\omega k H_0^2 ab}\frac{H_0^2}{2g\delta}\left[a + b + \frac{\pi^2 k^2}{\gamma^4}\left(\frac{n^2}{a} + \frac{m^2}{b}\right)\right]$$

$$= \frac{2\delta}{\gamma^2 kab}\left[(a+b)\gamma^4 + \pi^2 k^2\left(\frac{n^2}{a} + \frac{m^2}{b}\right)\right] \tag{Ex 9.3.12}$$

with $\gamma^2 = \left(n\pi/a\right)^2 + \left(m\pi/b\right)^2$.

Similar methods may be employed to find the resistive losses in a cavity. It should be evident that the electromagnetic energy contained in a cavity is proportional to the volume, but for given energy density, the losses increase in proportion to the surface area, leading us to conclude that the Q of the cavity increases linearly with the dimensions of the cavity. (The Q, or quality, of an oscillator may be defined as $2\pi\times$ the energy stored divided by the energy loss per cycle.)

9.3 Dielectric Waveguides (Optical Fibers)

Metallic waveguides are not alone in providing low-loss conduits for electromagnetic radiation. Dielectric waveguides have in recent years gained great use as wideband, low-loss carriers of electromagnetic waves for communication. Metallic waveguides, being dispersive and useful only at microwave frequencies, have rather limited information-carrying capacity compared to optical fibers operating at typically 2×10^{14} Hz.

In its simplest form, a large (compared to the wavelength, λ) diameter fiber can be thought of confining the wave by successive total internal reflections. It will be evident that the wave is not truly confined as the evanescent wave penetrates the medium surrounding the fiber. A homogeneous fiber of this type is very dispersive as each reflection gives a phase change to the wave; in addition, waves travelling at different angles have different longitudinal velocities along the fiber. Some improvements can be made by giving the fiber a radial refractive index gradient, but the cost and lack of flexibility of thick fibers makes them less than desirable. The current emphasis is on "monomode" fibers whose radius is of the order of the wavelength of the waves it carries. In these fibers, much of the electromagnetic field can reside outside the fiber, making the fiber much more a guide than a conduit.

The relationships (9–23) and (9–24) between the longitudinal and transverse fields did not depend on anything but the form $e^{i(kz-\omega t)}$ of the travelling wave. However, the boundary conditions that the fields inside a dielectric fiber must satisfy lead to somewhat greater complication than the metal interface boundary conditions.

The wave equation for $E_{0,z}$ and$H_{0,z}$ is

$$\left[\nabla_t^2 + (\mu_1\varepsilon_1\omega^2 - k^2)\right]\begin{pmatrix} E_{0,z} \\ H_{0,z} \end{pmatrix} = 0 \quad \text{inside the fiber} \tag{9–68}$$

and

$$\left[\nabla_t^2 + (\mu_0\varepsilon_0\omega^2 - k^2)\right]\begin{pmatrix} E_{0,z} \\ H_{0,z} \end{pmatrix} = 0 \quad \text{outside} \tag{9–69}$$

A cladding other than vacuum is easily accommodated by replacing ε_0 and μ_0 by ε_2 and μ_2. The wave equations above must now be solved subject to the boundary conditions that $B_{0,\perp}$, $E_{0,\parallel}$, $D_{0,\perp}$, and $H_{0,\parallel}$ all be continuous across the interface.

We substitute $+\gamma^2$ for $\mu_1\varepsilon_1\omega^2 - k^2$ and, to avoid an oscillatory solution and hence a radial outflow of energy (recall that oscillatory fields lead to radiation) outside the cylinder, we must choose $\mu_0\varepsilon_0\omega^2 - k^2 = -\beta^2$.

Because of the more involved boundary conditions, the fields do not generally separate into TE and TM modes except in special circumstances such as azimuthal symmetry. For simplicity we consider first the propagation of a wave having no φ dependence, along a circular cross section fiber of radius a.

Letting

$$\psi = \begin{cases} E_{0,z} & \text{for TM modes} \\ H_{0,z} & \text{for TE modes} \end{cases} \tag{9–70}$$

we express (9–68) and (9–69) in polar coordinates and recognize the resulting equation as Bessel's equation and the modified Bessel equation, respectively, having well-behaved solutions

$$\psi = \begin{cases} \mathrm{J}_0(\gamma r) & r \le a \\ A\mathrm{K}_0(\beta r) & r \ge a \end{cases} \tag{9–71}$$

where only a single arbitrary constant, A, is required to match the interior and exterior solutions. The transverse fields are then readily obtained. Inside the cylinder

$$\begin{aligned} H_{0,r} &= \frac{ik}{\gamma^2}\frac{\partial H_{0,z}}{\partial r} & \qquad H_{0,\varphi} &= \frac{i\varepsilon_1\omega}{\gamma^2}\frac{\partial E_{0,z}}{\partial r} \\ E_{0,r} &= \frac{ik}{\gamma^2}\frac{\partial E_{0,z}}{\partial r} & \qquad E_{0,\varphi} &= -\frac{i\mu_1\omega}{\gamma^2}\frac{\partial H_{0,z}}{\partial r} \end{aligned} \tag{9–72}$$

Outside the cylinder, similar relations hold with $-\beta^2$ replacing γ^2. Explicitly, noting that $\mathrm{J}_0'(z) = -\mathrm{J}_1(z)$ and $\mathrm{K}_0'(z) = -\mathrm{K}_1(z)$, the fields for the TE modes are:

$$\left.\begin{aligned} H_{0,z} &= \mathrm{J}_0(\gamma r) \\ H_{0,r} &= -\frac{ik}{\gamma}\mathrm{J}_1(\gamma r) \\ E_{0,\varphi} &= \frac{i\mu_1\omega}{\gamma}\mathrm{J}_1(\gamma r) \end{aligned}\right\} r \le a \qquad \left.\begin{aligned} H_{0,z} &= A\mathrm{K}_0(\beta r) \\ H_{0,r} &= \frac{ikA}{\beta}\mathrm{K}_1(\beta r) \\ E_{0,\varphi} &= -\frac{i\mu_0\omega A}{\beta}\mathrm{K}_1(\beta r) \end{aligned}\right\} r \ge a \tag{9–73}$$

At the boundary $r = a$, the fields must satisfy

$$\begin{aligned} B_\perp = B_r \text{ is continuous} &\Rightarrow -\frac{ik\mu_1\mathrm{J}_1(\gamma a)}{\gamma} = \frac{ik\mu_0 A\mathrm{K}_1(\beta a)}{\beta} \\ H_\parallel = H_z \text{ is continuous} &\Rightarrow \mathrm{J}_0(\gamma a) = A\mathrm{K}_0(\beta a) \\ E_\parallel = E_\varphi \text{ is continuous} &\Rightarrow \frac{i\omega\mu_1\mathrm{J}_1(\gamma a)}{\gamma} = -\frac{i\omega\mu_0 A\mathrm{K}_1(\beta a)}{\beta} \end{aligned} \tag{9–74}$$

The first and the last of these conditions merely repeat one another. Eliminating the constant A from the first two equations, we obtain

$$\frac{\mu_1\mathrm{J}_1(\gamma a)}{\mu_0\gamma\mathrm{J}_0(\gamma a)} + \frac{\mathrm{K}_1(\beta a)}{\beta\mathrm{K}_0(\beta a)} = 0 \tag{9–75}$$

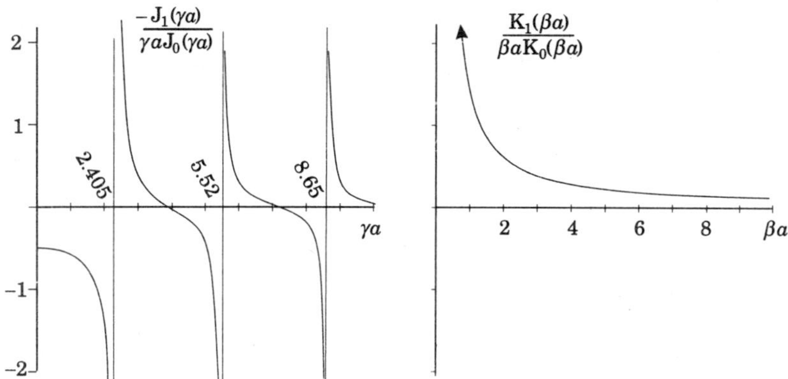

Figure 9.6: $-J_1(\gamma a)/\gamma a J_0(\gamma a)$ and $K_1(\beta a)/\beta a K_0(\beta a)$ plotted as a function of their arguments.

while from the definition of γ and β, $\gamma^2 + \beta^2 = (\varepsilon_1\mu_1 - \varepsilon_0\mu_0)\omega^2$ implies that $|\gamma|$ and $|\beta|$ are each less than $\omega\sqrt{\varepsilon_1\mu_1 - \varepsilon_0\mu_0}$. Considering $\beta = \beta(\gamma)$, we can solve the equations graphically. As an intermediate step, we first sketch $-J_1(\gamma a)/\gamma J_0(\gamma a)$ and $K_1(\beta a)/\beta K_0(\beta a)$ as a function of γa and βa, respectively, as shown in Figure 9.6.

Clearly, J_1/J_0 diverges and changes sign at roots of J_0. Moreover, since $J_1(x) \to x/2$ for small x, the ratio $-J_1(x)/xJ_0(x)$ will go to $-\frac{1}{2}$ for small x.

The modified Bessel function $K_n(x)$ behaves like a decreasing exponential at large z, but as $z \to 0$, $K_n(z) \to \frac{1}{2}(n+1)!(2/z)^n$ for $n \geq 1$, while $K_0(z) \to -\ln(z/2)$. Then $K_1(z)/zK_0(z) \to 1/z^2[-\ln(z/2)] \to \infty$ as $z \to 0$.

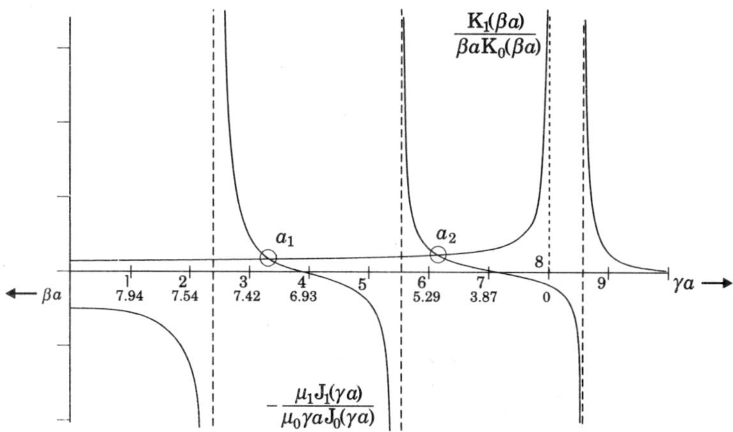

Figure 9.7: The intersection of the curves mark the roots corresponding to mode eigenvalues $\gamma_{n,0}$ for the $TE_{n,0}$ modes of a cylindrical dielectric waveguide.

To solve (9–75) we superimpose the two graphs of Figure 9.6 with a common argument γa. As γa increases to a maximum of $\omega a\sqrt{\varepsilon_1\mu_1 - \varepsilon_0\mu_0}$, βa decreases to zero. We plot both $-\mathrm{J}_1/\gamma\mathrm{J}_0$ and $\mathrm{K}_1/\beta\mathrm{K}_0$ as a function of γa in Figure 9.7. The graph is plotted here for $\omega a = 8$, sufficiently large that two roots of $\mathrm{J}_1/\gamma\mathrm{J}_0 + \mathrm{K}_1/\beta\mathrm{K}_0 = 0$ exist. Corresponding to these two roots are two modes having $\gamma a = \alpha_1$ and $\gamma a = \alpha_2$ (and, of course, corresponding β).

If the value of $\gamma a_{max} = \omega a\sqrt{\varepsilon_1\mu_1 - \varepsilon_0\mu_0}$ is less than 2.405, the first root of J_0, then no solutions exist. The cutoff frequency for $\mathrm{TE}_{n,0}$ modes is given by $\omega a = \alpha_n$. At this frequency, $\beta = 0$, giving $k = \omega\sqrt{\mu_0\varepsilon_0}$, the free space value. Below this frequency, β becomes imaginary, making K_0 and K_1 into Bessel functions that oscillate radially and decrease as $1/\sqrt{r}$. The fiber no longer acts as a guide, but instead as an antenna.

For frequencies well above cutoff, the curves of Figure 9.7 intersect near the roots of J_1. The dispersion relation then gives $\mu_1\varepsilon_1\omega^2 = \gamma^2 + k^2 \approx k^2$. The wave propagates at the speed of light appropriate (c/n) to the dielectric. The dispersion relation outside the dielectric gives $\beta^2 = k^2 - \mu_0\varepsilon_0\omega^2 \approx k^2(1 - 1/n^2)$, suggesting that the wave dies off in a distance of order k^{-1} away from the fiber.

As we will see in the next section, monomode communication fibers are normally operated in the $\mathrm{HE}_{1,1}$ mode which has no low frequency cut-off. Diminishing the difference of refractive index of the cladding and the fiber lowers the cut-off frequency for the TE and TM modes, so that the need for microscopic fibres to eliminate the higher order modes is relaxed.

⋆ 9.3.1 HE Modes

The most general modes propagating along dielectrics are called HE modes. They have both $E_z \neq 0$ and $H_z \neq 0$. Although algebraically somewhat more cumbersome than TE and TM modes, the $\mathrm{HE}_{1,1}$ mode is of special interest because it has no low frequency cutoff and very little dispersion.

For a cylindrical fiber of radius a, we have for $r \leq a$ as general wave solution

$$E_{0,z} = A_m\mathrm{J}_m(\gamma r)e^{im\varphi} \quad \text{and} \quad H_{0,z} = B_m\mathrm{J}_m(\gamma r)e^{im\varphi} \tag{9–76}$$

leading to the following transverse field components:

$$E_{0,r} = \left[\frac{ik}{\gamma}A_m\mathrm{J}'_m(\gamma r) - \frac{\omega m\mu_1}{\gamma^2 r}B_m\mathrm{J}_m(\gamma r)\right]e^{im\varphi} \tag{9–77}$$

$$H_{0,r} = \left[\frac{ik}{\gamma}B_m\mathrm{J}'_m(\gamma r) + \frac{\omega m\varepsilon_1}{\gamma^2 r}A_m\mathrm{J}_m(\gamma r)\right]e^{im\varphi} \tag{9–78}$$

$$E_{0,\varphi} = \left[-\frac{km}{\gamma^2 r}A_m\mathrm{J}_m(\gamma r) - \frac{i\omega\mu_1}{\gamma}B_m\mathrm{J}'_m(\gamma r)\right]e^{im\varphi} \tag{9–79}$$

$$H_{0,\varphi} = \left[-\frac{km}{\gamma^2 r}B_m\mathrm{J}_m(\gamma r) + \frac{i\omega\varepsilon_1}{\gamma}A_m\mathrm{J}'_m(\gamma r)\right]e^{im\varphi} \tag{9–80}$$

Similarly, for $r > a$, we choose again the exponentially decaying solutions:

$$E_{0,z} = C_m\mathrm{K}_m(\beta r)e^{im\varphi} \quad \text{and} \quad H_{0,z} = D_m\mathrm{K}_m(\beta r)e^{im\varphi} \tag{9–81}$$

where $-\beta^2 = \mu_0\varepsilon_0\omega^2 - k^2$. The associated transverse fields are

$$E_{0,r} = -\left[\frac{ik}{\beta}C_m\mathrm{K}'_m(\beta r) - \frac{\omega m\mu_0}{\beta^2 r}D_m\mathrm{K}_m(\beta r)\right]e^{im\varphi} \tag{9–82}$$

$$H_{0,r} = -\left[\frac{ik}{\beta}D_m\mathrm{K}'_m(\beta r) + \frac{\omega m\varepsilon_0}{\gamma^2 r}C_m\mathrm{K}_m(\beta r)\right]e^{im\varphi} \tag{9–83}$$

$$E_{0,\varphi} = -\left[-\frac{km}{\beta^2 r}C_m\mathrm{K}_m(\beta r) - \frac{i\omega\mu_0}{\beta}D_m\mathrm{K}'_m(\beta r)\right]e^{im\varphi} \tag{9–84}$$

$$H_{0,\varphi} = -\left[-\frac{km}{\beta^2 r}D_m\mathrm{K}_m(\beta r) + \frac{i\omega\varepsilon_0}{\beta}C_m\mathrm{K}'_m(\beta r)\right]e^{im\varphi} \tag{9–85}$$

The boundary conditions at $r = a$ impose the following equalities:

$$E_z(a_-) = E_z(a_+): \quad \Rightarrow \quad A_m\mathrm{J}_m(\gamma a) = C_m\mathrm{K}_m(\beta a) \tag{9–86}$$

$$H_z(a_-) = H_z(a_+): \quad \Rightarrow \quad B_m\mathrm{J}_m(\gamma a) = D_m\mathrm{K}_m(\beta a) \tag{9–87}$$

$$E_\varphi(a_-) = E_\varphi(a_+): \quad \Rightarrow \quad \left[-\frac{km}{\gamma^2 a}A_m\mathrm{J}_m(\gamma a) - \frac{i\omega\mu_1}{\gamma}B_m\mathrm{J}'_m(\gamma a)\right] = \left[\frac{km}{\beta^2 a}C_m\mathrm{K}_m(\beta a) + \frac{i\omega\mu_0}{\beta}D_m\mathrm{K}'_m(\beta a)\right] \tag{9–88}$$

$$H_\varphi(a_-) = H_\varphi(a_+): \quad \Rightarrow \quad \left[-\frac{km}{\gamma^2 a}B_m\mathrm{J}_m(\gamma a) + \frac{i\omega\varepsilon_1}{\gamma}A_m\mathrm{J}'_m(\gamma a)\right] = \left[\frac{km}{\beta^2 a}D_m\mathrm{K}_m(\beta a) - \frac{i\omega\varepsilon_0}{\beta}C_m\mathrm{K}'_m(\beta a)\right] \tag{9–89}$$

$$D_r(a_-) - D_r(a_+): \quad \Rightarrow \quad \varepsilon_1\left[\frac{ik}{\gamma}A_m\mathrm{J}'_m(\gamma a) - \frac{m\omega\mu_1}{\gamma^2 a}B_m\mathrm{J}_m(\gamma a)\right] = -\varepsilon_0\left[\frac{ik}{\beta}C_m\mathrm{K}'_m(\beta a) - \frac{m\omega\mu_0}{\beta^2 a}D_m\mathrm{K}_m(\beta a)\right] \tag{9–90}$$

$$B_r(a_-) = B_r(a_+): \quad \Rightarrow \quad \mu_1\left[\frac{ik}{\gamma}B_m\mathrm{J}'_m(\gamma a) + \frac{m\omega\varepsilon_1}{\gamma^2 a}A_m\mathrm{J}_m(\gamma a)\right] = -\mu_0\left[\frac{ik}{\beta}D_m\mathrm{K}'_m(\beta a) + \frac{m\omega\varepsilon_0}{\beta^2 a}C_m\mathrm{K}_m(\beta a)\right] \tag{9–91}$$

Dropping the subscripts m and arguments γa and βa for the sake of brevity, we reduce these six equations to a somewhat more tractable pair. From (9–88), we write

$$-\frac{km}{a}\left(\frac{A\mathrm{J}}{\gamma^2} + \frac{C\mathrm{K}}{\beta^2}\right) = i\omega\left(\frac{\mu_1}{\gamma}B\mathrm{J}' + \frac{\mu_0}{\beta}D\mathrm{K}'\right) \tag{9–92}$$

but, from (9–86), $A\mathrm{J} = C\mathrm{K}$ and, from (9–87), $B\mathrm{J} = D\mathrm{K}$. We therefore substitute for C and D in (9–92) to obtain

$$-\frac{mkA}{a}\left(\frac{1}{\gamma^2} + \frac{1}{\beta^2}\right) = i\omega B\left(\frac{\mu_1}{\gamma}\frac{\mathrm{J}'}{\mathrm{J}} + \frac{\mu_0}{\beta}\frac{\mathrm{K}'}{\mathrm{K}}\right) \tag{9–93}$$

Similarly, from (9–89),

$$-\frac{km}{a}\left(\frac{BJ}{\gamma^2}+\frac{DK}{\beta^2}\right)=-i\omega\left(\frac{\varepsilon_1}{\gamma}AJ+\frac{\varepsilon_0}{\beta}CK'\right) \tag{9–94}$$

or

$$\frac{mkB}{a}\left(\frac{1}{\gamma^2}+\frac{1}{\beta^2}\right)=i\omega A\left(\frac{\varepsilon_1}{\gamma}\frac{J'}{J}+\frac{\varepsilon_0}{\beta}\frac{K'}{K}\right) \tag{9–95}$$

In the same way, (9–90) becomes

$$\frac{m\omega B}{a}\left(\frac{\varepsilon_1\mu_1}{\gamma^2}+\frac{\varepsilon_0\mu_0}{\beta^2}\right)=ikA\left(\frac{\varepsilon_1}{\gamma}\frac{J'}{J}+\frac{\varepsilon_0}{\beta}\frac{K'}{K}\right) \tag{9–96}$$

while (9–91) becomes

$$\frac{m\omega A}{a}\left(\frac{\varepsilon_1\mu_1}{\gamma^2}+\frac{\varepsilon_0\mu_0}{\beta^2}\right)=-ikB\left(\frac{\mu_1}{\gamma}\frac{J'}{J}+\frac{\mu_0}{\beta}\frac{K'}{K}\right) \tag{9–97}$$

The remaining constants may now be eliminated. Multiplying (9–93) and (9–95), we have

$$\frac{m^2k^2AB}{a^2}\left(\frac{1}{\gamma^2}+\frac{1}{\beta^2}\right)^2=\omega^2AB\left(\frac{\mu_1}{\gamma}\frac{J'}{J}+\frac{\mu_0}{\beta}\frac{K'}{K}\right)\left(\frac{\varepsilon_1}{\gamma}\frac{J'}{J}+\frac{\varepsilon_0}{\beta}\frac{K'}{K}\right) \tag{9–98}$$

Alternatively, we might have multiplied (9–96) and (9–97) to eliminate the constants A and B. The same equation, (9-98), would be obtained. We finally restore the arguments to the functions to make evident the equation to be solved:

$$\frac{m^2k^2}{\omega^2}\left(\frac{1}{(\gamma a)^2}+\frac{1}{(\beta a)^2}\right)^2$$

$$=\left[\frac{\mu_1}{\gamma a}\frac{J'(\gamma a)}{J(\gamma a)}+\frac{\mu_0}{\beta a}\frac{K'(\beta a)}{K(\beta a)}\right]\left[\frac{\varepsilon_1}{\gamma a}\frac{J'(\gamma a)}{J(\gamma a)}+\frac{\varepsilon_0}{\beta a}\frac{K'(\beta a)}{K(\beta a)}\right] \tag{9–99}$$

with $\beta^2=(\varepsilon_1\mu_1-\varepsilon_0\mu_0)\omega^2-\gamma^2=(n^2-1)\omega^2/c^2-\gamma^2$. The roots of this equation are the values γa can take. Finding the roots is rather laborious (note that k is also a function of γa).

For $m=0$, we recover the characteristic equations for TE and TM modes, namely

$$\frac{\mu_1}{\gamma a}\frac{J_0'(\gamma a)}{J_0(\gamma a)}+\frac{\mu_0}{\beta a}\frac{K_0'(\beta a)}{K_0(\beta a)}=0 \tag{9–100}$$

for TE modes and

$$\frac{\varepsilon_1}{\gamma a}\frac{J_0'(\gamma a)}{J_0(\gamma a)}+\frac{\varepsilon_0}{\beta a}\frac{K_0'(\beta a)}{K_0(\beta a)}=0 \tag{9–101}$$

for TM modes.

Numerical calculations by Elasser[20] give roots for $m=0$ and 1 as sketched in Figure 9.8 for a polystyrene rod ($\varepsilon=2.56\,\varepsilon_0$) in terms of the dimensionless variables kc/ω and $\omega a/\pi c=2a/\lambda_0$.

[20]Walter M. Elasser (1949) *Journal of Applied Physics, 20.*

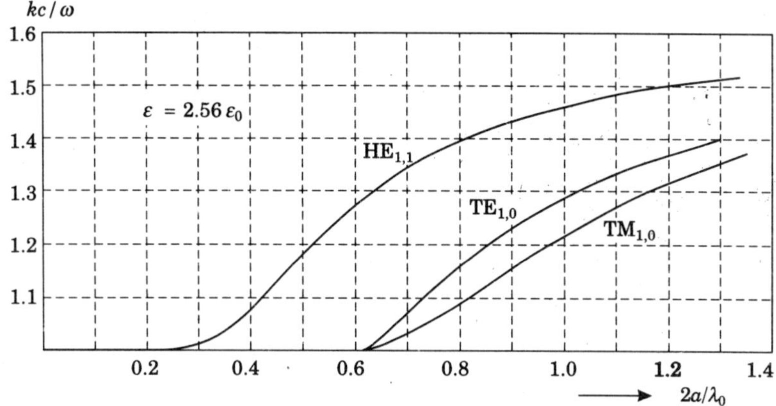

Figure 9.8: The effect of fiber diameter on wave speed in a polystyrene cylindrical rod for the $HE_{1,1}$, $TE_{1,0}$, and $TM_{1,0}$ modes. (Data from Elasser).

The $HE_{1,1}$ mode is unique in that it has no low-frequency cutoff. For small ωa, the phase velocity $\omega/k = c$. β then becomes very small, and most of the wave travels outside the dielectric. The dispersion, indicated by the slope of the curve, is also very low, and since most of the energy resides outside the fiber, losses can be made very small. If a cladding other than vacuum is used, the speed, the losses, and the dispersion of the wave will correspond to those of the cladding.

The dimensions of the fiber required to sustain only this mode in vacuum are of order one-tenth of a wavelength, too small even to be seen. Practical fibers of this sort will need to be clad in a second dielectric of macroscopic dimensions to gain mechanical strength. Choosing a cladding with refractive index close to that of the core raises the TE and TM cut-off frequencies so that only the $HE_{1,1}$ mode can be sustained in fiber cores of several micron diameter.[21]

---

[21] Much more information on optical fibers at this and more advanced levels can be obtained from the collection of *IEEE* reprints: P.J.B. Clarricoats, ed. (1975) *Optical Fibre Waveguides*, Peter Perigrinus Ltd.

Exercises and Problems

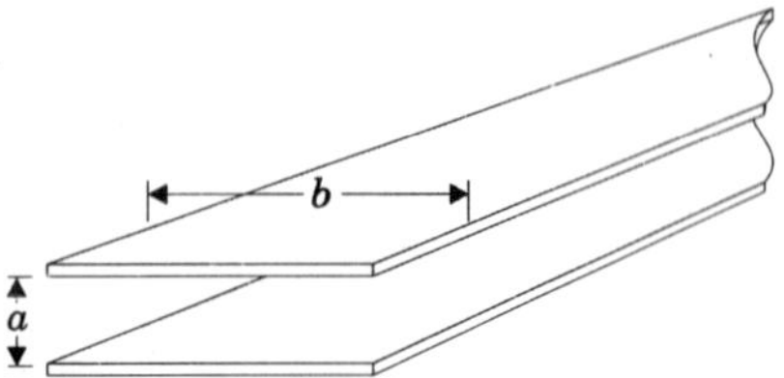

Figure 9.9: A parallel plate waveguide.

9-1 Calculate the fields for a $TM_{m,n}$ mode in a rectangular metal waveguide of dimensions $a \times b$. Find the dispersion relation and the cutoff frequency.

9-2 Find the TE modes of a rectangular cavity of dimensions $a \times b \times c$. In particular, find the three lowest resonant frequencies.

9-3 A pair of air-spaced parallel plates (Figure 9.9) form a waveguide. Find the modes of propagation in such a guide.

9-4 Find the wavelength of 12-GHz microwaves travelling in the $TE_{0,1}$ mode in an X-band waveguide (2.28×1.01 cm cross section).

9-5 Calculate the attenuation of the $TE_{1,0}$ mode in a copper wall waveguide (a) when the frequency is twice the cutoff frequency and (b) when the frequency is $1.05\times$ the cutoff frequency.

9-6 Find the TE modes of a rectangular cross section dielectric waveguide as well as the associated cutoff frequencies.

9-7 Use the results on spherical waves to obtain the resonant modes of a spherical cavity with perfectly conducting walls. Find the three lowest TE mode frequency as well as the two lowest nondegenerate TM mode frequencies. (Some numerical work may be required to obtain the latter.)

9-8 Find the Q of a rectangular cavity of dimensions $a \times b \times c$ when operated in a $TM_{1,0,1}$ mode. Assume the skin depth is 0.5 μm.

9-9 Obtain the ($m = 0$) TM mode characteristic equation for electromagnetic waves in a dielectric cylinder of radius a directly, as was done for the TE mode.

9-10 The space between two concentric, conducting cylinders forms a coaxial waveguide. Show that there are TEM modes of propagation for such a waveguide. Are there also TE and TM modes that have a longitudinal field?

9-11 Obtain the lowest frequency resonant mode of the cavity formed by two concentric, spherical conductors. The earth and the ionosphere form such a cavity with resonant frequencies of 8 Hz, 14 Hz, 20 Hz, $\cdots$; the modes are known as Schumann resonances.

9-12 Calculate the impedance of a coaxial line carrying a wave in the TEM mode. Assume the dielectric between the inner conductor and the shield has $\mu = \mu_0$ and $\varepsilon = 1.7\varepsilon_0$ and take an outer to inner radius ratio of 5.

9-13 The *impedance*, Z, of a medium may be deduced from

$$|S| = E^2/Z$$

Deduce that the impedance of a rectangular waveguide for TE waves is given by

$$Z = \sqrt{\frac{\mu}{\varepsilon}}\frac{\lambda}{\lambda_0}$$

9-14 A coaxial cylindrical resonator is formed in the space between two coaxial conducting cylinders with radii R_1 and R_2 sealed by two conducting planes, one at $z = 0$ and another at $z = L$. Find the TE modes and TM mode characteristic equations for modes of such a cavity.

9-15 A communications optical fiber has a 10 μm diameter core with refractive index $n_c = 1.465$ and cladding of index 1.420. Below the cut-off frequency for transverse modes, the fiber will support only the $\text{HE}_{1,1}$ mode. Find the cut-off frequency for $\text{TE}_{1,0}$ modes.

Chapter 10

Electromagnetic Radiation

10.1 The Inhomogeneous Wave Equation

When variable currents carry charges from one locality to another, causing a temporally changing charge density, the resulting time-dependent fields propagate outward at the finite velocity of light. We might therefore anticipate that an observer some distance from the varying charge distribution would sense temporally varying electric and magnetic fields, each decreasing as r^{-2} or faster, lagging a time $t = r/c$ behind the source. These fields, known as the induction fields, carry a diminishing energy (integrated over an enclosing sphere) as the distance to the source is increased. The vacuum wave equation, however, predicts the existence of electromagnetic waves whose fields diminish as r^{-1}, leading to a distance-independent energy transport.

It is in fact the temporal and consequential spatial variation of the potential that leads to the *radiation field.* Whereas differentiating a $1/r$ potential gives rise to a $1/r^2$ field, a potential of the form e^{ikr}/r gives rise to both a $1/r^2$ and a $1/r$ component in the field. It is this latter part that corresponds to the radiation field.

In order to relate these radiation fields to their sources, we will need to solve the inhomogeneous wave equations for the potentials, which in the Lorenz gauge, (3–56, 57), read:

$$\nabla^2 V - \mu\varepsilon\frac{\partial^2 V}{\partial t^2} = -\frac{\rho(\vec{r},t)}{\varepsilon} \tag{10–1}$$

and

$$\nabla^2 \vec{A} - \mu\varepsilon\frac{\partial^2 \vec{A}}{\partial t^2} = -\mu\vec{J}(\vec{r},t) \tag{10–2}$$

The electric and magnetic fields may then be obtained from the solutions of these equations as $\vec{E} = -\vec{\nabla}V - \partial\vec{A}/\partial t$ and $\vec{B} = \vec{\nabla}\times\vec{A}$. The two equations (10–1) and (10–2) are not independent, since ρ and $\vec{J}$ are related by the continuity equation.

In practice it is sufficient to evaluate $\vec{A}$ since $\vec{B}$ may be found directly as its curl, and outside the source, $\vec{E}$ may be obtained from $\vec{B}$. The problem of solving (10–1, 2) will now be considered.

10.1.1 Solution by Fourier Analysis

A commonly useful method of solving linear differential equations with a time dependence such as (10–1) or (10–2) is to frequency analyze the disturbance and the solution and temporarily deal with only one frequency component, thereby eliminating the time dependence. After the single-frequency solution has been found, the time dependent solution may be resynthesized by summing the frequency components. Both wave equations (3–56, 57) have the same form for V or any one Cartesian component of $\vec{A}$:

$$\nabla^2\psi(\vec{r},t) - \frac{1}{c^2}\frac{\partial^2\psi(\vec{r},t)}{\partial t^2} = -f(\vec{r},t) \tag{10–3}$$

We assume that the source function $f(\vec{r},t)$'s temporal behavior can be frequency analyzed, that is,

$$f(\vec{r},t) = \frac{1}{\sqrt{2\pi}}\int_{-\infty}^{\infty} f_\omega(\vec{r})e^{-i\omega t}d\omega \tag{10–4}$$

where the Fourier transform [22] $f_\omega(\vec{r})$ is given by the inverse transformation:

$$f_\omega(\vec{r}) = \frac{1}{\sqrt{2\pi}}\int_{-\infty}^{\infty} f(\vec{r},t)e^{i\omega t}dt \tag{10–5}$$

In similar fashion we frequency analyze the solution $\psi(\vec{r},t)$ to write

$$\psi(\vec{r},t) = \frac{1}{\sqrt{2\pi}}\int_{\infty}^{\infty} \psi_\omega(\vec{r})e^{-i\omega t}d\omega \tag{10–6}$$

with inverse

$$\psi_\omega(\vec{r}) = \frac{1}{\sqrt{2\pi}}\int_{-\infty}^{\infty} \psi(\vec{r},t)e^{i\omega t}dt \tag{10–7}$$

Substituting (10–4) and (10–6) into the wave equation (10–3), we obtain

$$\frac{1}{\sqrt{2\pi}}\int_{-\infty}^{\infty}\left[\nabla^2\psi_\omega(\vec{r}) + \frac{\omega^2}{c^2}\psi_\omega(\vec{r}) + f_\omega(\vec{r})\right]e^{-i\omega t}d\omega = 0 \tag{10–8}$$

which can be satisfied for all values of t only when the bracketed term vanishes. The Fourier components, which depend only on the spatial coordinates, must therefore satisfy

$$\nabla^2\psi_\omega(\vec{r}) + k^2\psi_\omega(\vec{r}) = -f_\omega(\vec{r}) \tag{10–9}$$

We can synthesize the solution of this equation by the superposition of unit point source solutions, one at each point $\vec{r}\,'$ of the source f, [i.e., solutions with $f = \delta(\vec{r}-\vec{r}\,')$]. Each unit point source solution $\mathrm{G}(\vec{r},\vec{r}\,')$ then satisfies the equation

$$\nabla^2\mathrm{G}(\vec{r},\vec{r}\,') + k^2\mathrm{G}(\vec{r},\vec{r}\,') = -\,\delta(\vec{r}-\vec{r}\,') \tag{10–10}$$

[22]Many authors do not distribute the $1/\sqrt{2\pi}$ equally between the transform and its inverse, instead giving one a unit coefficient before one integral and the other a $1/2\pi$ coefficient.

and the frequency component ψ_ω of the total solution may be found by summing all point source solutions with the appropriate weight $f_\omega(\vec{r}\,')$:

$$\psi_\omega(\vec{r}) = \int f_\omega(\vec{r}\,')\mathrm{G}(\vec{r},\vec{r}\,')d^3r' \tag{10–11}$$

The perceptive reader will have recognized G as the Green's function to the (inhomogeneous) *Helmholtz* equation (10–9). To find G, we note that the solution must be spherically symmetric about $\vec{r}\,'$. Denoting the distance from the source by $R = |\vec{r} - \vec{r}\,'|$, we find that at points other than $R = 0$, G must satisfy

$$\frac{1}{R}\frac{d^2}{dR^2}(R\mathrm{G}) + k^2\mathrm{G} = 0 \tag{10–12}$$

The solution of (10–12) is straightforward:

$$\frac{d^2}{dR^2}(R\mathrm{G}) + k^2(R\mathrm{G}) = 0 \quad \Rightarrow \quad R\mathrm{G} = Ae^{\pm ikR} \quad \Rightarrow \quad \mathrm{G} = \frac{Ae^{\pm ikR}}{R}. \tag{10–13}$$

The constant of integration A must be chosen to be consistent with the magnitude of the source inhomogeneity. To evaluate A, we integrate (10–10) over a small sphere of radius a, centered on $R = 0$ (or $\vec{r} = \vec{r}\,'$) with G given by (10–13).

$$\int \left[\nabla^2\mathrm{G}(\vec{r},\vec{r}\,') + k^2\mathrm{G}(\vec{r},\vec{r}\,')\right] d^3r = -\int \delta(R)d^3r \tag{10–14}$$

As $R \to 0$, $\mathrm{G} \to A/R$. Then for small R, the second term on the left-hand side of the equation becomes

$$\int_{sphere} k^2\mathrm{G}\,d^3r \quad \to \quad \int \frac{k^2A}{R}d^3r = \int_0^a \frac{k^2A}{R}4\pi R^2dR = 2\pi k^2Aa^2 \tag{10–15}$$

which vanishes as the radius a of the sphere decreases to zero. With the help of $\nabla^2(1/R) = -4\pi\delta(R)$, we write the remaining terms of (10–14) in the limit of vanishing a as

$$\lim_{a\to 0}\int -4\pi\delta(R)Ad^3r = \lim_{a\to 0}\int -\delta(R)d^3r \tag{10–16}$$

from which we conclude that $A = 1/(4\pi)$. Thus $\mathrm{G} = e^{\pm ikR}/4\pi R$. We now retrace our steps and synthesize ψ_ω

$$\psi_\omega(\vec{r}) = \frac{1}{4\pi}\int f_\omega(\vec{r}\,')\frac{e^{\pm ikR(\vec{r},\vec{r}\,')}}{R(\vec{r},\vec{r}\,')}d^3r' \tag{10–17}$$

and proceed to rebuild the time-dependent solution $\psi(\vec{r},t)$ by taking the inverse Fourier transform of (10–17).

$$\psi(\vec{r},t) = \frac{1}{4\pi}\frac{1}{\sqrt{2\pi}}\int_{-\infty}^{\infty}\int \frac{f_\omega(\vec{r}\,')e^{\pm ikR(\vec{r},\vec{r}\,')}}{R(\vec{r},\vec{r}\,')}d^3r'e^{-i\omega t}d\omega$$

$$= \frac{1}{4\pi}\int \frac{1}{\sqrt{2\pi}}\int_{-\infty}^{\infty} \frac{f_\omega(\vec{r}\,')e^{-i(\omega t \mp kR)}}{R(\vec{r},\vec{r}\,')}\,d\omega\,d^3r' \tag{10–18}$$

With the definition $t' = t \pm kR/\omega = t \pm R/c$, we may rewrite this result as

$$\psi(\vec{r},t) = \frac{1}{4\pi}\int \left[\frac{1}{\sqrt{2\pi}}\int_{-\infty}^{\infty} \frac{f_\omega(\vec{r}\,')e^{-i\omega t'}d\omega}{R(\vec{r},\vec{r}\,')}\right] d^3r'$$

$$= \frac{1}{4\pi}\int \frac{f(\vec{r}\,',t')d^3r'}{|\vec{r}-\vec{r}\,'|} \tag{10–19}$$

The general solution of (10–3) is then

$$\psi(\vec{r},t) = \frac{1}{4\pi}\int \frac{f(\vec{r}\,',t \pm R/c)}{|\vec{r}-\vec{r}\,'|}d^3r' \tag{10–20}$$

Mathematically, both the + and the − sign give valid solutions. Physically, the term with the − sign (the *retarded solution*) states that the current potential at $\vec{r}$ corresponds to that which was created by the sources a travel time R/c earlier. The term with the + sign (the *advanced solution*) says that the current potential depends on the behavior of the source in the future (at $t' = t + R/c$). For the time being, at least, we discard the advanced solution as unphysical.

A retarded potential may be visualized as arising in the following manner. Consider an observer located at $\vec{r}$. We might think of the observer surrounded by a succession of information-collecting spheres contracting toward $\vec{r}$ at velocity c (Figure 10.1). As a sphere passes a charged region of space, the then-existing charge and current density (at t') is divided by R and added to the potential already accumulated. When the sphere reaches $\vec{r}$ (at t), the accrued potential corresponds to that which the observer experiences at that particular time.

It is conventional to denote quantities evaluated at the retarded time by square brackets, [ ]. With this shorthand, the potentials become:

$$\vec{A}(\vec{r},t) = \frac{\mu_0}{4\pi}\int \frac{[\vec{J}(\vec{r}\,')]}{|\vec{r}-\vec{r}\,'|}d^3r' \tag{10–21}$$

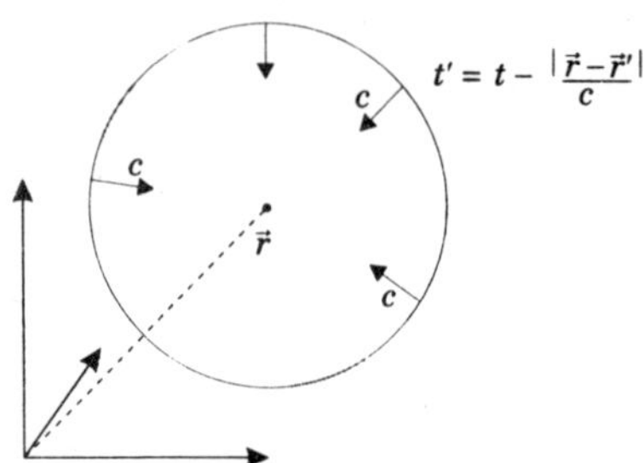

Figure 10.1: An information collecting sphere passes over collections of charge located at $\vec{r}\,'$ at $t' = t - R/c$ to converge on $\vec{r}$ at t.

and

$$V(\vec{r},t) = \frac{1}{4\pi\varepsilon_0}\int \frac{[\rho(\vec{r}\,')]}{|\vec{r}-\vec{r}\,'|}d^3r' \tag{10–22}$$

As we noted before, ρ and $\vec{J}$ are intimately connected through the continuity equation, so the two expressions are not independent of each other.

We have assumed $\vec{r}\,'$ fixed in this derivation. When the sources are in motion, R becomes a function of time, a circumstance we will postpone to consider in more detail later in this chapter.

10.1.2 Green's Function for the Inhomogeneous Wave Equation

It is fairly simple to find a Green's function for the inhomogeneous wave equation (10–3). Since time is involved, G will be $\mathrm{G}(\vec{r},\vec{r}\,';t,t')$ and must satisfy

$$\left(\nabla^2 - \frac{1}{c^2}\frac{\partial^2}{\partial t^2}\right)\mathrm{G}(\vec{r},\vec{r}\,';t,t') = -\delta(\vec{r}-\vec{r}\,')\delta(t-t') \tag{10–23}$$

In other words, G is the potential at $(\vec{r},t)$ due to a point charge at $\vec{r}\,'$ which is turned on for an infinitesimal interval at t'. In the previous section we have already solved the problem for a single frequency point source term $\delta(\vec{r}-\vec{r}\,')e^{-i\omega\tau}$, with τ arbitrary. The result was

$$\psi(\vec{r},\tau) = \frac{e^{\pm ikR}}{4\pi R}e^{-i\omega\tau} \tag{10–24}$$

(the travel delay is all contained in the $e^{\pm ikR}$ term). In vacuum we write $k = \omega/c$, and with this substitution, ψ becomes

$$\psi(\vec{r},\tau) = \frac{e^{\pm i\omega R/c}}{4\pi R}e^{-i\omega\tau} = \frac{e^{-i\omega(\tau\mp R/c)}}{4\pi R} \tag{10–25}$$

The δ function $\delta(t-t')$ in the differential equation above may be expressed as

$$\delta(t-t') = \frac{1}{2\pi}\int_{-\infty}^{\infty} e^{-i\omega(t-t')}d\omega \tag{10–26}$$

Picking the arbitrary τ to be $t-t'$, we see that the δ function is merely the sum, $\int e^{-i\omega\tau}d\omega$, of varying ω source terms. We therefore immediately write the solution G as the sum of the corresponding ψ:

$$\mathrm{G}(R,\tau) = \mathrm{G}(\vec{r},\vec{r}\,';t,t') = \frac{1}{2\pi}\int_{-\infty}^{\infty}\frac{e^{-i\omega(\tau\mp R/c)}d\omega}{4\pi R}$$

$$= \frac{\delta(\tau\mp R/c)}{4\pi R} = \frac{\delta\left(t-t'\mp\dfrac{|\vec{r}-\vec{r}\,'|}{c}\right)}{4\pi|\vec{r}-\vec{r}\,'|} \tag{10–27}$$

As before, we choose the retarded solution (with the $-$ sign) in order to avoid violating causality. The solution to the inhomogeneous wave equation in the absence of boundaries may now be written

$$\psi(\vec{r},t) = \frac{1}{4\pi}\int\int \frac{f(\vec{r}\,',t')\delta\left(t - t' - \dfrac{|\vec{r}-\vec{r}\,'|}{c}\right)}{|\vec{r}-\vec{r}\,'|} d^3r'dt' \tag{10–28}$$

When $\vec{r}\,'$ is stationary (not a function of t'), we easily recover our earlier solution (10–20) by integrating over t'.

10.2 Radiation from a Localized Oscillating Source

With a system whose charges and currents vary in time, we can frequency analyze the time dependence and handle each Fourier component separately. We therefore lose no generality by restricting our consideration to potentials, fields, and radiation from localized systems that vary sinusoidally in time. We take

$$\rho(\vec{r}\,',t) = \rho(\vec{r}\,')e^{-i\omega t} \tag{10–29}$$

and

$$\vec{J}(\vec{r}\,',t) = \vec{J}(\vec{r}\,')e^{-i\omega t} \tag{10–30}$$

where $\vec{J}$ and ρ are required to satisfy $\vec{\nabla}\cdot\vec{J} = i\omega\rho$ by the continuity equation. The potential arising from the charge distribution is then

$$\begin{aligned} V(\vec{r})e^{-i\omega t} &= \frac{1}{4\pi\varepsilon_0}\int \frac{\rho(\vec{r}\,')e^{-i\omega\left(t-|\vec{r}-\vec{r}\,'|/c\right)}}{|\vec{r}-\vec{r}\,'|} d^3r' \\ &= \frac{e^{-i\omega t}}{4\pi\varepsilon_0}\int \frac{\rho(\vec{r}\,')e^{ik|\vec{r}-\vec{r}\,'|}}{|\vec{r}-\vec{r}\,'|} d^3r' \end{aligned} \tag{10–31}$$

or

$$V(\vec{r}) = \frac{1}{4\pi\varepsilon_0}\int \frac{\rho(\vec{r}\,')e^{ik|\vec{r}-\vec{r}\,'|}}{|\vec{r}-\vec{r}\,'|} d^3r' \tag{10–32}$$

where we have again set $k = \omega/c$. Similarly, we find

$$\vec{A}(\vec{r}) = \frac{\mu_0}{4\pi}\int \frac{\vec{J}(\vec{r}\,')e^{ik|\vec{r}-\vec{r}\,'|}}{|\vec{r}-\vec{r}\,'|} d^3r' \tag{10–33}$$

The integrals (10–32) and (10–33) are generally intractable, and various approximations must be employed. We separate the problem into three special regions of interest:

(a) In the *near* zone, $k|\vec{r}-\vec{r}\,'| \ll 1$ (or $R \ll \lambda$) $\Rightarrow e^{ik|\vec{r}-\vec{r}\,'|} \approx 1$.

(b) In the *far* (radiation) zone, $r \gg \lambda$ and $r \gg r'$, and the denominator $|\vec{r}-\vec{r}\,'|$ is taken to be independent of r' although the argument of the complex exponential is not.

(c) In the *intermediate* zone r, r', and λ are all of the same order.

The consequences engendered by each of these approximations will now be considered. We write the appropriate approximations only for the vector potential $\vec{A}$; the generalization to V is obvious.

Because the radiation fields are frequently observed from a distance large compared to the source dimensions, it is convenient to place the source roughly at the origin and to express the potentials in the spherical polar coordinates of the observer.

(a) In the near zone (also called the induction zone) we set $e^{ik|\vec{r}-\vec{r}\,'|} = 1$, giving

$$\vec{A}(\vec{r}) = \frac{\mu_0}{4\pi}\int \frac{\vec{J}(\vec{r}\,')}{|\vec{r}-\vec{r}\,'|}d^3r' \tag{10–34}$$

(The amplitude of $\vec{A}$ is the static solution.) Recalling the expansion

$$\frac{1}{|\vec{r}-\vec{r}\,'|} = \sum_{\ell,m}\frac{4\pi}{2\ell+1}\frac{r_<^\ell}{r_>^{\ell+1}}\mathrm{Y}_\ell^{*m}(\theta',\varphi')\mathrm{Y}_\ell^m(\theta,\varphi) \tag{10–35}$$

we can express the induction zone result in spherical polars when $r > r'$ as

$$\vec{A}(\vec{r}) = \mu_0\sum_{\ell,m}\frac{1}{2\ell+1}\frac{\mathrm{Y}_\ell^m(\theta,\varphi)}{r^{\ell+1}}\int \vec{J}(\vec{r}\,')r'^\ell\mathrm{Y}_\ell^{*m}(\theta',\varphi')d^3r' \tag{10–36}$$

(b) In the far zone, $|\vec{r}-\vec{r}\,'| \approx r - (\vec{r}\cdot\vec{r}\,')/r + \cdots$. Taking $1/|\vec{r}-\vec{r}\,'|$ to be constant over the region of integration, we replace it by $1/r$ to get

$$\vec{A}(\vec{r}) = \frac{\mu_0}{4\pi}\frac{e^{ikr}}{r}\int \vec{J}(\vec{r}\,')e^{-ik(\vec{r}\cdot\vec{r}\,')/r}d^3r' \tag{10–37}$$

If, in addition, the source dimensions r' are small compared to a wavelength λ, then $kr' \ll 1$, and we can expand the exponential to obtain

$$\vec{A}(\vec{r}) = \frac{\mu_0}{4\pi}\frac{e^{ikr}}{r}\sum_\ell\frac{(-ik)^\ell}{\ell!}\int \vec{J}(\vec{r}\,')\left(\frac{\vec{r}\cdot\vec{r}\,'}{r}\right)^\ell d^3r' \tag{10–38}$$

This latter case includes the important example of an atom of typical dimensions 10^{-10}m radiating visible light of wavelength 5×10^{-7}m, observed at macroscopic distances.

(c) In the intermediate zone, we must use an exact expansion of $e^{ik|\vec{r}-\vec{r}\,'|}/|\vec{r}-\vec{r}\,'|$. The exact spherical polar expansion of this term is

$$\frac{e^{ik|\vec{r}-\vec{r}\,'|}}{|\vec{r}-\vec{r}\,'|} = 4\pi ik\sum_{\ell,m}\mathrm{h}_\ell^{(1)}(kr_>)\mathrm{j}_\ell(kr_<)\mathrm{Y}_\ell^m(\theta,\varphi)\mathrm{Y}_\ell^{*m}(\theta,'\varphi') \tag{10–39}$$

where $h_\ell^{(1)} \equiv j_\ell + in_\ell$ is a spherical Hankel function[23] of the first kind. With this expansion, the general expression for the vector potential when $r > r'$ becomes

$$\vec{A}(\vec{r}) = i\mu_0 k \sum_{\ell,m} h_\ell^{(1)}(kr) Y_\ell^m(\theta,\varphi) \int \vec{J}(\vec{r}\,') j_\ell(kr') Y_\ell^{*m}(\theta',\varphi') d^3r' \qquad (10\text{–}40)$$

If, in addition, the source dimensions are small compared to the wavelength ($kr' \ll 1$), we can approximate $j_n(kr')$ as $(kr')^n/(2n+1)!!$ [24] and the vector potential $\vec{A}$ may be written

$$\vec{A}(\vec{r}) = i\mu_0 k \sum_{\ell,m} h_\ell^{(1)}(kr) Y_\ell^m(\theta,\varphi) \int \frac{(kr')^\ell}{(2\ell+1)!!} \vec{J}(\vec{r}\,') Y_\ell^{*m}(\theta',\varphi') d^3r' \quad (10\text{–}41)$$

In each of the cases the source terms have been untangled from the field point terms. If the appropriate integral over the source (independent of the field point) can be evaluated, the potential is obtained at all points in space. If the source dimensions are small, the sums in (10–36), (10–38), and (10–41) each converge rapidly. We examine the first few terms of (10–41) in detail.

10.2.1 Electric Dipole Radiation

If only the first ($\ell = 0$) term of the expansion (10-41) for $\vec{A}(\vec{r})$ is retained, (we have assumed $kr' \ll 1$) we obtain

$$\vec{A}(\vec{r}) = \frac{\mu_0}{4\pi} \frac{e^{ikr}}{r} \int \vec{J}(\vec{r}\,') d^3r' \qquad (10\text{–}42)$$

when $h_0^{(1)}(kr)$ is replaced by its explicit form. This expression, in fact, holds also under the far field expansion, as is easily seen from (10–38).

To demonstrate that this term arises from the dipole component of the charge distribution, we would like to express the field potentials in terms of the electric charge distribution rather than the current density, a procedure that must clearly involve the continuity equation. This consideration motivates attempting to express the integrand as a function involving the divergence of $\vec{J}$.

Let us consider $\vec{\nabla}' \cdot [x'\vec{J}(\vec{r}\,')]$:

$$\begin{aligned}\vec{\nabla}' \cdot [x'\vec{J}(\vec{r}\,')] &= \vec{\nabla}' x' \cdot \vec{J}(\vec{r}\,') + x'[\vec{\nabla}' \cdot \vec{J}(\vec{r}\,')] \\ &= \hat{\imath} \cdot \vec{J} + x'(\vec{\nabla}' \cdot \vec{J}) \\ &= J_x + x'(\vec{\nabla}' \cdot \vec{J}) \end{aligned} \qquad (10\text{–}43)$$

[23] The first few spherical Hankel functions are explicitly

$$h_0^{(1)}(x) = \frac{-ie^{ix}}{x}, \quad h_1^{(1)}(x) = \frac{-e^{ix}}{x}\left(1 + \frac{i}{x}\right), \quad h_2^{(1)}(x) = \frac{ie^{ix}}{x}\left(1 + \frac{3i}{x} - \frac{3}{x^2}\right), \quad \text{etc.}$$

[24] The double factorial $(2\ell + 1)!!$ conventionally means $(2\ell+1)(2\ell-1)(2\ell-3)\cdots$ continued until either 0 or 1 is reached.

We therefore find that $J_x = \vec{\nabla}' \cdot (x'\vec{J}) - x'(\vec{\nabla}' \cdot \vec{J})$.

Integrating J_x over a volume large enough to contain all currents, we get

$$\begin{aligned}\int J_x(\vec{r}\,')d^3r' &= \int \vec{\nabla}' \cdot (x'\vec{J})d^3r' - \int x'(\vec{\nabla}' \cdot \vec{J})d^3r' \\ &= \oint x'\vec{J}(\vec{r}\,') \cdot d\vec{S} + \int x'\frac{\partial \rho}{\partial t}d^3r' \\ &= -i\omega \int x'\rho(\vec{r}\,')d^3r' \end{aligned} \tag{10–44}$$

We generalize this for the other two components of $\vec{J}$ and obtain the desired relation

$$\int \vec{J}d^3r' = -i\omega \int \vec{r}\,'\rho(\vec{r}\,')d^3r' = -i\omega\vec{p} \tag{10–45}$$

Thus

$$\vec{A}(\vec{r}) = \frac{-i\omega\mu_0}{4\pi}\frac{e^{ikr}}{r}\vec{p}_0 \tag{10–46}$$

where $\vec{p} = \vec{p}_0 e^{-i\omega t}$ is the electric dipole moment of the charge distribution.

We find the magnetic induction field of the oscillating dipole by taking the curl of the vector potential:

$$\begin{aligned}\vec{B}(\vec{r}) = \vec{\nabla} \times \vec{A}(\vec{r}) &= \frac{-i\omega\mu_0}{4\pi}\vec{\nabla}\left(\frac{e^{ikr}}{r}\right) \times \vec{p}_0 \\ &= \frac{-i\omega\mu_0}{4\pi}\left(\frac{ik}{r} - \frac{1}{r^2}\right)e^{ikr}\hat{r} \times \vec{p}_0 \\ &= \frac{\omega k\mu_0}{4\pi}\left(1 - \frac{1}{ikr}\right)\frac{e^{ikr}}{r}\hat{r} \times \vec{p}_0 \end{aligned} \tag{10–47}$$

In principle, we could now calculate $V(\vec{r})$ to evaluate $\vec{E}(\vec{r}) = -\vec{\nabla}V + i\omega\vec{A}$, but in practice, outside the source, where $\vec{J} = 0$, the electric field $\vec{E}$ is more easily obtained from $\vec{\nabla} \times \vec{B} = \mu_0\varepsilon_0\partial\vec{E}/\partial t = -i\omega/c^2\vec{E}$. Although it is somewhat laborious, we evaluate $\vec{E}$ by this method:

$$\begin{aligned}\vec{E}(\vec{r}) &= \frac{ic^2}{\omega}\vec{\nabla} \times \vec{B} = \frac{ic}{k}\vec{\nabla} \times \vec{B} \\ &= \frac{ic\omega\mu_0}{4\pi}\left[\vec{\nabla}\left(\frac{e^{ikr}}{r} - \frac{e^{ikr}}{ikr^2}\right) \times (\hat{r} \times \vec{p}_0) + \left(\frac{e^{ikr}}{r} - \frac{e^{ikr}}{ikr^2}\right)\vec{\nabla} \times (\hat{r} \times \vec{p}_0)\right] \\ &= \frac{ic\omega\mu_0}{4\pi}\left[ik\frac{e^{ikr}}{r}\hat{r} \times (\hat{r} \times \vec{p}_0) - \frac{e^{ikr}}{r^2}\hat{r} \times (\hat{r} \times \vec{p}_0) - \frac{e^{ikr}}{r^2}\hat{r} \times (\hat{r} \times \vec{p}_0)\right. \\ &\qquad \left. + \frac{2e^{ikr}}{ikr^3}\hat{r} \times (\hat{r} \times \vec{p}_0) + \frac{e^{ikr}}{ikr^2}(ikr - 1)\vec{\nabla} \times (\hat{r} \times \vec{p}_0)\right] \end{aligned} \tag{10–48}$$

To simplify this expression, we expand $\hat{r} \times (\hat{r} \times \vec{p}_0) = (\hat{r} \cdot \vec{p}_0)\hat{r} - \vec{p}_0$ in all but the first term, and further, we expand

$$\vec{\nabla} \times (\hat{r} \times \vec{p}_0) = \hat{r}(\vec{\nabla} \cdot \vec{p}_0) - \vec{p}_0(\vec{\nabla} \cdot \hat{r}) + (\vec{p}_0 \cdot \vec{\nabla})\hat{r} - (\hat{r} \cdot \vec{\nabla})\vec{p}_0$$

$$= -\vec{p}_0(\vec{\nabla} \cdot \hat{r}) + (\vec{p}_0 \cdot \vec{\nabla})\hat{r} \tag{10–49}$$

Each of the two terms in the latter expression may be further simplified using

$$\vec{\nabla} \cdot \hat{r} = \vec{\nabla} \cdot \left(\frac{\vec{r}}{r}\right) = \vec{\nabla}\left(\frac{1}{r}\right) \cdot \vec{r} + \frac{\vec{\nabla} \cdot \vec{r}}{r} = -\frac{\vec{r}}{r^3} \cdot \vec{r} + \frac{3}{r} = \frac{2}{r} \tag{10–50}$$

and

$$(\vec{p}_0 \cdot \vec{\nabla})\hat{r} = p_{0,x}\frac{\partial}{\partial x}\hat{r} + p_{0,y}\frac{\partial}{\partial y}\hat{r} + p_{0,z}\frac{\partial}{\partial z}\hat{r}$$

$$= p_{0,x}\left(\frac{\hat{\imath}}{r} - \frac{x\vec{r}}{r^3}\right) + p_{0,y}\left(\frac{\hat{\jmath}}{r} - \frac{y\vec{r}}{r^3}\right) + p_{0,z}\left(\frac{\hat{k}}{r} - \frac{z\vec{r}}{r^3}\right)$$

$$= \frac{\vec{p}_0}{r} - \frac{(\vec{p}_0 \cdot \vec{r})\vec{r}}{r^3} \tag{10–51}$$

Thus we obtain

$$\vec{\nabla} \times (\hat{r} \times \vec{p}_0) = -\frac{2\vec{p}_0}{r} + \frac{\vec{p}_0}{r} - \frac{(\vec{p}_0 \cdot \vec{r})\vec{r}}{r^3} = -\frac{\vec{p}_0}{r} - \frac{(\vec{p}_0 \cdot \hat{r})\hat{r}}{r} \tag{10–52}$$

Gathering terms for $\vec{E}$, we have

$$\vec{E}(\vec{r}) = \frac{ck\omega\mu_0}{4\pi}\frac{e^{ikr}}{r}(\vec{p}_0 \times \hat{r}) \times \hat{r}$$

$$-\frac{ic\omega\mu_0}{4\pi}\frac{e^{ikr}}{r^3}\frac{1}{ik}\Big\{(2ikr - 2)\left[(\hat{r} \cdot \vec{p}_0)\hat{r} - \vec{p}_0\right] + (ikr - 1)\left[\vec{p}_0 + (\hat{r} \cdot \vec{p}_0)\hat{r}\right]\Big\}$$

$$= \frac{ck\omega\mu_0}{4\pi}\frac{e^{ikr}}{r}\hat{r} \times (\vec{p}_0 \times \hat{r}) + \frac{c^2\mu_0}{4\pi}\frac{e^{ikr}}{r^3}(1 - ikr)\left[3(\hat{r} \cdot \vec{p}_0)\hat{r} - \vec{p}_0\right]$$

$$= \frac{ck\omega\mu_0}{4\pi}\frac{e^{ikr}}{r}\hat{r} \times (\vec{p}_0 \times \hat{r}) + \frac{1}{4\pi\varepsilon_0}\frac{(1 - ikr)e^{ikr}}{r^3}\left[3(\hat{r} \cdot \vec{p}_0)\hat{r} - \vec{p}_0\right] \tag{10–53}$$

The magnetic induction field $\vec{B}$ is everywhere transverse to $\vec{r}$ as might have been anticipated from the current flow in the oscillating dipole. Referring to figure 10.2, we note that the current in the dipole has to flow along the dipole giving rise to a magnetic induction field perpendicular to the plane containing $\vec{r}$. The electric field $\vec{E}$ has in general a component along $\vec{r}$. The nonvanishing longitudinal component of the electric field will prove to be a general feature of electric multipole radiation. We conclude that an oscillating electric dipole generates TM waves. We will see that radiation arising from oscillating magnetic dipoles has a purely transverse electric field, and $\vec{B}$ will have a longitudinal component.

The terms of both $\vec{B}$ and $\vec{E}$ separate naturally into a term varying inversely as r that will dominate at large distances and a second term diminishing as r^{-2} and r^{-3}

for the magnetic and electric fields, respectively, that will dominate at sufficiently small r.

In the near field ($kr' \ll kr \ll 1$) the fields are

$$\vec{B}(\vec{r},t) = \frac{i\omega\mu_0}{4\pi r^2}(\hat{r}\times\vec{p})e^{i(kr-\omega t)} \quad \text{and} \quad \vec{E}(\vec{r},t) = \frac{3(\hat{r}\cdot\vec{p})\hat{r}-\vec{p}}{4\pi\varepsilon_0 r^3}e^{i(kr-\omega t)} \tag{10–54}$$

The near field $\vec{E}$ is just the "electrostatic" field of an electric dipole appropriately delayed [$e^{i(kr-\omega t)} = e^{-i\omega(t-r/c)}$] for travel, and $\vec{B}$ is just the delayed "static" field resulting from the current needed to make the dipole oscillate. We see then that when we are sufficiently close to the source, ie, a small fraction of a wavelength, the fields are just the travel delayed static fields arising from the charge and current distribution r/c earlier.

In the radiation zone ($kr \gg 1$), the fields take the form

$$\vec{B}(\vec{r}) = \frac{\omega k\mu_0}{4\pi}\frac{e^{ikr}}{r}\hat{r}\times\vec{p} \tag{10–55}$$

and

$$\vec{E}(\vec{r}) = \frac{c\omega k\mu_0}{4\pi}\frac{e^{ikr}}{r}\hat{r}\times(\vec{p}\times\hat{r}) = c\vec{B}\times\hat{r} \tag{10–56}$$

Both $\vec{E}$ and $\vec{B}$ are transverse to $\vec{r}$ in the radiation zone and decrease as r^{-1}. The $1/r$ decrease in fields strength is typical of the radiation field. As an aside, we note that Gauss' law dictates that no spherically symmetric $1/r$ fields can exist.

The energy flux carried by the harmonically oscillating radiation is given by the Poynting vector:

$$\langle\vec{S}\rangle = \frac{1}{2\mu_0}\langle\vec{E}_0\times\vec{B}_0^*\rangle = \frac{\mu_0\omega^2k^2c}{32\pi^2r^2}|\hat{r}\times\vec{p}_0|^2\hat{r}$$

$$= \frac{\mu_0\omega^4}{32\pi^2r^2c}|\hat{r}\times\vec{p}_0|^2\hat{r} = \frac{\mu_0\omega^4}{32\pi^2r^2c}(p_0^2\sin^2\theta)\hat{r} \tag{10–57}$$

where θ is the angle between the dipole axis and $\vec{r}$. The power emitted per solid angle is given by

$$\frac{d\mathbf{P}}{d\Omega} = \langle\vec{S}\rangle\cdot\hat{r}\,r^2 = \frac{\mu_0\omega^4p_0^2\sin^2\theta}{32\pi^2c} \tag{10–58}$$

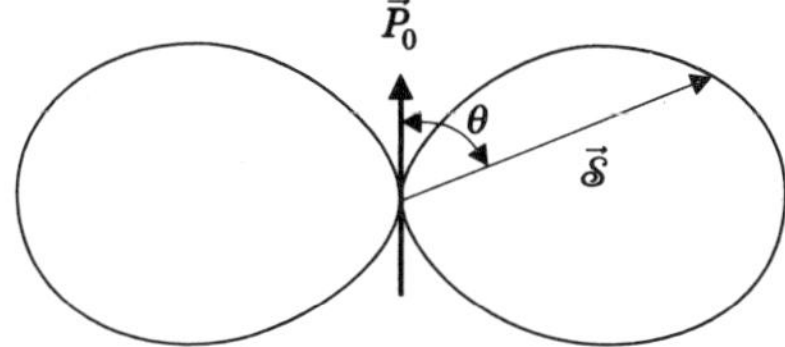

Figure 10.2: The radiation pattern of an oscillating electric dipole. The length of the arrow representing the Poynting vector is proportional to the radiation flux in the corresponding direction.

The angular distribution is illustrated in Figure 10.2.

The total power **P** emitted by the dipole is found by integrating (10–58) over the solid angle to obtain

$$\begin{aligned}\langle \mathbf{P} \rangle &= \frac{\mu_0 \omega^4 p_0^2}{32\pi^2 c} \int \sin^2\theta d\Omega \\ &= \frac{\mu_0 \omega^4 p_0^2}{32\pi^2 c} \int_0^{2\pi} \int_0^{\pi} \sin^3\theta d\theta d\varphi = \frac{\mu_0 \omega^4 p_0^2}{32\pi^2 c} \cdot \frac{8\pi}{3} = \frac{\mu_0 \omega^4 p_0^2}{12\pi c}\end{aligned} \tag{10–59}$$

EXAMPLE 10.1: Find the power radiated by a short rotating wire of length ℓ initially lying along the x axis centered at the origin carrying line charge density $\lambda = \lambda_0 x$ along its length. The wire rotates in the x-y plane about the z axis at angular frequency ω. ($\ell \ll c/\omega$)

Solution: The dipole moment of the wire at $t = 0$ is given by (Ex 2.2.2) as $\vec{p} = \frac{1}{12}\lambda_0 \ell^3 \hat{\imath}$. As the dipole rotates, its dipole moment at time t is

$$\vec{p}(t) = \frac{\lambda_0 \ell^3}{12}(\hat{\imath}\cos\omega t + \hat{\jmath}\sin\omega t) = \mathrm{Re}\Big(p(\hat{\imath} + i\hat{\jmath})e^{-i\omega t}\Big) \tag{Ex 10.1.1}$$

This expression can be interpreted as two linear oscillators executing harmonic oscillations 90° out of phase with one another. We compute the term $|\hat{r} \times \vec{p}_0|^2$ of (10–57) with $\vec{p}_0 = p(\hat{\imath} + i\hat{\jmath})$ and abbreviate the coefficients of $\hat{\imath}, \hat{\jmath}$, and $\hat{k}$ as A, B, C respectively

$$\hat{r} = \sin\theta\cos\varphi\,\hat{\imath} + \sin\theta\sin\varphi\,\hat{\jmath} + \cos\theta\,\hat{k} \equiv A\,\hat{\imath} + B\,\hat{\jmath} + C\,\hat{k} \tag{Ex 10.1.2}$$

then

$$|\hat{r} \times \vec{p}_0| = p(A\hat{\imath} + B\hat{\jmath} + C\hat{k}\,) \times (\hat{\imath} + i\,\hat{\jmath}\,) = p[\,iA\hat{k} - B\hat{k} + C\hat{\jmath} - iC\hat{\imath}\,] \tag{Ex 10.1.3}$$

whence

$$|\hat{r} \times \vec{p}_0|^2 = p^2(2C^2 + A^2 + B^2) = p^2(2\cos^2\theta + \sin^2\theta) \tag{Ex 10.1.4}$$

The angular distribution of power is then

$$\frac{d\mathbf{P}}{d\Omega} = \frac{\mu_0 \omega^4 p^2(2\cos^2\theta + \sin^2\theta)}{32\pi^2 c} = \frac{\mu_0 \omega^4 (2 - \sin^2\theta)}{32\pi^2 c}\left(\frac{\lambda_0 \ell^3}{12}\right)^2 \tag{Ex 10.1.5}$$

The rotating dipole radiates most strongly in the $\pm z$ direction. The total power is found by integrating (Ex 10.1.5) over the 4π solid angle, Focussing on just the angular part,

$$\int_0^{2\pi} \int_0^{\pi} (2 - \sin^2\theta)\sin\theta\, d\theta\, d\varphi = \frac{16\pi}{3} \tag{Ex 10.1.6}$$

The summarize, the total power emitted is

$$\mathbf{P} = \frac{\mu_0 \omega^4}{6\pi c}\left(\frac{\lambda_0 \ell^3}{12}\right)^2 \tag{Ex 10.1.7}$$

It might be noted that this result is exactly twice that obtained in (10–59) for the simple linear oscillator; a result that might have been anticipated from the fact that two oscillators in quadrature do not interfere.

10.2.2 Magnetic Dipole and Electric Quadrupole Radiation

If the first term of the vector potential (10–41) vanishes, or the source dimensions are not overwhelmingly small, the next ($\ell = 1$) term will contribute significantly to the field. This term will be seen to give rise to both electric quadrupole and magnetic dipole radiation.

Returning to the expansion of the vector potential (10–41)

$$\vec{A}(\vec{r}) = i\mu_0 k \sum_{\ell,m} \mathrm{h}_\ell^{(1)}(kr)\mathrm{Y}_\ell^m(\theta,\varphi) \int \vec{J}(\vec{r}\,')\mathrm{j}_\ell(kr')\mathrm{Y}_\ell^{*m}(\theta',\varphi')d^3r'$$

and using the summation identity for spherical harmonics (F–47)

$$\sum_m \mathrm{Y}_\ell^m(\theta,\varphi)\mathrm{Y}_\ell^{*m}(\theta',\varphi') = \frac{2\ell+1}{4\pi}\mathrm{P}_\ell(\cos\gamma) = \frac{2\ell+1}{4\pi}\mathrm{P}_\ell(\hat{r}\cdot\hat{r}\,') \tag{10–60}$$

(γ is the angle between $\vec{r}$ and $\vec{r}\,'$) we can write (10-41) as

$$\vec{A}(\vec{r}) = \mu_0 ik \sum_\ell \frac{2\ell+1}{4\pi}\mathrm{h}_\ell^{(1)}(kr) \int \vec{J}(\vec{r}\,')\mathrm{j}_\ell(kr')\mathrm{P}_\ell(\hat{r}\cdot\hat{r}\,')d^3r' \tag{10–61}$$

(Notice that we have eliminated the summation over m at the expense of reintroducing an $\vec{r}$ dependence into the integral.) As $\mathrm{P}_1(x) = x$, the $\ell = 1$ term yields

$$\vec{A}_1(\vec{r}) = \frac{3\mu_0 ik}{4\pi}\mathrm{h}_1^{(1)}(kr) \int \vec{J}(\vec{r}\,')\mathrm{j}_1(kr')(\hat{r}\cdot\hat{r}\,')d^3r' \tag{10–62}$$

For small kr', we approximate $\mathrm{j}_\ell(kr')$ by $(kr')^\ell/(2\ell+1)!!$, which gives for $\ell = 1$, $\mathrm{j}_1(kr') \approx \frac{1}{3}kr'$. Thus, the $\ell = 1$ term of the vector potential may be approximated as

$$\begin{aligned}\vec{A}_1(\vec{r}) &= \frac{\mu_0 ik}{4\pi}\mathrm{h}_1^{(1)}(kr) \int \vec{J}(\vec{r}\,')kr'(\hat{r}\cdot\hat{r}\,')d^3r' \\ &= \frac{-\mu_0 ik^2}{4\pi}\frac{e^{ikr}}{kr}\left(1+\frac{i}{kr}\right) \int \vec{J}(\vec{r}\,')(\hat{r}\cdot\vec{r}\,')d^3r' \\ &= \frac{-i\mu_0 ke^{ikr}}{4\pi r^2}\left(1-\frac{1}{ikr}\right) \int \vec{J}(\vec{r}\,')(\vec{r}\cdot\vec{r}\,')d^3r' \end{aligned} \tag{10–63}$$

Again we would like to relate this expression more directly to moments of the charge and current distributions. Since the integral in (10–63) evidently contains a moment of $\vec{J}$ (which implies also a second moment of ρ), we anticipate a decomposition in terms of a magnetic dipole and/or electric quadrupole. To remove the magnetic dipole contribution, we use the identity $\left[\vec{r}\,'\times\vec{J}(\vec{r}\,')\right]\times\vec{r} = (\vec{r}\,'\cdot\vec{r})\vec{J} - (\vec{r}\cdot\vec{J})\vec{r}\,'$

to rewrite the integrand in the not immediately obvious form

$$(\vec{r}\cdot\vec{r}\,')\vec{J} = \tfrac{1}{2}\left[(\vec{r}\,'\cdot\vec{r})\vec{J} + (\vec{r}\cdot\vec{J})\vec{r}\,' + (\vec{r}\,'\times\vec{J})\times\vec{r}\right] \tag{10–64}$$

The magnetic moment of a current distribution is generally (2–24)

$$\vec{m} = \tfrac{1}{2}\int \vec{r}\,'\times\vec{J}(\vec{r}\,')d^3r'$$

which lets us write (10–64) as

$$\vec{A}_1(\vec{r}) = \frac{-i\mu_0 k e^{ikr}}{4\pi r^2}\left(1 - \frac{1}{ikr}\right)\left\{\vec{m}\times\vec{r} + \tfrac{1}{2}\int\left[(\vec{r}\,'\cdot\vec{r})\vec{J} + (\vec{r}\cdot\vec{J})\vec{r}\,'\right]d^3r'\right\} \tag{10–65}$$

We put aside, for the moment, the integral in (10–65) and focus our attention on the magnetic dipole moment's contribution to the vector potential, namely

$$\vec{A}_M(\vec{r}) = \frac{i\mu_0 k e^{ikr}}{4\pi r}\left(1 - \frac{1}{ikr}\right)(\hat{r}\times\vec{m}) \tag{10–66}$$

Comparing this expression with those for the fields of the electric dipole, (10–47) and (10–53), we find that $\vec{A}_M$ has the same form as $\vec{B}$, in (10–47). Then $\vec{B}_M = \vec{\nabla}\times\vec{A}_M$ will have the same form as the electric dipole electric field, $\vec{E} = ic^2/\omega\vec{\nabla}\times\vec{B}$, previously found in (10-53). Making allowance for the different constants, we find [replacing $\vec{m}$ by $-i\omega\vec{p}$ to make the expression for $\vec{A}_M$ (10–66) correspond with that for $\vec{B}$ in (10–47)] $\vec{B} = 1/c^2\times$ (10–53) with $\vec{m}$ replacing $\vec{p}_0$ or

$$\vec{B}_M = \frac{k^2\mu_0}{4\pi}\frac{e^{ikr}}{r}\hat{r}\times(\vec{m}\times\hat{r}) + \frac{\mu_0}{4\pi}\frac{(1-ikr)e^{ikr}}{r^3}\left[3\hat{r}(\vec{m}\cdot\hat{r}) - \vec{m}\right] \tag{10–67}$$

The relation $\vec{\nabla}\times\vec{E} = i\omega\vec{B} = i\omega(\vec{\nabla}\times\vec{A})$ suggest $\vec{E} = i\omega\vec{A}$ (this can of course be verified by the rather tedious calculation of the curl of $\vec{B}$). Using this shortcut, we find

$$\vec{E}_M = \frac{-k\omega\mu_0}{4\pi}\left(1 - \frac{1}{ikr}\right)\frac{e^{ikr}}{r}(\hat{r}\times\vec{m}) \tag{10–68}$$

The radiation pattern will clearly be identical to that of the electric dipole shown in Figure 10.2. It will be noted that the electric field of magnetic dipole radiation is purely transverse to $\vec{r}$.

We return now to the remaining terms of (10–65), which we denote $\vec{A}_Q$, of the vector potential:

$$\vec{A}_Q = \frac{-\mu_0 ik e^{ikr}}{8\pi r^2}\left(1 - \frac{1}{ikr}\right)\int\left[(\vec{r}\cdot\vec{r}\,')\vec{J} + (\vec{r}\cdot\vec{J})\vec{r}\,'\right]d^3r' \tag{10–69}$$

The choice of subscript anticipates our conclusion that this component of the field arises from oscillating quadrupoles. The integral can again be "integrated by parts" to yield a term in div $\vec{J}$, which we will replace by $i\omega\rho$ by means of the continuity equation. The method is identical to that employed in Section 10.2.1. Consider

$$\begin{aligned}
\vec{\nabla}' \cdot \left[(\vec{r}\cdot\vec{r}\,')x'^{\alpha}\vec{J}\right] &= (\vec{r}\cdot\vec{r}\,')x'^{\alpha}(\vec{\nabla}'\cdot\vec{J}) + \vec{J}\cdot\vec{\nabla}'\left[(\vec{r}\cdot\vec{r}\,')x'^{\alpha}\right] \\
&= x_\beta x'^{\beta}x'^{\alpha}(\vec{\nabla}'\cdot\vec{J}) + J^{\gamma}\partial'_{\gamma}\left(x_\beta x'^{\beta}x'^{\alpha}\right) \\
&= x_\beta x'^{\beta}x'^{\alpha}(\vec{\nabla}'\cdot\vec{J}) + J^{\gamma}x_\beta\left(x'^{\alpha}\partial'_{\gamma}x'^{\beta} + x'^{\beta}\partial'_{\gamma}x'^{\alpha}\right) \\
&= x_\beta x'^{\beta}x'^{\alpha}(\vec{\nabla}'\cdot\vec{J}) + x_\beta J^{\beta}x'^{\alpha} + x_\beta x'^{\beta}J^{\alpha} \\
&= x_\beta x'^{\beta}x'^{\alpha}(\vec{\nabla}'\cdot\vec{J}) + (\vec{r}\cdot\vec{J})x'^{\alpha} + (\vec{r}\cdot\vec{r}\,')J^{\alpha} \qquad (10\text{–}70)
\end{aligned}$$

The latter two terms are just the α component of the integrand in (10–69). Concerning ourselves only with the α component for the moment, we have

$$\begin{aligned}
\int\left[(\vec{r}\cdot\vec{J})x'^{\alpha} + (\vec{r}\cdot\vec{r}\,')J^{\alpha}\right]d^3r' &= \int\vec{\nabla}'\cdot\left[(\vec{r}\cdot\vec{r}\,')x'^{\alpha}\vec{J}\right]d^3r' - \int x_\beta x'^{\beta}x'^{\alpha}(\vec{\nabla}'\cdot\vec{J})d^3r' \\
&= \oint x'^{\alpha}(\vec{r}\cdot\vec{r}\,')\vec{J}\cdot d\vec{S} - \int x_\beta x'^{\beta}x'^{\alpha}(i\omega\rho)d^3r' \\
&= -i\omega\int x'^{\alpha}x_\beta x'^{\beta}\rho(\vec{r}\,')d^3r' \qquad (10\text{–}71)
\end{aligned}$$

so that generalizing for all three components we obtain

$$\vec{A}_Q(\vec{r}) = \frac{-\mu_0\omega k e^{ikr}}{8\pi r^2}\left(1-\frac{1}{ikr}\right)\int \vec{r}\,'(\vec{r}\cdot\vec{r}\,')\rho(\vec{r}\,')\,d^3r' \qquad (10\text{–}72)$$

To relate $\vec{A}_Q$ to the quadrupole moment of the distribution, we consider the α component of $\vec{A}_Q$ and recall the definition of the quadrupole moment (2–16) to write:

$$\begin{aligned}
\left(\vec{A}_Q\right)^{\alpha} &= \frac{-\mu_0\omega k e^{ikr}}{8\pi r^2}\left(1-\frac{1}{ikr}\right)\frac{1}{3}x_\beta\int 3x'^{\beta}x'^{\alpha}\rho(\vec{r}\,')d^3r' \\
&= \frac{-\mu_0\omega k e^{ikr}}{24\pi r^2}\left(1-\frac{1}{ikr}\right)\left[x_\beta Q^{\alpha\beta} + x_\beta\int\delta^{\alpha\beta}r'^2\rho(\vec{r}\,')d^3r'\right] \\
&= \frac{-\mu_0\omega k e^{ikr}}{24\pi r^2}\left(1-\frac{1}{ikr}\right)\left[x_\beta Q^{\alpha\beta} + x^{\alpha}\int r'^2\rho(\vec{r}\,')d^3r'\right] \qquad (10\text{–}73)
\end{aligned}$$

The complete induction and radiation fields are rather complicated to write out. Instead we will restrict ourselves to the fields in the radiation zone, where they can be found from

$$\vec{B} = ik\hat{r}\times\vec{A} \quad \text{and} \quad \vec{E} = ikc(\hat{r}\times\vec{A})\times\hat{r} \qquad (10\text{–}74)$$

Since both fields involve the cross product of $\vec{r}$ with $\vec{A}$, the term with $\vec{r}$ multiplying the integral will make no contribution (only the α component is represented above) and may therefore be dropped from the expression for $\vec{A}$ without loss, leaving

$$\left(\vec{A}_Q\right)^{\alpha} = \frac{-\mu_0\omega k e^{ikr}}{24\pi r^2}x_\beta Q^{\alpha\beta} \qquad (10\text{–}75)$$

Abbreviating $X^\alpha = x_\beta Q^{\alpha\beta}$ (or $\vec{X} = \vec{r} \cdot \overleftrightarrow{Q}$) we compute the Poynting vector as

$$\langle \vec{S} \rangle = \frac{\vec{E}_0 \times \vec{B}_0}{2\mu_0} = \frac{\mu_0 \omega^2 k^4 c}{1152\pi^2 r^4} \left[(\hat{r} \times \vec{X}) \times \hat{r} \right] \times \left[\hat{r} \times \vec{X} \right]$$

$$= \frac{\mu_0 \omega^6}{1152\pi^2 r^4 c^3} \left[X^2 - (\vec{X} \cdot \hat{r})^2 \right] \hat{r} \qquad (10\text{–}76)$$

which leads to the angular power distribution

$$\frac{d\mathbf{P}}{d\Omega} = r^2 \langle \vec{S} \rangle \cdot \hat{r} = \frac{\mu_0 \omega^6}{1152\pi^2 r^2 c^3} \left[x_\beta Q^{\alpha\beta} Q_{\alpha\gamma} x^\gamma - \frac{x_\beta Q^{\alpha\beta} x_\alpha x^\gamma Q_{\gamma\delta} x^\delta}{r^2} \right] \qquad (10\text{–}77)$$

In order to find the total power emitted, this expression must be integrated over the solid angle. With x^α, x^β, x^γ, and x^δ chosen from $z = r\cos\theta$, $x = r \sin\theta \cos\varphi$, and $y = r\sin\theta \sin\varphi$, we evaluate the integrals (most easily done by expressing x, y, and z in terms of spherical harmonics):

$$\int_0^{4\pi} x_\alpha x^\beta d\Omega = \tfrac{4}{3}\pi r^2 \delta_\alpha^\beta \qquad (10\text{–}78)$$

$$\int_0^{4\pi} x_\alpha x_\beta x^\gamma x^\delta d\Omega = \tfrac{4}{15}\pi r^4 \left(\delta_{\alpha\beta}\delta^{\gamma\delta} + \delta_\alpha^\gamma \delta_\beta^\delta + \delta_\alpha^\delta \delta_\beta^\gamma \right) \qquad (10\text{–}79)$$

The term of (10–77) enclosed by square brackets integrated over the solid angle then becomes

$$\int_0^{4\pi} \left(Q^{\alpha\beta} Q_{\alpha\gamma} x_\beta x^\gamma - \frac{Q^{\alpha\beta} Q_{\gamma\delta} x_\alpha x_\beta x^\gamma x^\delta}{r^2} \right) d\Omega$$

$$= \tfrac{4}{3}\pi r^2 \left[Q_{\alpha\beta} Q^{\alpha\beta} - \tfrac{1}{5} \left(Q^{\alpha\beta}\delta_{\alpha\beta}\delta^{\gamma\delta} Q_{\gamma\delta} + Q_{\alpha\gamma}Q^{\alpha\gamma} + Q_{\alpha\delta}Q^{\alpha\delta} \right) \right] \qquad (10\text{–}80)$$

$\overleftrightarrow{Q}$ is a tensor with zero trace, $Q^{\alpha\beta}\delta_{\alpha\beta} = \sum_i Q_{ii}$ vanishes, leaving

$$\mathbf{P} = \frac{\mu_0 \omega^6}{1152\pi^2 c^3} \frac{4\pi}{3} \left(Q_{\alpha\beta} Q^{\beta\alpha} - \tfrac{2}{5} Q_{\alpha\gamma} Q^{\alpha\gamma} \right) = \frac{\mu_0 \omega^6}{1440\pi c^3} Q_{\alpha\beta} Q^{\alpha\beta} \qquad (10\text{–}81)$$

The power radiated by an electric quadrupole varies as the sixth power of ω compared to the fourth power for the dipole. This means that for very high frequency radiation such as that emitted by nuclei when they change internal energy states, quadrupole and even higher order radiation may dominate the decay.

EXAMPLE 10.2: Find the angular distribution of radiation as well as the total power radiated by the oscillating linear quadrupole composed of two equal charges $-q_0$ spaced at distance a from a central charge $2q_0$ oscillating in phase (illustrated in Figure 10.3).

Solution: The nonzero components of the quadrupole moment are (2–16)

$$Q_{xx} = Q_{yy} = 2a^2 q_0 \qquad \text{and} \qquad Q_{zz} = -4a^2 q_0 \qquad (\text{Ex } 10.2.1)$$

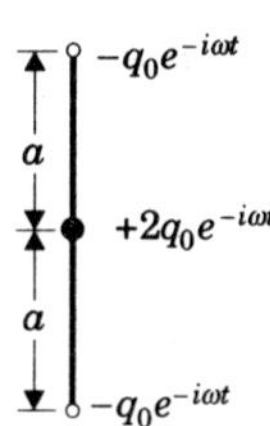

Figure 10.3: A linear oscillating quadrupole.

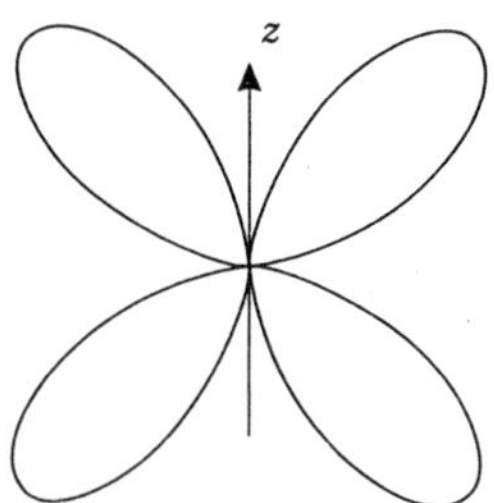

Figure 10.4: The radiation pattern of the linear quadrupole. The pattern is invariant under rotation about the quadrupole (z) axis.

The first term of the bracketed expression in (10-77) then becomes

$$Q_{xx}^2 x^2 + Q_{yy}^2 y^2 + Q_{zz}^2 z^2 = 4q_0^2 a^4(x^2+y^2+4z^2) = 4q_0^2 a^4 r^2(1+3\cos^2\theta) \quad \text{(Ex 10.2.2)}$$

while the second term becomes

$$\frac{1}{r^2}\left(Q_{xx}^2 x^4 + Q_{yy}^2 y^4 + Q_{zz}^2 z^4 + 2Q_{xx}Q_{yy}x^2y^2 + 2Q_{xx}Q_{zz}x^2z^2 + 2Q_{yy}Q_{zz}y^2z^2\right)$$

$$= \frac{4q_0^2a^4}{r^2}\left(x^4 + y^4 + 4z^4 + 2x^2y^2 - 4x^2z^2 - 4y^2z^2\right)$$

$$= \frac{4q_0^2a^4}{r^2}\left(x^2+y^2-2z^2\right)^2$$

$$= 4q_0^2a^4r^2\left(1-3\cos^2\theta\right)^2 \quad \text{(Ex 10.2.3)}$$

Combining the two terms, we obtain

$$\frac{d\mathbf{P}}{d\Omega} = \frac{\mu_0\omega^6}{1152\pi^2c^3}4a^4q_0^2\left(1+3\cos^2\theta - 1 + 6\cos^2\theta - 9\cos^4\theta\right)$$

$$= \frac{\mu_0\omega^6}{32\pi^2c^3}a^4q_0^2\cos^2\theta\sin^2\theta \quad \text{(Ex 10.2.4)}$$

The angular distribution is independent of the polar angle φ as would be anticipated from the geometry. A cross section in a plane containing the quadrupole gives the four-lobed pattern in Figure 10.4. The entire radiation pattern consists of the double cone obtained when the curve in Figure 10.4 is rotated about the z axis.

The mean total power radiated by the quadrupole may be obtained directly from (10–81) to yield

$$\langle\mathbf{P}\rangle = \frac{\mu_0\omega^6}{1440\pi c^3}\left(4a^4q_0^2 + 4a^4q_0^2 + 16a^4q_0^2\right) = \frac{\mu_0\omega^6}{60\pi c^3}a^4q_0^2 \quad \text{(Ex 10.2.5)}$$

This result could of course also have been obtained by integrating $d\mathbf{P}/d\Omega$ directly.

⋆ 10.2.3 Radiation by Higher Order Moments

The evaluation of higher multipole terms by the foregoing methods becomes progressively more difficult, particularly untangling the magnetic and electric moment contributions. Instead of continuing in this manner, we will pursue a more general approach. We will, using a method proposed by Bouwkamp and Casimir,[25] relate the spherical polar vacuum wave solution (3–67) and (3–68) to the source currents.

We briefly recall from Chapter 3 that the solutions of the vacuum wave equations for $\vec{E}$ and $\vec{B}$ were derived from a Debye potential

$$\psi_\ell^m(\vec{r}) = \left\{ \begin{array}{c} \mathrm{j}_\ell(kr) \\ \mathrm{n}_\ell(kr) \end{array} \right\} \mathrm{Y}_\ell^m(\theta,\varphi) \tag{10–82}$$

The TE fields were then obtained by taking

$$\vec{E}_{\mathrm{TE}} = ik\vec{\nabla}\times\vec{r}\psi \qquad \text{and} \qquad \vec{B}_{\mathrm{TE}} = \frac{1}{c}\vec{\nabla}\times(\vec{\nabla}\times\vec{r}\psi) \tag{10–83}$$

while the TM fields are given by

$$\vec{B}_{\mathrm{TM}} = \frac{-ik}{c}(\vec{\nabla}\times\vec{r}\psi) \qquad \text{and} \qquad \vec{E}_{\mathrm{TM}} = \vec{\nabla}\times(\vec{\nabla}\times\vec{r}\psi) \tag{10–84}$$

In order to obtain outgoing waves of the form e^{ikr}/r, the particular linear combination of $\mathrm{j}_\ell(kr)$ and $\mathrm{n}_\ell(kr)$ must be $\mathrm{h}_\ell^{(1)}(kr)$. Of special interest are the radial components of the fields which take the form

$$B_r^{\mathrm{TE}} = \frac{\ell(\ell+1)}{cr}\psi_\ell^m \qquad \text{and} \qquad E_r^{\mathrm{TM}} = \frac{\ell(\ell+1)}{r}\psi_\ell^m \tag{10–85}$$

If E_r and B_r (or equivalently, the more convenient H_r) are known, the entire field is determined. We will obtain the field components E_r and H_r generated by a harmonically oscillating source current distribution and so obtain the entire solution. If the source were more conveniently described as an oscillating charge distribution, the continuity equation would serve to relate the current to the charge density.

We begin, as always, with Maxwell's equations (8–10) to relate the fields to their sources. Let us start with the magnetic field. For harmonically varying fields, $\vec{H}$ is directly related to the currents by

$$\vec{\nabla}\times\vec{H} = \vec{J} - i\omega\varepsilon_0\vec{E} \tag{10–86}$$

while $\vec{E}$ is given by

$$\vec{\nabla}\times\vec{E} = i\mu_0\omega\vec{H} \tag{10–87}$$

As we want only the r component, we seek an inhomogeneous wave equation for $(\vec{r}\cdot\vec{H})$. To this end, we consider the vector identity[26]

$$\nabla^2(\vec{r}\cdot\vec{H}) = 2\vec{\nabla}\cdot\vec{H} + \vec{r}\cdot\nabla^2\vec{H} \tag{10–88}$$

[25] C.J. Bouwkamp and H.B.G. Casimir (1954) *Physica* XX, pp. 539-554.

[26] The identity is easily obtained using tensor notation:

$$\partial_i\partial^i(x_jH^j) = \partial_i(\delta_j^iH^j + x_j\partial^iH^j) = \partial_jH^j + \delta_{ij}\partial^iH^j + x_j\partial_i\partial^iH^j$$
$$= 2\partial_jH^j + x_j\partial_i\partial^iH^j = 2\vec{\nabla}\cdot\vec{H} + \vec{r}\cdot\nabla^2\vec{H}$$

The first term on the right hand side of (10–88) vanishes and we obtain an explicit expression for $\nabla^2\vec{H}$ using the vector identity (13)

$$\nabla^2\vec{H} \equiv -\vec{\nabla}\times(\vec{\nabla}\times\vec{H}) + \vec{\nabla}(\vec{\nabla}\cdot\vec{H})$$

whose last term of course must likewise vanish in vacuum. Substituting for curl $\vec{H}$ from equation (10–86) in the identity above we obtain

$$\begin{aligned}\nabla^2\vec{H} &= -\vec{\nabla}\times(\vec{J} - i\omega\varepsilon_0\vec{E})\\ &= -\vec{\nabla}\times\vec{J} + i\omega\varepsilon_0\vec{\nabla}\times\vec{E}\\ &= -\vec{\nabla}\times\vec{J} - \omega^2\varepsilon_0\mu_0\vec{H}\end{aligned} \tag{10–89}$$

When this result is inserted into (10–88), we obtain the desired wave equation:

$$\left(\nabla^2 + \frac{\omega^2}{c^2}\right)\left(\vec{r}\cdot\vec{H}\right) = -\vec{r}\cdot(\vec{\nabla}\times\vec{J}) \tag{10–90}$$

The solution to equation (10-89) is, according to (10–32),

$$(\vec{r}\cdot\vec{H}) = \frac{1}{4\pi}\int\frac{\vec{r}'\cdot\left[\vec{\nabla}'\times\vec{J}(\vec{r}')\right]e^{ik|\vec{r}-\vec{r}'|}}{|\vec{r}-\vec{r}'|}d^3r' \tag{10–91}$$

The same argument might be applied to $\vec{r}\cdot\vec{E}$, but unfortunately the divergence of $\vec{E}$ need not vanish. Instead, we apply the identity (10–88) to the quantity with vanishing divergence, $\vec{E} - \vec{J}/i\omega\varepsilon_0$. Following the tracks above,

$$\nabla^2\left[\vec{r}\cdot\left(\vec{E} - \frac{\vec{J}}{i\omega\varepsilon_0}\right)\right] = 2\vec{\nabla}\cdot\left(\vec{E} - \frac{\vec{J}}{i\omega\varepsilon_0}\right) + \vec{r}\cdot\nabla^2\left(\vec{E} - \frac{\vec{J}}{i\omega\varepsilon_0}\right) \tag{10–92}$$

The definition of the vector Laplacian (13) gives

$$\begin{aligned}\nabla^2\left(\vec{E} - \frac{\vec{J}}{i\omega\varepsilon_0}\right) &= -\vec{\nabla}\times\left[\vec{\nabla}\times\left(\vec{E} - \frac{\vec{J}}{i\omega\varepsilon_0}\right)\right] + \vec{\nabla}\left[\vec{\nabla}\cdot\left(\vec{E} - \frac{\vec{J}}{i\omega\varepsilon_0}\right)\right]\\ &= -\vec{\nabla}\times\left(\vec{\nabla}\times\vec{E}\right) + \vec{\nabla}\times\left(\vec{\nabla}\times\frac{\vec{J}}{i\omega\varepsilon_0}\right)\\ &= -\vec{\nabla}\times\left(i\omega\mu_0\vec{H}\right) + \vec{\nabla}\times\left(\vec{\nabla}\times\frac{\vec{J}}{i\omega\varepsilon_0}\right)\\ &= -i\mu_0\omega\left(\vec{J} - i\omega\varepsilon_0\vec{E}\right) + \vec{\nabla}\times\left(\vec{\nabla}\times\frac{\vec{J}}{i\omega\varepsilon_0}\right)\\ &= -\varepsilon_0\mu_0\omega^2\left(\vec{E} - \frac{\vec{J}}{i\omega\varepsilon_0}\right) + \vec{\nabla}\times\left(\vec{\nabla}\times\frac{\vec{J}}{i\omega\varepsilon_0}\right)\end{aligned} \tag{10–93}$$

The analogue of (10–90) becomes

$$\left(\nabla^2 + \frac{\omega^2}{c^2}\right)\left[\vec{r}\cdot\left(\vec{E} - \frac{\vec{J}}{i\omega\varepsilon_0}\right)\right] = \frac{-i}{\omega\varepsilon_0}\vec{r}\cdot\left[\vec{\nabla}\times(\vec{\nabla}\times\vec{J})\right] \tag{10–94}$$

with solution

$$\vec{r}\cdot\vec{E} = \frac{\vec{r}\cdot\vec{J}}{i\omega\varepsilon_0} + \frac{i}{4\pi\omega\varepsilon_0}\int \frac{\vec{r}\,'\cdot\left\{\vec{\nabla}'\times\left[\vec{\nabla}'\times\vec{J}(\vec{r}\,')\right]\right\}}{|\vec{r}-\vec{r}\,'|} e^{ik|\vec{r}-\vec{r}\,'|}d^3r' \tag{10–95}$$

Comparison of term under the integral with (3–46) we recognize the Laplacian of the solenoidal current $\nabla'^2\vec{J}_s(\vec{r}\,') = -\vec{\nabla}'\times[\vec{\nabla}'\times\vec{J}(\vec{r}\,')]$, so that we can abbreviate

$$\vec{r}\cdot\vec{E} = \frac{1}{i\omega\varepsilon_0}\left[\vec{r}\cdot\vec{J} + \frac{1}{4\pi}\int \frac{(\vec{r}\,'\cdot\nabla'^2\vec{J}_s)e^{ik|\vec{r}-\vec{r}\,'|}}{|\vec{r}-\vec{r}\,'|}d^3r'\right] \tag{10–96}$$

verifying that the radiation field depends only on the solenoidal current as discussed in 3.4.2. The longitudinal component of $\vec{J}$ in (10–91) has no curl and we can therefore, without loss, replace $\vec{J}$ by $\vec{J}_s$ as well.

We can cast (10-91) and (10–95) into the form of the vacuum fields by expanding $e^{ik|\vec{r}-\vec{r}\,'|}/|\vec{r}-\vec{r}\,'|$ in spherical polar coordinates. Taking r' as the lesser of r and r', equation (10–91) becomes

$$\vec{H}\cdot\vec{r} = ik\sum_{\ell,m}\mathrm{h}_\ell^{(1)}(kr)\mathrm{Y}_\ell^m(\theta,\varphi)\int \vec{r}\,'\cdot\left[\vec{\nabla}'\times\vec{J}(\vec{r}\,')\right]\mathrm{j}_\ell(kr')\mathrm{Y}_\ell^{*m}(\theta',\varphi')d^3r' \tag{10–97}$$

Comparing this to the result from the Debye potentials (3–67),

$$\vec{H}\cdot\vec{r} = \sum_{\ell,m} b_\ell^m\sqrt{\frac{\varepsilon_0}{\mu_0}}\ell(\ell+1)\mathrm{h}_\ell^{(1)}(kr)\mathrm{Y}_\ell^m(\theta,\varphi)$$

we find that the amplitude coefficient b_ℓ^m for TE wave generated by current $\vec{J}$ is

$$b_\ell^m = \frac{ik}{\ell(\ell+1)}\sqrt{\frac{\mu_0}{\varepsilon_0}}\int \vec{r}\,'\cdot[\vec{\nabla}'\times\vec{J}(\vec{r}\,')]\,\mathrm{j}_\ell(kr')\mathrm{Y}_\ell^{*m}(\theta',\varphi')d^3r' \tag{10–98}$$

The integral can be marginally simplified by using the identity

$$\vec{r}\,'\xi\cdot(\vec{\nabla}'\times\vec{J}) = \vec{J}\cdot(\vec{\nabla}'\times\vec{r}\,'\xi) + \vec{\nabla}'\cdot(\vec{J}\times\vec{r}\,'\xi)$$

to replace the integrand above. The $\vec{\nabla}'\cdot(\vec{J}\times\vec{r}\,'\xi)$ term in the volume integral may be converted to a surface integral surrounding the source, and since J is zero outside the source, that term vanishes. Our final form for the coefficient is then

$$b_\ell^m = \frac{ik}{\ell(\ell+1)}\sqrt{\frac{\mu_0}{\varepsilon_0}}\int \vec{J}(\vec{r}\,')\cdot\left[\vec{\nabla}'\times\vec{r}\,'\mathrm{j}_\ell(kr')\mathrm{Y}_\ell^{*m}(\theta',\varphi')\right]d^3r' \tag{10–99}$$

Similarly, from (10–95) we find that outside the source ($\vec{J}=0$),

$$\vec{E}\cdot\vec{r} = \frac{-k}{\omega\varepsilon_0}\sum_{\ell,m}\mathrm{h}_\ell^{(1)}(kr)\mathrm{Y}_\ell^m(\theta,\varphi)\int \vec{r}\,'\cdot\{\vec{\nabla}'\times[\vec{\nabla}'\times\vec{J}(\vec{r}\,')]\}\mathrm{j}_\ell(kr')\mathrm{Y}_\ell^{*m}(\theta',\varphi')d^3r'$$

$$= -\sqrt{\frac{\mu_0}{\varepsilon_0}} \sum_{\ell,m} h_\ell^{(1)}(kr) Y_\ell^m(\theta,\varphi) \int \vec{J} \cdot \left\{ \vec{\nabla}' \times \left[\vec{\nabla}' \times \vec{r}' j_\ell(kr') Y_\ell^{*m}(\theta',\varphi') \right] \right\} d^3r'$$

$$= \sum a_\ell^m \ell(\ell+1) \, h_\ell^{(1)}(kr) Y_\ell^m(\theta,\varphi) \tag{10–100}$$

where the latter form corresponds to the TM r-component of the electric field given in (3–68). This expression allows us to identify the amplitude coefficient for TM wave generation a_ℓ^m as

$$a_\ell^m = \frac{-1}{\ell(\ell+1)} \sqrt{\frac{\mu_0}{\varepsilon_0}} \int \vec{J} \cdot \left\{ \vec{\nabla}' \times \left[\vec{\nabla}' \times \vec{r}' j_\ell(kr') Y_\ell^{*m}(\theta',\varphi') \right] \right\} d^3r' \tag{10–101}$$

The radiation fields from an arbitrary source may now be written

$$\vec{E} = \sum_{\ell,m} a_\ell^m \vec{E}_{\ell,m}^{\,\mathrm{TM}}(\vec{r}) + b_\ell^m \vec{E}_{\ell,m}^{\,\mathrm{TE}}(\vec{r}) \tag{10–102}$$

and

$$\vec{H} = \sum_{\ell,m} a_\ell^m \vec{H}_{\ell,m}^{\,\mathrm{TM}}(\vec{r}) + b_\ell^m \vec{H}_{\ell,m}^{\,\mathrm{TE}}(\vec{r}) \tag{10–103}$$

If $j_\ell(kr')$ is replaced by its first order approximation $(kr')^n/(2n+1)!!$ for small kr', the multipole expansion results, with a_ℓ^m due to the electric 2^ℓ-pole moment and b_ℓ^m due to the magnetic 2^ℓ-pole moment. In this treatment it is unfortunately not transparent that H_r (or the TE coefficient, b_ℓ^m) derives directly from the currents, while E_r derives more directly from the charge distribution.

The "unit" fields have been defined so that the TE and TM fields give the same energy flux. Inspection of (10–99) and (10–101) reveals that the electric dipole-generated field coefficient, a_ℓ^m, has one more differentiation with respect to the source coordinates than b_ℓ^m, which corresponds very loosely to division by a distance of order the source dimension a. We infer that a given current gives rise to a magnetic multipole source term b_ℓ^m roughly ka ($\ll 1$) times that of the electric source term a_ℓ^m. With the harmonic time dependence assumed, $ka = \omega a/c \approx v/c$, where v is the velocity of the charge giving rise to the current. For optical radiation from atoms or molecules this leads to the expectation that magnetic dipole radiation will have a strength $v^2/c^2 \approx 10^{-4}$ compared to that of the electric dipole.

⋆ 10.2.4 Energy and Angular Momentum of the Multipole Fields

To compute the energy radiated by an oscillating source, it suffices to use the large kr asymptotic form of $h_\ell^{(1)}(kr) = (-i)^{\ell+1} e^{ikr}/(kr)$. Under this simplification, TM 2^ℓ-pole radiation (electric multipole) with unit amplitude is given by (3–66) as

$$\vec{H}^{\mathrm{TM}} = \frac{-ik}{\mu_0 c} \left\{ \vec{r} \times \vec{\nabla} \left[h_\ell^{(1)}(kr) Y_\ell^m(\theta,\varphi) \right] \right\}$$

$$= -ik \sqrt{\frac{\varepsilon_0}{\mu_0}} h_\ell^{(1)}(kr) \vec{r} \times \vec{\nabla} Y_\ell^m(\theta,\varphi)$$

$$= i^{-\ell}\sqrt{\frac{\varepsilon_0}{\mu_0}}\frac{e^{ikr}}{r}\vec{r}\times\vec{\nabla}\mathrm{Y}_\ell^m(\theta,\varphi) \tag{10–104}$$

while from the explicit expressions (3–68), the corresponding electric field is

$$\vec{E}^{\mathrm{TM}} \approx i^{-\ell}\frac{e^{ikr}}{r}\left[\frac{\partial}{\partial\theta}\mathrm{Y}_\ell^m(\theta,\varphi)\hat{\theta} + \frac{im}{\sin\theta}\mathrm{Y}_\ell^m(\theta,\varphi)\hat{\varphi}\right]$$

$$= i^{-\ell}e^{ikr}\vec{\nabla}\mathrm{Y}_\ell^m(\theta,\varphi)\ . \tag{10–105}$$

For magnetic multipole radiation (TE waves), the equivalent results are

$$\vec{E}^{\mathrm{TE}} = i^{-\ell}\frac{e^{ikr}}{r}\vec{r}\times\vec{\nabla}\mathrm{Y}_\ell^m(\theta,\varphi) \tag{10–106}$$

and

$$\vec{H}^{\mathrm{TE}} = i^{-\ell}\sqrt{\frac{\varepsilon_0}{\mu_0}}e^{ikr}\vec{\nabla}\mathrm{Y}_\ell^m(\theta,\varphi) \tag{10–107}$$

For either type, the Poynting vector, $\vec{\mathcal{S}}$, is given by

$$\langle\vec{\mathcal{S}}\rangle = \tfrac{1}{2}\mathrm{Re}(\vec{E}\times\vec{H}^*) = \tfrac{1}{2}\sqrt{\frac{\varepsilon_0}{\mu_0}}|E|^2\hat{r} = \tfrac{1}{2}\sqrt{\frac{\mu_0}{\varepsilon_0}}|H|^2\hat{r}$$

$$= \tfrac{1}{2}\sqrt{\frac{\varepsilon_0}{\mu_0}}\left|\vec{\nabla}\mathrm{Y}_\ell^m(\theta,\varphi)\right|^2\hat{r} \tag{10–108}$$

The total power, **P**, radiated by the source may be found by integrating (10–108) over the surface of a large enclosing sphere of radius R centered in the source:

$$\langle\mathbf{P}\rangle = \oint\langle\vec{\mathcal{S}}\rangle\cdot d\vec{S} = \frac{1}{2}\sqrt{\frac{\varepsilon_0}{\mu_0}}\oint\left|\vec{\nabla}\mathrm{Y}_\ell^m(\theta,\varphi)\right|^2\hat{r}\cdot d\vec{S}$$

$$= \frac{1}{2}\sqrt{\frac{\varepsilon_0}{\mu_0}}\left[\oint\vec{\nabla}\cdot(\mathrm{Y}_\ell^{m*}\vec{\nabla}\mathrm{Y}_\ell^m)dS - \int_{4\pi}|\,\mathrm{Y}_\ell^m\nabla^2\mathrm{Y}_\ell^m\,|R^2d\Omega\right] \tag{10–109}$$

The first integral vanishes because $\vec{\nabla}\mathrm{P}_\ell^m(\cos\theta)e^{i\varphi} \to \mathrm{P}_\ell^{m\pm1}(\cos\theta)e^{i\varphi}$, while the second may be simplified by $\nabla^2\mathrm{Y}_\ell^m = -\ell(\ell+1)\mathrm{Y}_\ell^m/R^2$ to yield

$$\langle\mathbf{P}\rangle = \frac{1}{2}\sqrt{\frac{\varepsilon_0}{\mu_0}}\ell(\ell+1)\int|\mathrm{Y}_\ell^m(\theta,\varphi)|^2\,d\Omega = \frac{1}{2}\sqrt{\frac{\varepsilon_0}{\mu_0}}\ell(\ell+1) \tag{10–110}$$

The angular momentum transported across a spherical surface of radius r during time dt is just the angular momentum contained in a shell of thickness $c\,dt$. In other words, the angular momentum lost from the volume surrounded by the shell during time dt is

$$d\vec{L} = \int_{shell}(\vec{r}\times\vec{\mathcal{P}})r^2cdtd\Omega$$

$$= c\int_{shell}\vec{r}\times\frac{\vec{E}\times\vec{H}}{c^2}r^2d\Omega dt \tag{10–111}$$

which, for sufficiently small dt, we may rewrite as

$$d\vec{L} = \frac{r^2}{c}\int\left[\vec{E}(\vec{r}\cdot\vec{H}) - \vec{H}(\vec{r}\cdot\vec{E})\right] d\Omega dt \tag{10–112}$$

To evaluate the integral, we specialize to TM radiation (electric multipole), although the TE results are identical. For TM radiation, $\vec{H}\cdot\vec{r} = 0$ and

$$\begin{aligned}\vec{E}\cdot\vec{r} &= \ell(\ell+1)\mathrm{h}_\ell^{(1)}(kr)\mathrm{Y}_\ell^m(\theta,\varphi)\\ &\approx \ell(\ell+1)i^{-\ell+1}\frac{e^{ikr}}{kr}\mathrm{Y}_\ell^m(\theta,\varphi)\end{aligned} \tag{10–113}$$

The φ component of $\vec{H}$, H_φ, contributes nothing to $\vec{L}$ since, using (F–32) to evaluate the derivative called for in (3–68) we find

$$\frac{\partial}{\partial\theta}\mathrm{P}_\ell^m(\cos\theta) = \tfrac{1}{2}\mathrm{P}_\ell^{m+1}(\cos\theta) - \tfrac{1}{2}(\ell+m)(\ell-m+1)\mathrm{P}_\ell^{m-1}(\cos\theta) \tag{10–114}$$

which, when multiplied by $\mathrm{Y}_\ell^m(\theta,\varphi)$ in the expression for E_r or E_φ, leads to a vanishing integral. From (3–68) and the asymptotic form $\mathrm{h}_\ell^{(1)}(kr) = (-i)^{\ell+1}e^{ikr}/(kr)$ we have

$$H_\theta \approx \frac{km}{\mu_0 c\sin\theta}(-i)^{\ell+1}\frac{e^{ikr}}{kr}\mathrm{Y}_\ell^m(\theta,\varphi) \tag{10–115}$$

Reverting to Cartesian coordinates,

$$H_z = -H_\theta\sin\theta \approx -m\sqrt{\frac{\varepsilon_0}{\mu_0}}\,i^{-\ell+1}\frac{e^ikr}{r}\mathrm{Y}_\ell^m(\theta,\varphi) \tag{10–116}$$

Thus the time average loss rate of angular momentum about the z axis due to radiation with the aid of (10–113) and (10–116) is found to be

$$\begin{aligned}\left\langle\frac{dL_z}{dt}\right\rangle &= \frac{r^2}{c}\int H_z(\vec{r}\cdot\vec{E})d\Omega\\ &= \frac{1}{2}\frac{m}{kc}\sqrt{\frac{\varepsilon_0}{\mu_0}}\ell(\ell+1)\int\left|\mathrm{Y}_\ell^m(\theta,\varphi)\right|^2 d\Omega = \frac{1}{2}\frac{m}{kc}\sqrt{\frac{\varepsilon_0}{\mu_0}}\ell(\ell+1)\end{aligned} \tag{10–117}$$

The ratio of radiated angular momentum to radiated power $\mathbf{P} = dW/dt$ is

$$\frac{\langle dL_z/dt\rangle}{\langle dW/dt\rangle} = \frac{m}{kc} = \frac{m}{\omega} \tag{10–118}$$

a relationship consistent with the requirements of quantum mechanics. According to quantum mechanics, atoms radiate photons with energy $\hbar\omega$ while changing the z component of angular momentum by $\hbar$. When radiating via higher moments, angular momentum is radiated in multiples of $\hbar$. (This treatment does not distinguish between spin and orbital angular momentum, and caution in the interpretation is advised.)

10.2.5 Radiation from Extended Sources

When the dimensions of the radiating system are not small compared to the wavelength of radiation, the multipole expansion of the potential is not valid and the integrals (10–21) or (10–22) for the potential must be evaluated directly. To make this discussion more concrete, let us consider the radiation arising from the thin, linear, center-fed antenna illustrated in Figure 10.5. We choose the z axis to lie along the antenna.

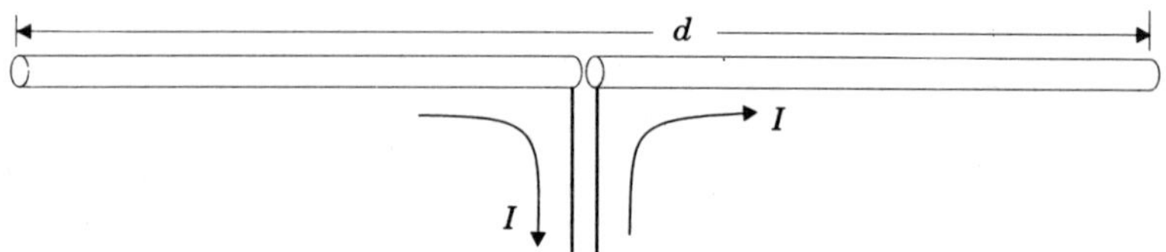

Figure 10.5: A center-fed linear antenna carrying symmetric current $I(z)$.

The antenna of length d is split by a small gap at its midpoint where each half is supplied by current $\pm I_0 e^{-i\omega t}$. To deduce the magnitude of the current on any point of the antenna, we neglect radiation damping. The current must be symmetric about the gap in the middle, and further, it must vanish at the ends. If the antenna were short, we would expect the entire right side to be uniformly charged to one polarity while the other side would be uniformly charged with the opposite polarity. For the one-dimensional problem, the continuity equation then states $\partial J/\partial z = -\partial\rho/\partial t$ = constant. Thus J would be of the form $(const.) \times (\frac{1}{2}d - |z|)\delta(x)\delta(y)$. For a longer antenna, we expect the charge density, and therefore also the current, to oscillate, still maintaining the endpoint boundary condition of $J = 0$ (Figure 10.6). We therefore take

$$\vec{J}(\vec{r},t) = \hat{k} I \sin\left(\tfrac{1}{2}kd - k|z|\right)\delta(x)\delta(y)e^{-i\omega t} \qquad \text{for } |z| \le \tfrac{1}{2}d \tag{10–119}$$

The supply current will evidently be $I_0 e^{-i\omega t} = I e^{-i\omega t}\sin\frac{1}{2}kd$. The vector potential due to an oscillating current is in general given by

$$\vec{A}(\vec{r}) = \frac{\mu_0}{4\pi}\int \frac{\vec{J}\left(\vec{r}\,', t - \frac{|\vec{r}-\vec{r}\,'|}{c}\right)}{|\vec{r}-\vec{r}\,'|} d^3r'$$

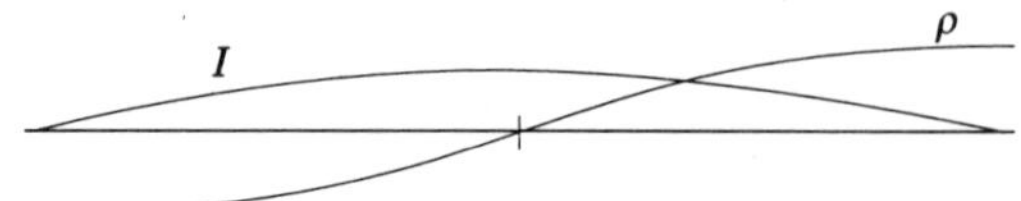

Figure 10.6: The figure shows a snapshot of the current and charge distribution along the linear antenna. The same current runs along the antenna at equal distances each side of center.

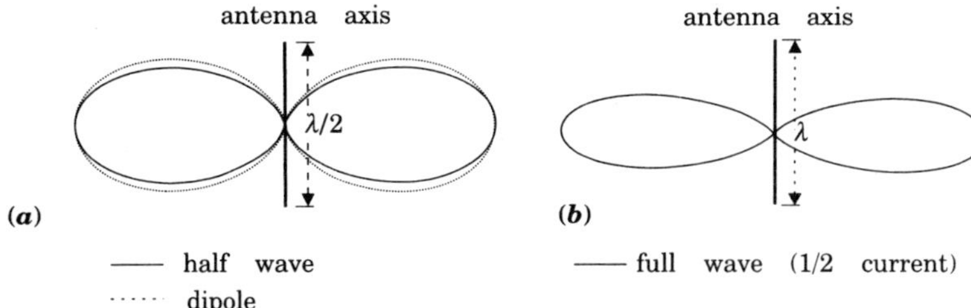

Figure 10.7: The radiation pattern from a half wave (left) and full wave (right) antenna. The dipole radiation pattern is superimposed on that for the half wave antenna for comparison.

$$= \frac{\mu_0}{4\pi}\int \vec{J}(\vec{r}\,',t)\frac{e^{ik|\vec{r}-\vec{r}\,'|}}{|\vec{r}-\vec{r}\,'|}d^3r' \tag{10–120}$$

In the radiation zone, we may approximate

$$\frac{e^{ik|\vec{r}-\vec{r}\,'|}}{|\vec{r}-\vec{r}\,'|} \quad \text{by} \quad \frac{e^{ikr}}{r}\cdot e^{-ik(\vec{r}\cdot\vec{r}\,')/r} = \frac{e^{ikr}}{r}e^{-ikz'\cos\theta} \tag{10–121}$$

With this approximation, the vector potential may be written

$$\begin{aligned}\vec{A}_0(\vec{r}) &= \frac{\mu_0}{4\pi}\frac{e^{ikr}}{r}\hat{k}\int_{-d/2}^{d/2} I\sin\left(\tfrac{1}{2}kd - k|z'|\right)e^{-ikz'\cos\theta}dz' \\ &= 2\frac{\mu_0}{4\pi}\frac{e^{ikr}}{kr}\hat{k}I\left[\frac{\cos\left(\frac{1}{2}kd\cos\theta\right)-\cos\frac{1}{2}kd}{\sin^2\theta}\right]\end{aligned} \tag{10–122}$$

In the radiation zone,

$$\vec{B} = ik\hat{r}\times\vec{A} \quad \Rightarrow \quad |\vec{B}_0| = k\sin\theta|\vec{A}_0| \tag{10–123}$$

and

$$\vec{E} = ick(\hat{r}\times\vec{A})\times\hat{r} \quad \Rightarrow \quad |\vec{E}_0| = ck\sin\theta|\vec{A}_0| \tag{10–124}$$

giving an angular power distribution

$$\frac{d\mathbf{P}}{d\Omega} = \frac{r^2}{2\mu_0}|\vec{E}\times\vec{B}^*| = \frac{2\mu_0}{(4\pi)^2}I^2c\left|\frac{\cos\left(\frac{1}{2}kd\cos\theta\right)-\cos\frac{1}{2}kd}{\sin\theta}\right|^2 \tag{10–125}$$

The angular distribution evidently depends on kd. For $kd \ll \pi$, one obtains, as should be anticipated, the dipole radiation pattern. For the special values $kd = \pi$ (half wave) and $kd = 2\pi$ (full wave) the angular distributions become

$$\frac{d\mathbf{P}}{d\Omega} = \frac{2\mu_0 cI^2}{(4\pi)^2}\frac{\cos^2\left(\frac{1}{2}\pi\cos\theta\right)}{\sin^2\theta} \qquad kd = \pi \tag{10–126}$$

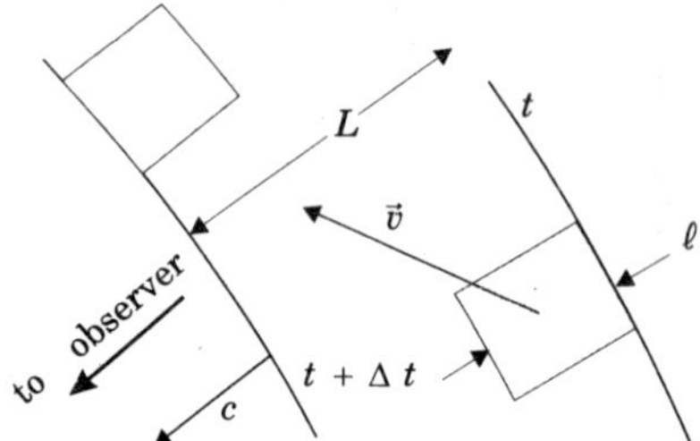

Figure 10.8: An information collecting sphere passes over the moving charge distribution.

and

$$\frac{d\mathbf{P}}{d\Omega} = \frac{8\mu_0 c I^2}{(4\pi)^2} \frac{\cos^4\left(\frac{1}{2}\pi\cos\theta\right)}{\sin^2\theta} \qquad kd = 2\pi \tag{10–127}$$

As can be seen from Figure 10.7, the half wave pattern resembles that of a dipole, but that of the full wave is considerably narrower.

The integrals over Ω, needed to find the total power, are most easily performed numerically, yielding

$$\mathbf{P} = \frac{I^2}{2}\frac{\mu_0 c}{4\pi} \times \begin{cases} 2.438 & \text{for} \quad kd = \pi \\ 6.70 & \text{for} \quad kd = 2\pi \end{cases} \tag{10–128}$$

10.3 The Liénard-Wiechert Potentials

The radiation field of a moving point charge is of particular importance because although we have pretended that charge densities are smooth, we know that charge is lumpy, and any real description of the world should, for instance, picture currents as moving point charges. The fact that for a moving point charge, the position vector $\vec{r}\,'$ is a function of time complicates matters considerably.

To acquire a naive understanding, let us begin with the calculation of the scalar potential arising from a moving finite-size lump of charge distribution. According to (10–22), the potential at $(\vec{r}, t)$ is given by

$$V(\vec{r}, t) = \frac{1}{4\pi\varepsilon_0}\int \frac{\rho\left(\vec{r}\,', t - \frac{|\vec{r}-\vec{r}\,'|}{c}\right)}{|\vec{r}-\vec{r}\,'|} d^3r' \tag{10–129}$$

Imagine that situated at $\vec{r}$, we collect information about the field by means of the information-collecting sphere to which we alluded earlier. For brevity we set $R = |\vec{r}-\vec{r}\,'|$ and $\hat{n}' = (\vec{r}-\vec{r}\,')/R$. The sphere, as it passes each point $\vec{r}\,'$ in space at $t' = t - R/c$, notes the charge density, divides by R, and sums the results as it converges on $\vec{r}$ at t.

Let us go through the motions of the sphere passing through a moving charge distribution. To be definite, let us suppose that the charge distribution has radial thickness ℓ and moves in a direction shown in Figure 10.8. If the charge has a component of its movement toward the observer as shown, the sphere will spend more time passing over it than if it were stationary and hence will count it more heavily in the final sum than if it were stationary. Because of the charge's movement, the sphere dwells on the charge for a time $\Delta t = L/c$ instead of the time ℓ/c it would have if the charge did not move. The contribution to the potential would then be

$$V = \frac{1}{4\pi\varepsilon_0}\frac{q}{R}\frac{L}{\ell} \tag{10–130}$$

To compute L/ℓ we note that during the time interval Δt during which the shell overlaps the charge, the shell travels distance $c\Delta t = L$ while the charge travels radial distance $(\vec{v}\cdot\hat{n}')\Delta t = L - \ell$. Eliminating Δt, we find

$$\frac{L}{\ell} = \frac{1}{1 - \dfrac{\vec{v}\,'\cdot\hat{n}'}{c}} = \frac{1}{1 - \vec{\beta}'\cdot\hat{n}'} \tag{10–131}$$

where we have defined $\vec{\beta}' = \vec{v}\,'/c$. In the limit as $\ell \to 0$, we can assign a unique value to the retarded velocity $\vec{v}\,'$ and direction $\hat{n}'$. The potential we find is then

$$V(\vec{r},t) = \frac{1}{4\pi\varepsilon_0}\frac{q}{\left(1 - \vec{\beta}'\cdot\hat{n}'\right)|\vec{r}-\vec{r}\,'|} \tag{10–132}$$

The vector potential $\vec{A}$ may be immediately written as a generalization of (10–132)

$$\vec{A}(\vec{r},t) = \frac{\mu_0}{4\pi}\frac{q\vec{v}\,'}{\left(1 - \vec{\beta}'\cdot\hat{n}'\right)|\vec{r}-\vec{r}\,'|} \tag{10–133}$$

Since (10–132) and (10–133) make no reference to the dimensions of the charge, they must hold equally for extended and point charges (merely take the limit as $\ell \to 0$).

The discerning reader may have noticed a subtle change in notation. Instead of the prime ($'$) being used to denote the charge's position or velocity, it has taken on the meaning of the source charge's retarded position, velocity, time, and so on.

⋆ 10.3.1 The Liénard-Wiechert Potentials Using Green's Functions

We now turn to a somewhat more formal derivation of (10-132) and (10–133). Consider the potential we construct from the time-dependent Green's function (10–27),

$$V(\vec{r},t) = \frac{1}{4\pi\varepsilon_0}\int \frac{\rho(\vec{r}\,',\tau)\delta\left[t - \left(\tau + \dfrac{|\vec{r}-\vec{r}\,'|}{c}\right)\right]}{|\vec{r}-\vec{r}\,'|}d^3r'\,d\tau \tag{10–134}$$

The charge density of a moving point charge q is given by $\rho(\vec{r}\,',\tau) = q\delta\big(\vec{r}\,' - \vec{r}_q(\tau)\big)$, where the charge is located at position $\vec{r}_q(\tau)$ at the time τ. Then

$$V(\vec{r},t) = \frac{1}{4\pi\varepsilon_0}\int \frac{q\delta\left[\vec{r}\,' - \vec{r}_q(\tau)\right]\delta\left(t - \tau - \dfrac{|\vec{r}-\vec{r}\,'|}{c}\right)}{|\vec{r}-\vec{r}\,'|}d^3r'\,d\tau \tag{10–135}$$

The integration over space is easily performed to give

$$V(\vec{r},t) = \frac{q}{4\pi\varepsilon_0}\int \frac{\delta\left(t-\tau-\dfrac{|\vec{r}-\vec{r}_q(\tau)|}{c}\right)}{|\vec{r}-\vec{r}_q(\tau)|}d\tau \tag{10–136}$$

The δ function in the integral above may have its argument simplified by means of the identity (C–7)

$$\delta\big(g(\tau)\big) = \sum_i \frac{\delta(\tau-\tau_i)}{|dg/d\tau|_{\tau=\tau_i}}$$

where the τ_i are the roots of $g(\tau)$. In our case $g(\tau) = t - |\vec{r}-\vec{r}_q(\tau)|/c - \tau$, giving

$$\left|\frac{dg}{d\tau}\right| = \left|-1 - \frac{d}{d\tau}\frac{|\vec{r}-\vec{r}_q(\tau)|}{c}\right| = \left|-1 + \frac{(\vec{r}-\vec{r}_q)}{c|\vec{r}-\vec{r}_q|}\cdot\frac{d\vec{r}_q}{d\tau}\right|$$

$$= \left|-1 + \frac{\vec{v}(\tau)\cdot\hat{n}(\tau)}{c}\right| = 1 - \vec{\beta}(\tau)\cdot\hat{n}(\tau) \tag{10–137}$$

Then

$$\delta\left(t - \frac{|\vec{r}-\vec{r}_q(\tau)|}{c} - \tau\right) = \frac{\delta(\tau-\tau_i)}{(1-\vec{\beta}\cdot\hat{n})|_{\tau=\tau_i}} \tag{10–138}$$

where $\tau_i = t - |\vec{r}-\vec{r}_q(\tau)|/c$, is the retarded time, t'. With this help, (10–136) is easily integrated over τ to yield

$$V(\vec{r},t) = \frac{q}{4\pi\varepsilon_0}\frac{1}{|\vec{r}-\vec{r}_q(t')|\left[1-\vec{\beta}(t')\cdot\hat{n}(t')\right]} \tag{10–139}$$

In identical fashion the vector potential may be obtained

$$\vec{A}(\vec{r},t) = \frac{\mu_0}{4\pi}\frac{q\vec{v}(t')}{|\vec{r}-\vec{r}_q(t')|\left[1-\vec{\beta}(t')\cdot\hat{n}(t')\right]} = \frac{1}{4\pi\varepsilon_0 c}\frac{q\vec{\beta}'}{|\vec{r}-\vec{r}\,'|\left(1-\vec{\beta}'\cdot\hat{n}'\right)} \tag{10–140}$$

It will be seen that (10–139) and (10–140) coincide with (10–132) and (10–133) obtained by more elementary means.

10.3.2 The Fields of a Moving Point Charge

Computing the fields $\vec{B} = \vec{\nabla}\times\vec{A}$ and $\vec{E} = -\vec{\nabla}V - \partial\vec{A}/\partial t$ from (10–139) and (10–140) is rather laborious because the retarded time and hence also the particle's retarded position coordinates depend on the field coordinates through the variable field travel delay, $|\vec{r}-\vec{r}\,'|/c$. For sophisticated readers, Jackson[27] obtains a covariant Green's function and then efficiently calculates the fields as the components of a covariant tensor. Because of the importance of the results, we will perform the indicated differentiation of (10–139) and (10–140) by elementary means, but postpone it to

[27] J.D. Jackson (1998) *Classical Electrodynamics*, 3rd. ed. John Wiley & Sons, New York.

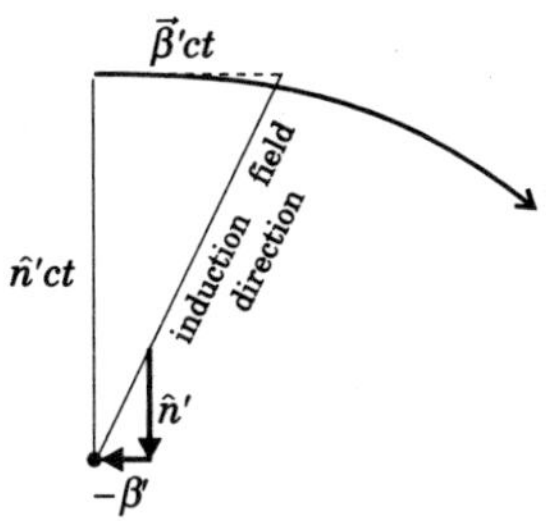

Figure 10.9: The induction (non-radiative) field of a moving point charge is detected by the observer to point to/from the position which the particle is extrapolated to occupy at time t, not its location at the retarded time, t', when the field was emitted.

the end of the chapter where at least a quick look is recommended to get the flavor of the labor. For now we merely quote the results:

$$\vec{E}(\vec{r},t) = \frac{q}{4\pi\varepsilon_0}\left[\frac{\hat{n}' - \vec{\beta}'}{\gamma'^2 R'^2 \xi'^3} + \frac{-\vec{a}'_\perp + \hat{n}' \times (\vec{a}' \times \vec{\beta}')}{c^2 R' \xi'^3}\right] \tag{10–141}$$

and

$$\vec{B}(\vec{r},t) = \frac{\hat{n}'}{c} \times \vec{E}(\vec{r},t) \tag{10–142}$$

where $\gamma'^2 \equiv 1/(1-\beta'^2)$, $\xi' \equiv 1 - \vec{\beta}' \cdot \hat{n}'$, and $\vec{a}'_\perp$ is the component of the retarded acceleration perpendicular to $\hat{n}'$.

The fields divide fairly naturally into a $1/R'^2$ part, the induction field, and a $1/R'$ part, the radiation field. It is interesting to note that the electric induction field points in the direction $\hat{n}' - \vec{\beta}'$; this vector points not from the retarded position, but from the extrapolated present position where the charge will be at the time the field is received if it continues undisturbed with its retarded velocity. The magnetic induction field similarly has a direction appropriate to a moving charge at the extrapolated current (i.e. at the time of observation) position as shown in Figure 10.9. The remainder of our effort in this chapter will be spent on the radiation fields

$$\vec{E}_R = \frac{q}{4\pi\varepsilon_0}\left[\frac{-\vec{a}'_\perp + \hat{n}' \times (\vec{a}' \times \vec{\beta}')}{c^2 R' \xi'^3}\right] \tag{10–143}$$

and

$$\vec{B}_R = \frac{\hat{n}'}{c} \times \vec{E}_R \tag{10–144}$$

We initially consider only the radiation from accelerated charges whose speed is small compared to that of light.

10.3.3 Radiation from Slowly Moving Charges

When the velocity of the charges is small compared to c, the term involving β' will be negligible and $\xi' = 1$. The radiation fields become

$$\vec{E}_R = \frac{q}{4\pi\varepsilon_0}\frac{-\vec{a}'_\perp}{c^2 R'} \quad \text{and} \quad \vec{B}_R = \frac{q}{4\pi\varepsilon_0}\frac{\hat{n}' \times (-\vec{a}'_\perp)}{c^3 R'} \tag{10–145}$$

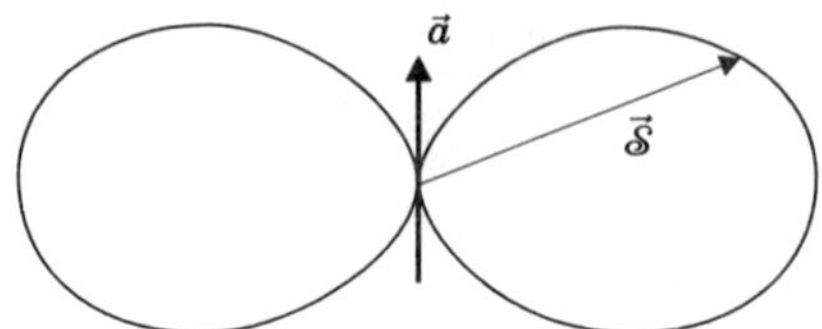

Figure 10.10: A nonrelativistic charge radiates most strongly in a direction at right angles to its acceleration.

yielding the Poynting vector

$$\vec{S} = \frac{\vec{E}\times\vec{B}}{\mu_0} = \frac{q^2}{(4\pi\varepsilon_0)^2}\frac{(a'_\perp)^2\hat{n}'}{\mu_0 c^5 R'^2}$$

$$= \frac{q^2(a'_\perp)^2\hat{n}'}{(4\pi)^2\varepsilon_0 c^3 R'^2} \qquad (10\text{–}146)$$

and, if $\hat{n}'$ makes angle θ with $\vec{a}\,'$, $a'_\perp = a'\sin\theta$, giving

$$\vec{S} = \frac{q^2}{(4\pi)^2\varepsilon_0}\frac{(a')^2\sin^2\theta}{c^3 R'^2}\hat{n}' \qquad (10\text{–}147)$$

The angular distribution of radiation from the accelerated charge may be written

$$\frac{d\mathbf{P}}{d\Omega} = \frac{q^2 a'^2\sin^2\theta}{(4\pi)^2\varepsilon_0 c^3} \qquad (10\text{–}148)$$

where Ω is the solid angle as seen by a stationary point at the source. The radiation pattern is illustrated in Figure 10.10. The total power radiated is found by integrating over the solid angle to be

$$\mathbf{P} = \frac{q^2 a'^2}{(4\pi)^2\varepsilon_0 c^3}\int_0^{2\pi}\int_0^{\pi}\sin^2\theta\sin\theta d\theta d\varphi = \frac{q^2 a'^2}{8\pi\varepsilon_0 c^3}\int_0^{\pi}\sin^3\theta d\theta$$

$$= \frac{q^2 a'^2}{6\pi\varepsilon_0 c^3} \qquad (10\text{–}149)$$

10.3.4 Thomson Scattering

Consider a free electron subjected to an oscillating field from a passing electromagnetic wave. The electron will accelerate and therefore radiate, but not in the same direction as the incident (plane) wave. The scattering of coherent electromagnetic waves by free nonrelativistic electrons is known as Thomson scattering. The cross section for scattering, σ, the area over which the scatterer may be thought to intercept all radiation to remove it from the incident beam, may be defined as

$$\sigma = \frac{\text{Power radiated}}{\text{Power incident/Unit area}} \qquad (10\text{–}150)$$

If an incident plane wave has electric field $\vec{E}$, then the electron's acceleration will be $\vec{a} = e\vec{E}/m$, leading to radiated power

$$\mathbf{P}_R = \frac{e^2}{6\pi\varepsilon_0 c^3}\left(\frac{eE}{m}\right)^2 \tag{10–151}$$

The incident power per unit area is $\vec{S} = E^2/\mu_0 c$ (note that we have used the instantaneous $\vec{E}$ and $\vec{S}$) so that the Thomson scattering cross section $\sigma_{\rm T}$ is

$$\sigma_{\rm T} = \frac{\mu_0 c}{6\pi\varepsilon_0 c^3}\frac{e^4}{m^2} = \frac{e^4/m^2c^4}{6\pi\varepsilon_0^2} \tag{10–152}$$

This result is usually expressed in terms of the "classical radius" of the electron $r_0 \equiv e^2/4\pi\varepsilon_0 m_0 c^2$ (the radius of a spherical shell of charge whose self-energy is m_0c^2). In terms of r_0 ($\approx 2.82\times 10^{-15}$ m for an electron), the Thomson cross section is

$$\sigma_{\rm T} = \frac{8\pi}{3}r_0^2 = 6.67\times 10^{-29}\ \mathrm{m}^2 \tag{10–153}$$

EXAMPLE 10.2: The sun is a highly ionized plasma with average free electron density about $10^{24}\mathrm{cm}^{-3}$. (*a*) Find the mean free path for electromagnetic radiation in the sun. (*b*) Estimate the time required for electromagnetic radiation to diffuse from the core of the sun to the outside ($R_\odot \approx 7\times 10^{10}$ cm).

Solution: (*a*) The mean free path is given by $\ell = 1/n\sigma = 1.50$ cm.

(*b*) To find the diffusion time, we consider the radiation's path as a random walk in three dimensions. Suppose that after N steps the packet of radiation under consideration is at distance D_N from the center of the sun. Then, after one more step of length ℓ at random angle θ with respect to D_N, the mean square displacement will be

$$\begin{aligned}\langle D_{N+1}^2\rangle &= \langle D_N^2 + \ell^2 + 2D_N\ell\cos\theta\rangle \\ &= \langle D_N^2\rangle + \ell^2 + 2D_N\ell\langle\cos\theta\rangle \\ &= \langle D_N^2\rangle + \ell^2\end{aligned} \tag{Ex 10.2.1}$$

The mean number of steps N required to reach R is $N = R^2/\ell^2$. Each step requires time ℓ/c, so that the total time required is

$$\frac{\ell}{c}\frac{R^2}{\ell^2} = \frac{R^2}{\ell c} = \frac{49\times 10^{20}\mathrm{cm}^2}{1.50\mathrm{cm}\times 3\times 10^{10}\mathrm{cm/s}} = 1.09\times 10^{11}\ \mathrm{s} \tag{Ex 10.2.2}$$

which is about 3500 years. This is actually an underestimate, as the density at the center of the sun is considerably larger that the average density. In any case, it is evident that it will take thousands of years before the luminosity of the sun reflects any change in conditions in the core.

10.3.5 Radiation by Relativistic Charges

When the velocity of a charged particle is not much less than c, it is more convenient to write the radiation field (10-143) as

$$\vec{E}_R = \frac{q}{4\pi\varepsilon_0}\frac{\hat{n}' \times \left[(\hat{n}' - \vec{\beta}') \times \dot{\vec{\beta}}'\right]}{cR'\xi'^3} \tag{10–154}$$

The component of the Poynting vector along $\hat{n}'$ is then

$$\begin{aligned}\vec{S}\cdot\hat{n}' = \frac{\vec{E}\times\vec{B}}{\mu_0}\cdot\hat{n}' &= \left(\frac{q}{4\pi\varepsilon_0}\right)^2 \frac{\left|\hat{n}' \times [(\hat{n}' - \vec{\beta}') \times \dot{\vec{\beta}}']\right|^2}{\mu_0 c^3 R'^2\xi'^6} \\ &= \frac{q^2}{(4\pi)^2\varepsilon_0 c}\frac{\left|\hat{n}' \times [(\hat{n}' - \vec{\beta}') \times \dot{\vec{\beta}}']\right|^2}{R'^2\xi'^6}\end{aligned} \tag{10–155}$$

a rather complicated expression. We will consider the cases of acceleration perpendicular to the velocity and acceleration parallel to the velocity separately. Before doing so, however, it's worth pointing out that $\vec{S}\cdot\hat{n}'$ is the energy flux moving in direction $\hat{n}'$, detected by an *observer* at $\vec{r}$. The energy will generally have been emitted during an earlier time interval $\Delta t'$ different in duration from the interval Δt during which it is observed, meaning that the reception rate will differ from the emission rate. It would probably be more sensible to consider, instead of the rate that radiation is received, the rate that the energy is emitted.

Suppose that at time t' the charge is at distance $R(t')$ from the observer, while at $t' + \Delta t'$ it is at distance $R(t' + \Delta t')$. The radiation emitted by the charge at t' arrives at the observer at time $t = t' + R(t')/c$, while the radiation emitted at $t' + \Delta t'$ arrives at $t + \Delta t = t' + \Delta t' + R(t' + \Delta t')/c$. Calculating the time interval during which the radiation arrives, we get

$$\begin{aligned}\Delta t &= t' + \Delta t' + \frac{R(t'+\Delta t')}{c} - t' - \frac{R(t')}{c} \\ &= \Delta t' + \frac{R(t'+\Delta t') - R(t')}{c} \\ &= \Delta t'\left[1 + \frac{1}{c}\frac{R(t'+\Delta t') - R(t')}{\Delta t'}\right]\end{aligned} \tag{10–156}$$

In the limit of small $\Delta t'$, this becomes simply $\Delta t = \Delta t'(1 - \vec{\beta}'\cdot\hat{n}') = \xi'\Delta t'$. Thus the energy received by the observer was in fact emitted at a rate ξ' greater than it is received. The power *radiated* (rather than received) per solid angle is given by

$$\frac{d\mathbf{P}(t')}{d\Omega} = R'^2\xi'(\vec{S}\cdot\hat{n}) = \frac{q^2}{(4\pi)^2\varepsilon_0 c}\frac{\left|\hat{n}' \times [(\hat{n}' - \vec{\beta}') \times \dot{\vec{\beta}}']\right|^2}{(1 - \hat{n}'\cdot\vec{\beta}')^5} \tag{10–157}$$

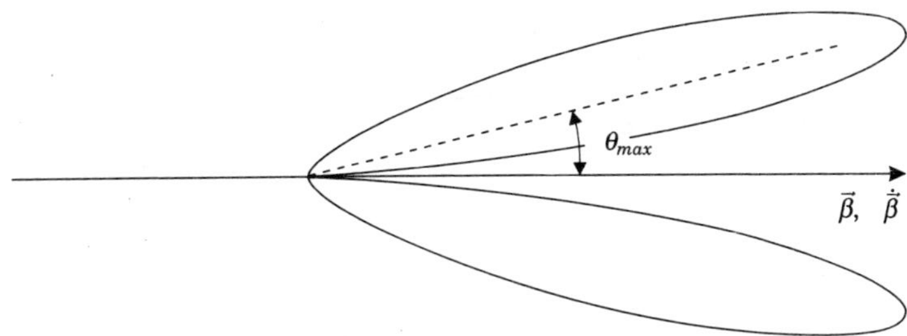

Figure 10.11: The radiation pattern of a relativistic charge with $v = 0.9c$, accelerated along its line of motion.

We proceed now to calculate the power emitted by charged particles accelerated along their line of motion.

Velocity and acceleration parallel: Let $\hat n'$ make angle θ' with $\vec\beta'$ (and $\dot{\vec\beta}'$) then $\hat n'\cdot\vec\beta' = \beta'\cos\theta'$ and

$$(\hat n' - \vec\beta')\times\dot{\vec\beta}' = \hat n'\times\dot{\vec\beta}' = \dot\beta'\sin\theta'(\hat n'\times\hat{\dot\beta}') \quad\Rightarrow\quad |\hat n'\times(\hat n'\times\dot{\vec\beta}')| = \dot\beta'\sin\theta' \qquad (10\text{–}158)$$

so that (10–157) becomes

$$\frac{d\mathbf{P}}{d\Omega} = \frac{q^2\dot\beta'^2\sin^2\theta'}{(4\pi)^2\varepsilon_0 c(1-\beta'\cos\theta')^5} \qquad (10\text{–}159)$$

For $\beta \ll 1$, this reproduces our nonrelativistic result (10–148), but as $\beta\to 1$, the pattern elongates in the direction of motion as sketched in Figure 10.11 for $\beta = 0.9$. The pattern can clearly be envisaged a that of Figure 10.10 with the lobes swept forward in the direction of the particle's motion. As β approaches unity, the angle of maximum emission, $\theta_{max}\to 1/(2\gamma)$, while the peak power density is proportional to γ^8!

Acceleration Perpendicular to Velocity: We choose a coordinate system in which, at t', $\vec\beta$ lies along the z axis and $\dot{\vec\beta}$ lies along the x axis. (For brevity, we drop the prime on the variables). In Cartesian coordinates, when the observer has angular coordinates θ and φ, $\vec\beta = (0,0,\beta)$, $\dot{\vec\beta} = (\dot\beta,0,0)$, and $\hat n = (\sin\theta\cos\varphi,\, \sin\theta\sin\varphi,\, \cos\theta)$. The term in the numerator of (10–154) then becomes

$$\begin{aligned}\hat n\times\left[(\hat n-\vec\beta)\times\dot{\vec\beta}\right] &= (\hat n\cdot\dot{\vec\beta})(\hat n-\vec\beta) - [\hat n\cdot(\hat n-\vec\beta)]\dot{\vec\beta}\\ &= \dot\beta\sin\theta\cos\varphi(\hat n-\vec\beta) - (1-\beta\cos\theta)\dot{\vec\beta}\end{aligned} \qquad (10\text{–}160)$$

which, when squared, gives

$$\begin{aligned}\left|\hat n\times\left[(\hat n-\vec\beta)\times\dot{\vec\beta}\right]\right|^2 &= (n^2 - 2\vec\beta\cdot\hat n + \beta^2)\dot\beta^2\sin^2\theta\cos^2\varphi\\ &\quad + (1-\beta\cos\theta)^2\dot\beta^2 - 2\dot\beta(1-\beta\cos\theta)(\hat n-\vec\beta)\cdot\dot{\vec\beta}\sin\theta\cos\varphi\end{aligned}$$

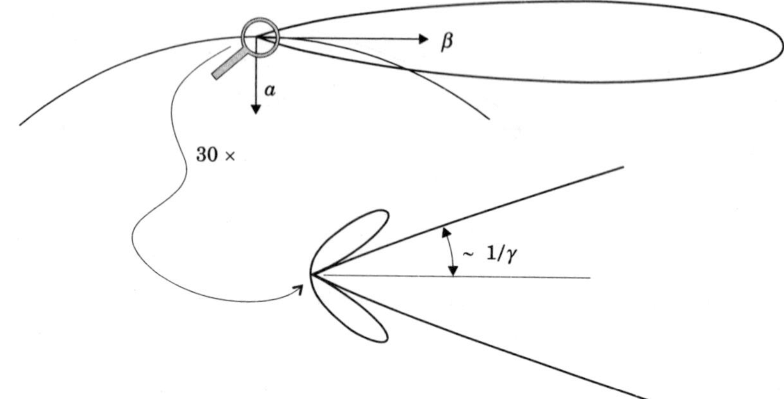

Figure 10.12: The radiation pattern from a charged particle with $v = 0.9c$, accelerated perpendicular to its velocity.

$$
\begin{aligned}
&= (1-2\beta\cos\theta+\beta^2)\dot\beta^2\sin^2\theta\cos^2\varphi \\
&\qquad\qquad + (1-\beta\cos\theta)^2\dot\beta^2 - 2\dot\beta^2(1-\beta\cos\theta)\sin^2\theta\cos^2\varphi \\
&= \dot\beta^2(1-\beta\cos\theta)^2\left[1+\frac{\sin^2\theta\cos^2\varphi(1-2\beta\cos\theta+\beta^2-2+2\beta\cos\theta)}{(1-\beta\cos\theta)^2}\right] \\
&= \dot\beta^2(1-\beta\cos\theta)^2\left[1-\frac{\sin^2\theta\cos^2\varphi}{\gamma^2(1-\beta\cos\theta)^2}\right] \qquad (10\text{–}161)
\end{aligned}
$$

The angular power distribution becomes

$$
\frac{d\mathbf{P}}{d\Omega} = \frac{q^2\dot\beta^2}{(4\pi)^2\varepsilon_0 c(1-\beta\cos\theta)^3}\left[1-\frac{\sin^2\theta\cos^2\varphi}{\gamma^2(1-\beta\cos\theta)^2}\right] \qquad (10\text{–}162)
$$

The total power emitted may be obtained by integrating $d\mathbf{P}/d\Omega$ or, as we will do in the next chapter, by Lorentz transforming the instantaneous rest frame result to yield

$$
\mathbf{P}(t') = \frac{q^2\dot\beta^2\gamma^4}{6\pi\varepsilon_0 c} \qquad (10\text{–}163)
$$

In fact, for arbitrary orientations of β and $\dot{\vec\beta}$, we will show (11–72) that

$$
\mathbf{P} = \frac{q^2\gamma^6}{6\pi\varepsilon_0 c}\left[(\dot\beta)^2 - |\vec\beta\times\dot{\vec\beta}|^2\right] \qquad (10\text{–}164)
$$

The radiation from a particle accelerated perpendicular to its velooity (circular motion) has an angular distribution vaguely like that of a particle accelerated along its velocity. It has a long narrow lobe facing forward (perpendicular to the acceleration), while the lobe facing against the velocity in the low speed limit is much smaller and swept forward, as illustrated in Figure 10.12. The power per solid angle

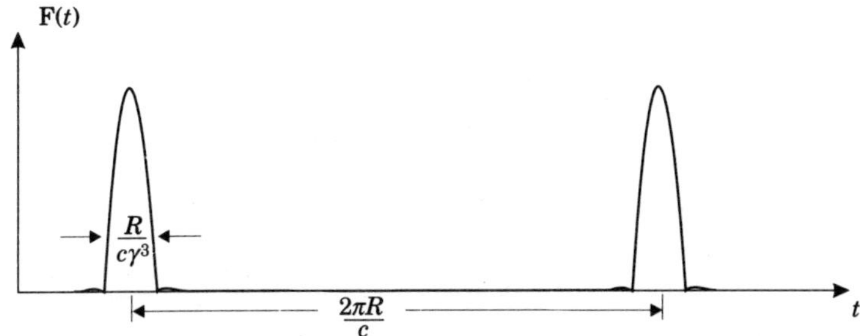

Figure 10.13: The qualitative time dependence of the power received from a charged particle in relativistic circular motion. The small bumps on each side of the main peak are due to the small, swept forward second lobe.

decreases to 0 when $\gamma^2(1-\beta\cos\theta)^2 = \sin^2\theta\cos^2\varphi$. (10–62) maximizes as a function of azimuthal angle when $\cos^2\varphi = 0$, ie. in a direction perpendicular to the plane of the orbit. The narrowing of the radiation lobes is referred to as the headlight effect.

10.3.6 Synchrotron Radiation

An important case of a charged particle accelerated in a direction perpendicular to its motion is provided by particles accelerated in a large circular trajectory in a synchrotron. Electrons or protons accelerated in a synchrotron may reach energies of more than a thousand times their rest mass energy and are therefore extremely relativistic. Let us, without getting embroiled in the mathematics, try to obtain a qualitative appreciation of the spectrum of synchrotron radiation.

As $\beta \to 1$, $(1-\beta\cos\theta) \to \frac{1}{2}(\gamma^{-2}+\theta^2)$ and the angular power distribution (10–162), illustrated in Figure 10.12, may be found to have an angular (full) width $\Delta\theta = 2/\gamma$ when observed in the plane of the orbit. An observer in the plane containing $\vec{v}$ and $\vec{a}$ will be illuminated by a pulse of electromagnetic radiation each time the headlight beam sweeps over him or her. If the particle travels at close to the speed of light, this repetition time is close to $T = 2\pi R/c = 2\pi/\omega_0$.

The pulse that eventually reaches the receiver is emitted during a time $\Delta t' = R\Delta\theta/c = 2R/\gamma c$. For a particle moving in the direction of the observer, the observer receives the pulse during a time $\Delta t = (1-\vec{\beta}\cdot\hat{n})\Delta t' = (1-\beta)2R/\gamma c$. When $\beta \approx 1$, $1-\beta$ may be approximated as $1/2\gamma^2$, giving the pulse received a width (in time)

$$\Delta t = \frac{R}{c\gamma^3} \tag{10–165}$$

Qualitatively, then, we expect a pulse $\mathrm{F}(t)$ of width $\sim \Delta t$ repeated at intervals $T = 2\pi/\omega_0 = 2\pi R/c$. The time dependence of the power received is sketched qualitatively in Figure 10.13.

Choosing the time origin at the center of a peak, we express the power as a Fourier series:

$$\mathrm{F}(t) = \frac{d\mathbf{P}}{d\Omega} = \sum_n A_n \cos n\omega_0 t \tag{10–166}$$

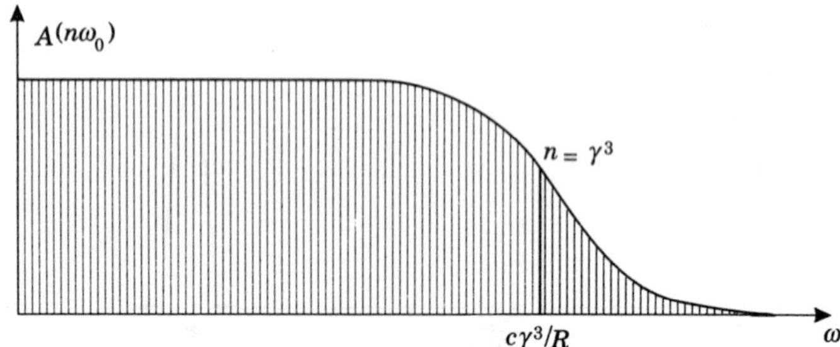

Figure 10.14: The qualitative spectrum of synchrotron radiation.

with $\omega_0 = c/R$ and A_n given by

$$A_n = \frac{2}{T}\int_{-T/2}^{T/2} \mathrm{F}(t)\cos n\omega_0 t\, dt \tag{10–167}$$

For terms with $n\omega_0 < c\gamma^3/R$, the cosine does not vary appreciably from 1 in the region of nonzero F. The Fourier coefficients A_n with $n < \gamma^3$ are thus all more or less equal. As $1/n\omega_0$ becomes smaller than the duration of the pulse, the cosine term will begin to oscillate within the pulse width, leading to rapidly decreasing integrals. We conclude that harmonics up to $n = \gamma^3$ occur in the power spectrum with approximately equal strength, as illustrated qualitatively in Figure 10.14.

For 1 GeV electrons in a 5-m radius cyclotron, $\gamma = 10^3\,\text{MeV}/.5\ \text{MeV} = 2\times 10^3$ and $\omega_0/2\pi \simeq 10^7\,\text{Hz}$. The spectrum extends past $10^{17}\,\text{Hz}$ ($\lambda = 3\times10^{-9}\,\text{m}$), well into the x-ray region. In recent years, specially built synchrotrons have been fitted with "wigglers," or "undulators", to increase their emission and to tune it to particular harmonics.

The lobes on the sides of the main peak are responsible for a slight increase in the amplitude of Fourier coefficients before the amplitude drops off again.

To obtain the exact results, the fields or potentials, rather than the power, must be frequency analyzed giving a qualitatively similar, but not quite so flat spectrum as the arguments above would indicate.

10.3.7 Bremstrahlung and Cherenkov radiation

We will not attempt anything but a brief qualitative discussion of bremstrahlung and Cherenkov radiation as both of these require a detailed knowledge of the medium through which a charged particle travels.

Bremstrahlung, German for "braking radiation" is emitted when a charged particle is brought rapidly to rest inside a material by interaction with other charged particles in the material. Although it is tempting to ascribe the radiation to the deceleration of the particle, a massive particle interacting with electrons in the medium will produce far greater acceleration of the electrons than it itself undergoes. It is the combined radiation from the medium and the incident particle that constitutes bremstrahlung.

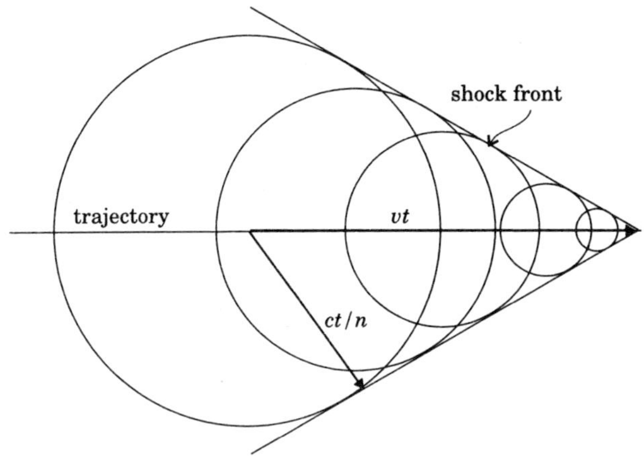

Figure 10.15: Cherenkov radiation comes from emitters coherently excited in the medium by the passage of a superluminal charged particle.

Cherenkov radiation is emitted when a charged particle travels through a medium at a speed greater than the speed of light in the medium (c/n). Again it is not the charged particle, which may be travelling at fairly constant speed, that emits the radiation but the medium. As the particle travels through the medium electrons in its neighborhood experience a rapidly varying electric field which elicits significant acceleration of the electrons. As the superluminal particle proceeds through the medium, it advances at $v \geq c/n$ while in its wake a disturbance spreads outward with velocity c/n, creating a shock cone resembling the bow wave a ship. The cone angle is readily determined from Figure 10.15. to be given by $\sin\theta = c/(nv)$. The particles on the shock front are excited coherently and therefore emit coherently in a direction perpendicular to the shock front. It will also be noted that the responding particles will be accelerated in a plane containing the trajectory of the impinging particle and the responding particle. We therefore anticipate that the radiation will be polarized along $a_\perp$, or in the plane containing the trajectory and the observer.

Cherenkov radiation is useful in determining the speed of relativistic particles as the cone angle, or alternatively the angle of the emitted radiation gives the speed in terms of the medium's refractive index.

10.4 Differentiating the Potentials

As mentioned, the differentiation of the potentials (10–132) and (10–133) to produce the fields $\vec{B} = \vec{\nabla} \times \vec{A}$ and $\vec{E} = -\vec{\nabla}V - \partial\vec{A}/\partial t$ is rather laborious and although not inherently difficult, requires considerable care. Abbreviating when convenient—$\vec{R}' = \vec{r} - \vec{r}\,'$, $\vec{\beta}' = \vec{v}\,'/c$, and $\xi' = 1 - \vec{\beta}\,' \cdot \hat{n}'$ —we begin by finding the gradient of

$$V(\vec{r}, t) = \frac{q}{4\pi\varepsilon_0 R'(t')\xi'(t')} \tag{10–168}$$

The position and velocity R' and β' of a moving particle are functions of t', but the retarded time t' depends explicitly on the field point coordinates and has a nonzero gradient; we write

$$\vec{\nabla}V = -(\vec{\nabla}V)\Big|_{t'} - \frac{\partial V}{\partial t'}\vec{\nabla}t' \tag{10–169}$$

To evaluate this expression we compute first a number of recurring derivatives.

$\boxed{\dfrac{\partial t'}{\partial t}}$:

$$t' = t - \frac{|\vec{r}-\vec{r}\,'|}{c} \quad \Rightarrow \quad \frac{\partial t'}{\partial t} = 1 - \frac{1}{c}\frac{\partial t'}{\partial t}\frac{d}{dt'}|\vec{r}-\vec{r}\,'|$$

$$= 1 + \frac{\vec{v}\,'\cdot\hat{n}}{c}\frac{\partial t'}{\partial t}$$

yielding

$$\frac{\partial t'}{\partial t} = \frac{1}{1-\vec{\beta}'\cdot\hat{n}\,'} = \frac{1}{\xi'} \tag{10–170}$$

a result we have already encountered in (10–156).

$\boxed{\dfrac{\partial \hat{n}'}{\partial t'}}$:

$$\frac{\partial \hat{n}'}{\partial t'} = \frac{\partial}{\partial t'}\left(\frac{\vec{r}-\vec{r}\,'}{|\vec{r}-\vec{r}\,'|}\right) = \frac{\dfrac{\partial \vec{R}}{\partial t'}}{R'} - \frac{\vec{R}'\dfrac{\partial R'}{\partial t'}}{R'^2}$$

$$= \frac{-\vec{v}\,'}{R'} + \frac{\vec{R}'(\vec{v}\,'\cdot\hat{n}')}{R'^2} = \frac{1}{R'}\left[(\vec{v}\,'\cdot\hat{n}')\hat{n}' - \vec{v}\,'\right]$$

$$= \frac{\hat{n}'\times(\hat{n}'\times\vec{v}\,')}{R'} \tag{10–171}$$

$\boxed{\dfrac{\partial \xi'}{\partial t'}}$:

$$\frac{\partial \xi'}{\partial t'} = \frac{\partial}{\partial t'}\left(1 - \frac{\vec{v}\,'\cdot\hat{n}'}{c}\right) = -\frac{\vec{a}\,'\cdot\hat{n}'}{c} - \vec{\beta}'\cdot\frac{\hat{n}'\times(\hat{n}'\times\vec{v}\,')}{R'} \tag{10–172}$$

$\boxed{\dfrac{\partial (R'\xi')}{\partial t'}}$:

$$\frac{\partial (R'\xi')}{\partial t'} = -\vec{v}\,'\cdot\hat{n}'\xi' - R'\frac{\vec{a}\,'\cdot\hat{n}'}{c} - \vec{\beta}'\cdot[\hat{n}'\times(\hat{n}'\times\vec{v}\,')]$$

$$= -R'\frac{\vec{a}\,'\cdot\hat{n}'}{c} - \vec{v}\,'\cdot\left(\hat{n}' - \frac{\vec{v}\,'\cdot\hat{n}'}{c}\hat{n}' + \frac{\vec{v}\,'\cdot\hat{n}'}{c}\hat{n}' - \frac{\vec{v}\,'}{c}\right)$$

$$= -R'\left(\frac{\vec{a}\,'\cdot\hat{n}'}{c}\right) - \vec{v}\,'\cdot\left(\hat{n}' - \frac{\vec{v}\,'}{c}\right) \tag{10–173}$$

$\boxed{\vec{\nabla}t'}$:

$$\vec{\nabla}t' = \vec{\nabla}\left(t - \frac{|\vec{r}-\vec{r}\,'|}{c}\right) = -\frac{1}{c}\left(\vec{\nabla}|\vec{r}-\vec{r}\,'| + \frac{\partial}{\partial t'}|\vec{r}-\vec{r}\,'|\vec{\nabla}t'\right)$$

$$= -\frac{1}{c}\frac{\vec{R}}{R} + \frac{\vec{v}\,'\cdot\hat{n}'}{c}\vec{\nabla}t' = -\frac{\hat{n}'}{c} + (\vec{\beta}'\cdot\hat{n}')\vec{\nabla}t'$$

$$\Rightarrow \quad \vec{\nabla}t' = \frac{-\hat{n}'/c}{1-\vec{\beta}'\cdot\hat{n}'} = -\frac{\hat{n}'}{\xi' c} \tag{10–174}$$

We will also need the gradient of various quantities calculated at constant t' (at constant t', the source position $\vec{r}\,'$ and $\vec{\beta}'$, are fixed).

$\boxed{\vec{\nabla}R'\Big|_{t'}}$:

$$\left(\vec{\nabla}|\vec{r}-\vec{r}\,'|\right)\Big|_{t'} = \frac{\vec{r}-\vec{r}\,'}{|\vec{r}-\vec{r}\,'|} = \hat{n}' \tag{10–175}$$

$\vec{\nabla}\xi'\Big|_{t'}$

$$\vec{\nabla}\xi'\Big|_{t'} = \hat{e}^i\partial_i\left[1 - \frac{\beta_j'(x^j - x^{j\prime})}{\sqrt{(x_j - x_j')(x^j - x^{j\prime})}}\right]$$
$$= \hat{e}^i\left[\frac{-\beta_j'\delta_i^j}{R'} + \frac{\beta_j'\left(x^j - x^{j\prime}\right)\left(x_i - x_i'\right)}{R'^3}\right]$$
$$= \frac{-\vec{\beta}'}{R'} + \frac{(\vec{\beta}'\cdot\vec{R}')\vec{R}'}{R'^3} = -\frac{\vec{\beta}'}{R'} + \frac{(\vec{\beta}'\cdot\hat{n}')\hat{n}'}{R'} \tag{10–176}$$

We are now ready to continue with our evaluation of the gradient of $-V$. We begin with $-\vec{\nabla}V\big|_{t'}$.

$$-\vec{\nabla}V\Big|_{t'} = -\vec{\nabla}\frac{1}{4\pi\varepsilon_0}\frac{q}{R'\xi'}\Big|_{t'} = \frac{q}{4\pi\varepsilon_0}\left(\frac{\xi'\vec{\nabla}R' + R'\vec{\nabla}\xi'}{R'^2\xi'^2}\right)$$
$$= \frac{q}{4\pi\varepsilon_0}\left[\frac{\hat{n}'(1 - \vec{\beta}'\cdot\hat{n}') - \vec{\beta}' + (\vec{\beta}'\cdot\hat{n}')\hat{n}'}{R'^2\xi'^2}\right]$$
$$= \frac{q}{4\pi\varepsilon_0}\frac{\hat{n}' - \vec{\beta}'}{R'^2\xi'^2} \tag{10–177}$$

We continue with the second term of the negative gradient:

$$-\frac{\partial V}{\partial t'} = -\frac{q}{4\pi\varepsilon_0}\frac{\partial}{\partial t'}\frac{1}{R'\xi'}$$
$$= \frac{q}{4\pi\varepsilon_0 R'^2\xi'^2}\left(-\frac{(\vec{a}'\cdot\hat{n}')R'}{c} - \vec{v}'\cdot\hat{n}' + \frac{v'^2}{c}\right) \tag{10–178}$$

Combining the two terms, we compute $-\vec{\nabla}V = -\vec{\nabla}V\Big|_{t'} + \dfrac{\hat{n}'}{\xi' c}\dfrac{\partial V}{\partial t'}$:

$$-\vec{\nabla}V = \frac{q}{4\pi\varepsilon_0\xi'^3R'^2}\left[(\hat{n}' - \vec{\beta}')\xi' + \xi'\frac{\hat{n}'}{\xi' c}\left(\frac{\vec{a}'\cdot\hat{n}'}{c}R' + \vec{v}'\cdot\hat{n}' - \frac{v'^2}{c}\right)\right]$$
$$= \frac{q}{4\pi\varepsilon_0\xi'^3}\left[\frac{\hat{n}'(1 - \beta'^2) - \vec{\beta}'(1 - \vec{\beta}'\cdot\hat{n}')}{R'^2} + \frac{\hat{n}'(\vec{a}'\cdot\hat{n}')}{R'c^2}\right] \tag{10–179}$$

Next we calculate $-\partial\vec{A}/\partial t$:

$$-\frac{\partial\vec{A}}{\partial t} = -\frac{\partial\vec{A}}{\partial t'}\frac{\partial t'}{\partial t} = -\frac{1}{\xi'}\frac{\partial\vec{A}}{\partial t'} = \frac{-q}{4\pi\varepsilon_0 c\xi'}\frac{\partial}{\partial t'}\left(\frac{\vec{\beta}'}{R'\xi'}\right)$$
$$= \frac{q}{4\pi\varepsilon_0 c\xi'}\left[\frac{-\vec{a}'/c}{R'\xi'} + \frac{\vec{\beta}'}{R'^2\xi'^2}\frac{\partial(R'\xi')}{\partial t'}\right]$$
$$= \frac{q}{4\pi\varepsilon_0\xi'^3}\left[\frac{-\xi'\vec{a}'}{R'c^2} + \vec{\beta}'\frac{-R'(\vec{a}'\cdot\hat{n}')/c^2 - \vec{\beta}'\cdot(\hat{n}' - \vec{\beta}')}{R'^2}\right]$$

$$= \frac{q}{4\pi\varepsilon_0\xi'^3}\left[\frac{(-\vec{a}'\cdot\hat{n}')\vec{\beta}-\xi'\vec{a}'}{R'c^2}+\frac{\left(\beta'^2-\vec{\beta}'\cdot\hat{n}'\right)\vec{\beta}'}{R'^2}\right] \tag{10–180}$$

We can now evaluate the electric field,

$$\vec{E} = \frac{q}{4\pi\varepsilon_0\xi'^3}\left[\frac{\hat{n}'(1-\beta'^2)-\vec{\beta}'(1-\vec{\beta}'\cdot\hat{n}')+\vec{\beta}'(\beta'^2-\vec{\beta}'\cdot\hat{n}')}{R'^2}\right.$$

$$\left.+\frac{\hat{n}'(\vec{a}'\cdot\hat{n}')-\vec{a}'(1-\vec{\beta}'\cdot\hat{n}')-(\vec{a}'\cdot\hat{n}')\vec{\beta}'}{R'c^2}\right]$$

$$= \frac{q}{4\pi\varepsilon_0\xi'^3}\left[\frac{(\hat{n}'-\vec{\beta}')(1-\beta'^2)}{R'^2}+\frac{-\vec{a}'_\perp+\hat{n}'\times(\vec{a}'\times\vec{\beta}')}{R'c^2}\right] \tag{10–181}$$

where $\vec{a}'_\perp = \vec{a}' - (\vec{a}'\cdot\hat{n}')\hat{n}'$ is the component of the retarded acceleration perpendicular to $\hat{n}'$. This is the result we earlier quoted as (10–141).

The magnetic induction field of a moving point charge is found in similar fashion from $\vec{\nabla}\times\vec{A}$:

$$\vec{\nabla}\times\vec{A}\,[\vec{r},t'(\vec{r})] = (\vec{\nabla}\times\vec{A})\Big|_{t'} - \frac{\partial\vec{A}}{\partial t'}\times\vec{\nabla}t' \tag{10–182}$$

Evaluating the first term, we get

$$(\vec{\nabla}\times\vec{A})\Big|_{t'} = \frac{q}{4\pi\varepsilon_0 c}\vec{\nabla}\times\frac{\vec{\beta}'}{R'\xi'}\bigg|_{t'} = \frac{q}{4\pi\varepsilon_0 c}\vec{\nabla}\left(\frac{1}{R'\xi'}\right)\times\vec{\beta}'\bigg|_{t'}$$

$$= \frac{-q}{4\pi\varepsilon_0 c}\frac{\hat{n}'-\vec{\beta}'}{R'^2\xi'^2}\times\vec{\beta}' = \frac{-q}{4\pi\varepsilon_0 c}\frac{\hat{n}'\times\vec{\beta}'}{R'^2\xi'^2} \tag{10–183}$$

where we have used the fact that $\vec{\nabla}\times t'$ vanishes to eliminate the term in $\vec{\nabla}\times\vec{\beta}'$. Next we obtain $-\partial\vec{A}/\partial t'$:

$$-\frac{\partial\vec{A}}{\partial t'} = -\frac{q}{4\pi\varepsilon_0 c}\frac{\partial}{\partial t'}\left(\frac{\vec{\beta}'}{R'\xi'}\right)$$

$$= \frac{-q}{4\pi\varepsilon_0 c}\left\{\frac{\vec{a}'/c}{R'\xi'} - \frac{\vec{\beta}'}{R'^2\xi'^2}\left[-R'\frac{\vec{a}'\cdot\hat{n}'}{c} - \vec{v}'\cdot(\hat{n}'-\vec{\beta}')\right]\right\}$$

$$= \frac{q}{4\pi\varepsilon_0\xi'^2}\left[\frac{-\vec{a}'(1-\vec{\beta}'\cdot\hat{n}')-\vec{\beta}'(\vec{a}'\cdot\hat{n}')}{R'c^2} - \frac{\vec{\beta}'(\vec{\beta}a'\cdot\hat{n}'-\beta'^2)}{R'^2}\right] \tag{10–184}$$

The numerator of the first term may be expressed as $-\vec{a}'+\hat{n}'\times(\vec{a}'\times\vec{\beta}')$ in order to bear greater similarity to the equivalent term in the electric field expression. Finally we find $\vec{B}$.

$$\vec{B} = \frac{q}{4\pi\varepsilon_0\xi'^2}\left\{\frac{-\hat{n}'\times\vec{\beta}'}{R'^2c}+\left[\frac{-\vec{a}'+\hat{n}'\times(\vec{a}'\times\vec{\beta}')}{R'c^2}+\frac{\vec{\beta}'(\vec{\beta}'\cdot\hat{n}'-\beta'^2)}{R'^2}\right]\times\frac{-\hat{n}'}{\xi'c}\right\}$$

$$= \frac{q}{4\pi\varepsilon_0 \xi'^3} \left[\frac{(1-\beta'^2)\vec{\beta}' \times \hat{n}'}{R'^2 c} + \hat{n}' \times \frac{-\vec{a}' + \hat{n}' \times (\vec{a}' \times \vec{\beta}')}{R' c^3} \right]$$

$$= \frac{\hat{n}'}{c} \times \vec{E} \tag{10–185}$$

a result we stated without proof in (10–142).

It will be noted that nowhere in the foregoing derivations have we assumed any harmonic time dependence, yet a radiation field arises, apparently due only to an arbitrary movement of the sources and the finite speed of propagation of the fields.

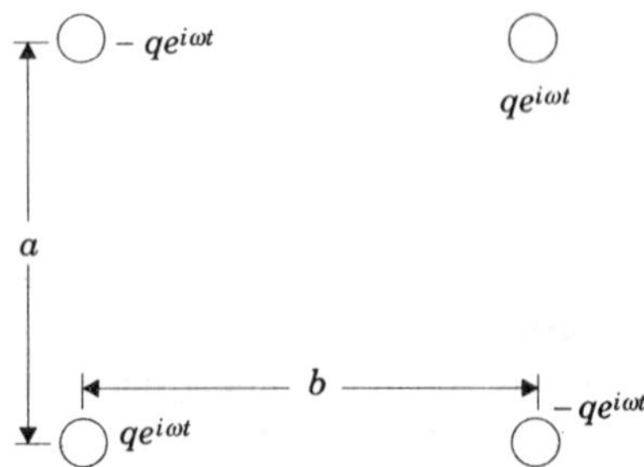

Figure 10.16: The electric quadrupole moment of problem 10-1.

Exercises and Problems

10-1 An electric quadrupole consisting of four equal charges (Figure 10.16), placed on the corners of a rectangle of sides a and b, oscillates with diagonally opposite charges in phase and contiguous charges 180° out of phase. Find the radiation pattern as well as the total power emitted by the quadrupole.

10-2 The power emitted by an electric dipole may be written

$$\left\langle \frac{dW}{dt} \right\rangle = \frac{|\ddot{p}|^2}{4\pi\varepsilon_0(3c^3)}$$

Find a similar expression for the power emitted by an oscillating magnetic dipole and electric quadrupole. Generalize the latter to obtain a guess for the power emitted by gravitational radiation from an oscillating mass quadrupole. Why would you expect this result to be similar? What would lead to differences?

10-3 A bar magnet with magnetization $\vec{M}$ parallel to its long axis is rotated about an axis perpendicular to and bisecting the magnet. Find the power emitted by the magnet if $\ell \ll c/\omega$.

10-4 In *nuclear magnetic resonance* (NMR for short) the magnetic moments of hydrogen nuclei in a sample rotate about the direction of the flux density $\vec{B}_0$. Find the rate at which a single atom would classically radiate energy. Why should this result be invalid? Under which conditions might it still be valid?

10-5 Assume that an atom emits radiation at 500 nm. Estimate the rate of emission of (a) electric dipole, (b) electric quadrupole, and (c) magnetic dipole radiation expected on a classical basis. What if the radiation had wavelength 50 nm?

10-6 Suppose that a hypothetical neutron star ($m = 2 \times 10^{30}$ kg, $R = 10^4$ m) has its neutrons aligned in a direction perpendicular to its axis of rotation. Find the rate that the neutron loses energy by dipole radiation. How long would it take for a neutron star rotating at $300/2\pi$ Hz to lose 9/10 of its initial rotational energy?

10-7 Find the impedance of a center-fed half wave antenna as well as that of a full wave antenna.

10-8 Consider a short ($\lambda \gg d$) linear antenna supplied at its center with current $I_0 = Ie^{-i\omega t} \sin \frac{1}{2}kd$.

(a) Find the dipole moment of the antenna, assuming that each half has a uniform charge density.

(b) Calculate the power radiated per solid angle by treating the antenna as an electric dipole.

(c) Calculate the power per solid angle emitted by applying a small d approximation to eliminate the cosines in equation (10–125).

10-9 A half wave antenna is parallel to a large conducting sheet at a distance $\frac{1}{4}\lambda$ from the sheet. Find the angular distribution of the power radiated by the antenna and the impedance of the antenna.

10-10 Assume that the electron in a hydrogen atom moves nonrelativistically in a circular Bohr orbit of radius r. Consider the circular motion as the superposition of two linear oscillations in orthogonal directions, having the same frequency but oscillating $\frac{1}{2}\pi$ out of phase.

(a) Show that the power radiated is

$$\mathbf{P} = \frac{q^6}{96\pi^3\varepsilon_0^3 c^3 m^2 r^4}$$

(b) If we assume that $\mathbf{P}/\omega \ll W = q^2/8\pi\varepsilon_0 r$, the orbit remains circular as it decays. How long would it take classically for the orbit to shrink from r_0 to 0? Insert the appropriate value of r_0 (for the 1S state) to obtain the classical lifetime of the hydrogen atom.

10-11 A hypothetical two-electron Bohr atom has its electrons counter-rotating in the same circular orbit as shown in Figure 2.9 (b). Find the rate of power emission.

10-12 Redo part (a) of problem (10-10) using the expression for radiation from a nonrelativistic accelerated particle, (10–149).

10–13 Calculate the energy emitted classically by a charged particle in an elliptical orbit with major axis A, minor axis B and period T.

10-14 Show that for a particle accelerated in a direction parallel to its velocity, $d\mathbf{P}/d\Omega$ is maximum at $\theta = 1/2\gamma$ when β approaches unity. (Hint: Set $\beta = 1 - \epsilon$, $\epsilon \ll 1$.) Show also that

$$\left.\frac{d\mathbf{P}}{d\Omega}\right|_{\theta_{\max}} \to \alpha\gamma^8$$

as $\beta \to 1$, and evaluate α.

10-15 Show that for a particle accelerated in a direction perpendicular to its velocity, $v \approx c$, the "headlight cone" width tends to $2/\gamma$. Find also how the power emitted in a direction parallel to the motion varies with γ.

Chapter 11

The Covariant Formulation

11.0 Covariance

Since our early use of special relativity to deduce the magnetic interaction between two currents, we have largely disregarded the requirements of relativity. In the last chapter there was in fact a tacit assumption that the fields travel with velocity c independent of motion of the source or the observer. In this chapter we recast the electromagnetic field equations in manifestly covariant form and use the four-vector forms to deduce the Lorentz transformation laws for the potentials, the fields, the magnetization, and polarization. We will also use relativistic covariance to find the energy and momentum radiated by relativistic charges.

By covariant, we mean that under special relativistic transformations (one frame moving with constant velocity with respect to another), equations of physics remain formally invariant. A quick review of four-tensors used in special relativity is strongly recommended before starting this section. To avoid confusion between the Lorentz factor $1/\sqrt{1-\beta^2}$ for a particle in motion and that for a frame transformation, we denote the former as γ, and identify the Lorentz factor for the frame transformation as Γ. The appendix on tensors (Section B.5) illustrates the use (as well as the utility) of this slightly nonstandard notation.

Four vectors X^μ have[28] a time-like component x^0 and three spatial components x^1, x^2, x^3 that we will frequently abbreviate as a three-vector $\vec{x}$, so we write $(x^0, \vec{x})$. Points in space-time are also known as events and have contravariant components (ct, x^1, x^2, x^2). The *interval* between two events is given by $dS^2 = \sqrt{(cdt)^2 - (dx)^2 - (dy)^2 - (dz)^2}$ giving metric tensor $g_{0\mu} = \delta_{0\mu}$ and $g_{i,j} = -\delta_{i,j}$ (i, j = 1, 2, 3) There seems to be no unanimity in the literature over the sign of the elements of the metric tensor for special relativity. We will use the signs stated as that gives a real positive interval between causally related events; roughly 30% of the literature uses the signs reversed. So long as one or the other is consistently used, no net change results. When using tensor notation in relativity, we adopt the

[28] Note the ambiguity of this notation as X^μ can either represent the μ component of X or the entire 4-vector, the context should make it clear which is intended.

standard convention that Greek indices (superscripts and subscripts) range from 0 to 3 while Roman letter indices range from 1 to 3. The Roman letters t, x, y, and z are of course exempt from this convention. Specific components of 3-vectors are normally labelled by subscripts x, y and z. In this chapter, the prime is reserved for Lorentz transformed quantities rather than source coordinates.

11.1 The Four-Potential and Coulomb's Law

Fundamental to our discussion of electromagnetic theory has been the requirement of conservation of charge. By this we mean in the present context that the charge contained in an isolated volume with boundaries all observers agree on will be independent of the volume's velocity with respect to an observer. Experimental evidence for this assertion is offered by the overall neutrality of atoms, independent of the speed of their electrons. Because observers travelling at different speed deduce different volumes contained within the boundaries, the charge *density* in the volume must necessarily depend on the volume's speed. In particular, an observer finds the length of the volume with relative speed v shortened by the factor $1/\gamma$ in the direction of the volume's motion. The charge density inferred by the observer is therefore $\rho = \gamma\rho_0$, where ρ_0 is the "proper" charge density, that observed in the rest frame of the charge.

Motivated by the notion that currents are charges in motion, we define the four-current density as the proper charge density ρ_0 times the four-velocity V^μ (it will be recalled that a constant times a four vector remains a four vector)

$$J^\mu = \rho_0 V^\mu = \rho_0\gamma(c, \vec{v}) = (\rho c, \vec{J}) \tag{11–1}$$

and the "four-potential"

$$\Phi^\mu = \left(\frac{V}{c}, \vec{A}\right) \tag{11–2}$$

That Φ is actually a four-vector (in other words it transforms like a first rank Lorentz tensor under frame transformations) remains to be demonstrated.

From the classical formulation of the theory, each component of Φ satisfies

$$\Phi^\mu = \frac{\mu_0}{4\pi}\int \frac{[J^\mu]d^3r}{R} \tag{11–3}$$

(The prime to differentiate the source coordinates from the field point coordinates is temporarily dropped as we will need it to denote frame transformations.) To demonstrate that the integral behaves as a four-vector, we note that the retarded current $[J^\mu]$ lies on the light cone (in other words, the argument $[J^\mu]$ has 0 interval from $\vec{r}$) of for all coincident observers independent of their motion; therefore all observers include the same sources into the summation (integral 11-3). It remains to show that d^3r/R is an invariant (scalar).

We assign coordinates $(0, 0, 0, 0)$ to the observer and (ct, x, y, z) to the retarded event for which $R = -ct$ and d^3r are to be evaluated. R is assumed to make angle θ with the x axis. A second observer, whose frame, coincident with the first at $t' = t = 0$, moves with a Lorentz factor Γ in the x direction with respect to the

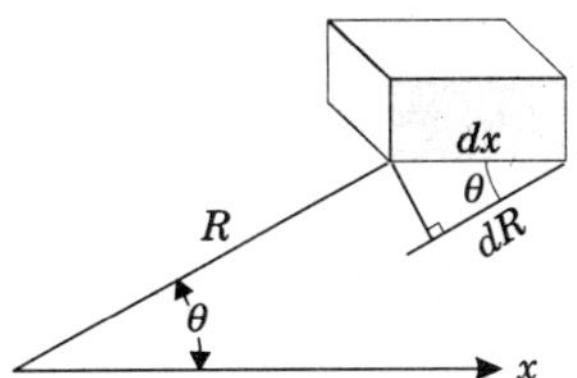

Figure 11.1: Fields from the far end of dr^3 are emitted a time $dt = dR/c$ earlier than those from the near end of the volume.

first observer, locates the elementary volume at $R' = -ct'$. The second observer's distance from the source at the retarded time is

$$\begin{aligned} R' &= -ct' = -\Gamma(ct - \beta x) \\ &= R\Gamma\left(1 + \frac{\beta x}{R}\right) \\ &= R\Gamma(1 + \beta\cos\theta) \end{aligned} \tag{11–4}$$

To get the relationship between the stationary and moving volume elements d^3r and d^3r', it is tempting but wrong to try $d^3r' = \Gamma d^3r$ since this applies only if the whole volume is measured at the same time. The far end of dx must in fact be measured at a time $dt = dR/c$ earlier than the near end (Figure 11.1). Taking d^3r' $= dx'\,dy'\,dz'$, we find that the transverse sides, dy' and dz' are invariant but that $dx' = \Gamma(dx - \beta cdt)$. Taking the line of sight depth $cdt = -dR = -dx\cos\theta$, we obtain $dx' = \Gamma dx(1 + \beta\cos\theta)$. We find then $d^3r' = \Gamma(1 + \beta\cos\theta)d^3r$, from which we infer that $d^3r'/R' = d^3r/R$ is invariant. We conclude that $[J^\mu]d^3r/R$ is a four-vector, and since the result of adding four-tensors (even at distinct points if the metric tensor $g_{\mu\nu}$ doesn't vary) is a four-tensor, Φ is a four-vector. The covariant form of the four-potential is $\Phi_\mu = (V/c, -\vec{A})$.

It is instructive consider an alternative approach to proving $(V/c, \vec{A})$ a four-vector. Recall the wave equation for V and $\vec{A}$ in the Lorenz gauge (3–56, 57)

$$\nabla^2 V - \frac{1}{c^2}\frac{\partial^2 V}{\partial t^2} = -\frac{\rho}{\varepsilon_0} = -\mu_0\rho c^2 \tag{11–5}$$

and

$$\nabla^2\vec{A} - \frac{1}{c^2}\frac{\partial^2\vec{A}}{\partial t^2} = -\mu_0\vec{J} \tag{11–6}$$

We divide (11–5) by c to obtain

$$\nabla^2(V/c) - \frac{1}{c^2}\frac{\partial^2(V/c)}{\partial t^2} = -\mu_0 c\rho \tag{11–7}$$

and recognize the right hand side of (11–6) and (11–7) as the space-like and time-like components of the four vector $\mu_0 J^\nu$ so that we combine (11–6) and (11–7) into

$$\left(\nabla^2 - \frac{1}{c^2}\frac{\partial^2}{\partial t^2}\right)\left(\frac{V}{c}, \vec{A}\right)^\mu = -\mu_0 J^\mu \tag{11–8}$$

The d'Alambertian operator, $\nabla^2 - (1/c^2)\partial^2/\partial t^2$, frequently abbreviated as $\Box$, is the scalar operator $\partial_\mu\partial^\mu$, so that we can conclude that $(V/c,\ \vec{A})^\mu$ constitutes a contravariant four-vector

11.2 The Electromagnetic Field Tensor

Because the vector potential is the spacelike component of the four-potential, we would expect the magnetic induction field to be given by the four-dimensional curl of the four-potential. Further, as we have seen, Lorentz transformations do not change the order of a tensor, but they do transform electric to magnetic fields and conversely, the electric field will also need to be contained within the tensor produced by curling Φ. With this motivation, we define the *electromagnetic field tensor* as

$$F_{\mu\nu} \equiv \frac{\partial}{\partial x^\mu}\Phi_\nu - \frac{\partial}{\partial x^\nu}\Phi_\mu = \partial_\mu\Phi_\nu - \partial_\nu\Phi_\mu \tag{11–9}$$

Performing the differentiations, we have explicitly

$$F_{\mu\nu} = \begin{pmatrix} 0 & \frac{E_x}{c} & \frac{E_y}{c} & \frac{E_z}{c} \\ -\frac{E_x}{c} & 0 & -B_z & B_y \\ -\frac{E_y}{c} & B_z & 0 & -B_x \\ -\frac{E_z}{c} & -B_y & B_x & 0 \end{pmatrix} \tag{11–10}$$

The full contravariant form of the field tensor $F^{\mu\nu}$ is

$$F^{\mu\nu} = g^{\mu\sigma}g^{\nu\lambda}F_{\sigma\lambda} = \begin{pmatrix} 0 & -\frac{E_x}{c} & -\frac{E_y}{c} & -\frac{E_z}{c} \\ \frac{E_x}{c} & 0 & -B_z & B_y \\ \frac{E_y}{c} & B_z & 0 & -B_x \\ \frac{E_z}{c} & -B_y & B_x & 0 \end{pmatrix} \tag{11–11}$$

With the aid of the contravariant field tensor we write the manifestly covariant equation

$$\partial_\mu F^{\mu\nu} = \mu_0 J^\nu \tag{11–12}$$

expressing two of Maxwell's equations:

$$\vec{\nabla}\cdot\vec{E} = \frac{\rho}{\varepsilon_0} \quad \text{and} \quad (\vec{\nabla}\times\vec{B}) - \frac{1}{c^2}\frac{\partial\vec{E}}{\partial t} = \mu_0\vec{J} \tag{11–13}$$

The remaining Maxwell equations are obtained from the somewhat less elegant third-rank tensor equation

$$\partial_\sigma F_{\mu\nu} + \partial_\mu F_{\nu\sigma} + \partial_\nu F_{\sigma\mu} = 0 \tag{11–14}$$

with μ, ν, and σ any three distinct numbers from 0 to four. In terms of the fields (11–14) translates as

$$\vec{\nabla} \cdot \vec{B} = 0 \qquad \text{and} \qquad \vec{\nabla} \times \vec{E} = -\frac{\partial \vec{B}}{\partial t} \tag{11–15}$$

Alternatively, one can construct the *Dual Tensor*, $G_{\mu\nu} = \frac{1}{2}\epsilon_{\mu\nu\sigma\tau}F^{\sigma\tau}$ which[29] produces $F_{\sigma\tau}$ above with $\vec{E}$ and $\vec{B}$ interchanged. The equations (11–15) can then be expressed as $\partial_\mu G^{\mu\nu} = 0$ (see problem 11-5).

Under velocity changes, the components of $\vec{E}$ and $\vec{B}$ must transform as components of a second-rank tensor, namely

$$(F^{\mu\nu})' = \alpha_\sigma^{\mu'}\alpha_\rho^{\nu'}F^{\sigma\rho} \tag{11–16}$$

with $\alpha_\rho^{\nu'}$ the coefficients of the Lorentz transformation. We take the x axis along the direction of relative motion of the frames. The nonzero coefficients of the transformation, $\alpha_\sigma^{\mu'}$ to a frame moving with velocity βc in the x direction are $\alpha_0^{0'} = \alpha_1^{1'} = \Gamma$, $\alpha_1^{0'} = \alpha_0^{1'} = -\beta\Gamma$, and $\alpha_2^{2'} = \alpha_3^{3'} = 1$ (Appendix B.5). We have, for example,

$$\begin{aligned}
\frac{E_x'}{c} &= (F^{10})' = \alpha_\sigma^{1'}\alpha_\rho^{0'}F^{\sigma\rho} = \alpha_\sigma^{1'}\left(\alpha_0^{0'}F^{\sigma 0} + \alpha_1^{0'}F^{\sigma 1}\right) \\
&= \alpha_\sigma^{1'}\Gamma\left(F^{\sigma 0} - \beta F^{\sigma 1}\right) \\
&= \Gamma\alpha_0^{1'}\left(F^{00} - \beta F^{01}\right) + \Gamma\alpha_1^{1'}\left(F^{10} - \beta F^{11}\right) \\
&= \Gamma^2\beta^2 F^{01} + \Gamma^2 F^{10} = \Gamma^2 F^{10} - \Gamma^2\beta^2 F^{10} \\
&= \frac{E_x}{c}
\end{aligned} \tag{11–17}$$

and

$$\begin{aligned}
\frac{E_y'}{c} &= (F^{20})' = \alpha_\sigma^{2'}\alpha_\rho^{0'}F^{\sigma\rho} = \alpha_\sigma^{2'}\left(\alpha_0^{0'}F^{\sigma 0} + \alpha_1^{0'}F^{\sigma 1}\right) \\
&= \alpha_\sigma^{2'}\left(\Gamma F^{\sigma 0} - \Gamma\beta F^{\sigma 1}\right) = \Gamma F^{20} - \Gamma\beta F^{21} \\
&= \Gamma\left(\frac{E_y}{c} - \frac{v_x B_z}{c}\right)
\end{aligned} \tag{11–18}$$

The other components follow in the same fashion. The transformations can be conveniently written in terms of the familiar three-fields as

$$\begin{aligned}
E_{\|}' &= E_{\|} & B_{\|}' &= B_{\|} \\
\vec{E}_\perp' &= \Gamma\left(\vec{E}_\perp + \vec{v}\times\vec{B}_\perp\right) & \vec{B}_\perp' &= \Gamma\left(\vec{B}_\perp - \frac{\vec{v}}{c^2}\times\vec{E}_\perp\right)
\end{aligned} \tag{11–19}$$

[29] The tensor $\epsilon_{\mu\nu\rho\sigma}$ is the four-dimensional analogue of the Levi-Cevita symbol. It is the totally antisymmetric tensor having elements +1 when μ, ν, ρ, and σ are an even permutation of 0, 1, 2, and 3; −1 for any odd permutation; and 0 if any two or more indices are equal.

11.2.1 Magnetization and Polarization

Because the components of the electric and magnetic fields are the components of a second-rank four-tensor, $F_{\mu\nu}$, we would anticipate that the components of $\vec{D} = \varepsilon_0\vec{E} + \vec{P}$ and $\vec{H} = \vec{B}/\mu_0 - \vec{M}$ would also be the components of a second-rank tensor, say $H_{\mu\nu}$. We construct such a tensor as $H_{\mu\nu} = \varepsilon_0 c^2 F_{\mu\nu} + P_{\mu\nu}$, where $P_{\mu\nu}$ is a second-rank tensor containing the magnetization and polarization. The 01 element gives

$$H_{01} = \varepsilon_0 c E_x + P_{01} \tag{11–20}$$

requiring P_{01} to be cP_x. Similarly, the 13 element gives

$$H_{13} = \varepsilon_0 c^2 B_y + P_{13} = \frac{B_y}{\mu_0} + P_{13} = H_y \tag{11–21}$$

which identifies P_{13} as $-M_y$. Continuing in the same fashion for the other elements, we construct the tensors $H_{\mu\nu}$ and $P_{\mu\nu}$ as follows.

$$\begin{pmatrix} 0 & cD_x & cD_y & cD_z \\ -cD_x & 0 & -H_z & H_y \\ -cD_y & H_z & 0 & -H_x \\ -cD_z & -H_y & H_x & 0 \end{pmatrix} = \frac{F_{\mu\nu}}{\mu_0} + \begin{pmatrix} 0 & cP_x & cP_y & cP_z \\ -cP_x & 0 & M_z & -M_y \\ -cP_y & -M_z & 0 & M_x \\ -cP_z & M_y & -M_x & 0 \end{pmatrix} \tag{11–22}$$

Maxwell's equations for $\vec{D}$ and $\vec{H}$ in a material medium may now be written in terms of $H^{\mu\nu}$ as

$$\partial_\mu H^{\mu\nu} = J^\nu \tag{11–23}$$

The transformation laws for magnetizations and polarizations of a moving medium follow immediately.

$$\begin{aligned} P'_{\parallel} &= P_{\parallel} & M'_{\parallel} &= M_{\parallel} \\ \vec{P}'_{\perp} &= \Gamma\left(\vec{P}_{\perp} - \frac{\vec{v}}{c^2} \times \vec{M}_{\perp}\right) & \vec{M}'_{\perp} &= \Gamma\left(\vec{M}_{\perp} + \vec{v} \times \vec{P}_{\perp}\right) \end{aligned} \tag{11–24}$$

EXAMPLE 11.1: A dielectric sphere of radius a and permittivity ε is placed in an initially uniform field $E_0\hat{k}$ and rotates about the x axis with angular velocity $\omega \ll c/a$. Find the magnetic induction field outside the sphere.

Solution: From (Ex 7.7.13), the electric field inside the sphere is

$$\vec{E} = \frac{3\varepsilon_0}{\varepsilon_1 + 2\varepsilon_0}\vec{E}_0 \tag{Ex 11.1.1}$$

while the relation $\vec{D} = \varepsilon_0\vec{E} + \vec{P} = \varepsilon_1\vec{E}$ gives $\vec{P} = (\varepsilon_1 - \varepsilon_0)\vec{E}$; hence

$$\vec{P} = \frac{3\varepsilon_0(\varepsilon_1 - \varepsilon_0)}{\varepsilon_1 + 2\varepsilon_0}\vec{E}_0 \tag{Ex 11.1.2}$$

A molecular dipole within the sphere at position $\vec{r}\,'$ has velocity $\vec{\omega} \times \vec{r}\,'$. Therefore, according to (11-24), the magnetization of the rotating sphere is (note the change in sign because the dipoles are moving rather than the observer)

$$\begin{aligned} \vec{M}' &= (\vec{\omega} \times \vec{r}\,') \times \vec{P} \\ &= -(\vec{r}\,' \cdot \vec{P})\vec{\omega} + (\vec{\omega} \cdot \vec{P})\vec{r}\,' \\ &= -\frac{3\,\varepsilon_0(\varepsilon_1 - \varepsilon_0)z' E_0 \omega \hat{\imath}}{\varepsilon_1 + 2\varepsilon_0} \end{aligned} \tag{Ex 11.1.3}$$

In Example 7.10 we found the scalar magnetic potential from a sphere carrying precisely such a magnetization. Adapting that result, we have

$$V_m(r > a) = -\frac{3\varepsilon_0(\varepsilon_1 - \varepsilon_0)\omega E_0 a^5 z x}{5(\varepsilon_1 + 2\varepsilon_0)r^5} \tag{Ex 11.1.4}$$

The magnetic induction field is then $\vec{B}(r > a) = -\mu_0 \vec{\nabla} V_m$

$$\vec{B}(\vec{r} > a) = \frac{3\mu_0\varepsilon_0(\varepsilon_1 - \varepsilon_0)\omega a^5 E_0}{5(\varepsilon_1 + 2\varepsilon_0)r^5}\left(z\hat{\imath} + x\hat{k} - \frac{5xz\vec{r}}{r^2}\right) \tag{Ex 11.1.5}$$

The interior field is equally easily found using the interior scalar potential given in example 7.10. A solution for arbitrary ω would include a radiation field.

11.3 Invariants

A number of invariants (zero-rank tensors) are easily obtained from the four-vectors and four-tensors so far obtained. We begin with the divergence of the four-current:

$$\partial_\nu J^\nu = \partial_\nu \partial_\mu H^{\mu\nu} \tag{11–25}$$

The order of differentiation in the last expression may be interchanged, and μ and ν, being dummy indices, may be exchanged. The result of these manipulations is that $\partial_\nu \partial_\mu H^{\mu\nu} = \partial_\mu \partial_\nu H^{\mu\nu} = \partial_\nu \partial_\mu H^{\nu\mu}$. But $H^{\mu\nu} = -H^{\nu\mu}$, which leads to the conclusion that

$$\partial_\nu J^\nu = 0 \tag{11–26}$$

which, of course is just an expression of the continuity equation

$$\frac{\partial \rho}{\partial t} + \vec{\nabla} \cdot \vec{J} = 0$$

The frame independence of the divergence of the four-potential,

$$\partial_\mu \Phi^\mu = \vec{\nabla} \cdot \vec{A} + \frac{1}{c^2}\frac{\partial V}{\partial t} \tag{11–27}$$

expresses the invariance of gauge choice. In the Lorenz gauge it is taken to be zero.

Other invariants are $J_\mu \Phi^\mu = \rho V - \vec{J} \cdot \vec{A}$, the interaction energy density between the field and the charge; $F^{\mu\nu}F_{\mu\nu} = 2(B^2 - E^2/c^2)$; and $\epsilon_{\mu\nu\rho\sigma}F^{\mu\nu}F^{\rho\sigma} = -\vec{E} \cdot \vec{B}/c$.

The wave equation is easily obtained from Maxwell's equation in covariant form, as shown below.

$$\begin{aligned}\partial_\mu F^{\mu\nu} &= \partial_\mu\left(\partial^\mu \Phi^\nu - \partial^\nu \Phi^\mu\right) = \mu_0 J^\nu \\ &= \partial_\mu \partial^\mu \Phi^\nu - \partial^\nu\left(\partial_\mu \Phi^\mu\right) = \mu_0 J^\nu\end{aligned} \tag{11–28}$$

which, after setting $\partial_\mu \Phi^\mu = 0$ to obtain the Lorenz gauge, becomes

$$\frac{1}{c^2}\frac{\partial^2}{\partial t^2}\Phi^\nu - \nabla^2 \Phi^\nu = \mu_0 J^\nu \tag{11–29}$$

11.4 The Stress-Energy-Momentum Tensor

The Lorentz (3-)force per unit volume, $\vec{f} = \rho\vec{E} + \vec{J}\times\vec{B}$, is the spacelike component of the four-force density $K^\mu = F^{\mu\nu}J_\nu$. Expressed in terms of three-vectors, this body force is $K^\mu = \left(\vec{\beta}\cdot\vec{f},\ \vec{f}\right)$. The zeroth component of K is $1/c$ times the power per unit volume expended by the electromagnetic field.

In general, we expect to be able to write the force per unit volume as the divergence of the stress. Labelling the four-stress tensor as $\overleftrightarrow{\eta}$, we anticipate

$$K^\nu = -\partial_\mu \eta^{\mu\nu} \tag{11–30}$$

To construct $\overleftrightarrow{\eta}$ we start from the Lorentz force equation $K_\mu = F_{\mu\nu}J^\nu$ and replace J^ν by derivatives of the fields using $\partial_\sigma F^{\sigma\nu} = \mu_0 J^\nu$. Thus

$$K_\mu = \frac{1}{\mu_0}F_{\mu\nu}\partial_\sigma F^{\sigma\nu} = \frac{1}{\mu_0}\left[\partial_\sigma\left(F_{\mu\nu}F^{\sigma\nu}\right) - F^{\sigma\nu}\partial_\sigma F_{\mu\nu}\right] \tag{11–31}$$

The first term is already of the required form. The second term can be expanded to

$$F^{\sigma\nu}\partial_\sigma F_{\mu\nu} = \tfrac{1}{2}F^{\sigma\nu}\partial_\sigma F_{\mu\nu} + \tfrac{1}{2}F^{\sigma\nu}\partial_\sigma F_{\mu\nu} \tag{11–32}$$

and, exchanging the dummy indices ν and σ on the second term, we get

$$\begin{aligned}F^{\sigma\nu}\,\partial_\sigma F_{\mu\nu} &= \tfrac{1}{2}F^{\sigma\nu}\partial_\sigma F_{\mu\nu} + \tfrac{1}{2}F^{\nu\sigma}\partial_\nu F_{\mu\sigma} \\ &= \tfrac{1}{2}F^{\sigma\nu}\partial_\sigma F_{\mu\nu} + \tfrac{1}{2}F^{\sigma\nu}\partial_\nu F_{\sigma\mu} \\ &= \tfrac{1}{2}F^{\sigma\nu}\left(\partial_\sigma F_{\mu\nu} + \partial_\nu F_{\sigma\mu}\right)\end{aligned} \tag{11–33}$$

Using $\partial_\sigma F_{\mu\nu} + \partial_\nu F_{\sigma\mu} + \partial_\mu F_{\nu\sigma} = 0$, we can replace the term in parentheses with $-\partial_\mu F_{\nu\sigma}$, leading to

$$F^{\sigma\nu}\partial_\sigma F_{\mu\nu} = -\tfrac{1}{2}F^{\sigma\nu}\partial_\mu F_{\nu\sigma} = \tfrac{1}{2}F^{\sigma\nu}\partial_\mu F_{\sigma\nu} = \tfrac{1}{4}\partial_\mu\left(F^{\sigma\nu}F_{\sigma\nu}\right) \tag{11–34}$$

Finally, replacing the dummy index σ by λ in the right hand term to avoid confusion with the σ in ∂_σ

$$\begin{aligned}K_\mu &= \frac{1}{\mu_0}\left[\partial_\sigma\left(F_{\mu\nu}F^{\sigma\nu}\right) - \tfrac{1}{4}\partial_\mu\left(F^{\lambda\nu}F_{\lambda\nu}\right)\right] \\ &= \frac{1}{\mu_0}\partial_\sigma\left[F_{\mu\nu}F^{\sigma\nu} - \tfrac{1}{4}\delta_\mu^\sigma\left(F^{\lambda\nu}F_{\lambda\nu}\right)\right]\end{aligned} \tag{11–35}$$

We have succeeded in writing the four-force density in a form similar to that of (11–30):

$$K_\mu = -\partial_\sigma \eta_{\mu}^{\cdot\sigma} \tag{11–36}$$

with

$$\eta_{\mu}^{\cdot\sigma} = -\frac{1}{\mu_0}\left[F_{\mu\nu}F^{\sigma\nu} - \tfrac{1}{4}\delta_\mu^\sigma\left(F^{\lambda\nu}F_{\lambda\nu}\right)\right] \tag{11–37}$$

Rather than immediately converting η to fully contravariant form, we use the somewhat more symmetric expression offered by equation (11–36) to compute the components of η explicitly in terms of the fields. The evaluation is not difficult but is a bit lengthy, as there are 16 elements to evaluate. We begin with

$$F^{\lambda\nu}F_{\lambda\nu} = 2\left(B^2 - \frac{E^2}{c^2}\right) \tag{11–38}$$

giving

$$\eta_{\mu}^{\cdot\sigma} = -\frac{1}{\mu_0}\left[F_{\mu\nu}F^{\sigma\nu} - \tfrac{1}{2}\delta_\mu^\sigma\left(B^2 - \frac{E^2}{c^2}\right)\right] \tag{11–39}$$

Referring to (11–10) and (11–11), we write the components of η as

$$\eta_0^{\cdot 0} = -\frac{1}{\mu_0}\left[F_{0\nu}F^{0\nu} - \frac{1}{2}\left(B^2 - \frac{E^2}{c^2}\right)\right]$$

$$= -\frac{1}{\mu_0}\left(- \frac{E^2}{c^2} - \frac{1}{2}B^2 + \frac{1}{2}\frac{E^2}{c^2}\right) = \frac{1}{2}\left(\varepsilon_0 E^2 + \frac{B^2}{\mu_0}\right) \tag{11–40}$$

$$\eta_1^{\cdot 1} = -\frac{1}{\mu_0}\left[F_{1\nu}F^{1\nu} - \frac{1}{2}\left(B^2 - \frac{E^2}{c^2}\right)\right] = -\frac{1}{\mu_0}\left(- \frac{E_x^2}{c^2} + B_z^2 + B_y^2 - \frac{B^2}{2} + \frac{E^2}{2c^2}\right)$$

$$= \frac{1}{\mu_0}\left[\frac{E_x^2}{c^2} + B_x^2 - \frac{1}{2}\left(B^2 + \frac{E^2}{c^2}\right)\right] \tag{11–41}$$

Similarly,

$$\eta_2^{\cdot 2} = \frac{1}{\mu_0}\left[\frac{E_y^2}{c^2} + B_y^2 - \frac{1}{2}\left(B^2 + \frac{E^2}{c^2}\right)\right] \tag{11–42}$$

$$\eta_3^{\cdot 3} = \frac{1}{\mu_0}\left[\frac{E_z^2}{c^2} + B_z^2 - \frac{1}{2}\left(B^2 + \frac{E^2}{c^2}\right)\right] \tag{11–43}$$

$$\eta_1^{\cdot 0} = -\frac{F_{1\nu}F^{0\nu}}{\mu_0} = -\frac{1}{\mu_0}\left(\frac{B_z E_y}{c} - \frac{B_y E_z}{c}\right) = -\frac{1}{\mu_0}\left(\frac{\vec{E}\times\vec{B}}{c}\right)_x \tag{11–44}$$

$$\eta_2^{\cdot 0} = -\frac{F_{2\nu}F^{0\nu}}{\mu_0} = -\frac{1}{\mu_0}\left(\frac{\vec{E}\times\vec{B}}{c}\right)_y \tag{11–45}$$

$$\eta_3^{\cdot 0} = -\frac{F_{3\nu}F^{0\nu}}{\mu_0} = -\frac{1}{\mu_0}\left(\frac{\vec{E}\times\vec{B}}{c}\right)_z \tag{11–46}$$

$$\eta_{2}^{\cdot 1} = -\frac{F_{2\nu}F^{1\nu}}{\mu_0} = \frac{1}{\mu_0}\left(\frac{E_xE_y}{c^2} + B_xB_y\right) \tag{11–47}$$

$$\eta_{3}^{\cdot 1} = \frac{1}{\mu_0}\left(\frac{E_xE_z}{c^2} + B_xB_z\right) \tag{11–48}$$

$$\eta_{3}^{\cdot 2} = \frac{1}{\mu_0}\left(\frac{E_yE_z}{c^2} + B_yB_z\right) \tag{11–49}$$

The remaining elements may be found by raising and lowering the indices using the metric tensor $g_{\mu\nu}$. In particular,

$$\eta^{i}_{\cdot 0} = g^{ii}g_{00}\eta_{i}^{\cdot 0} = -\eta_{i}^{\cdot 0} \tag{11–50}$$

and

$$\eta^{i}_{\cdot j} = g^{ii}g_{jj}\eta_{i}^{\cdot j} = \eta_{i}^{\cdot j} \tag{11–51}$$

for i and j chosen from 1, 2, and 3.

The fully contravariant form is also easily obtained by raising the lower index:

$$\begin{aligned} \eta^{ij} &= g^{ii}\,\eta_{i}^{\cdot j} = -\eta_{i}^{\cdot j} & \qquad \eta^{00} &= \eta_{00} = \eta_{0}^{\cdot 0} \\ \eta^{0j} &= g^{00}\eta_{0}^{\cdot j} = \eta_{0}^{\cdot j} & \qquad \eta^{j0} &= g^{jj}\eta_{j}^{\cdot 0} = -\eta_{j}^{\cdot 0} \end{aligned} \tag{11–52}$$

The resulting tensor is explicitly

$$\eta^{\mu\nu} = \left(\begin{array}{c|c} \dfrac{\varepsilon_0 E^2}{2} + \dfrac{B^2}{2\mu_0} & \longleftarrow \dfrac{\vec{E}\times\vec{B}}{\mu_0 c} \longrightarrow \\ \cdots\cdots & \cdots\cdots \\ \uparrow & \\ \dfrac{\vec{E}\times\vec{B}}{\mu_0 c} & -\varepsilon_0 E^\mu E^\nu - \dfrac{B^\mu B^\nu}{\mu_0} \\ \downarrow & \qquad + \tfrac{1}{2}\delta^{\mu\nu}\left(\varepsilon_0 E^2 + \dfrac{B^2}{\mu_0}\right) \end{array}\right) \tag{11–53}$$

The matrix has been partitioned to isolate the first row and first column. The remaining 3×3 matrix in the lower right hand corner is the Maxwell stress tensor, $\overleftrightarrow{T}^{\rm M}$ we have previously encountered. We recognize the 00 element to be the energy density U. As we will see, the 3-vector $(\eta^{10}, \eta^{20}, \eta^{30})$ formed by the space-like elements of the first column is the $1/c$ times $\vec{S}$, the Poynting vector. It will also become clear the row vector $(\eta^{01}, \eta^{02}, \eta^{03})$ is c times the electromagnetic momentum density.

We pause to justify the assertions about these implications of (11–53). Let us consider the spacelike terms of K^μ:

$$K^j = -\partial_\nu \eta^{\nu j} = -\partial_0\eta^{0j} - \partial_i T^{ij} = -\frac{\partial}{\partial t}\left(\frac{\vec{E}\times\vec{B}}{\mu_0 c^2}\right)^j - \partial_i T^{ij} \tag{11–54}$$

$K^j = f^j$ is the force density (force/volume) exerted by the fields on charged particles and currents. Since f is in general the rate of change of momentum density of the particles, and $-\partial_i T^{ij}$ is the force on the particles due to the field, we conclude that the term $\eta^{0j} = c(\vec{D} \times \vec{B})^j$ is the j component of $c \times$ the momentum density of the field. This is not, of course, a new result.

The K^0 term $(\vec{f} \cdot \vec{v}/c)$ gives

$$\frac{\vec{f} \cdot \vec{v}}{c} = -\partial_\nu \eta^{\nu 0} = -\frac{\partial}{c\partial t}\left(\frac{\varepsilon_0 E^2}{2} + \frac{B^2}{2\mu_0}\right) - \vec{\nabla} \cdot \frac{(\vec{E} \times \vec{B})}{\mu_0 c} \tag{11–55}$$

In other words, the power absorbed by the charges is

$$\vec{f} \cdot \vec{v} = -\frac{\partial}{\partial t}\left(\frac{\varepsilon_0 E^2}{2} + \frac{B^2}{2\mu_0}\right) - \frac{1}{\mu_0}\vec{\nabla} \cdot (\vec{E} \times \vec{B}) \tag{11–56}$$

The first term, $\frac{1}{2}(\varepsilon_0 E^2 + B^2/\mu_0)$ $[= \frac{1}{2}(\vec{E} \cdot \vec{D} + \vec{B} \cdot \vec{H})]$, may be identified as the energy density of the fields and the second term is the energy flux. The K^0 term can of course also be written as $K^0 = J_\mu F^{\mu 0} = \vec{J} \cdot \vec{E}/c$, giving $\vec{J} \cdot \vec{E} = \vec{f} \cdot \vec{v}$.

Although there appears to be no difference between the first row and first column in vacuum, they do differ in a material medium. In a medium, the 0th row $\eta^{0j} = c(\vec{D} \times \vec{B})^j$ whereas the first column, $\eta^{j0} = c^{-1}(\vec{E} \times \vec{H})$. The asymmetry implies non-conservation of electromagnetic angular momentum in material media and has led to considerable debate about the adequacy of this description.[30]

If the volume of interest contains only fields, with no charges or currents, then we must have $\partial_\mu \eta^{\mu\nu} = 0$.

The energy and momentum of the total field may be obtained by integrating the densities over the volume to give the contravariant momentum four-vector

$$P^\mu = \frac{1}{c}\int \eta^{0\mu}\, dx_1\, dx_2 dx_3 \tag{11–57}$$

having components

$$P^0 = \frac{1}{c}\int \left(\frac{\varepsilon_0 E^2}{2} + \frac{B^2}{2\mu_0}\right) d^3r = \frac{W}{c} \tag{11–58}$$

and

$$P^i = \frac{1}{c}\int \eta^{0i} d^3r = \int \frac{\vec{E} \times \vec{B}}{\mu_0 c^2} d^3r \tag{11–59}$$

P^0 is $1/c$ times the energy of the field and P^i is the momentum of the field.

It is interesting to calculate the momentum of the field of a slowly moving point charge, such as a nonrelativistic electron. We write the energy-stress-momentum

[30]see for example *Analysis of the Abraham-Minkowski controversy by means of two simple examples* M.A. López-Marinõ and J.L. Jiménez, Foundations of Physics Letters, **17**, Number 1, February 2004. pp. 1-23

tensor for the field of an electron, which for simplicity is assumed to be spherically symmetric in the proper frame, in abbreviated form

$$\overleftrightarrow{\eta} = \begin{pmatrix} U_0 & 0 \\ 0 & \overleftrightarrow{T}^{\rm M} \end{pmatrix} \tag{11-60}$$

where U_0 is the electrostatic field energy density and $\overleftrightarrow{T}^{\rm M}$ is the Maxwell stress tensor. Transforming the elements of $\overleftrightarrow{\eta}$ to those in a slowly moving frame having velocity $-\beta c$ ($\beta \ll 1 \Rightarrow \Gamma \simeq 1$) along the i axis, we obtain to first order in β

$$\begin{aligned} U_0' = \eta^{00'} &= \alpha_\mu^{0'} \alpha_\nu^{0'} \eta^{\mu\nu} = \alpha_\mu^{0'} \left(\alpha_0^{0'} \eta^{\mu 0} + \alpha_i^{0'} \eta^{\mu i}\right) \simeq \alpha_\mu^{0'} \left(\eta^{\mu 0} + \beta \eta^{\mu i}\right) \\ &= \alpha_0^{0'} \eta^{00} + \alpha_0^{0'} \beta \eta^{0i} + \alpha_i^{0'} \eta^{i0} + \alpha_i^{0'} \beta \eta^{ii} \simeq \eta^{00} = U_0 \end{aligned} \tag{11-61}$$

$$\begin{aligned} \eta^{0i'} &= \alpha_\mu^{0'} \alpha_\nu^{i'} \eta^{\mu\nu} = \alpha_\mu^{0'} \left(\alpha_0^{i'} \eta^{\mu 0} + \alpha_j^{i'} \eta^{\mu j}\right) = \alpha_\mu^{0'} \left(\beta \eta^{\mu 0} + \eta^{\mu i}\right) \\ &= \alpha_0^{0'} \beta \eta^{00} + \alpha_j^{0'} \eta^{ji} = \beta (U_0 + \eta^{ii}) \end{aligned} \tag{11-62}$$

(as i is not a dummy index, no summation is implied in η^{ii}) and for $j \neq i$

$$\begin{aligned} \eta^{0j'} &= \alpha_\mu^{0'} \alpha_\nu^{j'} \eta^{\mu\nu} = \alpha_\mu^{0'} \left(\alpha_0^{j'} \eta^{\mu 0} + \alpha_\ell^{j'} \eta^{\mu \ell}\right) \\ &= \alpha_\mu^{0'} \eta^{\mu j} = \alpha_0^{0'} \eta^{0j} + \alpha_\ell^{0'} \eta^{\ell j} = \beta \eta^{ij} \end{aligned} \tag{11-63}$$

The energy of the field of the moving electron is then unchanged (to first order in v/c),

$$W' = \int U_0 d^3 r = W \tag{11-64}$$

and the $j \neq i$ component of momentum is

$$P^{j'} = \frac{1}{c} \int \eta^{0j'} d^3 r = \frac{v}{c^2} \int \varepsilon_0 E^i E^j d^3 r = 0 \tag{11-65}$$

since the integral of E^j vanishes. The i component leads to a more interesting result.

$$\begin{aligned} P^{i'} &= \frac{1}{c} \int \eta^{0i'} d^3 r = \frac{\beta}{c} \int \left[U_0 - \varepsilon_0 \left(E^i E^i - \tfrac{1}{2} E_0^2\right)\right] d^3 r \\ &= \frac{\beta}{c} \left[W - \varepsilon_0 \int \left(\tfrac{1}{3} E_0^2 - \tfrac{1}{2} E_0^2\right) d^3 r\right] = \frac{\beta}{c} \left(W + \tfrac{1}{3} W\right) = \frac{4}{3} \frac{W}{c^2} v^i \end{aligned} \tag{11-66}$$

More generally, the momentum associated with the field of a charged particle is

$$\vec{p} = \frac{4}{3} \frac{W}{c^2} \vec{v} \tag{11-67}$$

The inertial mass of the field appears to be $\frac{4}{3} W/c^2$(!) in conflict with the normal relativistic expectation that the inertial mass should be the energy/c^2. Lest the reader suspect that there is a serious error in the preceding derivation, we will obtain this result again later using quite different methods.

We now proceed to use the covariant formalism to generalize some earlier results in radiation by an accelerated charge.

11.5 Radiation from Relativistic Particles

In the previous chapter we calculated the power radiated by an accelerated particle in its own rest frame ($\beta = 0$) as

$$\mathbf{P} = \frac{q^2 a'^2}{6\pi\varepsilon_0 c^3} \tag{11–68}$$

Noting that $dW = \mathbf{P}dt$ and that both dW and dt are time-like components of a four-vector, we find that $\mathbf{P}$ must be a scalar under Lorentz transformations. From the general expression (10-143, 144) of the fields, $\mathbf{P}$ can depend only on $\vec{\beta}$ and $\dot{\vec{\beta}}$; no higher derivatives occur. We therefore seek an invariant that is proportional to a^2 at low speeds and contains derivatives no higher than $\dot{\beta}$. Consider the four-acceleration

$$A^\mu \equiv \frac{d}{d\tau}V^\mu = \gamma\Big(\gamma^3 c(\dot{\vec{\beta}}\cdot\vec{\beta}),\ \gamma^3(\dot{\vec{\beta}}\cdot\vec{\beta})\vec{v} + \gamma\dot{\vec{v}}\Big) \tag{11–69}$$

where we have used $V^\mu = \gamma(c, \vec{v})$ and $d/d\tau = \gamma d/dt$. A^μ clearly depends only on $\dot{\vec{\beta}}$ and $\vec{\beta}$, and, as $\beta \to 0$, $A^\mu \to (0, \dot{\vec{v}})$. Thus $A^\mu A_\mu \to -a^2$. We are led then, to writing the covariant form of the radiated power as

$$\mathbf{P} = -\frac{q^2}{6\pi\varepsilon_0 c}\frac{A_\mu A^\mu}{c^2} \tag{11–70}$$

In the frame where the particle has $\beta \neq 0$ we find

$$\begin{aligned}
\frac{A_\mu A^\mu}{c^2} &= \gamma^2\Big\{\gamma^6(\dot{\vec{\beta}}\cdot\vec{\beta})^2 - \Big|\gamma^3(\dot{\vec{\beta}}\cdot\vec{\beta})\vec{\beta} + \gamma\dot{\vec{\beta}}\Big|^2\Big\} \\
&= \gamma^6\left[(\dot{\vec{\beta}}\cdot\vec{\beta})^2\gamma^2(1-\beta^2) - 2(\dot{\vec{\beta}}\cdot\vec{\beta})^2 - (1-\beta^2)\dot{\beta}^2\right] \\
&= -\gamma^6\left\{\dot{\beta}^2 - \left[\beta^2\dot{\beta}^2 - (\vec{\beta}\cdot\dot{\vec{\beta}})^2\right]\right\} = -\gamma^6\left[\dot{\beta}^2 - |\vec{\beta}\times\dot{\vec{\beta}}|^2\right]
\end{aligned} \tag{11–71}$$

The total power radiated by a relativistic particle is then

$$\mathbf{P} = -\frac{q^2}{6\pi\varepsilon_0 c}\frac{A_\mu A^\mu}{c^2} = \frac{q^2}{6\pi\varepsilon_0 c}\gamma^6\left[\dot{\beta}^2 - (\vec{\beta}\times\dot{\vec{\beta}})^2\right] \tag{11–72}$$

The momentum radiated by the particle is now also easily obtained from kinematic considerations. It should be pointed out that the rate of momentum radiation is not the entire electromagnetic reaction force on the accelerated particle. There will also appear an inductive reaction force rather perversely labelled the *radiation reaction*.

The four-force acting on a particle is related to its four-momentum by

$$\frac{dP^\mu}{d\tau} = \gamma\frac{d}{dt}\left(\frac{W}{c}, \vec{p}\right) \tag{11–73}$$

where $\vec{p}$ is the three-momentum. In the instantaneous rest frame of the particle, $d\vec{p}'/dt' = \vec{F}'$. Transforming to a lab frame where the particle has velocity $\vec{\beta}c$ (the lab frame then has velocity $-\vec{\beta}c$) and taking the x axis along $\vec{\beta}$ for convenience, we obtain for the 0 component in the lab frame

$$\frac{dP^0}{d\tau} = \frac{\gamma}{c}\frac{dW}{dt} = \Gamma\frac{dP^{0'}}{d\tau} + \beta\Gamma\frac{dP^{1'}}{d\tau}$$

$$= \frac{\Gamma\gamma'}{c}\frac{dW'}{dt'} + \beta\Gamma\gamma'\frac{dp'_x}{dt'} = \frac{\Gamma\gamma'}{c}\frac{dW'}{dt'} + \beta\Gamma\gamma' F'_x \tag{11–74}$$

The x component of force in the lab frame is likewise given by

$$\frac{dP^1}{d\tau} = \gamma\frac{dp_x}{dt} = \Gamma\gamma'\frac{dp'_x}{dt'} + \beta\frac{\Gamma\gamma'}{c}\frac{dW'}{dt'} = \frac{\beta}{c}\gamma\frac{dW}{dt} + (1-\beta^2)\Gamma\gamma' F'_x \tag{11–75}$$

As the $'$ variables were in the particle's rest frame, we have $\gamma' = 1$ and $\gamma = \Gamma$. We can simplify dp_x/dt to read

$$\frac{dp_x}{dt} = \frac{\beta}{c}\frac{dW}{dt} + (1-\beta^2)F'_x \tag{11–76}$$

We have at this point no means to determine F' because the force needed to give the required acceleration needs to overcome not only the inertia offered by the particle's mass, but also the inertia of the changing fields. We postpone consideration of F' to the next chapter. Denoting the radiative momentum losses by the subscript *rad*, we generalize the first term of (11–76) for $\vec{\beta}$ in an arbitrary direction using (11–72)

$$\frac{d\vec{p}_{rad}}{dt} = \frac{\vec{\beta}}{c}\frac{dW}{dt} = \frac{-q^2\gamma^6}{6\pi\varepsilon_0 c^2}\vec{\beta}\left[\dot{\beta}^2 - |\vec{\beta}\times\dot{\vec{\beta}}|^2\right] \tag{11–77}$$

This part of the reaction force on an accelerated particle results in non-linear equations of motion which are generally rather intractable.

Both the radiative energy and momentum losses may be combined in the clearly covariant equation

$$\frac{dP^\mu}{d\tau} = \frac{q^2}{6\pi\varepsilon_0 c^5}(A_\nu A^\nu)\, V^\mu \tag{11–78}$$

We can relate approximately the radiated power to the three-force $d\vec{p}/dt$ required to accelerate the particle by neglecting the portion of the work going into radiated momentum and expressing the power radiated (11–70) in terms of the four-momentum rather than the four-acceleration. Thus we write $m_0 A^\mu = dP^\mu/d\tau$:

$$\mathbf{P} = -\frac{q^2}{6\pi\varepsilon_0}\frac{c}{(m_0c^2)^2}\frac{dP_\mu}{d\tau}\frac{dP^\mu}{d\tau}$$

$$= -\frac{q^2}{6\pi\varepsilon_0}\frac{c\gamma^2}{(m_0c^2)^2}\frac{dP_\mu}{dt}\frac{dP^\mu}{dt} \tag{11–79}$$

and using $P^\mu = (W/c,\, \vec{p}) = \left(\sqrt{m_0^2c^2+p^2},\, \vec{p}\right)$, we find that

$$\frac{dP^\mu}{dt} = \left(\frac{d(W/c)}{dt},\, \frac{d\vec{p}}{dt}\right) = \left(\frac{p\dfrac{d|p|}{dt}}{W/c},\, \frac{d\vec{p}}{dt}\right) \tag{11–80}$$

whence

$$\frac{dP^\mu}{dt}\frac{dP_\mu}{dt} = \frac{p^2\left(\dfrac{d|p|}{dt}\right)^2}{W^2/c^2} - \left(\frac{d\vec{p}}{dt}\right)^2 = \frac{\beta^2m^2c^2}{m^2c^2}\left(\frac{d|p|}{dt}\right)^2 - \left(\frac{d\vec{p}}{dt}\right)^2$$

$$= \beta^2\left(\frac{d|p|}{dt}\right)^2 - \left(\frac{d\vec{p}}{dt}\right)^2 \tag{11–81}$$

The expression for the radiated power from the accelerated charge may now be written in terms of the accelerating forces,

$$\mathbf{P} = \frac{q^2}{6\pi\varepsilon_0}\frac{c\gamma^2}{(m_0c^2)^2}\left[\left(\frac{d\vec{p}}{dt}\right)^2 - \beta^2\left(\frac{d|p|}{dt}\right)^2\right] \tag{11–82}$$

When the force is parallel to the momentum, $\left|\dfrac{d\vec{p}}{dt}\right| = \dfrac{d|p|}{dt}$ and

$$\mathbf{P} = \frac{q^2}{6\pi\varepsilon_0}\frac{c\gamma^2}{(m_0c^2)^2}(1-\beta^2)\left|\frac{dp}{dt}\right|^2 = \frac{q^2c}{6\pi\varepsilon_0(m_0c^2)^2}\left|\frac{dp}{dt}\right|^2 \tag{11–83}$$

When the accelerating force is perpendicular to the momentum, $\dfrac{d|p|}{dt} = 0$, leading to

$$\mathbf{P} = \frac{q^2\gamma^2}{6\pi\varepsilon_0m_0^2c^3}\left(\frac{d\vec{p}}{dt}\right)^2 \tag{11–84}$$

generally much larger than the previous term. We conclude that for a given accelerating force, perpendicular forces lead to radiation a factor γ^2 larger than parallel forces.

Exercises and Problems

11-1 Using the fact that the phase of a plane electromagnetic wave is independent of an observer's velocity, show that the four-component quantity $K^\mu \equiv (\omega/c, \vec{k})$ is a four-vector. Use the transformation properties of four-vectors to obtain the Doppler effect formula (frequency transformation) for motion (a) parallel to the wave vector $\vec{k}$ and (b) perpendicular to the wave vector.

11-2 Show that $A^\mu V_\mu$ is identically zero. A^μ is the four-acceleration $dV^\mu/d\tau$.

11-3 Demonstrate that $f^j = F^{j\nu}J_\nu$ is the Lorentz (3-)force density

11-4 Verify that Maxwell's equations in vacuum result from (11–12) and (11–14). Show that Maxwell's equations for $\vec{D}$ and $\vec{H}$ result from (11-23).

11-5 The *electromagnetic dual tensor* $G^{\mu\nu}$ is defined by $G^{\mu\nu} = \frac{1}{2}\epsilon^{\mu\nu\rho\sigma}F_{\rho\sigma}$. Find the components of this tensor, and show that $\partial_\mu G^{\mu\nu} = 0$ is equivalent to $\vec{\nabla}\cdot\vec{B} = 0$ and $\vec{\nabla}\times\vec{E} = -\partial\vec{B}/\partial t$.

11-6 Find the remaining components of $(F^{\mu\nu})'$ required to demonstrate (11-19).

11-7 Obtain the transformation laws for the vector potential and the scalar potential when the primed system moves with velocity $\vec{v}$ with respect to the unprimed system.

11-8 Show that the transformations (11–19) applied to the field of a stationary charge give the electric field of a uniformly moving charge as expressed by the non-radiative part of (10-141).

11-9 Show from elementary considerations (length contraction of two sides) that a moving rectangular current loop acquires an dipole moment, and determine the magnitude and direction of this electric dipole moment.

11-10 Deduce Maxwell's equations for fields in a moving dielectric having $\rho = \vec{J} = 0$, relating $\vec{E}$ and $\vec{B}$ to $\vec{P}$ and $\vec{M}$ and to each other. Assume that the velocity V is sufficiently small that we may take $\Gamma = 1$. (Note that $\vec{P}$ and $\vec{M}$ move along with the dielectric.)

11-11 Find the electric field appearing around a uniformly magnetized sphere rotating at angular frequency $\omega \ll c/a$ about a central axis parallel to the magnetization.

11-12 Develop the Stress-Energy-Momentum tensor for fields in a medium using $H^{\mu\nu}$ to relate the fields to the sources.

11-13 Derive the Lorentz transformation for a frame moving in an arbitrary direction.

11-14 Show that if $\vec{E}$ and $\vec{H}$ are perpendicular in one inertial frame, they will be perpendicular in all inertial frames.

11-15 If $\vec{E}$ is perpendicular to $\vec{B}$ but $|\vec{E}| \neq |c\vec{B}|$, then there is a frame where the field is either purely electric or purely magnetic. Find the velocity of that frame. What happens if $|\vec{E}| = |c\vec{B}|$?

11-16 Use Maxwell's equations in a moving medium (problem 11-10) to find the velocity of light in a moving transparent dielectric.

11-17 The energy of an electron in a linear accelerator is increased at a rate of 1 MeV/meter. Find the power radiated by such an electron. (Note that the radiated power is independent of the electron's velocity.) Is the radiated power a major portion of the power input?

11-18 Find the power radiated by a highly relativistic electron in a circular orbit in terms of its velocity v and its radius R. Evaluate this expression for a 10-GeV electron in a 20-m radius orbit and find the energy loss per revolution. Would it be easy to supply several times this energy to obtain a net acceleration at this velocity?

Chapter 12

Radiation Reaction—Electrodynamics

12.1 Electromagnetic Inertia

We have in the foregoing chapters synthesized a powerful, coherent, self-consistent theory of electromagnetism, satisfying relativistic covariance naturally. In this chapter we will see that this magnificent construction, when dealing with the interaction of charged particles with their fields, presents some very uncomfortable inconsistencies with the world as we understand it. In particular, the theory will be shown to violate causality, the notion that causes precede their effects.

On our first discussion of the fields, we agreed to exclude the sensing particle's own field from that experienced by the particle on the grounds that the particle cannot pull itself up by its own bootstraps. As we have seen, the radiation field of an accelerated point particle is not generally symmetric about the particle and appears able to exist on its own even after the particle is snuffed out.

To convince ourselves that electromagnetic fields can have an existence independent of their sources, we might consider the radiation by positronium, the atom formed by an electron and its antiparticle the positron. Such atoms emit radiation in much the same manner as hydrogen, but the atom annihilates in its ground state. It would be surprising if radiation, having escaped from the atom, suddenly ceased to exist when the atom did. In fact it continues to spread, totally unaffected by the atom's disappearance. (Of course, on its annihilation the electron and positron also emit two 512-KeV γ-rays that carry the lost mass energy.) Disappearance of these electromagnetic fields would cause a serious violation of energy conservation.

A less contrived rationale for considering a particle's interaction with its own field is offered by the example of an atom emitting an electromagnetic wave that is reflected by a mirror. It would be surprising indeed if the atom were prohibited from interacting with the reflected wave that would surely be indistinguishable from the wave emitted by another atom.

We are now forced to reexamine our belief that the particle cannot interact with its own field. If the field has an existence independent of the particle and it is asymmetric about the particle, there seems no good reason to exclude this field from those felt by the particle. As a preliminary, we make the following observations.

(a) When an electron is in uniform motion, its field will contribute to the momentum since for a small change in velocity, the momentum of both the field and the particle would change together.

(b) If an electron radiates as a result of being accelerated by an external force, that force must supply the energy and momentum of both the fields (induction and radiation fields) and the inertia of the "bare" particle. If the electron is to offer a resistance to acceleration corresponding to its own inertia as well as that of the fields, then the fields must in fact produce a *reaction* force on the electron, diminishing the acceleration produced by external forces.

Let us first consider the interaction of the electron with its induction field. A change in velocity leads to a changing magnetic field, which induces an electric field at the position of the electron. The electric field may be found from $\vec{E} = -\partial\vec{A}/\partial t$. A change in $\delta\vec{v}$ in the motion of the electron produces a corresponding change in the vector potential

$$\delta\vec{A} = \frac{\mu_0}{4\pi}\frac{q}{r}\delta\vec{v} \tag{12–1}$$

at a distance r from the electron. Unfortunately this gives an infinite result at the 'edge' of a point particle. To avoid the infinity, we might, for the sake of making some progress, postulate that the $1/r$ form of the vector potential breaks down at some minimum distance r_0 because of a supposed structure of the particle. As a rough approximation, we assume that the interaction of the particle with its field occurs entirely at this cutoff distance r_0.

The reaction force on the accelerated particle due to this changing field is

$$\vec{F} = q\vec{E} = -q\frac{\partial\vec{A}}{\partial t} = -\frac{\mu_0 q^2}{4\pi r_0}\frac{d\vec{v}}{dt} = -\frac{q^2}{4\pi\varepsilon_0 r_0 c^2}\frac{d\vec{v}}{dt} \tag{12–2}$$

The reaction is an exact analogy to the back EMF in an electrical circuit.

The form of the reaction is that of an inertial reaction

$$\vec{F} = -m_{em}\frac{d\vec{v}}{dt} \quad \text{with} \quad m_{em} = \frac{q^2}{4\pi\varepsilon_0 r_0 c^2} \tag{12–3}$$

A similar estimate of the "inertial mass" of the electron's field is obtained by equating $1/c^2$ times the energy of the field of a uniformly charged spherical shell of radius r_0 to m_{em} (Ex 4.1.2):

$$\frac{W}{c^2} = \frac{e^2}{8\pi\varepsilon_0 r_0 c^2} \tag{12–4}$$

These arguments yield only estimates of the inertial mass of the electron's field, since r_0 is entirely adjustable, depending on the model.

It is possible to establish the relationship between W and m_{em} independently of r_0. The momentum carried by arbitrary fields in vacuum is given by

$$\vec{p}_{em} = \int \frac{\vec{E}\times\vec{B}}{\mu_0 c^2} d^3r \tag{12–5}$$

The magnetic induction field may be found from (11–19) to yield for low velocities

$$\vec{B} = \frac{\vec{\beta} \times \vec{E}}{c} = \varepsilon_0 \mu_0 \vec{v} \times \vec{E} \tag{12–6}$$

The momentum of the field moving with the charged particle is then

$$\vec{p}_{em} = \int \frac{\varepsilon_0 \vec{E} \times (\vec{v} \times \vec{E})}{c^2} d^3 r = \varepsilon_0 \int \frac{E^2 \vec{v} - (\vec{v} \cdot \vec{E})\vec{E}}{c^2} d^3 r \tag{12–7}$$

The electric field, $\vec{E}$, must be axially symmetric about the direction of the velocity $\vec{v}$, which we take to be along the z axis. Therefore, writing $\vec{E} = E_z \hat{k} + \vec{E}_t$, $\vec{E}_t \cdot \vec{v}$ is seen to vanish and further $\int (E_z v)\vec{E}_t d\varphi$ vanishes so that the second term of the integral reduces to

$$\int (\vec{v} \cdot \vec{E})\vec{E} d^3 r = \int E_z^2 \vec{v}\, d^3 r \simeq \tfrac{1}{3} \int E^2 \vec{v}\, d^3 r \tag{12–8}$$

Hence

$$\vec{p}_{em} = \frac{4\vec{v}}{3c^2} \int \frac{\varepsilon_0 E^2}{2} d^3 r = \frac{4}{3} \frac{W}{c^2} \vec{v} \tag{12–9}$$

a curious result we have already encountered in the last chapter. Presumably, because we cannot obtain an electron without its field, the empirical mass is the sum of a "bare" mass and the inertia of its electromagnetic field. The factor of $\frac{4}{3}$ in the electromagnetic inertia is rather difficult to accommodate in the total mass because it conflicts with the mass energy equivalence relationship of special relativity.

12.2 The Reaction Force Needed to Conserve Energy

Although we have considered the kinematics of radiating electrons, we have largely ignored the dynamics of accelerated particles. In particular, it should be clear that when an accelerated particle emits radiation it must do so at the expense of mechanical energy. It is tempting to assume that the energy loss during any particular time interval will be small compared to the kinetic energy change due to acceleration of the particle, so that the radiation loss will represent only a small perturbation to the motion. If this is indeed the case, we can account for this radiation "drag" by a force called *radiation reaction.* This force however, will be able to account for only an average effect on the motion. It is important to recognize that this force, in spite of its name, is not responsible for the momentum carried by the radiation field. It is in fact an interaction of the electron with its own (retarded) induction field.

On average, the reaction force F_R must act like a drag force on the charged particle, opposing the velocity to decrease its kinetic energy. Equating the rate that the reaction force extracts energy from the electron to feed the radiation to the radiative loss rate, we write

$$\vec{F}_R \cdot \vec{v} = -\frac{q^2}{6\pi\varepsilon_0 c^3} \dot{v}^2 \tag{12–10}$$

On a small enough time scale, however, this cannot be the whole story for reasons we detail below.

When the emitted energy is large compared to the energy gained from the external accelerating force, the trajectory of the particle as well as the reaction force become rather ill defined. Let us briefly examine the energy gain and energy loss of a particle accelerated during a short time interval Δt. For simplicity we take the particle initially at rest. The gain in kinetic energy is then

$$\Delta W = \tfrac{1}{2}m(a\Delta t)^2 \tag{12–11}$$

while the loss to radiation is

$$\Delta W_R = \frac{q^2 a^2}{6\pi\varepsilon_0 c^3}\Delta t \tag{12–12}$$

For times Δt smaller than 2τ, with

$$\tau = \frac{q^2}{6\pi\varepsilon_0 m c^3} \tag{12–13}$$

($\tau \cong 6.26 \times 10^{-24}$ seconds for an electron), the radiative energy loss exceeds the change in kinetic energy of the particle. Thus, on this time scale the radiation reaction force becomes the dominant force, leading apparently to a particle accelerating itself without externally applied force. We can no longer use an average force to determine the motion. Since light travels a distance $c\tau \simeq 2 \times 10^{-15}$ m, the distance scale where radiation plays a dominant role in the behavior of an electron is 10^{-15} m. The fluctuations are suggestive of the perturbations introduced by the spontaneous emission of virtual photons in the Quantum Electrodynamics picture.

In spite of the short term fluctuations we can obtain the average reaction force by considering conservation of energy over a sufficiently long time or along a restricted trajectory. Providing that the energy of the induction fields is the same at the beginning and at the end of the averaging period, conservation of energy requires that the energy lost to the radiation drag force equal the energy radiated:

$$\int_{t_1}^{t_2} \vec{F}_R \cdot \vec{v}\,dt + \int_{t_1}^{t_2} \frac{q^2}{6\pi\varepsilon_0 c^3}\dot{v}^2 dt = 0 \tag{12–14}$$

Integrating the second term by parts gives

$$\int_{t_1}^{t_2} \left(\vec{F}_R - \frac{q^2\ddot{\vec{v}}}{6\pi\varepsilon_0 c^3}\right) \cdot \vec{v}\,dt + \frac{q^2}{6\pi\varepsilon_0 c^3}\vec{v}\cdot\dot{\vec{v}}\Big|_{t_1}^{t_2} = 0 \tag{12–15}$$

The integrated second term represents the short-term fluctuations we considered earlier. For periodic motion, or for acceleration occurring within a limited time, an appropriate choice of t_2 and t_1 will eliminate this term. We can then conserve the average energy by setting

$$\vec{F}_R = \frac{q^2}{6\pi\varepsilon_0 c^3}\ddot{\vec{v}} \tag{12–16}$$

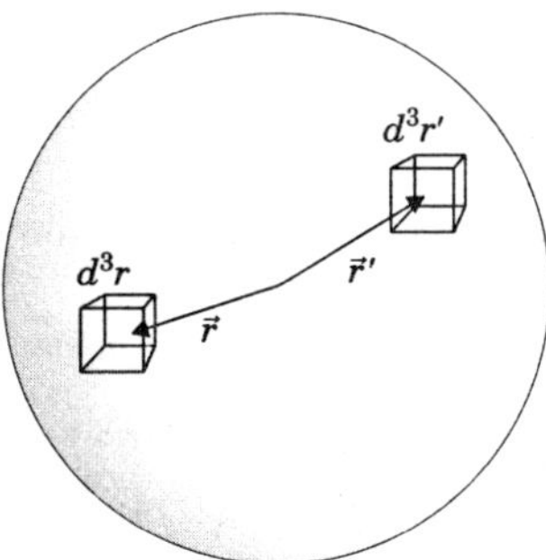

Figure 12.1: The force on a small volume within the ball of charge is calculated as the sum of forces the charge every other volume element d^3r' exercises on the charge in d^3r. Then the total force on the ball is obtained by summing the forces on each element.

as the reaction force. The radiated momentum term (11–77) is a factor of roughly β smaller than F_R and will be included in a full covariant expression. As we will see in section 12.4, equation (12–16) leads to an unstable equation of motion whose solution has a velocity that increases exponentially even in the absence of any external force.

12.3 Direct Calculation of Radiation Reaction: The Abraham-Lorentz Model

Although energy conservation certainly gives a requirement of a reaction force, it doesn't give much insight into how such a reaction force might arise. To gain such insight, we will calculate the force on a rigid, spherically symmetric charged ball due to all other parts of the ball (figure 12.1) when it is accelerated. We initially calculate the force that a small volume element d^3r experiences from all the other segments of the sphere and then sum over all segments. (A rigid extended particle is inconsistent with special relativity, but we ignore this.) It should be emphasized that this calculation in no way reflects the modern Quantum Electrodynamics understanding of radiation reaction.

The force on the element $\rho(\vec{r})d^3r$ is, in terms of the fields at its location,

$$d\vec{F}(\vec{r},t) = \left[\rho(\vec{r})\vec{E}(\vec{r},t) + \vec{J}(\vec{r}) \times \vec{B}(\vec{r},t)\right] d^3r \tag{12–17}$$

We may, without loss of generality, place ourselves in the instantaneous rest frame of the particle so that the force reduces to

$$d\vec{F}(\vec{r},t) = \rho(\vec{r})\vec{E}(\vec{r},t)d^3r \tag{12–18}$$

where the electric field $\vec{E}$ is calculated from $\vec{E} = -\vec{\nabla}V - \partial\vec{A}/\partial t$ and the scalar potential V and vector potential $\vec{A}$ are the retarded potentials due to the rest of the particle; that is,

$$V(\vec{r},t) = \frac{1}{4\pi\varepsilon_0}\int \frac{\rho(\vec{r}\,', t - R/c)}{R} d^3r' \tag{12–19}$$

and

$$\vec{A}(\vec{r},t) = \frac{\mu_0}{4\pi}\int \frac{\vec{J}(\vec{r}\,',t-R/c)}{R}d^3r' \tag{12–20}$$

where $R = |\vec{r}-\vec{r}\,'|$ is the distance from the source volume element, d^3r', to the field point $\vec{r}$, both inside the charged ball.

For R/c small, we expand $\rho(\vec{r},t-R/c)$ as a power series in R/c to obtain

$$\rho(\vec{r},t-R/c) = \sum_{n=0}^{\infty}\frac{1}{n!}\left(-\frac{R}{c}\right)^n \frac{\partial^n\rho(\vec{r},t)}{\partial t^n} \tag{12–21}$$

which, inserted into the integral for the retarded potential, gives

$$V(\vec{r},t) = \frac{1}{4\pi\varepsilon_0}\sum_{n=0}^{\infty}\frac{(-1)^n}{n!c^n}\int R^{n-1}\frac{\partial^n\rho(\vec{r}\,',t)}{\partial t^n}d^3r' \tag{12–22}$$

In identical fashion we expand the vector potential to obtain

$$\frac{\partial\vec{A}(\vec{r},t)}{\partial t} = \frac{\partial}{\partial t}\left[\frac{\mu_0}{4\pi}\sum_{n=0}^{\infty}\frac{(-1)^n}{n!c^n}\int R^{n-1}\frac{\partial^n\vec{J}(\vec{r}\,',t)}{\partial t^n}d^3r'\right] \tag{12–23}$$

The force on our selected volume element using $\vec{E} = -\vec{\nabla}V - \partial\vec{A}/\partial t$, is then

$$d\vec{F} = -\frac{\rho(\vec{r},t)d^3r}{4\pi\varepsilon_0}\sum_{n=0}^{\infty}\frac{(-1)^n}{n!c^n}\int\frac{\partial^n}{\partial t^n}\left[\rho(\vec{r}\,',t)\vec{\nabla}R^{n-1} + \frac{R^{n-1}}{c^2}\frac{\partial\vec{J}(\vec{r}\,',t)}{\partial t}\right]d^3r' \tag{12–24}$$

We consider the first two terms arising from the scalar potential separately. The $n = 1$ term vanishes identically as it involves $\vec{\nabla}R^0$. The $n = 0$ term arising from the scalar potential leads to a total force, which is just the electrostatic self-force. Labelling successive terms of the force with a subscript,

$$\begin{aligned}\vec{F}_0 &= \int d\vec{F}_0 = \int\rho(\vec{r},t)\int\frac{\rho(\vec{r}\,',t)\vec{R}}{4\pi\varepsilon_0 R^3}d^3r'd^3r \\ &= \int\rho(\vec{r},t)\vec{E}_S(\vec{r},t)d^3r = 0\end{aligned} \tag{12–25}$$

Eliminating these two terms and increasing by two the summation index on the terms contributed by the scalar potential, we rewrite the sum as

$$\begin{aligned}d\vec{F} = -\frac{\rho(\vec{r},t)d^3r}{4\pi\varepsilon_0}&\sum_{n=0}^{\infty}\frac{(-1)^n}{n!c^{n+2}}\frac{\partial^{n+1}}{\partial t^{n+1}} \\ &\int\left[\frac{\partial\rho(\vec{r}\,',t)}{\partial t}\frac{\vec{\nabla}R^{n+1}}{(n+1)(n+2)} + R^{n-1}\vec{J}(\vec{r}\,',t)\right]d^3r'\end{aligned} \tag{12–26}$$

The continuity equation may be used to replace $\partial\rho/\partial t$ by $-\vec{\nabla}\cdot\vec{J}$, allowing us to write

$$\frac{\partial\rho(\vec{r}\,',t)}{\partial t}\frac{\vec{\nabla}R^{n+1}}{(n+1)(n+2)} = -\frac{R^{n-1}\vec{R}}{n+2}\left[\vec{\nabla}'\cdot\vec{J}(\vec{r}\,',t)\right] \tag{12–27}$$

and with the aid of the vector relation (7), the α component of (12–27) may be rewritten as a divergence

$$\vec{\nabla}' \cdot \left[R_\alpha R^{n-1}\vec{J}\right] = \vec{\nabla}'\left[R_\alpha R^{n-1}\right] \cdot \vec{J} + R_\alpha R^{n-1}\left[\vec{\nabla}' \cdot \vec{J}\right]$$

$$= -R^{n-1}J_\alpha - (n-1)R_\alpha R^{n-3}\vec{R} \cdot \vec{J} + \left[R^{n-1}R_\alpha(\vec{\nabla}' \cdot \vec{J})\right] \quad (12\text{–}28)$$

to integrate the first term of the integral in (12–26) by parts after replacing $\partial\rho/\partial t$ by $-\vec{\nabla} \cdot \vec{J}$ one component at a time. We obtain

$$\int \frac{\partial \rho(\vec{r}\,', t)}{\partial t} \frac{\vec{\nabla} R^{n+1}}{(n+1)(n+2)} d^3r' = -\frac{1}{n+2} \int R^{n-1} \left[\vec{J} + (n-1)\frac{\vec{J} \cdot \vec{R}}{R^2}\vec{R}\right] d^3r' \quad (12\text{–}29)$$

The integral for $d\vec{F}$ then becomes

$$d\vec{F} = -\frac{\rho\, d^3r}{4\pi\varepsilon_0} \sum_{n=0}^{\infty} \frac{(-1)^n}{n!c^{n+2}} \int R^{n-1} \frac{\partial^{n+1}}{\partial t^{n+1}} \left[\frac{n+1}{n+2}\vec{J}(\vec{r}\,', t) - \frac{n-1}{n+2}\frac{\vec{J} \cdot \vec{R}}{R^2}\vec{R}\right] d^3r' \quad (12\text{–}30)$$

For a rigid body, $\vec{J}(\vec{r}\,', t) = \rho(\vec{r}\,', t)\vec{v}(t)$; therefore $\vec{J} \cdot \vec{R} = \rho\vec{v} \cdot \vec{R}$. Furthermore, only the component of $d\vec{F}$ parallel to $\vec{v}$ can survive the integration, so we write $\langle (\vec{J} \cdot \vec{R})\vec{R} \rangle = \rho[(\vec{R} \cdot \vec{v})^2/v^2]\vec{v}$ and recast the integral as

$$d\vec{F} = \frac{\rho d^3r}{4\pi\varepsilon_0} \sum_{0}^{\infty} \frac{(-1)^n}{n!c^{n+2}} \frac{\partial^{n+1}}{\partial t^{n+1}} \int R^{n-1}\rho(\vec{r}\,', t)\vec{v}(t) \left[\frac{n+1}{n+2} - \frac{n-1}{n+2}\frac{(\vec{R} \cdot \vec{v})^2}{R^2v^2}\right] d^3r' \quad (12\text{–}31)$$

At low velocity, the term $(\vec{R} \cdot \vec{v})^2$ may be replaced by its mean value $\frac{1}{3}(R^2v^2)$, which gives, for the bracketed term,

$$\frac{n+1}{n+2} - \frac{1}{3}\frac{n-1}{n+2} = \frac{2}{3} \quad (12\text{–}32)$$

The expression for the total force on the particle becomes finally

$$\vec{F} = -\frac{1}{6\pi\varepsilon_0} \sum_{n=0}^{\infty} \frac{(-1)^n}{n!c^{n+2}} \frac{\partial^{n+1}}{\partial t^{n+1}} \vec{v} \int d^3r\rho(\vec{r}, t) \int R^{n-1}\rho(\vec{r}\,', t)d^3r' \quad (12\text{–}33)$$

The first two terms in the sum may be evaluated explicitly to give

$$\vec{F}_0 = -\frac{\dot{\vec{v}}}{6\pi\varepsilon_0 c^2} \int\int \frac{\rho(\vec{r}, t)\rho(\vec{r}\,', t)}{R} d^3r\, d^3r' = -\frac{4}{3}\frac{W}{c^2}\dot{\vec{v}} \quad (12\text{–}34)$$

and

$$\vec{F}_1 = \frac{\ddot{\vec{v}}}{6\pi\varepsilon_0 c^3} \int\int \rho(\vec{r}, t)\rho(\vec{r}\,', t)d^3r\, d^3r' = \frac{q^2}{6\pi\varepsilon_0 c^3}\ddot{\vec{v}} \quad (12\text{–}35)$$

The higher order terms' order of magnitude may be approximated as

$$\vec{F}_n \sim \frac{\pm 1}{6\pi\varepsilon_0} \frac{b^{n-1}q^2}{n!c^{n+2}} \vec{v}^{(n+1)} \quad (12\text{–}36)$$

The first term is just the immediate generalization of $\vec{F}_R$ while we have encountered the second term (equation 11–78) as the momentum carried by the radiation field. This equation (12–46) is known as the Dirac radiation reaction equation. The equation can be rewritten in a more symmetric form using $\beta^\mu = V^\mu/c$ and

$$\frac{d^2}{d\tau^2}(\beta^\nu\beta_\nu) = 0 = \beta^\nu\frac{d^2\beta_\nu}{d\tau^2} + \beta_\nu\frac{d^2\beta^\nu}{d\tau^2} + 2\frac{d\beta_\nu}{d\tau}\frac{d\beta^\nu}{d\tau} \tag{12–47}$$

to obtain

$$m_0c\frac{d\beta^\mu}{d\tau} - \frac{q^2}{6\pi\varepsilon_0c^2}\beta^\nu\left(\beta_\nu\frac{d^2\beta^\mu}{d\tau^2} - \beta^\mu\frac{d^2\beta_\nu}{d\tau^2}\right) = F^\mu_{ext} \tag{12–48}$$

The covariant equation retains all the problems, such as the run-away solution, associated with the nonrelativistic equation.

12.6 Alternative Formulations

Early in the century, many workers, notably Einstein, remarking the similarity between gravity and electromagnetism, attempted to unify these forces in much the same way that electricity and magnetism are unified in Maxwell's theory. Before the advent of general relativity, Nordström proposed a theory adding a fifth dimension to flat Minkowski four-space. In this space he introduced a five-vector field, for which he wrote down Maxwell's equations, including a conserved five-current. It was possible to identify the first four components of the five-potential as the usual electromagnetic potential, while the fifth component could be identified with the (scalar) gravitational potential.

Kaluza, in the wake of Einstein's general theory of gravity, extended Einstein's theory to a five-dimensional space with the fifth dimension curled back on itself by the imposition of a cylindrical constraint. In this theory the electromagnetic potentials are identified with components of the metric tensor involving the fifth dimension. Klein related the periodicity of the fifth dimension to Planck's constant to obtain a primitive quantum theory of five-dimensional relativity. The resulting theory and higher dimensional generalizations are known as Kaluza-Klein theories. Related to these theories are the much more recent *string* theories, which hope to unify all forces.

The development of quantum mechanics provided considerable impetus to develop a quantum electromagnetic field theory. It was recognized as early as 1926 that the radiation field could be quantized. The formal equivalence of the field Hamiltonian to that of a harmonic oscillator led to quantum operators for the creation and annihilation of photons. In the early theory it was not the entire electromagnetic field that was quantized, but only the radiative part.

Quantum electrodynamics reestablishes the unity of electromagnetic theory by quantizing the entire field. The interaction of charged particles is ascribed to the emission of a photon by one charge followed by the absorption of that photon by the other, so that photons become the carriers of the field. The strength of electromagnetic interactions compared to quantum interactions is given by the dimensionless constant

$$\frac{e^2}{4\pi\varepsilon_0\hbar c} = \frac{1}{137.036} \tag{12–49}$$

known as the fine structure constant. The relatively small size of this constant allows the interaction with the field to be treated as a perturbation. The theory has had some resounding successes, including the prediction of the g-factor of the electron to some 12 significant figures and the *Lamb shift.*

Quantum electrodynamics was extended by Weinberg, Salam, and Glashow, who succeeded in unifying the weak interactions sometimes known as *Fermi coupling* with quantum electrodynamics in much the same sense that Maxwell unified electricity and magnetism to produce an electro-weak theory of interactions. The strong force theory, quantum chromodynamics, can apparently also be accommodated by the theory.

Remarkably, gravity, the weakest of all forces, has largely defied successful inclusion in what are usually called grand unified theories.

Exercises and Problems

12-1 Use the change of variable $y = (t' - t)/\tau$ to replace t' in the integral of equation (12–41) in order to more clearly express the dependence of $\vec{F}(t')$ on t.

12-2 Use the results of problem 12-1 to write the equation of motion of a one-dimensional charged harmonic oscillator subject to a force $-m\omega_0^2 x$. Assuming a solution of the form $x = x_0 e^{-\alpha t}$, obtain the characteristic equation for α. Solve the (cubic) equation when $\omega\tau \ll 1$.

12-3 Investigate the consequences of using (12–16) to describe the damping force on the harmonic oscillator of problem 12-2.

12-4 A point particle of mass m and charge e initially at rest is acted on by a force

$$F = m \times (3 \times 10^{23}\text{m/s}^2)$$

for a time 10^{-15}s. Find the resulting relativistic motion as a function of time (a) neglecting radiation damping, (b) using the differential equation of motion, and (c) using the integro-differential equation of motion.

Appendix A

Other Systems of Units

Throughout this book, we have used a system of units known as *Système International* (SI), or (rationalized) mksa (meter-kilogram-second-ampere). The advantage of these units is the fact that the measure of potential difference, the volt, and the measure of current, the ampere, are those in common use, leading to the customary electrical units of ohms, henrys, and farads, as well as the mks units of power, force, and energy.

Other systems—electrostatic (*esu*), electromagnetic (*emu*), and especially Gaussian (in several flavors)—are also frequently used. Most theoreticians (especially those who set $c = \hbar = 1$) will tell us it is trivial to convert from one system to another. Unfortunately, the changes involved are much more subtle than those in changing inches to centimeters or the like; the fields have different dimensions in various systems. The result is that the student (as well as the author) is frequently perplexed at the changes required to go from one system to another.

We comment briefly on the rationale behind each of the systems in common use.

The static theory was based entirely on two force laws: the force between two charges is given by Coulomb's law,

$$F_e = k_e \frac{qq'}{R^2} \tag{A-1}$$

and the force per unit length between two long parallel currents which is

$$\frac{F_m}{\ell} = 2k_m \frac{I_1 I_2}{R} \tag{A-2}$$

In SI units, $k_e = 1/4\pi\varepsilon_0 \simeq 9 \times 10^9$ N-m^2/C^2 and $k_m = \mu_0/4\pi = 10^{-7}$ N/A^2. Other systems of units make different choices for these force constants. Because electrical and magnetic forces are related by a Lorentz transformation, it is clear that these two constants cannot be chosen arbitrarily (if the current and charge are related by $dq/dt = I$). For all but a modified Gaussian system (which poses $dq/dt = cI$), the intuitive relation between the current and charge accumulation is maintained. Thus for SI, esu, emu, and Gaussian units we relate the two constants as follows.

Both F_e and F_m are forces and therefore must have the same dimensions in any system of units. Using $I = q/t$, we find k_e/k_m has dimensions of (distance)^2/(time)^2. The ratio is easily calculated in SI units to be $1/\mu_0\varepsilon_0 = c^2$ with $c = 2.9972458 \times 10^8$ m/s.

Whereas SI units of distance, mass, and length are mks (meters, kilograms, and seconds), all other systems use cgs units of centimeter, gram, and second as their basis.

A.1 ESU and EMU Systems

If one feels that Coulomb's law is the fundamental law, then it is natural to assign the value 1 to the constant k_e, implying that $k_m = 1/c^2$. The resulting force laws are then

$$F_e = \frac{qq'}{R^2} \qquad \text{and} \qquad \frac{F_m}{\ell} = \frac{2}{c^2}\frac{I_1 I_2}{R} \tag{A–3}$$

This definition leads to the *electrostatic system of units* (esu).

Applying the definition, we construe that two unit charges at 1 cm from each other exert a force of 1 dyne upon each other. Clearly this constitutes a definition of the unit of charge. The charge so defined is the *statcoulomb*. The corresponding unit of current is the *statampere*, and the unit of potential difference, the *statvolt*, is one erg/statcoulomb. Other units are similarly defined with the prefix *stat-* to distinguish them from the SI units. In practice, quantities measured in units of the esu system are frequently specified only as esu, making dimensional analysis a bit awkward. In this system, the magnetic induction field is defined by the Lorentz force law $\vec{F} = q(\vec{E} + \vec{v} \times \vec{B})$, or equivalently $\vec{\nabla} \times \vec{E} = -\partial\vec{B}/\partial t$.

To relate statcoulombs to coulombs, we note that two equal charges Q exert a force

$$F = \frac{8.9874 \times 10^9 Q^2 \text{N-m}^2/\text{C}^2}{R^2} \tag{A–4}$$

on each other. Converting to cgs units, we get

$$F = 8.9874 \times 10^9 \text{N-m}^2/\text{C}^2 \times (10^5\text{dyne/N}) \times (10^4\text{cm}^2/\text{m}^2)\frac{Q^2}{R^2}$$

$$= \frac{8.9874 \times 10^{18}\text{dyne-cm}^2/\text{C}^2 Q^2}{R^2} = \frac{1\ \text{dyne-cm}^2/\text{statcoul}^2 Q^2}{R^2} \tag{A–5}$$

which clearly requires 1 statcoulomb $\simeq 2.9979 \times 10^9$ (numerically equal to $10c$ in SI units) coulombs.

If, alternatively the current law, rather than Coulomb's law, is felt to be fundamental (we do in fact use this law to define the SI coulomb), then it is natural to write

$$F_e = \frac{c^2 qq'}{R^2} \qquad \text{and} \qquad \frac{F_m}{\ell} = \frac{2I_1 I_2}{R} \tag{A–6}$$

to define the *electromagnetic system of units* (emu). Two equal currents of 1 *abampere* spaced at a centimeter attract each other with a force of 2 dynes per centimeter of length. It is easily verified using the method above that 1 abampere = 10 amperes. The associated charge is then 1 abcoulomb = 1 abampere-second = 10 C.

Quantities measured in the emu system are all prefixed by *ab-* (for *absolute*). Again the units are frequently specified only as emu rather than abfarad or the like.

A.2 Gaussian Units

The Gaussian system is an unrationalized hybrid (*rationalized* refers to the inclusion of the 4π in the force law) of the emu and esu systems. Electric charge and quantities derived from it, such as current and dipole moment are measured in esu while the magnetic induction field is defined by the emu system. Faraday's law, relating $\vec{E}$ and $\vec{B}$ and the Lorentz force law, must be modified to account for these differing definitions. In Gaussian units, Maxwell's equations read

$$\begin{aligned} \vec{\nabla}\cdot\vec{D} &= 4\pi\rho & \vec{\nabla}\cdot\vec{B} &= 0 \\ \vec{\nabla}\times\vec{E} &= -\frac{1}{c}\frac{\partial\vec{B}}{\partial t} & \vec{\nabla}\times\vec{H} &= \frac{4\pi}{c}\vec{J}+\frac{1}{c}\frac{\partial\vec{D}}{\partial t} \end{aligned} \tag{A–7}$$

with $\vec{D}=\vec{E}+4\pi\vec{P}$ and $\vec{B}=\vec{H}+4\pi\vec{M}$. The definitions of $\vec{D}$ and $\vec{H}$ imply that in vacuum, $\vec{D}=\vec{E}$ and $\vec{H}=\vec{B}$ in the Gaussian system. This last equality leads to considerable confusion and misuse of $\vec{H}$ and $\vec{B}$.

The Lorentz force must also be modified in order to obtain the correct force law for moving charges. In the Gaussian system it becomes

$$\vec{F}=q(\vec{E}+\frac{\vec{v}}{c}\times\vec{B}) \tag{A–8}$$

Note that $\vec{B}$ in the Gaussian system is a different physical quantity from that in SI units.

For linear materials,

$$\vec{D}=\varepsilon\vec{E} \qquad \vec{B}=\mu\vec{H} \qquad \vec{J}=\sigma\vec{E} \qquad \vec{P}=\chi_e\vec{E} \qquad \vec{M}=\chi_m\vec{H} \tag{A–9}$$

(those who use Gaussian units uniformly use σ rather than g for the conductivity) so that $\varepsilon=1+4\pi\chi_e$ and $\mu=1+4\pi\chi_m$.

The vector fields $\vec{E}$, $\vec{D}$ and $\vec{P}$ as well as $\vec{B}$, $\vec{H}$, and $\vec{M}$ all have the same dimensions. Nonetheless, the units of each field have different names. $\vec{B}$ is measured in gauss, $\vec{H}$ and $\vec{M}$ in Oersted, and the flux Φ in gauss-cm^2 = Maxwell. Of course, statvolts and statamps are retained from esu, leading to electric and displacement fields measured in statvolt/cm.

Heaviside-Lorentz units are rationalized Gaussian units that banish the 4π from Maxwell's equations (but reimport it into Coulomb's law and the Biot-Savart law).

Finally, there is a modified version of the Gaussian system where the charge is measured in the statcoulombs of the esu system but current is measured in abamps of emu. In this system we can no longer maintain $I=dq/dt$; internal consistency requires $cI=dq/dt$, or more usefully in terms of the continuity equation,

$$\vec{\nabla}\cdot c\vec{J}+\frac{\partial\rho}{\partial t}=0 \tag{A–10}$$

It suffices to replace $\vec{J}$ by $c\vec{J}$ in Maxwell's equations so that Ampère's law becomes

$$\vec{\nabla} \times \vec{H} = 4\pi\vec{J} + \frac{1}{c}\frac{\partial \vec{D}}{\partial t} \tag{A-11}$$

Clearly this plethora of systems is bound to produce confusion, particularly as it is rarely stated which version of Gaussian units is being used.

Conversion from one system of units to another requires both a conversion of the formulas (for example, Coulomb's law in Gaussian units becomes $F = qq'/r^2$) and a substitution of numerical values appropriate to the new system. Table A.1 lists the required substitutions in the formulas involving the fields and related variables of the SI and the Gaussian systems.

SI system		Gaussian system
q, ρ, I, $\vec{J}$, $\vec{p}$, $\vec{P}$, $\cdots$	$\leftrightarrow$	$(4\pi\varepsilon_0)^{1/2}q$, ρ, $I, \vec{J}$, $\vec{p}$, $\vec{P}, \cdots$
$\vec{E}$, V,	$\leftrightarrow$	$(4\pi\varepsilon_0)^{-1/2}\vec{E}$, V
$\vec{D}$	$\leftrightarrow$	$(\varepsilon_0/4\pi)^{1/2}\vec{D}$
$\vec{B}$, $\vec{A}$, $\vec{M}$, $\vec{m}$, Φ	$\leftrightarrow$	$(\mu_0/4\pi)^{1/2}\vec{B}$, $\vec{A}$, $\vec{M}$, $\vec{m}$, Φ
$\vec{H}$, V_m	$\leftrightarrow$	$(4\pi\mu_0)^{-1/2}\vec{H}$, V_m
χ_e, χ_m	$\leftrightarrow$	$4\pi\chi_e$, $4\pi\chi_m$
$(\mu_0\varepsilon_0)^{-1/2}$	$\leftrightarrow$	c
L(inductance)	$\leftrightarrow$	$(4\pi\varepsilon_0)^{-1}L$
C(capacitance)	$\leftrightarrow$	$(4\pi\varepsilon_0)C$
g(conductivity)	$\leftrightarrow$	$4\pi\varepsilon_0 g$(usually denoted by σ)

Table A.1 Substitutions required in order to convert a formula from SI units to Gaussian or the converse.

Variable	SI	Gaussian
length	1 m	10^2 cm
mass	1 kg	10^3 g
force	1 N	10^5 dynes
energy	1 J	10^7 ergs
charge	1 C	3×10^9 statcoul
current	1 A	3×10^9 statamp
potential	1 V	1/300 statvolt
Electric field	1 V/m	$1/3\times10^{-4}$ statvolt/cm
Displacement field	1 C/m^2	$12\pi\times10^5$ statvolt/cm
Mag. field intensity	1 A/m	$4\pi\times10^{-3}$ oersted
Mag. induction field	1 T (=Wb/m^2)	10^4 gauss
Magnetic flux	1 Wb	10^8 maxwell
Magnetization	1 A/m	10^{-3} oersted
Polarization	1 C/m^2	3×10^5 statvolt/cm
Capacitance	1 F	9×10^{11} statfarad
Inductance	1 H	$1/9\times10^{-11}$ stathenry
Resistance	1 Ω	$1/9\times10^{-11}$ statohm

Table A.2 The numerical correspondence between Gaussian and SI units. Note that although $\vec{D}$ and $\vec{P}$ are measured in the same units, the conversion for them is not the same, differing by a factor of 4π. Similar remarks apply to $\vec{H}$ and $\vec{M}$. The conversion factors abbreviated as 3 should really be 2.99792458 and the 9's should be $(2.9972458)^2$.

The magnitudes of the quantities measured in each of the systems are related in Table A.2.

Appendix B

Vectors and Tensors

B.1 Vectors

It is customary in elementary physics courses to define vectors as quantities having both direction and magnitude, and it is tacitly assumed that these quantities obey the same geometry as line segments. Mathematicians, on the other hand, typically using an axiomatic approach, make little reference to geometry when speaking of vector spaces and vectors. It is not infrequent that a student after taking a linear algebra course concludes that the vectors of mathematics and those of physics have little in common. In fact, there are differences in that physicists usually[31] really mean a first-rank tensor when they speak of a vector. Let us begin by considering vectors as defined by mathematicians.

Definition: A *vector space* is a set V of elements called *vectors*, satisfying the following axioms:

A. To every pair of vectors, $\vec{x}$ and $\vec{y}$ in V, there exists a vector $(\vec{x}+\vec{y})$ such that

 (1) $\vec{x}+\vec{y}=\vec{y}+\vec{x}$
 (2) $\vec{x}+(\vec{y}+\vec{z})=(\vec{x}+\vec{y})+\vec{z}$
 (3) There exists a unique vector $\vec{0}$ such that $\vec{x}+\vec{0}=\vec{x}$
 (4) There exists a unique inverse $-\vec{x}$ for every $\vec{x}$ in V such that $\vec{x}+(-\vec{x})=\vec{0}$

B. To every pair α and $\vec{x}$, where α is a scalar and $\vec{x}$ is a vector in V, there exists a vector $\alpha\vec{x}$ in V called the product of α and $\vec{x}$, satisfying

 (1) $\alpha(\beta\vec{x})=(\alpha\beta)\vec{x}$
 (2) $1\cdot\vec{x}=\vec{x}$
 (3) $(\alpha+\beta)\vec{x}=\alpha\vec{x}+\beta\vec{x}$

It is worth noting that not every quantity characterized by magnitude and direction is a vector under this definition. To qualify, the quantity must also obey the laws of vector algebra. To illustrate, suppose a rigid body is rotated about

[31] This seems rather an excessive generalization; the state vectors of quantum mechanics and similar abstract vectors that have no geometric interpretation are frequently used in physics. In this appendix, we concern ourselves only with those vectors that behave like directed line segments.

some arbitrary axis. The rotation can be represented by a line segment of length proportional to the angle of rotation, directed along the axis of rotation. Directed line segments of this kind do *not* add like vectors. The addition is not commutative: a rotation about the z axis followed by one about the y axis is not the same as a rotation about the y axis followed by one about the z axis.

B.1.1 Bases and Transformations

The N vectors $\vec{A}_1$, $\vec{A}_2$, $\vec{A}_3$, ... , $\vec{A}_N$ are called *linearly independent* if and only if (frequently abbreviated by *iff*)

$$c_1\vec{A}_1 + c_2\vec{A}_2 + c_3\vec{A}_3 + \cdots + c_N\vec{A}_N = 0 \tag{B–1}$$

implies that each of the coefficients c_1, c_2, c_3, ..., c_N vanishes.

Two linearly dependent vectors are *collinear*. If $\vec{A}$ and $\vec{B}$ are linearly independent, then any vector $\vec{C}$ that can be expressed as a linear combination of $\vec{A}$ and $\vec{B}$ is said to be *coplanar* with $\vec{A}$ and $\vec{B}$ and has a unique expansion $\vec{C} = s\vec{A} + t\vec{B}$.

> Proof: Suppose $\vec{C}$ has another expansion, say $\vec{C} = \alpha\vec{A} + \beta\vec{B}$. Subtracting one from the other gives $(s-\alpha)\vec{A}+(t-\beta)\vec{B} = 0$. Since $\vec{A}$ and $\vec{B}$ are linearly independent, $s - \alpha = 0$ and $t - \beta = 0$; in other words, the expansion is the same ($\alpha = s$ and $\beta = t$).

This theorem is easily extended to three or more dimensions.

By a *basis* for a three-dimensional space we mean *any* set of three linearly independent vectors $\vec{e}_1$, $\vec{e}_2$, and $\vec{e}_3$ in that space. Given a basis $\vec{e}_1$, $\vec{e}_2$, and $\vec{e}_3$, every vector $\vec{A}$ in 3-space has a unique expansion in this basis: $\vec{A} = r\vec{e}_1 + s\vec{e}_2 + t\vec{e}_3$.

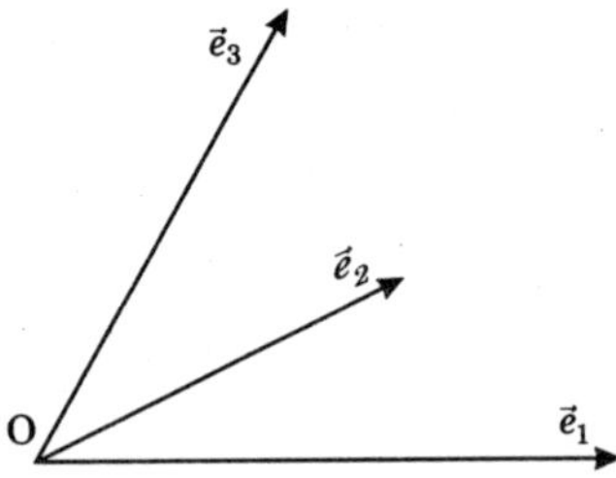

Figure B.1: Any three linearly independent vectors form a basis of a three dimensional vector space.

Suppose the basis vectors $\vec{e}_1$, $\vec{e}_2$, and $\vec{e}_3$ are all drawn from a common origin O, as in Figure B.1. The triad in general gives an *oblique* coordinate system. Any point (the terminus of a vector) can be specified as the set of three numbers (x^1, x^2, x^3) that give the expansion on the basis. When $\vec{e}_1$, $\vec{e}_2$, and $\vec{e}_3$ are all orthogonal, of unit magnitude and constant, the coordinate system is said to be *Cartesian*, and the coordinates of a point may then be written (x_1, x_2, x_3) for reasons that will become evident. We will also write the basis as $\hat{\imath}_1, \hat{\imath}_2, \hat{\imath}_3$ instead of $\vec{e}_1, \vec{e}_2, \vec{e}_3$ in orthonormal systems.

The great merit of vectors applied to physics is the fact that equations describing physical phenomena can be formulated without reference to any particular coordinate system. However, in actually carrying out calculations needed to solve given problems, one must usually eventually cast the problem into a form involving coordinates. This can be done by introducing a suitable coordinate system and then replacing the given vector (or tensor) equation by an equivalent system of scalar equations involving only the components of the vectors obeying the ordinary rules of arithmetic.

B.1.2 Transformation of Basis Vectors

Consider two distinct bases $\vec{e}_1$, $\vec{e}_2, \vec{e}_3$ and $\vec{e}_1'$, $\vec{e}_2'$, $\vec{e}_3'$ drawn from the same point O. Then any of the primed basis vectors can be expanded in terms of the unprimed basis. Let $\alpha_{i'}^1, \alpha_{i'}^2$, and $\alpha_{i'}^3$ be the expansion coefficients of $\vec{e}_i'$ in the unprimed basis; that is,

$$\begin{aligned} \vec{e}_1' &= \alpha_{1'}^1 \vec{e}_1 + \alpha_{1'}^2 \vec{e}_2 + \alpha_{1'}^3 \vec{e}_3 \\ \vec{e}_2' &= \alpha_{2'}^1 \vec{e}_1 + \alpha_{2'}^2 \vec{e}_2 + \alpha_{2'}^3 \vec{e}_3 \\ \vec{e}_3' &= \alpha_{3'}^1 \vec{e}_1 + \alpha_{3'}^2 \vec{e}_2 + \alpha_{3'}^3 \vec{e}_3 \end{aligned} \tag{B–2}$$

or, more concisely,

$$\vec{e}_i' = \sum_k \alpha_{i'}^k \vec{e}_k \tag{B–3}$$

The nine numbers $\alpha_{i'}^k$ are called the coefficients of the transformation.

Similarly, we let $\alpha_i^{k'}$ be the coefficients of the inverse transformation (note the location of the prime on the transformation coefficients; if one thinks of the prime as attached to the index of the basis vector, it becomes a bit more obvious):

$$\vec{e}_i = \sum \alpha_i^{k'} \vec{e}_k' \tag{B–4}$$

The $\alpha_{i'}^k$ and $\alpha_j^{\ell'}$ are simply related, as is demonstrated below:

$$\begin{aligned} \vec{e}_i' &= \alpha_{i'}^1 \vec{e}_1 + \alpha_{i'}^2 \vec{e}_2 + \alpha_{i'}^3 \vec{e}_3 \\ &= \alpha_{i'}^1 \left(\alpha_1^{1'} \vec{e}_1' + \alpha_1^{2'} \vec{e}_2' + \alpha_1^{3'} \vec{e}_3'\right) + \alpha_{i'}^2 \left(\alpha_2^{1'} \vec{e}_1' + \alpha_2^{2'} \vec{e}_2' + \alpha_2^{3'} \vec{e}_3'\right) \\ &\qquad + \alpha_{i'}^3 \left(\alpha_3^{1'} \vec{e}_1' + \alpha_3^{2'} \vec{e}_2' + \alpha_3^{3'} \vec{e}_3'\right) \\ &= \vec{e}_1' \sum_\ell \alpha_{i'}^\ell \alpha_\ell^{1'} + \vec{e}_2' \sum_\ell \alpha_{i'}^\ell \alpha_\ell^{2'} + \vec{e}_3' \sum_\ell \alpha_{i'}^\ell \alpha_\ell^{3'} \end{aligned} \tag{B–5}$$

Since any vector including $\vec{e}_i'$ has a unique expansion, we conclude that for $i = 1$ the first sum must equal unity, while the other two sums must vanish. Similarly for $i = 2$, the second sum equals unity and the first and third vanish. In exactly the same fashion we find

$$\vec{e}_i = \vec{e}_1 \sum_\ell \alpha_i^{\ell'} \alpha_{\ell'}^1 + \vec{e}_2 \sum_\ell \alpha_i^{\ell'} \alpha_{\ell'}^2 + \vec{e}_3 \sum_\ell \alpha_i^{\ell'} \alpha_{\ell'}^3 \tag{B–6}$$

The two sets of equations for the coefficients can be summarized by

$$\sum_\ell \alpha_{i'}^{\ell}\alpha_{\ell}^{j\,'} = \delta_{i'}^{j\,'} \qquad \text{and} \qquad \sum_\ell \alpha_i^{\ell'}\alpha_{\ell'}^{j} = \delta_i^j \tag{B-7}$$

where the *Kronecker delta* δ_i^j is 0 if $i \neq j$ and 1 if $i = j$.

B.1.3 Scalar and Vector Products

We expand the operations that we can perform on vectors beyond the addition provided for in the axioms by defining a *scalar* or *dot* product by

$$\vec{A}\cdot\vec{B} = |\vec{A}||\vec{B}\,|\cos(\vec{A},\vec{B}) \tag{B-8}$$

The magnitude $|\vec{A}\,| \equiv \sqrt{A^2}$ of a vector will be defined in B.2.1; for now we rely on our intuitive notions of length. In addition to a scalar product we also define the *vector* or *cross* product. The cross product has magnitude

$$|\vec{A}\times\vec{B}\,| = |\vec{A}||\vec{B}\,|\sin(\vec{A},\vec{B}) \tag{B-9}$$

and an orientation perpendicular to the plane defined by $\vec{A}$ and $\vec{B}$ in the direction that a right-hand screw advances in turning from $\vec{A}$ to $\vec{B}$ (through the smallest angle).

B.1.4 Reciprocal Bases

Consider the general problem of expanding an arbitrary vector $\vec{A}$ with respect to three linearly independent vectors $\vec{e}_1$, $\vec{e}_2$, and $\vec{e}_3$ that are neither orthogonal nor of unit length. Let the expansion coefficients of the vector $\vec{A}$ in this basis be A^1, A^2, and A^3 (the superscripts are *not* to be interpreted as exponents; the context should make it clear when the superscript is an exponent) so that $\vec{A}$ may be written

$$\vec{A} = A^1\vec{e}_1 + A^2\vec{e}_2 + A^3\vec{e}_3 \tag{B-10}$$

We will consider the problem of finding the expansion coefficients. With an orthonormal basis the solution consists of taking the dot product of $\vec{A}$ with each of the basis vectors; $A^i = \vec{A}\cdot\hat{\imath}_i$. When the basis is not orthogonal, the resolution of $\vec{A}$ is less obvious. Considerable simplification is brought to the problem by the introduction of the *reciprocal basis*.

Two bases $\vec{e}_1$, $\vec{e}_2$, $\vec{e}_3$ and $\vec{e}^{\,1}$, $\vec{e}^{\,2}$, $\vec{e}^{\,3}$ are said to be reciprocal if they satisfy $\vec{e}_i\cdot\vec{e}^{\,k} = \delta_i^k$. To construct the reciprocal basis from the ordinary one, we note that $\vec{e}^{\,1}$ must be perpendicular to $\vec{e}_2$ and $\vec{e}_3$. We therefore set $\vec{e}^{\,1} = m(\vec{e}_2\times\vec{e}_3)$. The requirement that $\vec{e}_1\cdot\vec{e}^{\,1} = 1$ implies that

$$m\vec{e}_1\cdot(\vec{e}_2\times\vec{e}_3) = 1 \tag{B-11}$$

leading to

$$\vec{e}^{\,1} = \frac{\vec{e}_2\times\vec{e}_3}{\vec{e}_1\cdot(\vec{e}_2\times\vec{e}_3)} \tag{B-12}$$

The other two reciprocal basis vectors are found by cyclically permuting the indices:

$$\vec{e}^{\,2} = \frac{\vec{e}_3 \times \vec{e}_1}{\vec{e}_1 \cdot (\vec{e}_2 \times \vec{e}_3)} \quad \text{and} \quad \vec{e}^{\,3} = \frac{\vec{e}_1 \times \vec{e}_2}{\vec{e}_1 \cdot (\vec{e}_2 \times \vec{e}_3)} \tag{B–13}$$

An orthonormal basis is its own reciprocal basis: $\hat{\imath}^1 = \hat{\imath}_1$, $\hat{\imath}^2 = \hat{\imath}_2$, and $\hat{\imath}^3 = \hat{\imath}_3$.

To return to the problem of finding the expansion coefficients of $\vec{A}$, we exploit the orthogonality of the basis with its reciprocal basis. Taking the dot product of $\vec{A}$ with $\vec{e}^{\,i}$, we find $\vec{A} \cdot \vec{e}^{\,i} = A^i$ so that generally

$$\vec{A} = (\vec{A} \cdot \vec{e}^{\,1})\vec{e}_1 + (\vec{A} \cdot \vec{e}^{\,2})\vec{e}_2 + (\vec{A} \cdot \vec{e}^{\,3})\vec{e}_3 \tag{B–14}$$

EXAMPLE B.1: An oblique coordinate system has basis $\vec{e}_1 = 2\hat{\imath} + 3\hat{\jmath} + \hat{k}$, $\vec{e}_2 = \hat{\imath} - \hat{\jmath} + \hat{k}$, and $\vec{e}_3 = \hat{\jmath} + \hat{k}$. Find the expansion of the vector $\vec{A} = 5\hat{\imath} + 6\hat{\jmath} + 7\hat{k}$ in this basis.

Solution: We first generate the reciprocal basis by the prescription above to obtain $\vec{e}^{\,1} = \frac{1}{6}(2\hat{\imath} + \hat{\jmath} - \hat{k})$, $\vec{e}^{\,2} = \frac{1}{6}(2\hat{\imath} - 2\hat{\jmath} + 2\hat{k})$ and $\vec{e}^{\,3} = \frac{1}{6}(-4\hat{\imath} + \hat{\jmath} + 5\hat{k})$ so that the components of $\vec{A}$ in this basis $(\vec{e}_i)$ are

$$\begin{aligned} A^1 &= (\vec{A} \cdot \vec{e}^{\,1}) = 1.5 \\ A^2 &= (\vec{A} \cdot \vec{e}^{\,2}) = 2 \\ A^3 &= (\vec{A} \cdot \vec{e}^{\,3}) = 3.5 \end{aligned} \tag{Ex B.1.1}$$

It is readily verified that $\sum_i A^i \vec{e}_i = \vec{A}$.

B.1.5 The Summation Convention

We will henceforth make free use of the following convention, used universally in contemporary physical and mathematical literature.

(1) Every index appearing once in an expression can take on the values 1, 2, and 3 (0, 1, 2, and 3 in special relativity). Thus A_i denotes any member of the set $\{A_1, A_2, A_3\}$; A_{ik} of the set $\{A_{11}, A_{12}, A_{13}, A_{21}, A_{22}, A_{23}, A_{31}, A_{32}, A_{33}\}$; and so on.

(2) If a free index appears twice in a term, once as superscript and once as subscript, summation over that index, allowing the index to take all its possible values, is implied. For example

$$\begin{aligned} A_i{}^i &\equiv A_1{}^1 + A_2{}^2 + A_3{}^3 \\ A_i B^i &\equiv A_1 B^1 + A_2 B^2 + A_3 B^3 \\ A_i B^k C^i &\equiv B^k (A_1 C^1 + A_2 C^2 + A_3 C^3) \end{aligned} \tag{B–15}$$

With this convention, $\vec{e}_i' = \alpha_{i'}^k \vec{e}_k$ and $\vec{e}_i = \alpha_i^{k'} \vec{e}_k'$. Clearly an expression such as $A_i B^i$ is independent of the letter chosen for summation index; for this reason such an index is often called a dummy index.

The summation is occasionally extended to repeated indices irrespective of their position. As we will see, this makes sense only for entities expressed in an orthonormal basis.

B.1.6 Covariant and Contravariant Components of a Vector

The vector $\vec{A} = A^i\vec{e}_i$ could equally well be expanded in the reciprocal basis as $\vec{A} = A_i\vec{e}^{\,i}$. The numbers A^i are called the *contravariant* components of $\vec{A}$, while the numbers A_i are its *covariant* components. These designations arise from the fact that under transformations of the coordinate system, the covariant components in the 'primed' system are obtained from the unprimed components by the same transformation that takes $\vec{e}_i \to \vec{e}_k^{\,\prime}$ whereas the contravariant components in the primed system are obtained by the inverse transformation that takes $\vec{e}_i^{\,\prime} \to \vec{e}_k$. To confirm this assertion, let $\vec{A} = A_i\vec{e}^{\,i} = A^i\vec{e}_i$ in the basis defined by $\{\vec{e}_i\}$, while $\vec{A} = A_i'\vec{e}^{\,i'} = A^{i'}\vec{e}_i^{\,\prime}$ in the basis defined by $\{\vec{e}_i{}'\}$. First we note that

$$\vec{e}_i^{\,\prime} \cdot \vec{e}^{\,k} = \alpha_{i'}^{k} \quad \text{and} \quad \vec{e}_i \cdot \vec{e}^{\,k'} = \alpha_i^{k'} \tag{B–16}$$

Using the above equations, we convert between the primed and unprimed system

$$A_i' = \vec{A} \cdot \vec{e}_i^{\,\prime} = A_k\vec{e}^{\,k} \cdot \vec{e}_i^{\,\prime} = \alpha_{i'}^{k} A_k \tag{B–17}$$

Similarly

$$A^{i'} = \vec{A} \cdot \vec{e}^{\,i'} = A^k\vec{e}_k \cdot \vec{e}^{\,i'} = \alpha_k^{i'} A^k \tag{B–18}$$

Comparing these transformation coefficients to those for the basis (B–3, 4), $\vec{e}_i^{\,\prime} = \alpha_{i'}^{k}\vec{e}_k$ and $\vec{e}_i = \alpha_i^{k'}\vec{e}_k^{\,\prime}$, confirms our assertion.

B.1.7 The Metric Tensor

Converting between covariant and contravariant components of a vector is easily done by means of the *metric tensor*, also known as the raising and lowering operator.

Consider the vector $\vec{A} = A^i\vec{e}_i$. Taking the dot product of $\vec{A}$ with $\vec{e}_j$, we obtain

$$(\vec{A} \cdot \vec{e}_j) = A^i\vec{e}_i \cdot \vec{e}_j \tag{B–19}$$

so that

$$A_j = g_{ij}A^i \tag{B–20}$$

where $g_{ij} \equiv \vec{e}_i \cdot \vec{e}_j$ is known as the metric tensor. This name comes from the fact that g relates physical displacements to changes in the coordinates. In analogous fashion, setting $g^{ij} = \vec{e}^{\,i} \cdot \vec{e}^{\,j}$ lets us write $A^j = g^{ij}A_i$. The definition of the reciprocal basis requires $g_i^j = \vec{e}_i \cdot \vec{e}^{\,j} = \delta_i^j$. It is also evident that $g_{ij} = g_{ji}$. For orthonormal systems, $g_{ij} = \delta_{ij}$.

The contravariant components g^{ik} of the metric tensor may also be expressed in terms of the covariant basis:

$$g^{i\ell} = \vec{e}^{\,i} \cdot \vec{e}^{\,\ell} = \frac{(\vec{e}_j \times \vec{e}_k) \cdot (\vec{e}_m \times \vec{e}_n)}{|\,(\vec{e}_1 \times \vec{e}_2) \cdot \vec{e}_3\,|^2} \tag{B–21}$$

where the triplets i, j, k and ℓ, m, n are each in cyclic order. We abbreviate the denominator as V^2, with V the volume of the parallelepiped spanned by the basis. With this abbreviation and the aid of (4),

$$g^{i\ell} = \frac{1}{V^2}\begin{vmatrix} \vec{e}_j \cdot \vec{e}_m & \vec{e}_j \cdot \vec{e}_n \\ \vec{e}_k \cdot \vec{e}_m & \vec{e}_k \cdot \vec{e}_n \end{vmatrix} = \frac{1}{V^2}\begin{vmatrix} g_{jm} & g_{jn} \\ g_{km} & g_{kn} \end{vmatrix} \tag{B–22}$$

Alternatively, $A_\ell = g_{i\ell}A^i$ can be written it as $A^i = \sum_\ell (g^{-1})_{i\ell}A_\ell$. The i, ℓ element of the inverse to a nonsingular matrix g is generally given by

$$(g^{-1})_{i\ell} = \frac{G^{i\ell}}{G} \tag{B–23}$$

where G is the determinant of the metric coefficient matrix ($G = \det g_{i\ell}$) and $G^{i\ell}$ is the cofactor of $g_{i\ell}$. The cofactor $G^{i\ell}$ may be written as the determinant of g_{ij} with the i-th row and ℓ-th column removed and multiplied by $(-1)^{i+\ell}$.

$$G^{i\ell} = (-1)^{i+\ell} \begin{vmatrix} g_{jm} & g_{jn} \\ g_{km} & g_{kn} \end{vmatrix} \tag{B–24}$$

Thus, returning to our expression for A^i above,

$$A^i = \frac{G^{i\ell}}{G} A_\ell \quad \Rightarrow \quad g^{i\ell} = \frac{G^{i\ell}}{G} \tag{B–25}$$

Comparison of the two expressions (B–22) and B–25) for $g^{i\ell}$ leads to the conclusion that $G = V^2$.

B.2 Tensors

Recall that a scalar is a quantity whose specification requires only one number. On the other hand, a vector requires three numbers; its components with respect to some basis. Scalars and vectors are both special cases of a more general object called a *tensor* of order n, whose specification in any three-dimensional coordinate system requires 3^n numbers called its components.

Of course, a tensor is more than just a collection of 3^n numbers (recall that not any set of three numbers constitute a vector). The key property of a tensor is the transformation law of its components under a change of coordinate system. The precise form of this transformation law is a consequence of the physical or geometric meaning of the tensor.

Suppose we have a law of physics involving components $a, b, c, \ldots$ of the various physical quantities. It is an empirical fact that the law has the same form when written in terms of the components a', b', c', ... of the same quantities with respect to a coordinate system shifted and rotated with respect to the first. In other words, a properly formulated physical law is invariant in form under displacements and rotations of the coordinate system. The operations under which we expect the laws of physics to be invariant correspond to the symmetries of space such as the isotropy and homogeneity of space. Under many conditions we also expect the laws to be invariant under reflection of the coordinate system, and special relativity also requires the laws to be invariant under velocity transformations of the reference frame.

B.2.1 Zero-Rank Tensor (Scalars)

We begin by refining our definition of scalars. By a scalar (zero-rank tensor) we will mean a quantity uniquely specified by a single number independent of the coordinate system and invariant under changes of the coordinate system.

are to be read if $A^{ij} \neq A^{ji}$. In particular a quantity like $g_{ij}A_k^{\cdot j} = A_{ki}$ whereas $g_{ij}A^{j}_{\cdot k} = A_{ik}$.

EXAMPLE B.4: Given two vectors, $\vec{A}$ and $\vec{B}$, show that the nine products of the components $T^{ij} = A^iB^j$ form a second-rank tensor.

Solution: In the transformed coordinate system

$$T^{ij'} = A^{i'}B^{j'} = (\alpha_\ell^{i'}A^\ell)(\alpha_k^{j'}B^k) = \alpha_\ell^{i'}\alpha_k^{j'}A^\ell B^k = \alpha_\ell^{i'}\alpha_k^{j'}T^{\ell k} \qquad \text{(Ex B.4.1)}$$

clearly it transforms as a second-rank contravariant tensor as shown in the second equality of (B–31).

EXAMPLE B.5: Show that the coefficients a_k^i of a linear transformation between the vector A^k and B^i, $B^i = a_k^iA^k$, form a second-rank mixed tensor.

Solution: We multiply both sides of the relation above by $\alpha_i^{m'}$ (and perform the implied summation) to obtain

$$\alpha_i^{m'}B^i = \alpha_i^{m'}a_k^iA^k \qquad \text{(Ex B.5.1)}$$

or

$$B^{m'} = \alpha_i^{m'}a_k^iA^k = \alpha_i^{m'}a_k^i\alpha_{j'}^kA^{j'} \qquad \text{(Ex B.5.2)}$$

In Σ', the linear transformation takes the form $B^{m'} = \left(a_j^m\right)'A^{j'}$. Comparing the two expressions we find

$$\left(a_j^m\right)' = \alpha_i^{m'}\alpha_{j'}^ka_k^i \qquad \text{(Ex B.5.3)}$$

which is precisely the transformation required of a second-rank mixed tensor.

EXAMPLE B.6: The Kronecker delta, δ_i^j, is a second-rank tensor. Clearly the definition of δ does not change with a change in coordinate system. Using the identity (B–7),

$$\left(\delta_i^j\right)' = \alpha_{i'}^k\alpha_k^{j'} = \alpha_{i'}^k\alpha_m^{j'}\delta_k^m \qquad \text{(Ex B.6.1)}$$

showing that δ_i^j is indeed a second-rank tensor. It has the same form in all coordinate systems and is therefore called *isotropic*.

Other examples of second-rank tensors include the moment of inertia tensor, the stress tensor, the quadrupole moment tensor, and many others.

B.2.4 Differentiation of Tensors

(a) Differentiation of a tensor with respect to a scalar produces another tensor of the same rank and type.

(b) Differentiation of a tensor with respect to the coordinates x^i produces a new tensor whose covariant order is increased by 1 (similar to taking a gradient) *provided* that the basis does not depend on the coordinates. Thus $\partial_i A^{k\ell m} = B_i^{\cdot k\ell m}$ and $\partial^i A^{jk\ell} = B^{ijk\ell}$ are tensors of the type and rank indicated by the indices.

The Fundamental Theorem: *The product of two tensors (including contraction as indicated by the indices) forms a tensor of the rank and type indicated by the indices.*

Quotient Rule: *When two of the three entities in an expression such as $A^i = C^{ij}B_j$ are tensors, then the third is also a tensor of the type and rank indicated by the indices.*

B.2.5 Pseudotensors

In general we expect physical laws to be invariant not only under rotations, translations, and stretching of the coordinate system, but also under inversion of the coordinates. In fact, a normal (polar) vector's components change sign under inversion, but those of axial vectors (produced as cross products) do not change sign. The components of an axial vector do not transform like those of an arrow. For this reason, the axial vector is called a *pseudotensor* of rank one. Extending this notion to tensors of rank r, one finds that the components of an rth rank tensor satisfy $P'_{i\ell m\ldots s} = (-1)^r P'_{i\ell m\ldots s}$ under inversion. By contrast, rth rank pseudo tensors have parity $(-1)^{r+1}$.

EXAMPLES:

(*a*) $(\vec{A}\times\vec{B})\cdot\vec{C}$ is a pseudoscalar. This particular pseudoscalar may be identified with the volume of a parallelepiped with edges of length A, B, and C.

(*b*) $\vec{A}\times\vec{B}$ is a pseudovector.

(*c*) $D_{ij} = (\vec{A}\times\vec{B})_i C_j$ is a second-rank pseudotensor.

(*d*) The entity with components $\epsilon_{ijk} = (\vec{e}_i\times\vec{e}_j)\cdot\vec{e}_k$, called a *Levi-Cevita* symbol, is pseudotensor of third-rank of particular importance. The vector i component of the cross-product $\vec{C} = \vec{A}\times\vec{B}$ is the contraction of ϵ_{ijk} with the second-rank tensor A^jB^k. This is easily shown by expanding

$$\vec{A}\times\vec{B} = (A^j\vec{e}_j)\times(B^k\vec{e}_k) = A^jB^k\vec{e}_j\times\vec{e}_k \tag{B–33}$$

The covariant i component of the cross product is found by taking the dot product with $\vec{e}_i$

$$\begin{aligned}(\vec{A}\times\vec{B})_i &= (\vec{A}\times\vec{B})\cdot\vec{e}_i = A^jB^k(\vec{e}_j\times\vec{e}_k)\cdot\vec{e}_i = (\vec{e}_i\times\vec{e}_j)\cdot\vec{e}_kA^jB^k \\ &= \epsilon_{ijk}A^jB^k\end{aligned} \tag{B–34}$$

In Cartesian coordinates,

$$\epsilon_{ijk} = \begin{cases} 1 & \text{for } i,\ j,\ k \text{ a cyclic permutation of } 1,\ 2, \text{ and } 3 \\ 0 & \text{for any two indices equal} \\ -1 & \text{for } i,\ j,\ k \text{ anticyclic}\end{cases} \tag{B–35}$$

An important identity for ϵ_{ijk}, easily proved from the definition, is $\epsilon_{ijk}\epsilon^{\ell mk} = \delta_i^\ell\delta_j^m - \delta_i^m\delta_j^\ell$. This identity may be used to great advantage in proving vector identities, as we will demonstrate in the following section.

B.3 Vector Identities

In the following list we derive the vector identities (1) to (13) found on the end pages of this book. When using Cartesian coordinates, no distinction need be made between covariant and contravariant components of a vector.

(1)
$$\begin{aligned}\left[\vec{A}\times(\vec{B}\times\vec{C})\right]^i &= \epsilon^{ijk}A_j\left(\vec{B}\times\vec{C}\right)_k\\ &= \epsilon^{ijk}A_j\epsilon_{k\ell m}B^\ell C^m\\ &= \epsilon^{ijk}\epsilon_{\ell mk}A_jB^\ell C^m\\ &= \left(\delta_\ell^i\delta_m^j - \delta_m^i\delta_\ell^j\right)A_jB^\ell C^m\\ &= A_jC^jB^i - A_jB^jC^i = (\vec{A}\cdot\vec{C})B^i - (\vec{A}\cdot\vec{B})C^i\end{aligned}$$

(2)
$$\begin{aligned}\left[(\vec{A}\times\vec{B})\times\vec{C}\right]^i &= \epsilon^{ijk}(\vec{A}\times\vec{B})_jC_k\\ &= \epsilon^{ijk}\epsilon_{jm\ell}A^mB^\ell C_k\\ &= \epsilon^{jki}\epsilon_{jm\ell}A^mB^\ell C_k\\ &= \left(\delta_m^k\delta_\ell^i - \delta_\ell^k\delta_m^i\right)A^mB^\ell C_k\\ &= B^iA^kC_k - A^iB^kC_k = B^i(\vec{A}\cdot\vec{C}) - A^i(\vec{B}\cdot\vec{C})\end{aligned}$$

(3)
$$\begin{aligned}\vec{A}\cdot(\vec{B}\times\vec{C}) &= A^k\left(\vec{B}\times\vec{C}\right)_k = A^k\epsilon_{k\ell m}B^\ell C^m\\ &= \epsilon_{mk\ell}A^kB^\ell C^m = C^m\epsilon_{mk\ell}A^kB^\ell\\ &= C^m\left(\vec{A}\times\vec{B}\right)_m = (\vec{A}\times\vec{B})\cdot\vec{C}\end{aligned}$$

(4)
$$\begin{aligned}(\vec{A}\times\vec{B})\cdot(\vec{C}\times\vec{D}) &= (\vec{A}\times\vec{B})_i(\vec{C}\times\vec{D})^i\\ &= \epsilon_{ijk}A^jB^k\epsilon^{i\ell m}C_\ell D_m\\ &= \left(\delta_j^\ell\delta_k^m - \delta_m^j\delta_k^\ell\right)A^jB^kC_\ell D_m\\ &= A^jC_jB^kD_k - A^jD_jB^kC_k\\ &= (\vec{A}\cdot\vec{C})(\vec{B}\cdot\vec{D}) - (\vec{A}\cdot\vec{D})(\vec{B}\cdot\vec{C})\end{aligned}$$

(5)
$$\left[\vec{\nabla}(\psi\xi)\right]_i = \partial_i(\psi\xi) = \xi\partial_i\psi + \psi\partial_i\xi$$

(6)
$$\begin{aligned}\left[\vec{\nabla}\times(\psi\vec{A})\right]^i &= \epsilon^{ijk}\partial_j(\psi A_k)\\ &= \epsilon^{ijk}(\partial_j\psi)A_k + \psi\epsilon^{ijk}\partial_jA_k\\ &= \left[(\vec{\nabla}\psi)\times\vec{A}\right]^i + \psi(\vec{\nabla}\times\vec{A})^i\end{aligned}$$

Although the identities are generally valid, the derivation of the following two identities involving a divergence is, strictly speaking, valid only in orthonormal coordinates as the general covariant divergence operators will not be discussed until section B.4.2.

(7) $$\begin{aligned}\vec{\nabla}\cdot(\psi\vec{A}) &= \partial_i(\psi A^i) = (\partial_i\psi)A^i + \psi\partial_i A^i \\ &= \vec{\nabla}\psi\cdot\vec{A} + \psi\vec{\nabla}\cdot\vec{A}\end{aligned}$$

(8) $$\begin{aligned}\vec{\nabla}\cdot(\vec{A}\times\vec{B}) &= \partial_i(\vec{A}\times\vec{B})^i = \partial_i\epsilon^{ijk}A_jB_k \\ &= \epsilon^{ijk}\big(\partial_iA_j\big)B_k + \epsilon^{ijk}\big(\partial_iB_k\big)A_j \\ &= \epsilon^{kij}\big(\partial_iA_j\big)B_k - \epsilon^{jik}\big(\partial_iB_k\big)A_j \\ &= \big(\vec{\nabla}\times\vec{A}\big)^kB_k - \big(\vec{\nabla}\times\vec{B}\big)^jA_j = \big(\vec{\nabla}\times\vec{A}\big)\cdot\vec{B} - \big(\vec{\nabla}\times\vec{B}\big)\cdot\vec{A}\end{aligned}$$

(9) $$\begin{aligned}\big[\vec{\nabla}(\vec{A}\cdot\vec{B})\big]_i &= \partial_i(A_jB^j) = B^j\partial_iA_j + A^j\partial_iB_j \\ &= B^j\big(\partial_iA_j - \partial_jA_i\big) + A^j\big(\partial_iB_j - \partial_jB_i\big) + B^j\partial_jA_i + A^j\partial_jB_i\end{aligned}$$

where we have used $B_j\partial_iA^j = B^j\partial_iA_j$ and added and subtracted the two rightmost terms. The first two terms vaguely resemble a curl; in fact, expanding

$$\begin{aligned}\big[\vec{A}\times(\vec{\nabla}\times\vec{B})\big]_i &= \epsilon_{ijk}A^j(\epsilon^{k\ell m}\partial_\ell B_m) \\ &= \epsilon_{kij}\epsilon^{k\ell m}A^j\partial_\ell B_m \\ &= \big(\delta_i^\ell\delta_j^m - \delta_i^m\delta_j^\ell\big)A^j\partial_\ell B_m \\ &= A^j\partial_iB_j - A^j\partial_jB_i\end{aligned}$$

we have the second term of the expansion of $\big[\vec{\nabla}(\vec{A}\cdot\vec{B})\big]_i$ above. Similarly,

$$\big[\vec{B}\times(\vec{\nabla}\times\vec{A})\big]_i = B^j\partial_iA_j - B^j\partial_jA_i$$

Making these replacements, we find

$$\big[\vec{\nabla}(\vec{A}\cdot\vec{B})\big]_i = [\vec{B}\times(\vec{\nabla}\times\vec{A})]_i + [\vec{A}\times(\vec{\nabla}\times\vec{B})]_i + (\vec{B}\cdot\vec{\nabla})A_i + (\vec{A}\cdot\vec{\nabla})B_i$$

(10) $$\begin{aligned}\big[\vec{\nabla}\times(\vec{A}\times\vec{B})\big]^i &= \epsilon^{ijk}\partial_j(\vec{A}\times\vec{B})_k = \epsilon^{ijk}\partial_j\epsilon_{k\ell m}A^\ell B^m \\ &= \epsilon^{kij}\epsilon_{k\ell m}\big(B^m\partial_jA^\ell + A^\ell\partial_jB^m\big) \\ &= \left(\delta_\ell^i\delta_m^j - \delta_m^i\delta_\ell^j\right)\big(B^m\partial_jA^\ell + A^\ell\partial_jB^m\big) \\ &= (\vec{B}\cdot\vec{\nabla})A^i - B^i(\vec{\nabla}\cdot\vec{A}) + A^i(\vec{\nabla}\cdot\vec{B}) - (\vec{A}\cdot\vec{\nabla})B^i\end{aligned}$$

(11) $$\begin{aligned}\vec{\nabla}\cdot(\vec{\nabla}\times\vec{A}) &= \partial_i(\vec{\nabla}\times\vec{A})^i \\ &= \partial_i\epsilon^{ijk}\partial_jA_k = \epsilon^{jik}\partial_j\partial_iA_k \\ &= -\epsilon^{ijk}\partial_j\partial_iA_k = -\partial_i\epsilon^{ijk}\partial_jA_k = -\vec{\nabla}\cdot(\vec{\nabla}\times\vec{A})\end{aligned}$$

where we have relabelled dummy indices i and j as j and i respectively in the second line and exchanged the order of differentiation in the third. We conclude that $\vec{\nabla}\cdot(\vec{\nabla}\times\vec{A})\equiv 0$. We prove $\vec{\nabla}\times(\vec{\nabla}V)=0$ in the same fashion:

$$\text{(12)}\qquad \begin{aligned}\left[\vec{\nabla}\times(\vec{\nabla}V)\right]^i &= \epsilon^{ijk}\partial_j(\vec{\nabla}V)_k \\ &= \epsilon^{ijk}\partial_j\partial_k V = \epsilon^{ikj}\partial_k\partial_j V \\ &= -\epsilon^{ijk}\partial_j\partial_k V = -\left[\vec{\nabla}\times(\vec{\nabla}V)\right]^i\end{aligned}$$

We conclude that $\vec{\nabla}\times(\vec{\nabla}V)\equiv 0$.

$$\text{(13)}\qquad \begin{aligned}\left[\vec{\nabla}\times(\vec{\nabla}\times\vec{A})\right]^i &= \epsilon^{ijk}\partial_j(\vec{\nabla}\times\vec{A})_k = \epsilon^{ijk}\partial_j\epsilon_{k\ell m}\partial^\ell A^m \\ &= \epsilon^{kij}\epsilon_{klm}\partial_j\partial^\ell A^m \\ &= (\delta^i_\ell\delta^j_m - \delta^i_m\delta^j_\ell)\partial_j\partial^\ell A^m \\ &= \partial^i(\partial_j A^j) - \partial_\ell\partial^\ell A^i = \partial^i(\vec{\nabla}\cdot\vec{A}) - \nabla^2 A^i\end{aligned}$$

B.4 Curvilinear Coordinates

Any three numbers q^1, q^2, q^3 uniquely specifying the position of a point in space are called the coordinates of the point in space. To illustrate, in cylindrical polar coordinates $q^1 = r = \sqrt{x_1^2+x_2^2}$, $q^2=\varphi=\tan^{-1}x_2/x_1$, and $q^3 = z = x_3$, where x_1, x_2, and x_3 are the Cartesian coordinates of the point.

The surfaces described when one of the coordinates is held constant while the others vary over their complete range are called *coordinate surfaces.* In the example above, the coordinate surfaces are

$r=a$	cylinder of radius a
$z=b$	plane parallel to the xy plane
$\varphi=\varphi_0$	half-infinite plane containing the z axis

The curves obtained when two of the coordinates are fixed (the intersection of two coordinate planes) are called coordinate curves. The coordinate curves for cylindrical coordinates are

$r=a$	$\varphi=b$	straight line parallel to the z axis
$\varphi=a$	$z=b$	straight line pointing radially outward at $z=b$
$r=a$	$z=b$	circle of radius a at $z=b$

B.4.1 Bases and Coordinate Axes

By a basis of a "generalized" coordinate system q^1, q^2, q^3 we mean any set of three vectors $\vec{e}_1$, $\vec{e}_2$, $\vec{e}_3$ of (locally) fixed length, pointing tangent to the positive direction of the coordinate curve. Thus $\vec{e}_1$ is tangent to the coordinate curve of q^1 (q^2 and q^3 fixed). The basis is said to be local, since it varies from point to point. In general, the basis vectors are neither of unit length nor mutually perpendicular. Coordinate systems whose basis vectors intersect at right angles are said to be orthogonal.

Over small distances the separation between two points can be found from $(dS)^2 = g_{ik}dq^i dq^k$, where g_{ik} is the *local metric tensor.* The element of arc length along a coordinate curve q^i is

$$(dS)_i = |\vec{e}_i|dq^i = \sqrt{g_{ii}}\,dq^i \quad \text{(no summation)} \tag{B–36}$$

while the element of area, $d\sigma_1$, on the coordinate surface q^1 = constant (this surface has normal proportional to $\vec{e}_1$) is

$$\begin{aligned} d\sigma_1 &= |\vec{e}_2 \times \vec{e}_3|\, dq^2 dq^3 \\ &= \sqrt{(\vec{e}_2 \times \vec{e}_3)\cdot(\vec{e}_2 \times \vec{e}_3)}dq^2 dq^3 \\ &= \sqrt{(\vec{e}_2\cdot\vec{e}_2)(\vec{e}_3\cdot\vec{e}_3) - (\vec{e}_2\cdot\vec{e}_3)(\vec{e}_2\cdot\vec{e}_3)}dq^2 dq^3 \\ &= \sqrt{g_{22}g_{33} - (g_{23})^2}\, dq^2 dq^3 \end{aligned} \tag{B–37}$$

Similarly,

$$d\sigma_2 = \sqrt{g_{11}g_{33} - (g_{13})^2}\, dq^1 dq^3 \quad \text{and} \quad d\sigma_3 = \sqrt{g_{11}g_{22} - (g_{12})^2}\, dq^1 dq^2 \tag{B–38}$$

In terms of the coordinates, the differential volume element is given by $\vec{e}_1 \cdot (\vec{e}_2 \times \vec{e}_3) dq_1 dq_2 dq_3 = \sqrt{G}\, dq_1 dq_2 dq_3$, where $G = \det(g_{ik})$ is the determinant of the metric tensor. In orthogonal systems, the only nonzero elements of g_{ij} are g_{11}, g_{22}, and g_{33}.

Suppose that the relationship between a system of generalized coordinates q^1, q^2, q^3 and an underlying system of Cartesian coordinates x_1, x_2, x_3 is given by the formulas

$$q^1 = q^1(x_1, x_2, x_3) \qquad q^2 = q^2(x_1, x_2, x_3) \qquad q^3 = q^3(x_1, x_2, x_3) \tag{B–39}$$

and

$$x_1 = x_1(q^1, q^2, q^3) \qquad x_2 = x_2(q^1, q^2, q^3) \qquad x_3 = x_3(q^1, q^2, q^3) \tag{B–40}$$

where the Jacobians

$$J = \det\left(\frac{\partial q^i}{\partial x_k}\right) \quad \text{and} \quad J^{-1} = \det\left(\frac{\partial x_i}{\partial q^k}\right)$$

are nonzero ($J = 1/r$ for cylindrical coordinates and $J = 1/(r\sin\theta)$ for spherical polars). We can write the position vector of an arbitrary point as $\vec{r} = \vec{r}(q^1, q^2, q^3) = x_1\hat{\imath}_1 + x_2\hat{\imath}_2 + x_3\hat{\imath}_3$. Then

$$d\vec{r} = \frac{\partial \vec{r}}{\partial q^1}dq^1 + \frac{\partial \vec{r}}{\partial q^2}dq^2 + \frac{\partial \vec{r}}{\partial q^3}dq^3 = \frac{\partial \vec{r}}{\partial q^i}dq^i \tag{B–41}$$

It follows that

$$(dS)^2 = d\vec{r}\cdot d\vec{r} = \frac{\partial \vec{r}}{\partial q^i}\cdot\frac{\partial \vec{r}}{\partial q^j}\,dq^i dq^j \tag{B-42}$$

We therefore choose the vectors of the local basis as

$$\vec{e}_i = \frac{\partial \vec{r}}{\partial q^i} \tag{B-43}$$

and the metric tensor in this basis is $g_{ij} = \dfrac{\partial \vec{r}}{\partial q^i}\cdot\dfrac{\partial \vec{r}}{\partial q^j}$.

EXAMPLE B.7: Find the local basis and metric tensor for spherical polar coordinates defined by

$$x_1 = r\sin\theta\cos\varphi \qquad x_2 = r\sin\theta\sin\varphi \qquad x_3 = r\cos\theta$$

Solution: The basis expressed in terms of the Cartesian basis is

$$\begin{aligned}
\vec{e}_r &= \frac{\partial \vec{r}}{\partial r} = \hat{\imath}_1\sin\theta\cos\varphi + \hat{\imath}_2\sin\theta\sin\varphi + \hat{\imath}_3\cos\theta\\
\vec{e}_\theta &= \frac{\partial \vec{r}}{\partial \theta} = \hat{\imath}_1 r\cos\theta\cos\varphi + \hat{\imath}_2 r\cos\theta\sin\varphi - \hat{\imath}_3 r\sin\theta\\
\vec{e}_\varphi &= \frac{\partial \vec{r}}{\partial \varphi} = \hat{\imath}_1 r\sin\theta\sin\varphi + \hat{\imath}_2 r\sin\theta\cos\varphi
\end{aligned} \tag{Ex B.7.1}$$

leading to a diagonal metric tensor with elements

$$\begin{aligned}
g_{rr} &= \vec{e}_r\cdot\vec{e}_r = \sin^2\theta\cos^2\varphi + \sin^2\theta\sin^2\varphi + \cos^2\theta = 1\\
g_{\theta\theta} &= \vec{e}_\theta\cdot\vec{e}_\theta = r^2(\cos^2\theta\cos^2\varphi + \cos^2\theta\sin^2\varphi + \sin^2\theta) = r^2\\
g_{\varphi\varphi} &= \vec{e}_\varphi\cdot\vec{e}_\varphi = r^2(\sin^2\theta\sin^2\varphi + \sin^2\theta\cos^2\varphi) = r^2\sin^2\theta
\end{aligned} \tag{Ex B.7.2}$$

B.4.2 Differentiation of Tensors in Curvilinear Coordinates

We have already shown that $\partial V/\partial q^i = F_i$ are the covariant components of a vector we call $\vec{\nabla}V$. When the basis varies with position in space, the entity obtained by differentiating the components of a tensor with respect to the coordinates is *not* generally a tensor. To be specific, let us consider the result of differentiating a vector with respect to the components.

The change in a vector field $\vec{A}$ in moving from $\{q_i\}$ to $\{q_i + dq_i\}$ is in general given by

$$d\vec{A} = \frac{\partial \vec{A}}{\partial q^k}\,dq^k \tag{B-44}$$

When $\vec{A}$ is expressed in terms of the local basis, the components of $\vec{A}$ at the new position must express not only the change in $\vec{A}$ but also the change in the basis resulting from the displacement to the new position. Algebraically, for $\vec{A} = A^i \vec{e}_i$, the partial derivative of $\vec{A}$ with respect to the coordinate q_k is

$$\frac{\partial \vec{A}}{\partial q^k} = \frac{\partial A^i}{\partial q^k}\vec{e}_i + A^i \frac{\partial \vec{e}_i}{\partial q^k} \tag{B–45}$$

The components of this partial derivative of $\vec{A}$, conventionally denoted $A^i_{\ .k}$, are the components of a second-rank tensor called the covariant derivative of $\vec{A}$ (the partial derivatives of the components are *not*, however, components of a tensor). The components of the covariant derivative are readily obtained by taking the dot product of $\partial \vec{A}/\partial q^k$ with a local basis vector or its reciprocal. Thus

$$\frac{\partial \vec{A}}{\partial q^k} \cdot \vec{e}_i \equiv A_{i,k} \tag{B–46}$$

are the covariant components of the covariant derivative of the vector $\vec{A}$. We repeat that the comma preceding the k in the subscript conventionally indicates covariant differentiation with respect to q^k. The mixed components of the covariant derivative are similarly obtained:

$$\frac{\partial \vec{A}}{\partial q^k} \cdot \vec{e}^{\,i} \equiv A^i_{\ ,k} \tag{B–47}$$

Expanding the components we have

$$\begin{aligned} A^j_{\ ,k} = \frac{\partial \vec{A}}{\partial q^k} \cdot \vec{e}^{\,j} &= \frac{\partial A^i}{\partial q^k}\vec{e}_i \cdot \vec{e}^{\,j} + A^i \frac{\partial \vec{e}_i}{\partial q^k} \cdot \vec{e}^{\,j} \\ &= \frac{\partial A^j}{\partial q^k} + A^i \left(\frac{\partial \vec{e}_i}{\partial q^k} \cdot \vec{e}^{\,j} \right) \end{aligned} \tag{B–48}$$

The term in parentheses is called a *Christoffel symbol of the second kind*[32] and is usually denoted by

$$\left\{ {}_{i}{}^{j}{}_{k} \right\} \equiv \frac{\partial \vec{e}_i}{\partial q^k} \cdot \vec{e}^{\,j} = \vec{e}^{\,j} \cdot \partial_k \vec{e}_i \tag{B–49}$$

so that we write

$$A^j_{\ ,k} = \frac{\partial A^j}{\partial q^k} + A^i \left\{ {}_{i}{}^{j}{}_{k} \right\} = \partial_k A^j + A^i \left\{ {}_{i}{}^{j}{}_{k} \right\} \tag{B–50}$$

The fully covariant form of the covariant derivative can be found in the same fashion:

$$\begin{aligned} A_{j,k} &= \partial_k (A_i \vec{e}^{\,i}) \cdot \vec{e}_j = (\vec{e}^{\,i} \cdot \vec{e}_j) \partial_k A_i + A_i \vec{e}_j \cdot \partial_k \vec{e}^{\,i} \\ &= \partial_k A_j + A_i \vec{e}_j \cdot \partial_k \vec{e}^{\,i} \end{aligned} \tag{B–51}$$

[32] Christoffel symbols of the first kind, $[i, jk] \equiv \vec{e}_i \cdot (\partial \vec{e}_j / \partial q^k)$, are also frequently encountered.

With the aid of $\partial_k(g_i^j) = 0 \Rightarrow e_j \cdot \partial_k \vec{e}^i = -\vec{e}^i \cdot \partial_k \vec{e}_j$, we rearrange the second term to obtain

$$A_{j,k} = \frac{\partial A_j}{\partial q^k} - \left\{ {i \atop j \quad k} \right\} A_i \tag{B–52}$$

That the Christoffel symbol is invariant under exchange of the two lower indices is seen from

$$\frac{\partial \vec{e}_i}{\partial q^k} = \frac{\partial}{\partial q^k}\frac{\partial \vec{r}}{\partial q^i} = \frac{\partial}{\partial q^i}\frac{\partial \vec{r}}{\partial q^k} = \frac{\partial \vec{e}_k}{\partial q^i} \tag{B–53}$$

The foregoing is easily extended to the covariant differentiation of higher order tensors. Differentiating a second rank tensor gives the following three possible forms:

$$F^{ik}_{,\ell} = \frac{\partial F^{ik}}{\partial q^\ell} + \left\{ {i \atop m \quad \ell} \right\} F^{mk} + \left\{ {k \atop m \quad \ell} \right\} F^{im} \tag{B–54}$$

$$F_{ik,\ell} = \frac{\partial F_{ik}}{\partial q^\ell} - \left\{ {m \atop i \quad \ell} \right\} F_{mk} - \left\{ {m \atop k \quad \ell} \right\} F_{im} \tag{B–55}$$

$$F^{i}_{\cdot k,\ell} = \frac{\partial F^{i}_{\cdot k}}{\partial q^\ell} + \left\{ {i \atop m \quad \ell} \right\} F^{m}_{\cdot k} - \left\{ {m \atop k \quad \ell} \right\} F^{i}_{\cdot m} \tag{B–56}$$

The Cristoffel symbols are *not* themselves tensors; nevertheless, the summation convention for repeated indices is used.

B.4.3 Differential Operators

The values of the differential operators, gradient, and curl (also sometimes called "rot") are frequently best expressed in terms of unit vectors in order that magnitudes can be computed from components. If the divergence and curl are to have physical meaning they will need to be made covariant. We define the "del" (or *nabla*) operator by

$$\vec{\nabla} \equiv \vec{e}^i \frac{\partial}{\partial q^i} \equiv \vec{e}^i \partial_i \tag{B–57}$$

The **gradient** of a scalar in terms of a unit basis is then

$$\vec{\nabla} V = \vec{e}^i \partial_i V = \frac{\hat{e}^i}{\sqrt{g_{ii}}}\frac{\partial V}{\partial q^i} \tag{B–58}$$

where $\hat{e}_i$ is a unit vector along $\vec{e}_i$ ($\hat{e}_i = \vec{e}_i/|\vec{e}_i|$ and $\hat{e}^i = \vec{e}^i/|\vec{e}^i| = \vec{e}^i|\vec{e}_i|$). We already know the components $\partial_i V$ to be components of a contravariant first rank tensor so no modification is required in non-cartesian frames. We denote the unit basis components of a vector with a superscript * so that $F^{*i} = (\vec{\nabla} V)^{*i} = \sqrt{g^{ii}}\partial V/\partial q^i$.

The **divergence** of a vector field $\vec{A}$ is the contracted covariant derivative of $\vec{A}$; that is,

$$\text{div}\vec{A} \equiv A^i_{\ ,i} = \partial_i A^i + \left\{ {i \atop i \quad j} \right\} A^j \tag{B–59}$$

The sum

$$\left\{ {i \atop i \quad j} \right\} = \vec{e}^i \cdot \frac{\partial \vec{e}_i}{\partial q^j} \tag{B–60}$$

may with some effort be shown to equal $\frac{1}{\sqrt{G}}\frac{\partial\sqrt{G}}{\partial q^j}$ where G is the determinant of the metric tensor g_{ij} so that

$$\vec{\nabla}\cdot\vec{A} = \frac{\partial A^j}{\partial q^j} + \frac{A^j}{\sqrt{G}}\frac{\partial}{\partial q^j}(\sqrt{G}) = \frac{1}{\sqrt{G}}\frac{\partial}{\partial q^j}\left(A^j\sqrt{G}\right) \tag{B–61}$$

To express this result in terms of the components of $\vec{A}$ in a unit basis it suffices to replace A^j in (B–61) by $A^{*j} = \sqrt{g^{jj}}A^j$, so that we may write

$$\vec{\nabla}\cdot\vec{A} = \frac{1}{\sqrt{G}}\frac{\partial}{\partial q^j}\left(A^{*j}\sqrt{g^{jj}G}\right) \tag{B–62}$$

The **curl** of a vector field $\vec{A}$ is

$$\vec{\nabla}\times\vec{A} = \vec{e}^{\,j}\partial_j\times\vec{A} = \vec{e}^{\,j}\times\partial_j\vec{A} = \vec{e}^{\,j}\times\vec{e}^{\,k}A_{k.j} \tag{B–63}$$

The cross product

$$\vec{e}^{\,j}\times\vec{e}^{\,k} = \frac{1}{\sqrt{G}}\times\begin{cases} \vec{e}_i & \text{if } i,\ j,\ k \text{ is a cyclic permutation of } 1,\ 2,\ 3 \\ -\vec{e}_i & \text{if } i,\ j,\ k \text{ is an anticyclic permutation of } 1,\ 2,\ 3 \\ 0 & \text{if two of the indices coincide} \end{cases} \tag{B–64}$$

Making use of this we write the i component of the curl as

$$\left(\vec{\nabla}\times\vec{A}\right)^i = \frac{1}{\sqrt{G}}\left(A_{k.\ell} - A_{\ell.k}\right) \tag{B–65}$$

where the indices i, k, and ℓ are taken in cyclic order. Writing

$$A_{k,\ell} = \frac{\partial A_k}{\partial q^\ell} - \left\{ {m \atop k\quad \ell} \right\} A_m \tag{B–66}$$

and

$$A_{\ell.k} = \frac{\partial A_\ell}{\partial q^k} - \left\{ {m \atop \ell\quad k} \right\} A_m \tag{B–67}$$

we find the Christoffel symbols cancel in the difference so that

$$\begin{aligned}\left(\vec{\nabla}\times\vec{A}\right)^i &= \frac{1}{\sqrt{G}}\left(\frac{\partial A_k}{\partial q^\ell} - \frac{\partial A_\ell}{\partial q^k}\right) \\ &\equiv \epsilon^{i\ell k}\partial_\ell A_k\end{aligned} \tag{B–68}$$

As we can see, the expression for the curl remains unchanged from that used in Cartesian coordinates. Again we may express these results in terms of the components with respect to a unit basis resulting in

$$\left(\vec{\nabla}\times\vec{A}\right)^{*i} = \frac{\sqrt{g_{ii}}}{\sqrt{G}}\left(\frac{\partial(\sqrt{g_{kk}}A_k^*)}{\partial q^\ell} - \frac{\partial(\sqrt{g_{\ell\ell}}A_\ell^*)}{\partial q^k}\right) \tag{B–69}$$

The **Laplacian** ∇^2 of a scalar field V is defined as the divergence of the gradient of V. Combining the expressions for div and grad, we find that

$$\begin{aligned}\nabla^2 V &= \frac{1}{\sqrt{G}}\partial_i\left[\sqrt{G}(\vec{\nabla}V)^i\right]\\ &= \frac{1}{\sqrt{G}}\partial_i\left(\sqrt{G}g^{ik}\partial_k V\right)\end{aligned} \tag{B–70}$$

EXAMPLE B.8: To solidify some of these notions let us construct the local basis, the metric tensor, and the Laplacian in *toroidal* coordinates.

Solution: The toroidal coordinates α, β, φ are related to the Cartesian coordinates as follows:

$$x = \frac{c\sinh\alpha\cos\varphi}{\cosh\alpha - \cos\beta} \qquad y = \frac{c\sinh\alpha\sin\varphi}{\cosh\alpha - \cos\beta} \qquad z = \frac{c\sin\beta}{\cosh\alpha - \cos\beta} \tag{Ex B.8.1}$$

with $0 \le \alpha \le \infty$, $0 \le \beta \le 2\pi$, and $0 \le \varphi \le 2\pi$.

In terms of cylindrical coordinates, $r = \sqrt{x^2+y^2}$, z, and $\varphi_c = \tan^{-1} y/x$, the relations may also be expressed as

$$r = \frac{c\sinh\alpha}{\cosh\alpha - \cos\beta} \qquad z = \frac{c\sin\beta}{\cosh\alpha - \cos\beta} \qquad \varphi_c = \varphi \tag{Ex B.8.2}$$

The coordinate surfaces are a torus for α = constant, two over-lapping spheres for β = constant, and a vertical plane for φ = constant, as will be shown below.

The ratio $z/r = \sin\beta/\sinh\alpha$ may be used to eliminate either α or β in the equations for r or z.

α **constant**: To eliminate $\cos\beta$ from the equation for r, we note that $\sin^2\beta = (z^2/r^2)\sinh^2\alpha$. Then

$$r^2\cos^2\beta = r^2(1-\sin^2\beta) = r^2 - z^2\sinh^2\alpha \tag{Ex B.8.3}$$

Using the relation between r and the toroidal coordinates, $r\cosh\alpha - r\cos\beta = c\sinh\alpha$, we replace $r^2\cos^2\beta$ in (Ex B.8.3) to obtain

$$(r\cosh\alpha - c\sinh\alpha)^2 = r^2 - z^2\sinh^2\alpha \tag{Ex B.8.4}$$

On expansion this becomes

$$r^2\left(\cosh^2\alpha - 1\right) - 2r\cosh\alpha\sinh\alpha + z^2\sinh^2\alpha = -c^2\sinh^2\alpha \tag{Ex B.8.5}$$

which, upon being divided by $\sinh^2\alpha$, gives

$$r^2 - 2rc\frac{\cosh\alpha}{\sinh\alpha} + z^2 = -c^2 \tag{Ex B.8.6}$$

We complete the square for terms in r in (Ex B.8.6) to obtain,

$$\left(r^2 - 2rc\coth\alpha + c^2\coth^2\alpha\right) + z^2 = c^2(\coth^2\alpha - 1) \tag{Ex B.8.7}$$

or

$$(r - c \coth \alpha)^2 + z^2 = \frac{c^2}{\sinh^2 \alpha} \quad \text{(Ex B.8.8)}$$

the equation of a circle centered at $r = c \coth \alpha$, having a radius $c/\sinh \alpha$. This same circle is produced for any value of the azimuthal angle φ. In other words, as we rotate r about the z axis, a torus centered on the z axis is swept out.

β **constant**: When β is constant we eliminate $\cosh \alpha$ from the z equation. We again use the relation $\sinh^2 \alpha = (r^2/z^2) \sin^2 \beta$ to write

$$\cosh^2 \alpha = 1 + \sinh^2 \alpha = \frac{r^2}{z^2} \sin^2 \beta + 1 \quad \text{(Ex B.8.9)}$$

or

$$z^2 \cosh^2 \alpha = r^2 \sin^2 \beta + z^2 \quad \text{(Ex B.8.10)}$$

Then, returning to (Ex B.8.1), $z \cosh \alpha = z \cos \beta + c \sin \beta$ may used to eliminate $z^2 \cosh^2 \alpha$ from (Ex B.8.10) and we obtain

$$r^2 \sin^2 \beta + z^2 = z^2 \cos^2 \beta + 2zc \sin \beta \cos \beta + c^2 \sin^2 \beta \quad \text{(Ex B.8.11)}$$

We gather the terms in z and divide by $\sin^2 \beta$ to get

$$\frac{z^2 - z^2 \cos^2 \beta - 2zc \cos \beta \sin \beta + r^2 \sin^2 \beta}{\sin^2 \beta} = z^2 - 2zc \cot \beta + r^2 = c^2 \quad \text{(Ex B.8.12)}$$

and complete the square to obtain

$$\left(z^2 - 2zc \cot \beta + c^2 \cot^2 \beta\right) + r^2 = c^2(1 + \cot^2 \beta) \quad \text{(Ex B.8.13)}$$

$$(z - c \cot \beta)^2 + r^2 = \frac{c^2}{\sin^2 \beta} \quad \text{(Ex B.8.14)}$$

the equation of a circle in the z-r plane centered at $z = c \cot \beta$, and of radius $c/\sin \beta$. Rotation about the z axis sweeps out a sphere centered at positive z for $\beta \in (-\pi/2, \pi/2)$ and negative z for remaining values of β.

To obtain the metric tensor, we first construct the local basis:

$$\vec{e}_\varphi = \frac{d\vec{r}}{d\varphi} = \frac{d}{d\varphi}(x\hat{\imath} + y\hat{\jmath} + z\hat{k}) = \left(\frac{c \sinh \alpha}{\cosh \alpha - \cos \beta}\right)(-\sin \varphi \hat{\imath} + \cos \varphi \hat{\jmath}) \quad \text{(Ex B.8.15)}$$

$$\vec{e}_\alpha = \left[\frac{c \cosh \alpha}{(\cosh \alpha - \cos \beta)} - \frac{c \sinh^2 \alpha}{(\cosh \alpha - \cos \beta)^2}\right](\cos \varphi \hat{\imath} + \sin \varphi \hat{\jmath}) - \frac{c \sin \beta \sinh \alpha \hat{k}}{(\cosh \alpha - \cos \beta)^2} \quad \text{(Ex B.8.16)}$$

$$\vec{e}_\beta = \frac{-c \sinh \alpha \sin \beta (\cos \varphi \hat{\imath} + \sin \varphi \hat{\jmath})}{(\cosh \alpha - \cos \beta)^2} + \left[\frac{c \cos \beta}{\cosh \alpha - \cos \beta} - \frac{c \sin^2 \beta}{(\cosh \alpha - \cos \beta)^2}\right] \hat{k} \quad \text{(Ex B.8.17)}$$

The orthogonality of the basis is readily verified and the metric tensor now follows in straightforward fashion. The $\varphi\varphi$ element may be written immediately, as

$$g_{\varphi\varphi} = \frac{c^2 \sinh^2 \alpha}{(\cosh\alpha - \cos\beta)^2}$$

while $g_{\alpha\alpha}$ may be computed as

$$\begin{aligned} g_{\alpha\alpha} &= \left[\frac{c\cosh\alpha}{(\cosh\alpha - \cos\beta)} - \frac{c\sinh^2\alpha}{(\cosh\alpha-\cos\beta)^2}\right]^2 + \frac{c^2\sin^2\beta\sinh^2\alpha}{(\cosh\alpha-\cos\beta)^4} \\ &= \frac{c^2}{(\cosh\alpha-\cos\beta)^2}\Bigg\{\cosh^2\alpha + \sinh^2\alpha \\ &\qquad \times \left[\frac{\sinh^2\alpha - 2\cosh\alpha(\cosh\alpha-\cos\beta)+\sin^2\beta}{(\cosh\alpha-\cos\beta)^2}\right]\Bigg\} \\ &= \frac{c^2}{(\cosh\alpha-\cos\beta)^2}(\cosh^2\alpha - \sinh^2\alpha) = \frac{c^2}{(\cosh\alpha-\cos\beta)^2} \end{aligned} \qquad \text{(Ex B.8.18)}$$

In an equally laborious fashion, it can be shown that

$$g_{\beta\beta} = \frac{c^2}{(\cosh\alpha-\cos\beta)^2} \qquad \text{(Ex B.8.19)}$$

while all the nondiagonal terms vanish.

The squared element of arc length is thus

$$(dS)^2 = \frac{c^2}{(\cosh\alpha-\cos\beta)^2}\left[(d\alpha)^2 + (d\beta)^2 + \sinh^2\alpha\,(d\varphi)^2\right] \qquad \text{(Ex B.8.20)}$$

and the Laplacian $\nabla^2 f$ becomes

$$\begin{aligned} \frac{(\cosh\alpha-\cos\beta)^3}{c^2\sinh\alpha}\Bigg\{\frac{\partial}{\partial\alpha}\left[\frac{\sinh\alpha}{(\cosh\alpha-\cos\beta)}\frac{\partial f}{\partial\alpha}\right] + \frac{\partial}{\partial\beta}\left[\frac{\sinh\alpha}{(\cosh\alpha-\cos\beta)}\frac{\partial f}{\partial\beta}\right] \\ + \frac{1}{(\cosh\alpha-\cos\beta)\sinh\alpha}\frac{\partial^2 f}{\partial\varphi^2}\Bigg\} \end{aligned} \qquad \text{(Ex B.8.21)}$$

Although not strictly a part of this discussion, we complete our consideration of toroidal coordinates by effecting the separation of variables required for the solution of Laplace's equation. The equation cannot be separated directly, but if we make the substitution $f = \sqrt{\cosh\alpha - \cos\beta}\,v$, Laplace's equation becomes

$$\frac{\partial^2 v}{\partial\alpha^2} + \frac{\partial^2 v}{\partial\beta^2} + \coth\alpha\frac{\partial v}{\partial\alpha} + \frac{v}{4} + \frac{1}{\sinh^2\alpha}\frac{\partial^2 v}{\partial\varphi^2} = 0 \qquad \text{(Ex B.8.22)}$$

which is separable. In fact, setting $v = \mathrm{A}(\alpha)\mathrm{B}(\beta)\Phi(\varphi)$, one obtains the equations

$$\frac{d^2\Phi}{d\varphi^2} + \mu^2\Phi = 0 \qquad \frac{d^2\mathrm{B}}{d\beta^2} + \nu^2\mathrm{B} = 0 \qquad \text{(Ex B.8.23)}$$

and

$$\frac{1}{\sinh\alpha}\frac{d}{d\alpha}\left(\sinh\alpha\frac{dA}{d\alpha}\right)-\left(\nu^2-\tfrac{1}{4}+\frac{\mu^2}{\sinh^2\alpha}\right)A=0 \qquad (\text{Ex B.8.24})$$

having solutions for the original function $f(\alpha,\beta,\varphi)$

$$f(\alpha,\beta,\varphi)=\sqrt{\cosh\alpha-\cos\beta}\left\{\begin{array}{l}P^{\mu}_{\nu-\frac{1}{2}}(\cosh\alpha)\\ Q^{\mu}_{\nu-\frac{1}{2}}(\cosh\alpha)\end{array}\right\}\left\{\begin{array}{c}\cos\nu\beta\\ \sin\nu\beta\end{array}\right\}\left\{\begin{array}{c}\cos\mu\varphi\\ \sin\mu\varphi\end{array}\right\} \quad (\text{Ex B.8.25})$$

Physical problems will usually impose periodic boundary conditions on Φ and B, leading to integral values for μ and ν.

B.5 Four-Tensors in Special Relativity

According to the tenets of special relativity the equations of physics should be invariant not only under changes in orientation or translation of the coordinate system but also under the changes in the velocity of the coordinate system. In particular, scalars must remain invariant under all such transformations. One important such scalar is c, the velocity of light, whose invariance requires including t, or more conveniently ct, as one of the coordinates of a space-time point.

If the components of a four-vector $\vec{X}=(x^0,x^1,x^2,x^3)$ are (ct,x,y,z), then in going from a frame Σ to another frame Σ', moving with velocity βc along the x axis, the vector $\vec{X}$ has contravariant components $X^{\mu'}=\alpha^{\mu'}_{\nu}X^{\nu}$ with

$$\alpha^{0'}_{0}=\Gamma=1/\sqrt{1-\beta^2},\quad \alpha^{1'}_{1}=\Gamma,\quad \alpha^{0'}_{1}=\alpha^{1'}_{0}=-\beta\Gamma,\quad \alpha^{2'}_{2}=\alpha^{3'}_{3}=1 \qquad (\text{B–71})$$

while all other transformation coefficients are zero. (The inverse transformations are found be reversing the sign of β.) All other 4-vectors, of course, obey the same transformation between inertial frames.

The metric defined by this basis $(\hat{t},\hat{\imath},\hat{\jmath},\hat{k})$ has nonzero elements $g_{11}=g_{22}=g_{33}=-1$ and $g_{00}=+1$. (A diminishing number of authors of relativity texts favor the opposite signs for the elements of g; so long as one is consistent, it makes no difference.) This metric gives, as fundamental measure, the *interval* between two space-time points (known as *events*), the scalar

$$(dS)^2=(cdt)^2-(dx)^2-(dy)^2-(dz)^2 \qquad (\text{B–72})$$

We use this scalar to define *proper time* τ (from the French *propre* meaning *own*), by $d\tau=dS/c$. Then the relation between a moving particle's proper time and that of a "stationary" observer is given by

$$\begin{aligned} d\tau &= \sqrt{(dt)^2-\frac{1}{c^2}\left[(dx)^2+(dy)^2+(dz)^2\right]} \\ &= dt\sqrt{1-\frac{1}{c^2}\left[\left(\frac{dx}{dt}\right)^2+\left(\frac{dy}{dt}\right)^2+\left(\frac{dz}{dt}\right)^2\right]} \end{aligned}$$

Exercises and Problems

B.1 Prove $B_j\partial_i A^j = B^j\partial_i A_j$.

B.2 Demonstrate that the last line of (Ex B.8.18) follows from the line before it.

B.3 Obtain the relativistic transformation law for a particle's momentum.

B.4 Show that $g_{ik}G^{ik} = G$ (no summation over i).

B.5 Find the form of the squared element of arc length, $(dS)^2$, in spherical polar coordinates.

B.6 Prove that metric tensor g_{ik} is a second rank covariant tensor.

B.7 Find the covariant components of $\vec{A}$ in the oblique coordinate system of example B.1.

B.8 Find the metric tensor to the basis in example B.1. Use this to show that $(\Delta S)^2 = g_{ij}(\Delta A^i)(\Delta A^j)$ comparing it to the $(\Delta S)^2$ found from the Cartesian form.

B.9 The angular momentum $\vec{L}$ with respect to the origin of a system on n particles is $\vec{L} = \sum_{j=1}^{n} m_{(j)}(\vec{r}_{(j)} \times \vec{v}_{(j)})$. If all the particles rotate as a rigid body, $\vec{v}_{(j)} = \vec{\omega} \times \vec{r}_{(j)}$. The angular momentum can then be expressed as $L^i = I^{ik}\omega_k$. Obtain the moment of inertia tensor I^{ik} and show it is a second rank symmetric tensor.

B.10 The *Strain* tensor for an object whose points are displaced by $\vec{U}(\vec{r})$ is given by $U^{ij} = \frac{1}{2}\left(\partial U^i/\partial x_j + \partial U^j/\partial x_i\right)$. Show that $\overleftrightarrow{U}$ is a second rank tensor.

B.11 Hyperbolic cylindrical coordinates (u, v, z) may be defined in terms of cartesian coordinates by

$$x = \sqrt{\rho + v}, \quad y = \sqrt{\rho - v}, \quad z = z$$

with $\rho = \sqrt{u^2 + v^2}$. Find the basis, the metric tensor, and the Laplacian in hyperbolic cylindrical coordinates.

B.12 Bispherical coordinates (η, θ, ϕ) are defined by

$$x = \frac{a \sin\theta \cos\phi}{\cosh\eta - \cos\theta}$$

$$y = \frac{a \sin\theta \sin\phi}{\cosh\eta - \cos\theta}$$

$$x = \frac{a \sinh\eta}{\cosh\eta - \cos\theta}$$

Show that the surface generated by holding η constant are spheres centered at $z = \pm a \coth\eta$. Find the metric tensor and obtain the Laplacian in bispherical coordinates.

Appendix C

The Dirac Delta Function

C.1 The Dirac Delta Function

The Dirac δ-function answers the need to describe quantities that exist only at a point, along a line, or on a sheet; in other words, quantities that do not extend over all dimensions. They are also frequently useful as approximations of real physical situations where quantities are non-vanishing in only a very thin layer or thin rod. Intuitively, the δ function in one dimension may be visualized as a function that is zero everywhere except the point where its argument vanishes. At this point the δ function is infinite, constrained only by the requirement that the area under the spike be unity. The δ function may also be taken as the zero-width limit of a normalized Gaussian, a normalized triangle, or even a normalized step function:

$$\delta(x) = \lim_{a\to 0}\begin{cases} \dfrac{1}{|a|} & \text{if x } \in \ (-a/2, a/2) \\ 0 & \text{if x } \notin \ (-a/2, a/2) \end{cases} \tag{C–1}$$

We will denote the step function of which we take the limit by $u_a(x)$. The representation of the δ function as the zero-width limit of well-behaved functions is particularly useful in deducing its properties.

The fundamental, defining property of the δ function is

$$\int_a^b f(x)\delta(x-c)dx = \begin{cases} f(c) & \text{if } c \in (a,b) \\ 0 & \text{if } c \notin (a,b) \end{cases} \tag{C–2}$$

A number of properties of the δ function follow immediately from this definition. Letting $f(x) = 1$, we have

$$\int_a^b \delta(x-c)dx = \begin{cases} 1 & \text{if c } \in (a,b) \\ 0 & \text{if c } \notin (a,b) \end{cases} \tag{C–3}$$

With the aid of the step function representation $u_a(x)$, we see immediately that if the argument x is replaced by cx, the rectangle narrows in width from a to

$a/|c|$ so that we expect the area under the spike to decrease by a factor $|c|$. More analytically, assuming that the region of integration includes the zero of the delta function argument, say b, we find that

$$\begin{aligned}\int f(x)\delta[c(x-b)]dx &= \lim_{a\to 0}\int f(x)u_a(cx-cb)dx \\ &= \lim_{a\to 0}\int_{b-a/2c}^{b+a/2c} f(x)\frac{1}{|a|}dx \\ &= \lim_{a\to 0}\frac{\overline{f}(x)a}{|ac|} = \frac{f(b)}{|c|}\end{aligned} \tag{C–4}$$

where we have restricted the range of integration to the interval where u_a does not vanish and used the mean value theorem to evaluate the integral. As a tends to zero, the mean value $\overline{f}$ of f over the interval of integration tends to $f(b)$, leading to the conclusion that

$$\delta[c(x-b)] = \frac{\delta(x-b)}{|c|} \tag{C–5}$$

We may generalize this result to the case when the argument of the δ function is itself a function, say $g(x)$. It is clear the $\delta[g(x)]$ has a nonzero value only when $g(x)$ vanishes, so that

$$\delta[g(x)] = \delta[g(x\approx r_1)] + \delta[g(x\approx r_2)] + \delta[g(x\approx r_3)] + \cdots \tag{C–6}$$

where r_1, r_2, r_3, ... are the roots of g.

In the neighborhood of one of these roots, say r_i, we may approximate $g(x)$ by $g(r_i)+g'(r_i)(x-r_i)$ so that $\delta(g(x\approx r_i) = \delta[g(r_i)+g'(r_i)(x-r_i)] = \delta[g'(r_i)(x-r_i)]$

$$\begin{aligned}\delta[g(x)] &= \delta[g'(r_1)(x-r_1)] + \delta[g'(r_2)(x-r_2)] + \delta[g'(r_3)(x-r_3)] + \cdots \\ &= \frac{\delta(x-r_1)}{|g'(r_1)|} + \frac{\delta(x-r_2)}{|g'(r_2)|} + \frac{\delta(x-r_3)}{|g'(r_3)|} + \cdots\end{aligned} \tag{C–7}$$

Example C.1: Expand $\delta(x^2-a^2)$.

Solution: The roots of x^2-a^2 are $x=\pm a$ and $g'(a)=2a$ and $g'(-a)=-2a$. We therefore conclude that

$$\delta(x^2-a^2) = \frac{\delta(x-a)}{2|a|} + \frac{\delta(x+a)}{2|a|} \tag{Ex C.1.1}$$

In Cartesian coordinates, the two- and three-dimensional δ functions are merely products of one-dimensional δ functions, but in non-Cartesian systems, some care is required. In spherical polars, for instance, the element of volume is $r^2drd(\cos\theta)d\varphi$, meaning that $\delta(\vec{r}-\vec{a})$ must take the form $r^{-2}\delta(r-a)\delta(\cos\theta-\cos\theta_a)\delta(\varphi-\varphi_a)$. More generally, the δ function $\delta(\vec{r}-\vec{r}\,')$ may be written in terms of coordinates (η_1, η_2, η_3)

$$\delta(\vec{r}-\vec{r}\,') = \frac{1}{|J(x_i,\eta_i)|}\delta(\eta_1-\eta_1')\delta(\eta_2-\eta_2')\delta(\eta_3-\eta_3') \tag{C–8}$$

where $J(x_i, \eta_i)$ is the Jacobian of the transformation relating the η_i to x_i. ($J_{ij} \equiv \partial x_i / \partial \eta_j$).

The charge density $\rho(\vec{r})$ of point charges q_i located at $\vec{r}_i$ is easily written in terms of the δ function

$$\rho(\vec{r}) = \sum q_i \delta(\vec{r} - \vec{r}_i) \tag{C–9}$$

Integrating ρ over any small volume containing $\vec{r}_i$ (it is assumed that only one of the charges' position vector lies in the volume) gives q_i, while a vanishing integral results when none of the r_i is contained.

Uniform line charges, say along the z axis, may similarly be represented in Cartesian coordinates by $\rho = \lambda\delta(x)\delta(y)$ with $\lambda = q/L$, the charge per unit length. This same line charge in spherical polar coordinates has the form

$$\rho(r, \theta, \varphi) = \frac{\lambda}{2\pi r^2}\delta(\cos\theta - 1) + \frac{\lambda}{2\pi r^2}\delta(\cos\theta + 1) \tag{C–10}$$

Integrating to obtain the charge contained in a spherical shell extending from a to b with $a, b \in (0, \frac{1}{2}L)$, we find

$$\begin{aligned} \int \rho d^3 r &= \int \frac{\lambda}{2\pi r^2}\left[\delta(\cos\theta - 1) + \delta(\cos\theta + 1)\right] r^2 dr d(\cos\theta) d\varphi \\ &= \frac{4\pi\lambda}{2\pi}\int_a^b dr = 2\lambda(b - a) \end{aligned} \tag{C–11}$$

as we would expect for the two segments.

An important identity involving δ functions is (26) on the inside back cover

$$\nabla^2 \frac{1}{|\vec{r} - \vec{r}\,'|} = -4\pi\delta(\vec{r} - \vec{r}\,') \tag{C–12}$$

To prove this identity, let us consider

$$\int_\tau f(\vec{r})\nabla^2 \frac{1}{|\vec{r} - \vec{r}\,'|} d^3 r \tag{C–13}$$

If $\vec{r}\,'$ is not in the region τ over which we are integrating, it is easily confirmed that $\nabla^2 1/|\vec{r} - \vec{r}\,'|$ vanishes identically. Any region τ that contains $\vec{r}$ may be subdivided into a small sphere of radius ρ centered on $\vec{r}\,'$ surrounded by surface Γ and the remaining volume where $\nabla^2|\vec{r} - \vec{r}\,'|^{-1}$ vanishes. Abbreviating $R = |\vec{r} - \vec{r}\,'|$, we have

$$\begin{aligned} \int_\tau f(\vec{r})\nabla^2 \frac{1}{R} d^3 r &= \int_{sphere} f(\vec{r})\vec{\nabla}\cdot\left(\vec{\nabla}\frac{1}{R}\right) d^3 r \\ &= \int_{sphere} f(\vec{r})\vec{\nabla}\cdot\left(\frac{-\vec{R}}{R^3}\right) d^3 r \end{aligned} \tag{C–14}$$

The mean value theorem allows us to replace $f(\vec{r})$ in the integral by $f(\vec{\xi})$ where $\vec{\xi}$ is some point in the sphere. Taking $f(\vec{\xi})$ outside the integral and applying the divergence theorem (20) to the remaining integral, we obtain

$$\int_\tau f(\vec{r})\nabla^2 \frac{1}{R} d^3 r = f(\vec{\xi})\int_\Gamma \frac{-\vec{R}\cdot d\vec{S}}{R^3}$$

$$= -f(\vec{\xi}) \int \frac{\rho^2 d\Omega}{\rho^2} = -f(\vec{\xi}) \int d\Omega$$

$$= -4\pi f(\vec{\xi}) \tag{C–15}$$

As ρ tends to 0, $\vec{\xi}$ must approach $\vec{r}\,'$, so that

$$\int f(\vec{r})\nabla^2 \frac{1}{R} d^3r = -4\pi f(\vec{r}\,') \tag{C–16}$$

Comparing this result with the definition (C–2) of the δ function, we conclude that

$$\nabla^2 \frac{1}{|\vec{r} - \vec{r}\,'|} = -4\pi\delta(\vec{r} - \vec{r}\,') \tag{C–17}$$

Appendix D

Orthogonal Function Expansions

D.1 Orthogonal Functions

Throughout this course it is frequently useful to express solutions of problems in terms of well-studied familiar functions. A power series expansion of some function might be offered as a typical example were it not for the fact that the coefficients of the truncated series will change as more terms are included. This defect is due to the lack of *orthogonality* of the monomials.

Consider a set of functions $\{U_n(x)\}$ defined for x in the range (a, b). Two of these function $U_i(x)$ and $U_j(x)$ are said to be orthogonal over (a, b) if (and only if):

$$\int_a^b U_i^*(x)\, U_j(x)\, dx = 0 \qquad \text{for } i \neq j \tag{D–1}$$

where U^* is the complex conjugate of U. The whole set is orthogonal over (a, b) if each function $U_n(x)$ is orthogonal to every other function $U_m(x)$ over this interval.

EXAMPLE D.1: Show that the set of function $\{U_n(x)\} = \{e^{inx}\}$ with m and n integers is orthogonal over the interval $(-\pi, \pi)$.

Solution: We compute the integral of (D–1)

$$\begin{aligned}
\int_{-\pi}^{\pi} e^{-inx} e^{imx} dx &= \int_{-\pi}^{\pi} e^{i(m-n)x} dx \\
&= \int_{-\pi}^{\pi} \cos(n-m)x + i \sin(m-n)x dx \\
&= \left.\frac{\sin(m-n)x}{(m-n)}\right|_{-\pi}^{\pi} - i \left.\frac{\cos(m-n)x}{(m-n)}\right|_{-\pi}^{\pi}
\end{aligned} \tag{Ex D.1.1}$$

As $\cos(r\pi) = \cos(-r\pi)$ for all r, the second term vanishes. The first term vanishes for all non-zero $(m - n)$ since $\sin(r\pi) = 0$. The $(m - n) = 0$ term is best evaluated from the initial integral:

$$\int_{-\pi}^{\pi} e^0 e^0 dx = 2\pi \tag{Ex D.1.2}$$

We conclude then that

$$\int_{-\pi}^{\pi} \mathrm{U}_n^*(x)\,\mathrm{U}_m(x)dx = \begin{cases} 0 & m \neq n \\ 2\pi & m = n \end{cases} \tag{Ex D.1.3}$$

proving the orthogonality of the set.

In general, $\int_a^b |\mathrm{U}_i(x)|^2 dx = \nu_i$; ν_i is called the normalization of the function. If $\nu_i = 1$ for all i, the orthogonal set is said to be orthonormal.

EXAMPLE D.2: Show that the set $\{\mathrm{U}_n(x)\} = \{\sin n\pi x\}$ with $n = 1, 2, 3\ldots$ is orthonormal over $(-1, 1)$

Solution: Substituting the form of U above into (D–1) we get

$$\int_{-1}^{1} \sin(n\pi x)\sin(m\pi x)dx = \delta_{mn} \tag{Ex D.2.1}$$

We shall also meet functions that are orthogonal over an interval only with a weighting factor $w(x)$. In other words, if $f_i(x)$ and $f_j(x)$ are orthogonal over (a, b) with weight factor $w(x)$ they satisfy

$$\int_a^b f_i^*(x)f_j(x)w(x)dx = 0 \qquad \text{for } i \neq j \tag{D–2}$$

Such solution tend to occur, for example, in problems with cylindrical symmetry, where the weighting factor w conveniently changes the dr to $r\,dr$, the appropriate factor for integrating over a cross-sectional surface. Other coordinate systems will similarly give rise to weighted orthogonal solutions.

Orthogonal functions occur in the analysis of many physical problems in normal mode vibrations, quantum mechanics, heat flow and eigenfunction problems in general.

D.1.1 Expansion of Functions in terms of an Orthogonal Set

Let us attempt to approximate the function $f(x)$ defined over (a, b) by a linear superposition of orthogonal functions (with unit weight factor), ie.

$$f(x) \approx \sum_{n=1}^{N} a_n \mathrm{U}_n(x) \tag{D–3}$$

The expansion coefficients may be evaluated as follows: Multiply both sides of the near equality (D–3) by $\mathrm{U}_m^*(x)$ and integrate the product from a to b.

$$\int_a^b f(x)\mathrm{U}_m^*(x)dx \approx \sum_{n=1}^{N} a_n \int_a^b \mathrm{U}_n(x)\mathrm{U}_m^*(x)dx$$

$$= a_1 \int_a^b \mathrm{U}_1 \mathrm{U}_m^* dx + a_2 \int_a^b \mathrm{U}_2 \mathrm{U}_m^* dx + a_3 \int_a^b \mathrm{U}_3 \mathrm{U}_m^* dx + \cdots$$

$$+ a_m \int_a^b \mathrm{U}_m \mathrm{U}_m^* dx + \cdots + a_N \int_a^b \mathrm{U}_N \mathrm{U}_m^* dx$$

$$= a_m \int_a^b \mathrm{U}_m \mathrm{U}_m^* dx = a_m \nu_m \quad \text{(D–4)}$$

Thus,

$$a_m \approx \frac{1}{\nu_m} \int_a^b f(x) \mathrm{U}_m^*(x) dx \quad \text{(D–5)}$$

It is not clear that the a_m found above necessarily provides the best choice to approximate $f(x)$ (The coefficients of a truncated Taylor series are not the best fit, for example.) We define the "square-error" of the approximation as

$$M_N \equiv \int_a^b \left| f(x) - \sum_{n=0}^{N} a_n \mathrm{U}_n(x) \right|^2 dx \quad \text{(D–6)}$$

Restricting ourselves to real functions $\mathrm{U}_n(x)$ for the moment, we produce the "least-square-error" by choosing a_n to minimize the square-error. To this end, we set $\partial M_n / \partial a_n = 0$ for every value of n. Differentiating (D–6) with respect to a_n we get, exploiting the linear independence of the U_n,

$$\int_a^b \left(-2\mathrm{U}_n(x) f(x) + 2a_n \left| \mathrm{U}_n(x) \right|^2 \right) dx = 0 \quad \text{(D–7)}$$

$\int 2a_n |U_n(x)|^2 dx = 2a_n \nu_n$, so that we may solve for a_n to get

$$a_n = \frac{1}{\nu_n} \int_a^b \mathrm{U}_n(x) f(x) dx \quad \text{(D–8)}$$

precisely the result in (D–5). A somewhat more laborious repetition of the above when U_n and a_n are complex yields with $a_n = p_n + iq_n$, $\partial M_N / \partial p_n = 0$ and $\partial M_N / \partial q_n = 0$

$$p_n = \mathrm{Re} \frac{1}{\nu_n} \int_a^b f(x) \mathrm{U}_n^*(x) dx \quad \text{(D–9)}$$

and

$$q_n = \mathrm{Im} \frac{1}{\nu_n} \int_a^b f(x) \mathrm{U}_n^*(x) dx \quad \text{(D–10)}$$

or, combining these results,

$$a_n = \frac{1}{\nu_n} \int_a^b f(x) \mathrm{U}_n^*(x) dx \quad \text{(D–11)}$$

In other words, the choice of coefficients made in (D–5) minimizes the integrated square error.

EXAMPLE D.3: Approximate $f(x) = x^3$ by a second degree polynomial using a Legendre polynomial series.

Solution: The Legendre polynomials $\mathrm{P}_n(x)$ are orthogonal over $(-1,1)$ with

$$\int_{-1}^{1} \mathrm{P}_m(x)\mathrm{P}_n(x)dx = \frac{2}{2n+1}\delta_{mn} \tag{Ex D.3.1}$$

and the first four polynomials are $\mathrm{P}_0(x) = 1$, $\mathrm{P}_1(x) = x$, $\mathrm{P}_2(x) = \frac{1}{2}(3x^2-1)$, $\mathrm{P}_3(x) = \frac{1}{2}(5x^3-3x)$. Let us consider the expansion

$$x^3 \approx \sum_0^2 a_n \mathrm{P}_n(x) \tag{Ex D.3.2}$$

The expansion coefficients are readily computed:

$$\begin{aligned} a_0 &= \tfrac{1}{2}\int_{-1}^{1} x^3\mathrm{P}_0(x)dx = 0 \qquad & a_1 &= \tfrac{3}{2}\int_{-1}^{1} x^3\mathrm{P}_1(x)dx = \tfrac{3}{5} \\ a_2 &= \tfrac{5}{2}\int_{-1}^{1} x^3\mathrm{P}_2(x)dx = 0 \qquad & a_3 &= \tfrac{7}{2}\int_{-1}^{1} x^3\mathrm{P}_3(x)dx = \tfrac{2}{5} \end{aligned} \tag{Ex D.3.3}$$

Including only terms up to P_2 we find $x^3 \approx \frac{3}{5}x$. For comparison, the Taylor expansion, $f(x) = f(0)+f'(0)x+\frac{1}{2}f''(0)x^2\ldots$ truncates to $x^3 \approx 0+3x^2\big|_0 x+\frac{1}{2}6x\big|_0 x^2 = 0$.

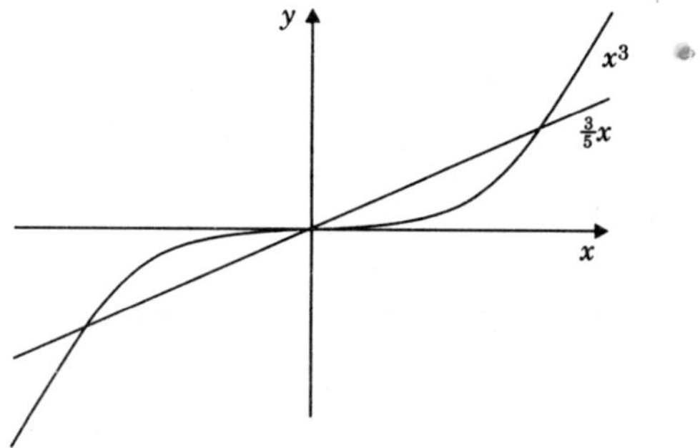

Figure C.1: The curve $y = x^3$ and the best-fit "quadratic".

Had we carried our expansion to include terms up to P_3 we would have obtained $x^3 \approx \frac{3}{5}x + \frac{2}{5}\cdot\frac{1}{2}(5x^3-3x) = x^3$ for an exact fit.

If any reasonably behaved (continuous, single valued, square integrable) function can be written as the limit of a sequence

$$f(x) = \lim_{N\to\infty} S_N(x) \quad \text{with} \quad S_N = \sum_{n=1}^{N} a_n \mathrm{U}_n(x) \tag{D–12}$$

the set $\{U_n\}$ is said to be *complete.* (Merely being infinite does not guarantee a set's completeness. For example $\{\sin n\pi x, n = 1, 2 \ldots \infty\}$ is not complete as all the U_n vanish at $x = 0$ and no function with $f(0) \neq 0$ can be constructed from a superposition.) If the set is complete over (a, b),

$$\begin{aligned} f(x) &= \sum_{n=1}^{\infty} a_n U_n(x) \\ &= \sum_{n=1}^{\infty} \left[\frac{1}{\nu_n} \int_a^b f(x') U_n^*(x') dx' \right] U_n(x) \\ &= \int_a^b \left[\sum_{n=1}^{\infty} \frac{1}{\nu_n} U_n(x) U^*{}_n(x') \right] f(x') dx' \end{aligned} \tag{D–13}$$

Comparing this to the definition of the Dirac δ function (C–2)

$$f(x) = \int_a^b \delta(x - x') f(x') dx' \qquad x \in (a, b)$$

we obtain the completeness relation

$$\sum_{n=1}^{\infty} \frac{1}{\nu_n} U_n(x) U_n^*(x') = \delta(x - x') \tag{D–14}$$

EXAMPLE D.4: Express the Dirac δ-function, $\delta(x - x')$, in terms of the set of complex exponentials $\{e^{inkx}; n \in (-\infty, \infty)\}$ orthogonal (and complete) over the interval $(a, a + L)$ with $k = 2\pi/L$.

Solution: The normalization of the functions above is easily obtained

$$\nu_m = \int_a^{a+L} e^{imkx} e^{-imkx} dx = \int_a^{a+L} dx = L \tag{Ex D.4.1}$$

Then, with the help of (D–14) we get

$$\delta(x - x') = \frac{1}{L} \sum_{m=-\infty}^{\infty} e^{imk(x-x')} = \frac{k}{2\pi} \sum_{m=-\infty}^{\infty} e^{imk(x-x')} \tag{Ex D.4.2}$$

For the case that the functions are orthogonal only with a weighting function $w(x)$, retracing the steps leading to (D–14) leads to

$$\sum \frac{w(x)}{\nu_n} U_n(x) U_n^*(x') = \delta(x - x') \tag{D–15}$$

It can be shown under fairly general conditions that the eigenvalue equation for the variable $y(x)$[32]

$$\frac{d}{dx}\left(f(x) \frac{dy}{dx} \right) - g(x) y = -\lambda w(x) y \tag{D–16}$$

[32]The problem is known as the Sturm-Liouville Problem. See for example S.M. Lea *Mathematics for Physicists*, Brookes/Cole (2004), Belmont, CA, USA

subject to vanishing Dirichlet or Neumann (or even a linear combination of these) boundary conditions has a complete set of eigenfunctions orthogonal with weight function $w(x)$ and real eigenvalues. Since most physical involving the Laplacian can be fit into form (D–16), we anticipate that the sets of eigenfunction will be orthogonal and complete.

Appendix E

Bessel Functions

E.1 Properties of Bessel Functions

Bessel functions occur frequently in the solutions of Laplace's equation in polar or cylindrical coordinates. They also occur in the context of integrals of cosines with trigonometric arguments. In this appendix we give some of the principal properties of Bessel functions as well as the related modified Bessel functions, Hankel functions, and spherical Bessel functions. The properties listed here are by no means exhaustive; for further properties as well as tables of values we refer the reader to *Handbook of Mathematical Functions*, edited by M. Abramowitz and I. Stegun and published by Dover Publications Inc., New York.

E.1.1 Differential Equation

The Bessel function (of the first kind), $\mathrm{J}_\nu(x)$, and Neumann function, $\mathrm{N}_\nu(x)$ [also frequently denoted $\mathrm{Y}_\nu(x)$ (we avoid this label as it would lead to confusion with the spherical harmonics) and called a Bessel function of the second kind], are solutions of the differential equation

$$x^2 f'' + xf' + (x^2 - \nu^2)f = 0 \tag{E–1}$$

where ν is a constant. In most physical systems, ν is required to be an integer by the azimuthal periodicity of the solution. The Neumann function, N_ν, diverges at the origin and is therefore frequently eliminated from the solution of boundary value problems.

The Bessel function may be found as a power series solution using recursion relations between the coefficients. The solution will be so obtained in section E.3 and may be written

$$\mathrm{J}_n(z) = \sum_{\lambda=0}^{\infty} \frac{(-1)^\lambda}{\lambda!(n+\lambda)!}\left(\frac{z}{2}\right)^{n+2\lambda} \tag{E–2}$$

when n is an integer, or as

$$\mathrm{J}_\nu(z) = \sum_{\lambda=0}^{\infty} \frac{(-1)^\lambda}{\lambda!\Gamma(\nu+\lambda+1)}\left(\frac{z}{2}\right)^{\nu+2\lambda} \tag{E–3}$$

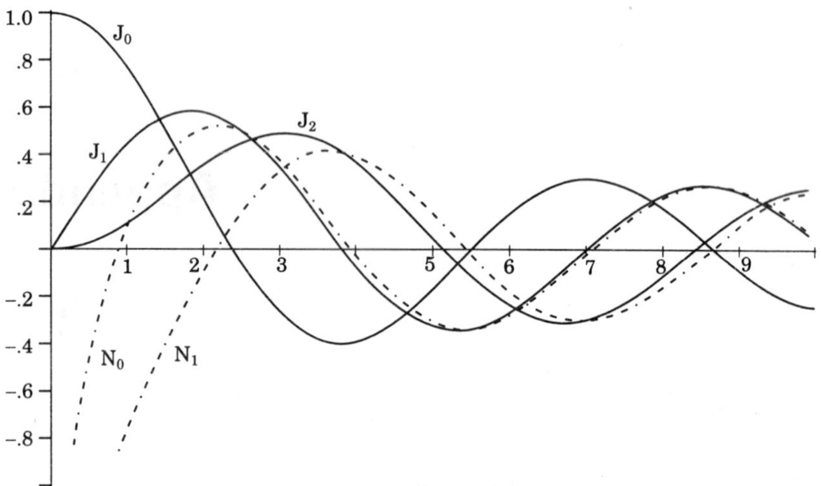

Figure E.1: The first few Bessel functions of integer order are plotted for arguments ranging from 0 to 10.

when ν is not an integer. The Neumann function may be generated from the Bessel function (of the first kind) by

$$\mathrm{N}_\nu(z) = \frac{\mathrm{J}_\nu(z)\cos(\nu\pi) - \mathrm{J}_{-\nu}(z)}{\sin(\nu\pi)} \tag{E–4}$$

The series solution is normalized so that $\mathrm{J}_0(0) = 1$. The first few integer-order Bessel function are plotted in Figure E.1. As the order increases J_n remains near zero for increasing intervals; J_{10}, for example, takes on significant values only when its argument exceeds 10.

E.1.2 Asymptotic Values: As $z \to \infty$, J_ν and N_ν tend asymptotically to 45° phase-shifted sines and cosines, with amplitude decreasing inversely as the square root of the argument:

$$\mathrm{J}_\nu(z) \to \sqrt{\frac{2}{\pi z}}\cos\left(z - \frac{\nu\pi}{2} - \frac{\pi}{4}\right) \quad \text{and} \quad \mathrm{N}_\nu(z) \to \sqrt{\frac{2}{\pi z}}\sin\left(z - \frac{\nu\pi}{2} - \frac{\pi}{4}\right) \tag{E–5}$$

as $z \to \infty$. Conversely, as the argument tends to zero,

$$\mathrm{J}_n(z) \to \frac{1}{n!}\left(\frac{z}{2}\right)^n \quad \text{and} \quad \begin{cases} \mathrm{N}_\nu(z) \to \dfrac{-\Gamma(\nu)}{\pi}\left(\dfrac{2}{z}\right)^\nu & \nu \neq 0 \\ \mathrm{N}_0(z) \to \dfrac{2}{\pi}\ln z \end{cases} \tag{E–6}$$

where the gamma function $\Gamma(\nu + 1)$ is the generalization of $\nu!$ for non-integral ν detailed following (E–45). As is evident from the power series solution (E–2), $\mathrm{J}_n(0) = 0$ when $n \neq 0$ and $\mathrm{J}_0(0) = 0$. The $\mathrm{J}_n(x)$ are bounded by ± 1.

E.1.3 Integral Forms

The Bessel functions of the first kind have the following integral representations:

$$\mathrm{J}_n(z) = \frac{1}{\pi}\int_0^\pi \cos(z\sin\theta - n\theta)d\theta \tag{E–7}$$

for integral n and

$$\mathrm{J}_\nu(z) = \frac{(\frac{1}{2}z)^\nu}{\sqrt{\pi}\,\Gamma(\nu+\frac{1}{2})}\int_{-1}^{1}(1-t^2)^{\nu-\frac{1}{2}}\cos zt\,dt \tag{E–8}$$

for $\mathrm{Re}(\nu) > -\frac{1}{2}$.

E.1.4 Explicit Forms:

$$\begin{aligned} &\mathrm{J}_{-n}(z) = (-1)^n \mathrm{J}_n(z) && \mathrm{N}_{-n}(z) - (-1)^n \mathrm{N}_n(z) \\ &\mathrm{J}_{\frac{1}{2}}(z) = \sqrt{\frac{2}{\pi z}}\sin z && \mathrm{J}_{-\frac{1}{2}}(z) = \sqrt{\frac{2}{\pi z}}\cos z \\ &\mathrm{J}_0'(z) = -\mathrm{J}_1(z) && \mathrm{N}_0'(z) = -\mathrm{N}_1(z) \end{aligned} \tag{E–9}$$

where $'$ denotes differentiation with respect to the argument.

E.1.5 Recursion Relations

The following relations hold for J or N or any linear combination:

$$\mathrm{J}_{\nu-1}(z) + \mathrm{J}_{\nu+1}(z) = \frac{2\nu}{z}\mathrm{J}_\nu(z) \tag{E–10}$$

$$\mathrm{J}_{\nu-1}(z) - \mathrm{J}_{\nu+1}(z) = 2\mathrm{J}_\nu'(z) \tag{E–11}$$

$$\frac{d}{dz}\left[z^\nu\ \mathrm{J}_\nu(z)\right] = z^\nu \mathrm{J}_{\nu-1}(z) \qquad \frac{d}{dz}\left[z^{-\nu}\mathrm{J}_\nu(z)\right] = -z^{-\nu}\mathrm{J}_{\nu+1}(z) \tag{E–12}$$

E.1.6 Generating Function and Associated Series:

$$e^{\frac{1}{2}z(t-1/t)} = \sum_{k=-\infty}^{\infty} \mathrm{J}_k(z)t^k \qquad (t \neq 0) \tag{E–13}$$

If we take $t = e^{i\theta}$, this becomes

$$e^{iz\sin\theta} = \sum_{k=-\infty}^{\infty} \mathrm{J}_k(z)e^{ik\theta} \tag{E–14}$$

whose real and imaginary parts may equated separately to obtain

$$\cos(z\sin\theta) = \mathrm{J}_0(z) + 2\sum_{k=0}^{\infty} \mathrm{J}_{2k}(z)\cos(2k\theta) \tag{E–15}$$

and
$$\sin(z\sin\theta) = 2\sum_{k=0}^{\infty} J_{2k+1}(z)\sin(2k+1)\theta \tag{E–16}$$

When θ is $\pi/2$, these series become

$$\cos z = J_0(z) - 2J_2(z) + 2J_4(z) - 2J_6(z) + \cdots \tag{E–17}$$

$$\sin z = 2J_1(z) - 2J_3(z) + 2J_5(z) - 2J_7(z) + \cdots \tag{E–18}$$

E.1.7 Addition Theorem

$$J_n(z_1 + z_2) = \sum_{m=-\infty}^{\infty} J_m(z_1)J_{n-m}(z_2) \tag{E–19}$$

E.1.8 Orthogonality

$$\int_0^1 J_\nu(\rho_{\nu m} r)J_\nu(\rho_{\nu n} r) r dr = \tfrac{1}{2}\left[J_{\nu+1}(\rho_{\nu m})\right]^2 \delta_{nm} \tag{E–20}$$

where $\rho_{\nu m}$ is the mth root of J_ν

$$\int_0^1 J_\nu(\rho'_{\nu m} r)J_\nu(\rho'_{\nu n} r) r dr = \tfrac{1}{2}\left(1 - \frac{\nu^2}{\rho'^2_{\nu n}}\right)\left[J_\nu(\rho'_{\nu n})\right]^2 \delta_{mn} \tag{E–21}$$

with $\rho'_{\nu n}$ the nth root of J'_ν.

E.1.9 Completeness:

$$\int_0^\infty J_\nu(kx)J_\nu(k'x) x dx = \frac{1}{k}\delta(k - k') \tag{E–22}$$

for k and $k' \geq 0$.

E.1.10 Roots

The first few roots of J_0, J_1, J_2, and J_3 are tabulated in Table E.1, and those of the corresponding J' are listed in Table E.2.

n	1	2	3	4
ρ_{0n}	2.40483	5.52008	8.65373	11.79153
ρ_{1n}	3.83171	7.01559	10.17347	13.32369
ρ_{2n}	5.13562	8.41724	11.61984	14.79595
ρ_{3n}	6.38016	9.76102	13.01520	16.22347

Table E.1 The nth root ρ_{in} of Bessel functions, J_i.

n	1	2	3	4
ρ'_{0n}	0.00000	3.38171	7.01559	10.17347
ρ'_{1n}	1.84118	5.33144	8.53632	11.70600
ρ'_{2n}	3.05424	6.70613	9.96947	13.17037
ρ'_{3n}	4.20119	8.01524	11.34592	14.58585

Table E.2 The nth root ρ'_{in} of Bessel function derivative, J'_i.

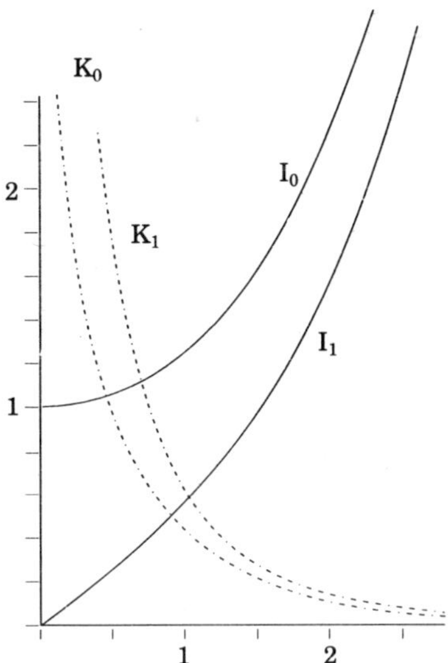

Figure E.2: The modified Bessel functions K and I diverge at 0 and ∞ respectively.

E.2 Related Functions

E.2.1 Hankel Functions

Hankel functions are the complex linear combinations of J and N:

$$\mathrm{H}_\nu^{(1)}(z) = \mathrm{J}_\nu(z) + i\mathrm{N}_\nu(z) \qquad \mathrm{H}_\nu^{(2)}(z) = \mathrm{J}_\nu(z) - i\mathrm{N}_\nu(z) \tag{E–23}$$

The asymptotic expansion of each is easily obtained from (E–5).

E.2.2 Modified Bessel Functions:

I_n and K_n are the solutions of the differential equation

$$x^2 f'' + xf' - (x^2 + n^2)f = 0 \tag{E–24}$$

and are related to Bessel functions by

$$\mathrm{I}_n(x) = -i^n \mathrm{J}_n(ix) \quad \text{and} \quad \mathrm{K}_n(x) = \frac{\pi i^{n+1}}{2} \mathrm{H}_n^{(1)}(ix) \tag{E–25}$$

Both these functions diverge: K_n diverges at 0 where $\mathrm{K}_0(z) \to -\ln z$ and $\mathrm{K}_\nu(z) \sim \frac{1}{2}\Gamma(\nu)(2/z)^\nu$ whereas $\mathrm{I}_\nu(z)$ diverges roughly exponentially as $z \to \infty$ as could have been predicted from the $\mathrm{J}_n(ix)$ form above. As $z \to 0$, $\mathrm{I}_\nu(z) \approx (\frac{1}{2}z)^\nu/\Gamma(\nu+1)$ for $(\nu \neq -1, -2, -3, \ldots)$. The derivatives of the modified Bessel functions satisfy $\mathrm{K}_0'(z) = -\mathrm{K}_1(z)$ and $\mathrm{I}_0'(z) = \mathrm{I}_1(z)$.

E.2.3 Spherical Bessel Functions

Spherical Bessel functions occur primarily as the radial functions in spherical boundary condition problems. They may be defined in terms of the Bessel functions of half-integral order:

$$j_\ell(z) = \sqrt{\frac{\pi}{2z}}\,\mathrm{J}_{\ell+\frac{1}{2}}(z) \quad \text{and} \quad n_\ell(z) = \sqrt{\frac{\pi}{2z}}\,\mathrm{N}_{\ell+\frac{1}{2}}(z) \tag{E–26}$$

Each has an asymptotic 1 over z envelope. Spherical Bessel functions are the solutions of

$$\frac{1}{z^2}\frac{d}{dz}\left(z^2\frac{d}{dz}\right)y + \left[1 - \frac{\ell(\ell+1)}{z^2}\right]y = 0 \tag{E–27}$$

Related to the spherical Bessel functions are the spherical Hankel functions, $\mathrm{h}_\ell^{(1)} \equiv \mathrm{j}_\ell + i n_\ell$ and $\mathrm{h}_\ell^{(2)} \equiv \mathrm{j}_\ell - i n_\ell$. The first few spherical Bessel functions are explicitly

$$\mathrm{j}_0(z) = \frac{\sin z}{z} \qquad \mathrm{j}_{-1}(z) = \frac{\cos z}{z} \qquad \mathrm{n}_0(z) = -\frac{\cos z}{z} \tag{E–28}$$

$$\mathrm{j}_1(z) = \frac{\sin z}{z^2} - \frac{\cos z}{z} \qquad \mathrm{n}_1(z) = -\frac{\cos z}{z^2} - \frac{\sin z}{z} \tag{E–29}$$

$$\mathrm{j}_2(z) = \left(\frac{3}{z^3} - \frac{1}{z}\right)\sin z - \frac{3}{z^2}\cos z \quad \mathrm{n}_2(z) = -\left(\frac{3}{z^3} - \frac{1}{z}\right)\cos z - \frac{3}{z^2}\sin z \tag{E–30}$$

E.3 Solution of Bessel's Equation

We will find a power series solution to Bessel's equation (E–1),

$$x^2y'' + xy' + (x^2 - \nu^2)y = 0$$

where ν is an arbitrary constant. We try a solution of the form

$$y = x^s\sum_{n=0}^{\infty} c_n x^n \qquad \text{with } c_0 \neq 0 \tag{E–31}$$

Substituting this, (E–31) into the equation (E–1) we calculate first the derivatives

$$y' = \sum_{n=0}^{\infty} c_n(n+s)x^{n+s-1} \tag{E–32}$$

$$y'' = \sum_{n=0}^{\infty} c_n(n+s)(n+s-1)x^{n+s-2} \tag{E–33}$$

which turn Bessel's equation into the algebraic equation

$$\sum_{n=0}^{\infty}\Big[c_n(n+s)(n+s-1)x^{n+s} + (n+s)c_nx^{n+s} + c_nx^{n+s+2} - \nu^2x^{n+s}\Big] = 0 \tag{E–34}$$

Gathering terms and factoring x^s from (E–34) we get

$$x^s \sum_{n=0}^{\infty} \Big\{ \big[(n+s)^2 - \nu^2\big] c_n x^n + c_n x^{n+2} \Big\} = 0 \tag{E–35}$$

The x^k are linearly independent functions, the vanishing of the sum requires the coefficient of each $x^k, k = 0, 1, 2, \ldots$ to vanish. For x^0 we obtain

$$c_0(s^2 - \nu^2)x^0 = 0 \qquad \Rightarrow \qquad s = \pm\nu \tag{E–36}$$

as we required that $c_0 \neq 0$. For x^1 (E–35) gives

$$c_1\big[(s+1)^2 - \nu^2\big]x^1 = 0 \tag{E–37}$$

which is compatible with $s = \pm\nu$ only if $c_1 = 0$ or $s = -\frac{1}{2}$. The remaining coefficients satisfy

$$\Big\{ c_n\big[(n+s)^2 - \nu^2\big] + c_{n-2} \Big\} x^n = 0 \tag{E–38}$$

or $c_n = -\dfrac{c_{n-2}}{(n+s)^2 - \nu^2}$. Assuming for the moment that $s \neq -\frac{1}{2}$,

$$\frac{c_n}{c_{n-2}} = -\frac{1}{(n+s-\nu)(n+s+\nu)} = -\frac{1}{n(n+2s)} \quad \text{with } s = \pm\nu \tag{E–39}$$

This *recursion* relation allows us to write each coefficient in terms of that preceding it by two, so that

$$\frac{c_2}{c_0} = -\frac{1}{2(2+2s)} = \frac{-1}{4(s+1)} \tag{E–40}$$

$$\frac{c_4}{c_0} = \frac{c_4}{c_2}\frac{c_2}{c_0} = \frac{-1}{4(4+2s)}\,\frac{-1}{4(s+1)} = \frac{1}{4\cdot 8(s+2)(s+1)} \tag{E–41}$$

$$\frac{c_6}{c_0} = \frac{c_6}{c_4}\frac{c_4}{c_0} = \frac{-1}{6(6+2s)}\,\frac{1}{4\cdot 8(s+1)(s+2)} = \frac{-1}{4\cdot 8\cdot 12(s+1)(s+2)(s+3)} \tag{E–42}$$

We write the solution in terms of c_0 as

$$y = c_0 x^s \left[1 - \frac{x^2}{4(s+1)} + \frac{x^4}{4\cdot 8(s+1)(s+2)} - \frac{x^6}{4\cdot 8\cdot 12(s+1)(s+2)(s+3)} + \cdots \right] \tag{E–43}$$

$$= \begin{cases} y_1 & \text{if } s = +|\nu| \\ y_2 & \text{if } s = -|\nu| \end{cases}$$

Thus we obtain two linearly independent solutions. Unfortunately when ν is an integer, the second solution, y_2 has a term with coefficient $1/(s + |s|)$, which for negative s leads to a divergent sum. We will find a second solution by alternate means, but first we consider the excluded case of $\nu^2 = \frac{1}{4}, s = -\frac{1}{2}$. In this case c_1 need not be zero and the recursion relations give

$$y = c_0 x^{-\frac{1}{2}} \left(1 - \frac{x^2}{2} + \frac{x^4}{4!} - \cdots \right) + c_1 x^{-\frac{1}{2}} \left(x - \frac{x^3}{3!} + \frac{x^5}{5!} \cdots \right)$$

$$= \frac{c_0}{\sqrt{x}} \cos x + \frac{c_1}{\sqrt{x}} \sin x \tag{E–44}$$

Returning to the $s \neq -\frac{1}{2}$, $s = +|\nu|$ case, we define $J_\nu(z)$, the *Bessel function of the first kind* by

$$J_\nu(z) = \frac{1}{\Gamma(\nu+1)}\left(\frac{z}{2}\right)^\nu \left[1 - \frac{1}{\nu+1}\left(\frac{z}{2}\right)^2 + \frac{1}{(\nu+1)(\nu+2)2!}\left(\frac{z}{2}\right)^4 - \cdots\right]$$

$$= \sum_{r=0}^{\infty} \frac{(-1)^r}{r!\Gamma(\nu+r+1)}\left(\frac{z}{2}\right)^{\nu+2r} \tag{E–45}$$

where $\Gamma(z+1) \equiv \int_0^\infty t^z e^{-t} dt$, is the gamma function, a generalization of the factorial, $z!$ to non integral z. It is readily shown that $\Gamma(z+1) = z\Gamma(z)$.

If $\nu = 0, 1, 2, 3 \ldots$ the J_n are Bessel functions of integer order.

E.3.1 The Second Solution

When ν is an integer, the second solution with $s = -\nu$ diverges. For nonintegral ν we define the Neumann function

$$N_\nu(z) \equiv \frac{J_\nu(z)\cos\nu\pi - J_{-\nu}(z)}{\sin\nu\pi} \tag{E–46}$$

The *Wronskian*

$$W(J_\nu, N_\nu) = \begin{vmatrix} J_\nu & N_\nu \\ J'_\nu & N'_\nu \end{vmatrix} = \frac{2}{\pi z} \neq 0 \tag{E–47}$$

which implies that N_ν and J_ν are linearly independent functions. Moreover, as N_ν is a linear combination of J_ν and $J_{-\nu}$, it solves Bessel's equation. Taking the limit $N_n(z) = \lim_{\nu\to n} N_\nu(z)$, we find it indeterminate, 0/0. However, using L'Hôpital's rule,

$$\frac{\partial}{\partial\nu}\left[J_\nu(z)\cos\nu\pi - J_{-\nu}(z)\right] = \frac{\partial J_\nu}{\partial\nu}\cos\nu\pi - \pi\sin\nu\pi J_\nu - \frac{\partial J_{-\nu}}{\partial\nu} \tag{E–48}$$

and

$$\frac{\partial}{\partial\nu}(\sin\nu\pi) = \pi\cos\nu\pi \tag{E–49}$$

together give

$$N_n(z) = \lim_{\nu\to n} N_\nu(z) = \frac{1}{\pi}\left(\frac{\partial J_\nu}{\partial\nu} - (-1)^n \frac{\partial J_{-\nu}}{\partial\nu}\right)_{\nu=n} \tag{E–50}$$

The limiting forms of the Neumann functions as $z \to 0$ are

$$N_0 = \frac{2}{\pi}\ln z \qquad N_\nu(z) = \frac{1}{\pi\Gamma(\nu)}\left(\frac{2}{z}\right)^\nu \tag{E–51}$$

whereas the limit as $z \to \infty$ is

$$N_\nu(z) \to \sqrt{\frac{2}{\pi z}}\sin\left(\frac{z - (\nu+\frac{1}{2})\pi}{2}\right) \tag{E–52}$$

E.3.2 Derivation of the Generating Function

The generating function for integer order Bessel functions is

$$e^{(z/2)(t-1/t)} = \sum_{n=-\infty}^{\infty} \mathrm{J}_n(z)t^n \tag{E–53}$$

The proof of this equality follows. We first expand the exponential as a power series

$$e^{(z/2)(t-1/t)} = \sum_{\mu=0}^{\infty} \frac{1}{\mu!}\left(\frac{z}{2}\right)^{\mu}\left(t-\frac{1}{t}\right)^{\mu} \tag{E–54}$$

and then use the binomial theorem to expand $(t-1/t)^{\mu}$

$$\left(t-\frac{1}{t}\right)^{\mu} = \sum_{\lambda=0}^{\infty}\binom{\mu}{\lambda} t^{\mu-\lambda}\left(\frac{-1}{t}\right)^{\lambda} \qquad \text{with } \binom{\mu}{\lambda} \equiv \frac{\mu!}{(\mu-\lambda)!\,\lambda!} \tag{E–55}$$

Thus

$$e^{(z/2)(t-1/t)} = \sum_{\mu=0}^{\infty}\sum_{\lambda=0}^{\infty}\frac{1}{\mu!}\frac{\mu!}{(\mu-\lambda)!\,\lambda!}\left(\frac{z}{2}\right)^{\mu} t^{\mu-\lambda}(-1)^{\lambda}t^{-\lambda}$$

$$= \sum_{\mu=0}^{\infty}\sum_{\lambda=0}^{\infty}\frac{(-1)^{\lambda}}{(\mu-\lambda)!\,\lambda!}\left(\frac{z}{2}\right)^{\mu} t^{\mu-2\lambda} \tag{E–56}$$

We substitute $n = \mu - 2\lambda$ as the summation variable for the first sum. For any value of μ as λ ranges from 0 to ∞, n varies from $-\infty$ to μ, thus as μ ranges to ∞ n varies from ∞ to $-\infty$.

$$e^{(z/2)(t-1/t)} = \sum_{n=-\infty}^{\infty}\left[\sum_{\lambda=0}^{\infty}\frac{(-1)^{\lambda}}{(n+\lambda)!\,\lambda!}\left(\frac{z}{2}\right)^{n+2\lambda}\right]t^n \tag{E–57}$$

Comparing the sum in square brackets with the sum (E–45) we conclude that this is the defining sum for $\mathrm{J}_n(z)$, hence

$$e^{(z/2)(t-1/t)} = \sum_{n=-\infty}^{\infty} \mathrm{J}_n(z)\,t^n \tag{E–58}$$

which completes the proof. We continue with a simple application of the generating function.

EXAMPLE E.1: Use the generating function to generate the addition theorem (E–19).

Solution: Consider

$$e^{\frac{1}{2}(z_1+z_2)(t-1/t)} = \sum_{n=-\infty}^{\infty} \mathrm{J}_n(z_1+z_2)t^n \tag{Ex E.1.1}$$

We can equally write the left hand side as a product of two exponentials

$$e^{\frac{1}{2}z_1(t-1/t)}e^{\frac{1}{2}z_2(t-1/t)} = \left[\sum_{m=-\infty}^{\infty} \mathrm{J}_m(z_1)t^m\right]\left[\sum_{\ell=-\infty}^{\infty} \mathrm{J}_\ell(z_2)t^\ell\right] \tag{Ex E.1.2}$$

and writing $m+\ell=n$ while replacing ℓ by n in the summation to get

$$\sum_{n=-\infty}^{\infty} \mathrm{J}_n(z_1+z_2)t^n = \sum_{n=-\infty}^{\infty}\left[\sum_{m=-\infty}^{\infty} \mathrm{J}_m(z_1)\mathrm{J}_{n-m}(z_2)\right]t^n \tag{Ex E.1.3}$$

Equating coefficients of t^n produces the desired result.

E.3.3 Spherical Bessel Functions

To solve the spherical Bessel equation (E–27) or equivalently,

$$\frac{d}{dz}\left(z^2\frac{dy}{dz}\right) + \left[z^2 - \ell(\ell+1)\right]y = 0 \tag{E–59}$$

we could start afresh on a power series solution but the substitution of $y = z^{-\frac{1}{2}}f$ converts the equation to Bessel's equation

$$z^2f'' + zf' + \left[z^2 - (\ell+\tfrac{1}{2})^2\right]f = 0 \tag{E–60}$$

leading to the conclusion that the solutions of (E–59) must be of the form $y = z^{-1/2}(c_1\mathrm{J}_{\ell+\frac{1}{2}} + c_2\mathrm{N}_{\ell+\frac{1}{2}})$. The customary normalization,

$$\mathrm{j}_n(z) = \sqrt{\frac{\pi}{2z}}\mathrm{J}_{n+\frac{1}{2}}(z) \quad \text{and} \quad \mathrm{n}_m(z) = \sqrt{\frac{\pi}{2z}}\mathrm{N}_{m+\frac{1}{2}}(z) \tag{E–61}$$

gives the trigonometric forms (E–28) etc. Generally, j_n, n_n, $\mathrm{h}_n^{(1)}$, or $\mathrm{h}_n^{(2)}$ may be found from the zeroth and first order functions using the recursion relation

$$f_{n+1}(z) = (2n+1)z^{-1}f_n(z) - f_{n-1}(z) \tag{E–62}$$

where f_n is any one of the four functions above. The derivatives may be evaluated from

$$nf_{n-1}(z) + (n+1)f_{n+1}(z) = (2n+1)\frac{d}{dz}f_n(z) \tag{E–63}$$

Appendix F

Legendre Polynomials and Spherical Harmonics

F.1 Legendre Functions

Legendre functions appear as solutions of the angular portion of Laplace's or Poisson's equation in spherical polar coordinates. Generally the argument of the functions will be $\cos\theta$ meaning, if the argument is x, it is restricted to lie in the range $x \in (-1, 1)$. The solution of the Legendre equation will be postponed to the end of this appendix, leaving the earlier part as a quick reference.

F.1.1 Differential Equation: The Legendre functions $\mathrm{P}_\nu^\mu(z)$ and $Q_\nu^\mu(z)$ of degree ν and order μ satisfy

$$(1-z^2)\frac{d^2f}{dz^2} - 2z\frac{df}{dz} + \left[\nu(\nu+1) - \frac{\mu^2}{1-z^2}\right] f(z) = 0 \tag{F-1}$$

This equation is more often encountered as

$$\left[\frac{1}{\sin\theta}\frac{\partial}{\partial\theta}\left(\sin\theta\frac{\partial}{\partial\theta}\right) + \nu\,(\nu+1) - \frac{\mu^2}{\sin^2\theta}\right] f(\theta) = 0 \tag{F-2}$$

having solutions $f(\theta) = a\mathrm{P}_\nu^\mu\,(\cos\theta) \;+\; b\,\mathrm{Q}_\nu^\mu\,(\cos\theta)$.

$\mathrm{Q}_\nu^\mu(z)$ is badly behaved at $z = \pm 1$, while $\mathrm{P}_\nu^\mu\,(z)$ is badly behaved at $z = \pm 1$ unless ν is an integer. For most physical applications μ is also an integer (because of the azimuthal periodicity of 2π). For $\mu = 0$, $\mathrm{P}_\ell^0\,(z) \equiv \mathrm{P}_\ell(z)$, a polynomial of degree ℓ and parity $(-1)^\ell$. The Legendre polynomial has the expansion

$$\mathrm{P}_\ell(z) = \sum_{r=0}^{[\frac{1}{2}\ell]} \frac{(-1)^r(2\ell-2r)!\,z^{\ell-2r}}{2^\ell(\ell-r)!\,(\ell-2r)!\,r!} = \frac{1}{2^\ell\ell!}\frac{d^\ell}{dz^\ell}\,(z^2-1)^\ell \tag{F-3}$$

where $[\frac{1}{2}\ell]$ means the largest integer less than (or equal to) $\frac{1}{2}\ell$. The (associated) Legendre functions of nonzero integral order m may be obtained from the Legendre polynomial using

$$\mathrm{P}_\ell^m(z) = (z^2-1)^{m/2}\frac{d^m}{dz^m}\mathrm{P}_\ell\,(z) \qquad Q_\ell^m(z) = (z^2-1)^{m/2}\frac{d^m}{dz^m}Q_\ell \tag{F-4}$$

F.1.2 Some Values:

$$|\mathrm{P}_\ell(x)| \le 1 \qquad \text{(F–5)} \qquad\qquad \mathrm{P}_\ell(1) = 1 \qquad \text{(F–6)}$$

$$\mathrm{P}_\ell(-x) = (-1)^\ell\, \mathrm{P}_\ell(x) \qquad \text{(F–7)} \qquad\qquad \mathrm{P}_{-(\ell+1)}(x) = \mathrm{P}_\ell(x) \qquad \text{(F–8)}$$

$$\mathrm{P}_\ell(0) = \begin{cases} 0 \text{ if } \ell \text{ is odd} \\[2ex] \dfrac{(-1)^{\ell/2}\,\ell!}{2^\ell\,(\ell/2)!\,(\ell/2)!} \text{ if } \ell \text{ is even} \end{cases} \qquad \text{(F–9)}$$

F.1.3 Explicit Forms

$$\mathrm{P}_0(z) = 1 \qquad \text{(F–10)} \qquad \mathrm{Q}_0(z) = \tfrac{1}{2}\ln\frac{z+1}{z-1} \qquad \text{(F–11)}$$

$$\mathrm{P}_1(z) = z \qquad \text{(F–12)} \qquad \mathrm{Q}_1(z) = \frac{z}{2}\ln\frac{z+1}{z-1} - 1 \qquad \text{(F–13)}$$

$$\mathrm{P}_2(z) = \tfrac{1}{2}(3z^2 - 1) \qquad \text{(F–14)} \qquad \mathrm{Q}_2(z) = \frac{3z^2-1}{4}\ln\frac{1+z}{1-z} - \frac{3z}{2} \qquad \text{(F–15)}$$

$$\mathrm{P}_3(z) = \tfrac{1}{2}(5z^3 - 3z) \qquad \text{(F–16)} \qquad \mathrm{Q}_\ell(z) = \tfrac{1}{2}\mathrm{P}_\ell(z)\ln\frac{z+1}{z-1} + f_\ell(z) \qquad \text{(F–17)}$$

$$\mathrm{P}_4(z) = \tfrac{1}{8}(35z^4 - 30z^2 + 3) \qquad \text{(F–18)}$$

$$\mathrm{P}_5(z) = \tfrac{1}{8}(63z^5 - 70z^3 + 15z) \qquad \text{(F–19)}$$

where $f_\ell z)$ is a polynomial of degree $\ell - 1$. In the expressions for Q_ℓ, the sign of the argument of the logarithm may be changed, as this generally adds $\pm\frac{1}{2}i\pi\,\mathrm{P}_\ell(z)$ to the solution. Since P_ℓ satisfies the same equation as Q_ℓ, the new form with $(1-z)$ replacing $(z-1)$ in the logarithm is an equally good solution.

Explicit expressions for the associated Legendre functions in terms of the argument $x = \cos\theta$ follow below:

$$\mathrm{P}_0(x) = 1 \qquad \text{(F–20)}$$

$$\mathrm{P}_1^1(x) = -(1-x^2)^{1/2} = -\sin\theta \qquad \text{(F–21)}$$

$$\mathrm{P}_2^1(x) = -3x\,(1-x^2)^{1/2} = -3\cos\theta\,\sin\theta \qquad \text{(F–22)}$$

$$\mathrm{P}_2^2(x) = 3\,(1-x^3) = 3\sin^2\theta \qquad \text{(F–23)}$$

$$\mathrm{P}_3^1(x) = -\tfrac{3}{2}(5\,x^2 - 1)\,(1-x^2)^{1/2} = -\tfrac{3}{2}\,(5\cos^2\theta - 1)\,\sin\theta \qquad \text{(F–24)}$$

$$\mathrm{P}_3^2(x) = 15\,x\,(1-x^2) = 15\cos\theta\,\sin^2\theta \qquad \text{(F–25)}$$

$$\mathrm{P}_3^3(x) = -15\,(1-x^2)^{3/2} = -15\sin^3\theta \qquad \text{(F–26)}$$

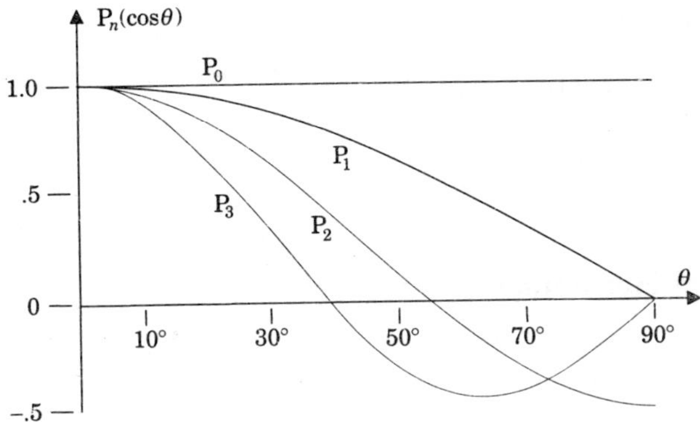

Figure F.1: The first few Legendre Polynomials as a function of $\cos\theta$.

The negative order Legendre functions may be obtained from those with positive order using

$$P_\ell^{-m}(z) = (-1)^m \frac{(\ell-m)!}{(\ell+m)!} P_\ell^m(z) \tag{F–27}$$

F.1.4 Recursion Relations:

$$(n+1)P_{n+1}(x) = (2n+1)xP_n(x) - nP_{n-1}(x) \tag{F–28}$$

$$(n+1)P_n(x) = \frac{d}{dx}P_{n-1}(x) - x\frac{d}{dx}P_n(x) \tag{F–29}$$

$$(2n+1)P_n(x) = \frac{d}{dx}P_{n+1}(x) - \frac{d}{dx}P_{n-1}(x) \tag{F–30}$$

$$\sqrt{1-x^2}\,P_\ell^{m+1}(x) = (\ell-m)xP_\ell^m(x) - (m+\ell)P_{\ell-1}^m(x) \tag{F–31}$$

$$\sqrt{1-x^2}\frac{d}{dx}P_\ell^m(x) = \tfrac{1}{2}(\ell+m)(\ell-m+1)P_\ell^{m-1}(x) - \tfrac{1}{2}P_\ell^{m+1}(x) \tag{F–32}$$

$$(\ell-m+1)P_{\ell+1}^m(x) = (2\ell+1)xP_\ell^m(x) - (m+\ell)P_{\ell-1}^m(x) \tag{F–33}$$

F.1.5 Generating Function

The generating function for Legendre polynomials:

$$(1-2xt+t^2)^{-1/2} = \sum_{n=0}^{\infty} P_n(x)t^n \qquad |x| \le 1 \tag{F–34}$$

gives rise to the useful result

$$\frac{1}{|\vec{r}-\vec{r}\,'|} = \frac{1}{r_>}\sum_{n=0}^{\infty}\left(\frac{r_<}{r_>}\right)^n P_n(\cos\theta) \tag{F–35}$$

F.1.6 Orthogonality

$$\int_{-1}^{1} \mathrm{P}_\ell^m(x)\mathrm{P}_k^m(x)\,dx = \frac{2}{2\ell+1}\frac{(\ell+m)!}{(\ell-m)!}\,\delta_{k\ell} \tag{F-36}$$

$$\int_{-1}^{1} \frac{\mathrm{P}_\ell^m(x)\,\mathrm{P}_\ell^n(x)}{1-x^2}\,dx = \frac{(\ell+m)!}{m(\ell-m)!}\,\delta_{mn} \tag{F-37}$$

F.2 Spherical Harmonics

The spherical harmonics defined by

$$\mathrm{Y}_\ell^m(\theta,\varphi) \equiv \sqrt{\frac{(2\ell+1)\,(\ell-m)!}{4\pi(\ell+m)!}}\,\mathrm{P}_\ell^m(\cos\theta)\,e^{im\varphi} \qquad |m|\le\ell \tag{F-38}$$

are the well-behaved solutions to

$$\left\{\frac{1}{\sin\theta}\frac{\partial}{\partial\theta}\left(\sin\theta\,\frac{\partial}{\partial\theta}\right) + \left[\ell\,(\ell+1) + \frac{m^2}{\sin^2\theta}\,\frac{\partial^2}{\partial\varphi^2}\right]\right\} f(\theta,\varphi) = 0 \tag{F-39}$$

F.2.1 Values and Relations:

$$\mathrm{Y}_\ell^{*\,m}(\theta,\varphi) = (-1)^m\,\mathrm{Y}_\ell^{-m}(\theta,\varphi) \tag{F-40}$$

$$\mathrm{Y}_\ell^\ell(\theta,\varphi) = (-1)^\ell\,\frac{\sqrt{(2\ell)!}}{2^\ell\ell!}\sqrt{\frac{2\ell+1}{4\pi}}\,\sin^\ell\theta\,e^{i\ell\varphi} \tag{F-41}$$

Defining the operator

$$L_\pm \equiv \pm e^{\pm i\varphi}\left(\frac{\partial}{\partial\theta} \pm i\cot\theta\frac{\partial}{\partial\varphi}\right) \tag{F-42}$$

we have

$$L_\pm \mathrm{Y}_\ell^m = \sqrt{(\ell\mp m)(\ell\pm m+1)}\mathrm{Y}_\ell^{m\pm1} \tag{F-43}$$

F.2.2 Explicit Forms

$\mathrm{Y}_0^0 = \sqrt{\frac{1}{4\pi}}$

$$\mathrm{Y}_1^0 = \sqrt{\tfrac{3}{4\pi}}\cos\theta \qquad \mathrm{Y}_1^{\pm1} = \mp\sqrt{\tfrac{3}{8\pi}}\sin\theta e^{\pm i\varphi} \tag{F-44}$$

$\mathrm{Y}_2^0 = \sqrt{\frac{5}{16\pi}}(3\cos^2\theta - 1) \quad \mathrm{Y}_2^{\pm1} = \mp\sqrt{\frac{15}{8\pi}}\sin\theta\cos\theta e^{\pm i\varphi} \quad \mathrm{Y}_2^{\pm2} = \sqrt{\frac{15}{32\pi}}\sin^2\theta e^{\pm 2i\varphi}$

F.2.3 Orthogonality

$$\int_0^{2\pi}\!\!\int_0^{\pi} \mathrm{Y}_\ell^m(\theta,\varphi)\,\mathrm{Y}_{\ell'}^{*m'}(\theta,\varphi)\sin\theta\,d\theta\,d\varphi = \delta_{l\,l'}\,\delta_{m\,m'} \tag{F-45}$$

F.2.4 Completeness:

$$\sum_{l=0}^{\infty}\sum_{m=-l}^{\ell} \mathrm{Y}_\ell^{*m}(\theta',\varphi')\mathrm{Y}_\ell^m(\theta,\varphi) = \delta(\varphi-\varphi')\,\delta(\cos\theta-\cos\theta') \tag{F–46}$$

F.2.5 Addition Theorem

$$\frac{4\pi}{(2\ell+1)}\sum_{m=-\ell}^{\ell} \mathrm{Y}_\ell^{*\,m}(\theta',\varphi')\mathrm{Y}_\ell^m(\theta,\ \varphi) = \mathrm{P}_\ell(\cos\gamma) \tag{F–47}$$

where γ is the angle between (θ',φ') and (θ,φ).

F.3 Solution of the Legendre Equation

A subset of the solutions of Legendre's equation (F–1) may be found by setting μ equal to zero so that the equation becomes

$$(1-x^2)\frac{d^2f}{dx^2} - 2x\frac{df}{dx} + \nu(\nu+1)f = 0 \tag{F–48}$$

which can be solved with a power series solution. Setting

$$f(x) = \sum_{n=0}^{\infty} c_n x^n \tag{F–49}$$

we evaluate the terms to get

$$\nu(\nu+1)f = \sum_{n=0}^{\infty}\nu(\nu+1)c_n x^n, \qquad -2x\frac{df}{dx} = -\sum_{n=1}^{\infty} 2nc_n x^n \tag{F–50}$$

$$(1-x^2)\frac{d^2f}{dx^2} = \sum_{n=2}^{\infty} c_n n(n-1)x^{n-2} - \sum_{n=2}^{\infty} c_n n(n-1)x^n \tag{F–51}$$

Collecting the terms, we obtain

$$\sum_{n=2}^{\infty} c_n n(n-1)x^{n-2} - \sum_{n=2}^{\infty} c_n n(n-1)x^n - \sum_{n=1}^{\infty} 2nc_n x^n + \nu(\nu+1)\sum_{n=0}^{\infty} c_n x^n = 0 \tag{F–52}$$

As the x^n are linearly independent functions, the coefficient of each power of x must vanish. Thus for x^0,

$$c_2\cdot 2\cdot 1 + \nu(\nu+1)c_0 = 0 \quad\Rightarrow\quad c_2 = \frac{-\nu(\nu+1)}{2}c_0 \tag{F–53}$$

For x^1,

$$c_3\cdot 3\cdot 2 - 2c_1 + \nu(\nu+1)c_1 = 0 \quad\Rightarrow\quad c_3 = \frac{2-\nu(\nu+1)}{3\cdot 2} = \frac{(2+\nu)(1-\nu)}{3\cdot 2}c_1 \tag{F–54}$$

and for $x^n (n \geq 2)$,

$$(n+2)(n+1)c_{n+2} - n(n-1)c_n - 2nc_n + \nu(\nu+1)c_n = 0$$

$$\Rightarrow \quad c_{n+2} = \frac{n(n+1) - \nu(\nu+1)}{(n+2)(n+1)} c_n = \frac{(n+1+\nu)(n-\nu)}{(n+1)(n+2)} c_n \tag{F–55}$$

We evaluate all the even coefficients in terms of c_0 to get

$$\frac{c_1}{c_0} = -\frac{\nu(\nu+1)}{2}, \quad \frac{c_4}{c_0} = \frac{\nu(\nu+1)}{2} \frac{(\nu+3)(\nu-2)}{3 \cdot 4},$$

$$\frac{c_6}{c_0} = -\frac{\nu(\nu+1)}{2} \frac{(\nu+3)(\nu-2)}{3 \cdot 4} \frac{(\nu+5)(\nu-4)}{5 \cdot 6}, \quad \text{etc.} \tag{F–56}$$

Similarly we find for the odd coefficients

$$\frac{c_3}{c_1} = -\frac{(\nu+2)(\nu-1)}{2 \cdot 3}, \quad \frac{c_5}{c_1} = \frac{(\nu+2)(\nu-1)}{2 \cdot 3} \frac{(\nu+4)(\nu-3)}{4 \cdot 5}, \quad \text{etc.} \tag{F–57}$$

so that our solution may be written

$$\begin{aligned} f = c_0 \Bigg[1 - \frac{\nu(\nu+1)}{2} x^2 + \frac{(\nu+1)(\nu+3)\nu(\nu-2)}{4!} x^4 \\ - \frac{(\nu+1)(\nu+3)(\nu+5)\nu(\nu-2)(\nu-4)}{6!} x^6 + \cdots \Bigg] \\ + c_1 \left[x - \frac{(\nu+2)(\nu-1)}{3!} x^3 + \frac{(\nu+2)(\nu+4)(\nu-1)|(\nu-3)}{5!} x^5 - \cdots \right] \end{aligned} \tag{F–58}$$

If ν is an integer, then we conclude from (F–55) that $c_{\nu+2} = 0$, implying $c_{\nu+4} = c_{\nu+6} = \ldots = 0$. Thus, for an even integer ν the even series becomes a polynomial of degree ν, whereas for odd ν the odd series becomes a polynomial of degree ν. Whichever of the two series terminates, it is after appropriate normalization defined to be the Legendre polynomial $\mathrm{P}_\nu(x)$ while the remaining series is the associated Legendre function $\mathrm{Q}_\nu(x)$.

The behavior of the infinite series at $x = \pm 1$ is worth investigating. The ratio of successive terms,

$$\frac{c_{n+2}}{c_n} = \frac{(n+\nu+1)(n-\nu)}{(n+1)(n+2)} \to -\left(1 - \frac{\nu^2}{n^2}\right) \tag{F–59}$$

as $n \to \infty$. Thus, the infinite series will not converge at $x = \pm 1$, leading to the exclusion of Q_ν, whether ν is an integer or not, as well as P_ν for non-integral ν, from the solution of physical problems that include $x = 1$ in the domain. For the usual application where $x = \cos\theta$ this means that whenever the points at $\theta = 0$ or $\theta = \pi$ are included only the Legendre polynomials will be included in the solution.

Including its normalization, the Legendre Polynomial is defined by

$$\mathrm{P}_\ell(x) = \sum_{r=0}^{[\frac{1}{2}\ell]} \frac{(-1)^r}{2^\ell r!} \frac{(2\ell - 2r)! x^{\ell - 2r}}{(\ell - r)!(\ell - 2r)!} \tag{F–60}$$

where $[\frac{1}{2}\ell] = \frac{1}{2}\ell$ when ℓ is even and $[\frac{1}{2}\ell] = \frac{1}{2}(\ell - 1)$ when ℓ is odd. To verify the equivalence of this series to the one above, we need merely verify that the ratio c_{n+2}/c_n satisfies (F–55). Abbreviating the series to

$$\mathrm{P}_\ell(x) = \sum_{r=0}^{[\frac{1}{2}\ell]} c_{\ell-2r} x^{\ell-2r} \tag{F–61}$$

We note that if $c_n = c_{\ell-2r}$, then $c_{n+2} = c_{\ell-2(r-1)}$, hence $\dfrac{c_{n+2}}{c_n} = \dfrac{c_{\ell-2(r-1)}}{c_{\ell-2r}}$, or

$$\frac{c_{n+2}}{c_n} = \frac{(-1)^{r-1}}{2^\ell (r-1)!} \frac{(2\ell-2r+2)!}{(\ell-r+1)!\,(\ell-2r+2)!} \cdot \frac{2^\ell(\ell-r)!\,r!}{(-1)^r(2\ell-2r)!}$$

$$= -\frac{(2\ell-2r+2)(2\ell-2r+1)r}{(\ell-r+1)(\ell-2r+2)(\ell-2r+1)} = \frac{(-2r)(2\ell-2r+1)}{(\ell-2r+2)(\ell-2r+1)} \tag{F–62}$$

Finally we replace $-2r$ by $(n-\ell)$ to obtain (F–55).

F.3.1 Derivation of the Generating Function

Throughout this book we make considerable use of the generating function for Legendre polynomials. In this section we will demonstrate that

$$(1 - xt + t^2)^{-1/2} = \sum_{n=0}^{\infty} \mathrm{P}_n(x) t^n \qquad |x \leq 1 \tag{F–63}$$

To begin, we expand $\mathrm{F}(x,t) = (1 - 2xt + t^2)^{-1/2}$ with $|x| \leq 1$, $|\,t\,| \leq 1$ using the binomial theorem

$$(1+\epsilon)^\alpha = 1 + \alpha\epsilon + \frac{\alpha(\alpha+1)\epsilon^2}{2!} + \frac{\alpha(\alpha+1)(\alpha+2)\epsilon^3}{3!} + \cdots$$

and specialize to $\alpha = -\frac{1}{2}$ to obtain

$$(1-\epsilon)^{-1/2} = 1 + \frac{1}{2}\epsilon + \frac{1}{2!}\frac{1}{2}\frac{3}{2}\epsilon^2 + \frac{1}{3!}\frac{1}{2}\frac{3}{2}\frac{5}{2}\epsilon^3 + \cdots$$

$$= \sum_{n=0}^{\infty} \frac{1}{2^n} \frac{(2n-1)!!}{n!} \epsilon^n = \sum_{n=0}^{\infty} \frac{(2n)!}{2^n n!(2n)!!} \epsilon^n$$

$$= \sum_{n=0}^{\infty} \frac{(2n)!\epsilon^n}{2^n n! 2n(2n-2)(2n-4)(2n-6)\cdots} = \sum_{n=0}^{\infty} \frac{(2n)!\epsilon^n}{2^{2n}(n!)^2} \tag{F–64}$$

Applying this, we write F as

$$[1 - t(2x-t)]^{-1/2} = \sum_{n=0}^{\infty} \frac{(2n)!\,t^n(2x-t)^n}{2^{2n}(n!)^2} \tag{F–65}$$

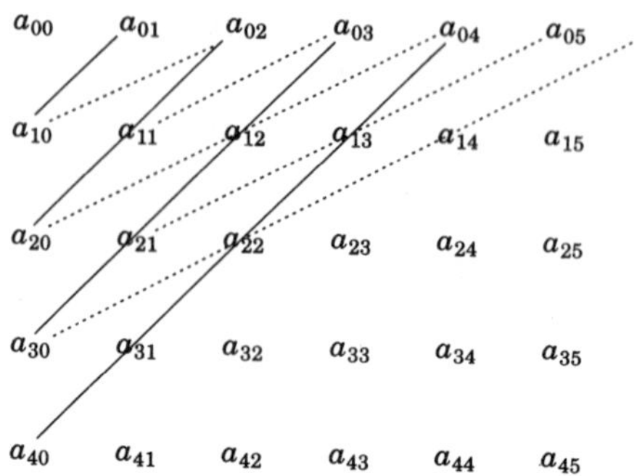

Figure F.2: The array of elements may be summed horizontally, vertically or along several different diagonals as indicated by the solid and dotted lines.

We expand $(2x - t)^n$ in turn using the binomial theorem

$$(2x - t)^n = \sum_{k=0}^{n} \frac{(-1)^k n!\,(2x)^{n-k} t^k}{k!\,(n-k)!} \tag{F–66}$$

the expression in (F–65) may then be written

$$(1 - 2xt + t^2)^{-1/2} = \sum_{n=0}^{\infty} \sum_{k=0}^{n} \frac{(-1)^k (2n)!\,(2x)^{n-k} t^{n+k}}{2^{2n} n!\,k!\,(n-k)!} \tag{F–67}$$

This convergent series may be summed with its terms grouped in a number of different orders. Consider the sum of elements $a_{k,n}$ of a rectangular array. The sum, written as the sum of diagonal elements connected by solid lines in Figure F.2

$$\sum_{n=0}^{\infty} \sum_{k=0}^{n} a_{k,n-k} = a_{0,0} + (a_{0,1} + a_{1,0}) + (a_{0,2} + a_{1,1} + a_{2,0}) + \cdots \tag{F–68}$$

may equally well be written as the sum

$$\sum_{n=0}^{\infty} \sum_{k=0}^{[\frac{1}{2}n]} a_{k,n-2k} = a_{0,0} + a_{0,1} + (a_{0,2} + a_{1,0}) + (a_{0,3} + a_{1,1}) + (a_{0,4} + a_{1,2} + a_{2,0}) + (a_{0,5} + a_{1,3} + a_{2,1}) \cdots \tag{F–69}$$

as indicated buy the dotted lines in the diagram. We rearrange the sum in (F–67) in the same manner. Identifying the terms of (F–67) by

$$a_{k,n-k} = \frac{(-1)^k (2n)!\,(2x)^{n-k} t^{n+k}}{2^{2n} k!\,n!\,(n-k)!} \tag{F–70}$$

we obtain $a_{k,n-2k}$ on replacing n by $n - k$ in (F–67), so that we find

$$(1 - 2xt + t^2)^{-1/2} = \sum_{n=0}^{\infty} \left[\sum_{k=0}^{[\frac{1}{2}n]} \frac{(-1)^k (2n-2k)!\,x^{n-2k}}{2^n k!\,(n-k)!\,(n-2k)!} \right] t^n$$

$$= \sum_{n=0}^{\infty} \mathrm{P}_n(x)t^n \tag{F–71}$$

In addition to giving the very useful expansion for $1/|\vec{r}-\vec{r}'|$ the generating function may be used to generate most of the recursion relations for the Legendre polynomials

EXAMPLE F.1: Differentiate the generating function with respect to t to find the recursion formula relating P_{n+1} to P_n and P_{n-1}.

Solution: Differentiating $\mathrm{F} = (1-2xt+t^2)^{-1/2}$ with respect to t as instructed

$$\frac{\partial \mathrm{F}}{\partial t} = \frac{x-t}{(1-2xt+t^2)^{3/2}} = \frac{x-t}{1-2xt+t^2}\mathrm{F}(x,t) \tag{Ex F.1.1}$$

In other words,

$$(x-t)\mathrm{F}(x,t) = (1-2xt+t^2)\frac{\partial \mathrm{F}}{\partial t} \tag{Ex F.1.2}$$

We replace F by its generating function expansion (F–71) to get

$$(x-t)\sum_{n=0}^{\infty} \mathrm{P}_n(x)t^n = (1-2xt+t^2)\sum_{n=1}^{\infty} \mathrm{P}_n(x)nt^{n-1} \tag{Ex F.1.3}$$

and equate the coefficients of t^n to obtain

$$x\mathrm{P}_n(x) - \mathrm{P}_{n-1}(x) = (n+1)\mathrm{P}_{n+1}(x) - 2nx\mathrm{P}_n(x) + (n-1)\mathrm{P}_{n-1}(x) \tag{Ex F.1.4}$$

which, after collecting terms, becomes

$$(n+1)\mathrm{P}_{n+1}(x) = (2n+1)x\mathrm{P}_n(x) - n\mathrm{P}_{n-1}(x) \tag{Ex F.1.5}$$

Using the result above, (Ex F.1.5), we can generate all the Legendre polynomials starting from P_0 and P_1, for example,

$$2\mathrm{P}_2(x) = 3x\mathrm{P}_1(x) - \mathrm{P}_0(x) \quad \Rightarrow \quad \mathrm{P}_2(x) = \tfrac{1}{2}(3x^2-1) \tag{F–72}$$

Had we differentiated F with respect to x instead of t, we would have obtained a relation between the Legendre Polynomials and their derivatives.

F.3.2 Associated Legendre Functions

Recall that we started out to solve

$$(1-x^2)\frac{d^2\mathrm{P}}{dx^2} - 2x\frac{d\mathrm{P}}{dx} + \left[n(n+1) - \frac{m^2}{1-x^2}\right]\mathrm{P} = 0 \tag{F–73}$$

but postponed the $m \neq 0$ case. With some effort it may be shown by induction that

$$\mathrm{P}_n^m(x) = (-1)^m(1-x^2)^{m/2}\frac{d^m\mathrm{P}_n(x)}{dx^m} \tag{F–74}$$

κ dielectric constant
L inductance
$\vec{L}$ angular momentum
ℓ length, counting number
$\vec{\mathcal{L}}$ angular momentum density
λ line charge density
λ wavelength
m mass
$\vec{m}$ magnetic moment
$\vec{M}$ magnetization
μ magnetic permeability
N_ℓ Neumann function
n_ℓ spherical Neumann function
$\hat{n}$ normal
ν frequency
$\mathcal{O}(h)$ terms of order h
$\mathbf{P}$ power
$\vec{P}$ polarization
P_ℓ^m Legendre function
P_{ij} coefficient of potential
$\vec{p}$ dipole moment
$\vec{p}$ momentum
$\vec{\mathcal{P}}$ momentum density
$\wp$ pressure
q, Q charge
$\overleftrightarrow{Q}$ quadrupole moment tensor
Q_{ij} i, j component of the quadrupole moment
R resistance
$\Re$ reluctance
$\vec{r}$ position vector
ρ charge density
ρ cylindrical polar radius
$\vec{S}$ surface
S entropy
$\vec{\mathcal{S}}$ Poynting vector
σ surface charge density
σ Stefan Boltzmann constant
$\overleftrightarrow{T}$ Maxwell stress tensor
T temperature
τ volume
U energy density
V electric potential
V_m magnetic scalar potential
V^μ four-velocity
$\vec{v}$ velocity
W energy
Ω solid angle, ohm
ω angular frequency
χ electric susceptibility
χ_m magnetic susceptibility
Y_ℓ^m spherical harmonic
Z impedance

Index

D

N

O

P

Useful Vector Identities

$$\vec{A} \times (\vec{B} \times \vec{C}) = (\vec{A} \cdot \vec{C})\vec{B} - (\vec{A} \cdot \vec{B})\vec{C} \tag{1}$$

$$(\vec{A} \times \vec{B}) \times \vec{C} = (\vec{A} \cdot \vec{C})\vec{B} - (\vec{B} \cdot \vec{C})\vec{A} \tag{2}$$

$$\vec{A} \cdot (\vec{B} \times \vec{C}) = (\vec{A} \times \vec{B}) \cdot \vec{C} \tag{3}$$

$$(\vec{A} \times \vec{B}) \cdot (\vec{C} \times \vec{D}) = \begin{vmatrix} \vec{A} \cdot \vec{C} & \vec{A} \cdot \vec{D} \\ \vec{B} \cdot \vec{C} & \vec{B} \cdot \vec{D} \end{vmatrix} \tag{4}$$

$$\vec{\nabla}(\psi\xi) = \xi\vec{\nabla}\psi + \psi\vec{\nabla}\xi \tag{5}$$

$$\vec{\nabla} \times (\psi\vec{A}) = \vec{\nabla}\psi \times \vec{A} + \psi\vec{\nabla} \times \vec{A} \tag{6}$$

$$\vec{\nabla} \cdot (\psi\vec{A}) = \vec{\nabla}\psi \cdot \vec{A} + \psi\vec{\nabla} \cdot \vec{A} \tag{7}$$

$$\vec{\nabla} \cdot (\vec{A} \times \vec{B}) = \vec{B} \cdot (\vec{\nabla} \times \vec{A}) - \vec{A} \cdot (\vec{\nabla} \times \vec{B}) \tag{8}$$

$$\vec{\nabla}(\vec{A} \cdot \vec{B}) = (\vec{A} \cdot \vec{\nabla})\vec{B} + (\vec{B} \cdot \vec{\nabla})\vec{A} + \vec{A} \times (\vec{\nabla} \times \vec{B}) + \vec{B} \times (\vec{\nabla} \times \vec{A}) \tag{9}$$

$$\vec{\nabla} \times (\vec{A} \times \vec{B}) = \vec{A}(\vec{\nabla} \cdot \vec{B}) - \vec{B}(\vec{\nabla} \cdot \vec{A}) + (\vec{B} \cdot \vec{\nabla})\vec{A} - (\vec{A} \cdot \vec{\nabla})\vec{B} \tag{10}$$

$$\vec{\nabla} \cdot (\vec{\nabla} \times \vec{A}) \equiv 0 \tag{11}$$

$$\vec{\nabla} \times (\vec{\nabla} V) \equiv 0 \tag{12}$$

$$\vec{\nabla} \times (\vec{\nabla} \times \vec{A}) = \vec{\nabla}(\vec{\nabla} \cdot \vec{A}) - \nabla^2\vec{A} \tag{13}$$

$$(\vec{\nabla} \times \vec{A})^i = \epsilon^{ijk}\partial_j A_k \qquad \text{(The summation convention is assumed)} \tag{14}$$

$$\epsilon_{ijk}\epsilon^{i\ell m} = \delta_j^\ell\delta_k^m - \delta_j^m\delta_k^\ell \tag{15}$$

$$\int_a^b \vec{\nabla}\psi \cdot d\vec{\ell} = \psi(b) - \psi(a) \tag{16}$$

$$\oint \psi d\vec{\ell} = \int d\vec{S} \times \vec{\nabla}\psi \tag{17}$$

$$\oint \vec{A} \cdot d\vec{\ell} = \int (\vec{\nabla} \times \vec{A}) \cdot d\vec{S} \text{ — Stokes' theorem} \tag{18}$$

$$\oint \psi d\vec{S} = \int \vec{\nabla}\psi d^3r \tag{19}$$

$$\oint \vec{A} \cdot d\vec{S} = \int \vec{\nabla} \cdot \vec{A} d^3r \text{ — Divergence theorem} \tag{20}$$

$$\oint d\vec{S} \times \vec{A} = \int (\vec{\nabla} \times \vec{A}) d^3r \tag{21}$$

$$\oint (\psi\vec{\nabla}\xi - \xi\vec{\nabla}\psi) \cdot d\vec{S} = \int (\psi\nabla^2\xi - \xi\nabla^2\psi) d^3r \text{ — Green's theorem} \tag{22}$$

Miscellaneous Properties of Dirac δ Functions

$$\delta(x - c) = 0 \quad \text{for } x \neq c \tag{23}$$

$$\int_a^b f(x)\delta(x - c)\, dx = f(c) \quad \text{for } c \in (a,\, b) \tag{24}$$

$$\delta[f(x)] = \sum_i \frac{1}{|f'(r_i)|}\delta(x - r_i) \qquad \text{where } r_i \text{ is a root of } f \tag{25}$$

$$\nabla^2 \frac{1}{|\vec{r} - \vec{r}\,'|} = -4\pi\delta(\vec{r} - \vec{r}\,') \tag{26}$$

$$\int_a^b f(x)\,\delta'(x - c)\, dx = f'(c) \quad \text{for } c \in (a,\, b) \tag{27}$$

Frequently Used Integrals

$$\int \frac{dx}{(a^2+x^2)^{3/2}} = \frac{x}{a^2\sqrt{a^2+x^2}} \quad (28) \qquad \int \frac{x\,dx}{(a^2+x^2)^{3/2}} = \frac{-1}{\sqrt{a^2+x^2}} \quad (29)$$

$$\int \frac{x\,dx}{\sqrt{a^2+x^2}} = \sqrt{a^2+x^2} \quad (30) \qquad \int \frac{dx}{\sqrt{a^2+x^2}} = \ln\left(x+\sqrt{a^2+x^2}\right) \quad (31)$$

$$\int \frac{x^2\,dx}{(a^2+x^2)^{3/2}} = \frac{-x}{\sqrt{a^2+x^2}} + \ln\left(x+\sqrt{a^2+x^2}\right) \quad (32)$$

Binomial Theorem

$$(1+\epsilon)^\alpha = 1 + \alpha\epsilon + \frac{\alpha(\alpha-1)}{2!}\epsilon^2 + \frac{\alpha(\alpha-1)(\alpha-2)}{3!}\epsilon^3 + \ \ldots \quad (33)$$

Vector Derivatives in Curvilinear Coordinates

Cylindrical Coordinates

$$\vec{\nabla} = \hat{r}\frac{\partial}{\partial r} + \frac{\hat{\varphi}}{r}\frac{\partial}{\partial \varphi} + \hat{k}\frac{\partial}{\partial z} \quad (34)$$

$$\vec{\nabla}\times\vec{A} = \left(\frac{1}{r}\frac{\partial A_z}{\partial \varphi} - \frac{\partial A_\varphi}{\partial z}\right)\hat{r} + \left(\frac{\partial A_r}{\partial z} - \frac{\partial A_z}{\partial r}\right)\hat{\varphi} + \frac{1}{r}\left[\frac{\partial}{\partial r}(rA_\varphi) - \frac{\partial A_r}{\partial \varphi}\right]\hat{k} \quad (35)$$

$$\vec{\nabla}\cdot\vec{A} = \frac{1}{r}\frac{\partial}{\partial r}\left(rA_r\right) + \frac{1}{r}\frac{\partial A_\varphi}{\partial \varphi} + \frac{\partial A_z}{\partial z} \quad (36)$$

$$\nabla^2\psi = \frac{1}{r}\frac{\partial}{\partial r}\left(r\frac{\partial \psi}{\partial r}\right) + \frac{1}{r^2}\frac{\partial^2\psi}{\partial \varphi^2} + \frac{\partial^2\psi}{\partial z^2} \quad (37)$$

Spherical Polar Coordinates

$$\vec{\nabla} = \hat{r}\frac{\partial}{\partial r} + \frac{\hat{\theta}}{r}\frac{\partial}{\partial \theta} + \frac{\hat{\varphi}}{r\sin\theta}\frac{\partial}{\partial \varphi} \quad (38)$$

$$\vec{\nabla}\times\vec{A} = \frac{\hat{r}}{r\sin\theta}\left[\frac{\partial(\sin\theta A_\varphi)}{\partial\theta} - \frac{\partial A_\theta}{\partial\varphi}\right] + \frac{\hat{\theta}}{r}\left[\frac{1}{\sin\theta}\frac{\partial A_r}{\partial\varphi} - \frac{\partial(rA_\varphi)}{\partial r}\right] + \frac{\hat{\varphi}}{r}\left[\frac{\partial(rA_\theta)}{\partial r} - \frac{\partial A_r}{\partial\theta}\right] \quad (39)$$

$$\vec{\nabla}\cdot\vec{A} = \frac{1}{r^2}\frac{\partial(r^2A_r)}{\partial r} + \frac{1}{r\sin\theta}\frac{\partial(\sin\theta\,A_\theta)}{\partial\theta} + \frac{1}{r\sin\theta}\frac{\partial A_\varphi}{\partial\varphi} \quad (40)$$

$$\nabla^2\psi = \frac{1}{r^2}\frac{\partial}{\partial r}\left(r^2\frac{\partial\psi}{\partial r}\right) + \frac{1}{r^2\sin\theta}\frac{\partial}{\partial\theta}\left(\sin\theta\frac{\partial\psi}{\partial\theta}\right) + \frac{1}{r^2\sin^2\theta}\frac{\partial^2\psi}{\partial\varphi^2} \quad (41)$$

$$= \frac{1}{r}\frac{\partial^2}{\partial r^2}(r\psi) + \frac{1}{r^2\sin\theta}\frac{\partial}{\partial\theta}\left(\sin\theta\frac{\partial\psi}{\partial\theta}\right) + \frac{1}{r^2\sin^2\theta}\frac{\partial^2\psi}{\partial\varphi^2} \quad (42)$$